Ultra-High Performance Concrete

Ultra-high performance concrete (UHPC) is an advanced cement-based composite material with compressive strength of over 120 MPa, high toughness, and superior durability. Since its development in the early 1990s, UHPC has attracted great interest worldwide due to its advantages. This book covers material selection and mixture design methods for developing UHPC, as well as performance and applications of UHPC, including fresh and hardened properties, setting and hardening, dimensional stability, static and dynamic properties, durability, long-term properties, and self-healing properties.

A range of potential applications and case studies are presented to illustrate how UHPC meets requirements for lightweight, high-rise, large-span, heavy-load bearing, fast-construction, and highly durable structures in civil and construction engineering. Also introduced is a typical new concrete, seawater sea-sand UHPC, which avoids the use of freshwater and river sand in marine construction.

The first book to fully cover the design, performance, and applications of UHPC, which is ideal for concrete technologists, designers, contractors, and researchers.

Ultra-High Performance Concrete

Design, Performance, and Application

Caijun Shi, Zemei Wu and
Nemkumar Banthia

CRC Press
Taylor & Francis Group
Boca Raton London New York

CRC Press is an imprint of the
Taylor & Francis Group, an **informa** business

Cover image: Yichang Bridge Group

First edition published 2024
by CRC Press
4 Park Square, Milton Park, Abingdon, Oxon, OX14 4RN

and by CRC Press
2385 NW Executive Center Drive, Suite 320, Boca Raton FL 33431

British Library Cataloguing-in-Publication Data
A catalogue record for this book is available from the British Library

ISBN: 978-1-032-06732-2 (hbk)
ISBN: 978-1-032-06733-9 (pbk)
ISBN: 978-1-003-20360-5 (ebk)

DOI: 10.1201/9781003203605

Typeset in Sabon
by Newgen Publishing UK

Contents

Preface

Ultra-high performance concrete (UHPC) is an advanced cement-based composite material with compressive strength of over 120 MPa, high toughness, and superior durability. Since its development in the early 1990s, UHPC has attracted great interest worldwide due to its advantages.

During the past three decades, a large number of papers on UHPC materials, structures, and applications have been published. Several series of symposiums or conferences on UHPC materials and structures have been held in Germany, USA, and China. Given the widespread application of UHPC materials, this book provides an in-depth summary of the historical evolution, scientific design principles, raw material selection, mixture design methods, fresh and hardened properties, and some typical up-to-date studies and applications.

This book first tracks the advances in cement-based materials with time and explores its characterization, design principles, and preparation technologies to uncover the mystery of the superior performance of UHPC. Special attention is then focused on material selection and mixture design methods for developing UHPC. Performances of UHPC, including fresh properties, setting and hardening, dimensional stability, static and dynamic properties, durability, and self-healing properties, are put forward to help readers understand UHPC thoroughly. New types of UHPC, i.e., seawater sea-sand UHPC, which addresses the challenges associated with the shortages of freshwater and river sand in producing durable concrete suitable for marine constructions, are also introduced. Since UHPC can meet the requirements of lightweight, high rise, large span, heavy load-bearing, fast construction, and high durability for civil and construction engineering, potential applications and relevant case studies in various engineering aspects are finally presented.

Combining with the related literature, most chapters in this book come from the research results of the authors under the financial support of two major projects funded by the National Science Foundation of China (NSFC) and the Ministry of Science and Technology of China. The authors have successfully organized three international conferences and two national conferences in China on "UHPC Materials and Structures" during the last ten years. This book cannot be completed without the hard work and contribution of the members of the research group at Hunan University. The authors would like to acknowledge the contributions from those graduate students and postdoctoral fellows, including Dr. Jianhui Liu, Dr. Chaohui Zhang, Dr. Yifan Zhao, Dr. Li Yang, Min Zhou, Dr. Yangxue Ou, Sailong Hou, Yihan Ma, Dr. Yiwei Liu, Hongwei Zhang, and Jing Xie.

This book is dedicated to a group of users composed of universities and testing laboratories, building material companies and industries, material scientists and experts, building and infrastructure authorities, designers, and civil engineers. It can be a part of a textbook or reference book for 4th-year undergraduate or graduate courses in universities.

About the authors

Caijun Shi is a chair professor in the College of Civil Engineering at Hunan University, China, and also an adjunct professor at the University of British Columbia. His research interests include characterization and utilization of industrial by-products and waste materials, design and testing of cement and concrete materials, development and evaluation of cement additives and concrete admixtures, and solid and hazardous waste management. One of his inventions—self-sealing/self-healing barrier—has been used as a municipal landfill liner in the world's largest landfill site in South Korea. He developed preconditioned technology to overcome the barrier to effective CO_2 curing of concrete products with CO_2 and proposed a method on "Design of High-Performance Concrete with Multiple Performance Requirements", which has been used for several large construction projects in China. His patented cement grinding aids have been widely used by several state-owned companies. He is the Editor-in-Chief of the *Journal of Sustainable Cement-Based Materials* and Coeditor of several peer-reviewed journals. His awards include the Outstanding Leadership Award for Graduate Supervisor Group, Research Excellence Award, Outstanding Overseas Chinese Contribution Award, Outstanding Supervisor for master students, and Outstanding Supervisor for Ph.D. students. He has been ranked as the No. 1 most-cited Chinese Scholar in the Building and Construction Sector since 2015. He has authored *Alkali-Activated Cements and Concretes*, *Rheology of Cement-based Materials*, and *Transport and Interactions of Chlorides in Cement-Based Materials*, which are also published by CRC Press. In recognizing his contributions to research in waste management and concrete technology, he was elected as a fellow of International Energy Foundation in 2001, American Concrete Institute in 2007, and RILEM in 2016.

Zemei Wu is a professor in the College of Civil Engineering at Hunan University, China. She obtained her Ph.D. degree from Missouri University of Science and Technology in 2018 and subsequently gained postdoctoral research experience there. Her research interests mainly focus on novel and sustainable construction materials, including ultra-high performance concrete, concrete nanotechnology, and recycled concrete aggregate. She has authored/coauthored more than 70 journal papers, 10 international conference papers, and 2 book chapters. She serves members of several technical committees, including voting member of Cement Grouting (ACI-552), and associate members of Silica Fume (ACI-232) and Ultra-High Performance Concrete (ACI-239). She has been invited as reviewer of more than 10 peer-reviewed journals and was recognized as Outstanding Reviewer for the *Journal of Cement and Concrete Composites* (2018), the *Journal of Construction and Building Materials* (2018), and the *Journal of Cleaner Production* (2018).

Nemkumar (Nemy) Banthia is a University Killam Professor, distinguished university scholar, and Canada Research Chair at the University of British Columbia. He is a materials scientist with a focus on ultra-high performance carbon-neutral building materials.

Dr. Banthia has graduated over 75 doctoral and post-doctoral students, holds 9 patents, and has published nearly 400 journal papers. He serves on the editorial boards of nine international journals and is the Editor-in-Chief of the *Journal of Cement and Concrete Composites*—a journal with the 2022 Impact Factor of 10.5.

His awards include the Wason Medal of American Concrete Institute, Solutions Through Research Award of the BC Innovation Council, Wolfson Merit Award of the Royal Society of the UK, Killam Research Prize from the Killam Foundation, Horst Leipholz Medal of the Canadian Society for Civil Engineering, Mufti Medal of Excellence of the International Society for Health Monitoring of Infrastructure, and Global Citizenship Award of alumni-UBC. His other awards include Distinguished Alumni Award of IIT-Delhi; Jacob Biely Faculty Research Prize from UBC; Dean's Medal of Distinction from the Faculty of Applied Science, UBC; and a University Killam Professor appointment from UBC.

Dr. Banthia is a fellow of the American Concrete Institute, Canadian Society for Civil Engineering, Indian Concrete Institute, Canadian Academy of Engineering (CAE), Indian National Academy of Engineering (INAE), and the Royal Society of Canada. He is one of Top 25 'Most Cited in the field of Construction & Building Materials' as per Platinum H-Index.

Chapter 1

Historical evolution of ultra-high performance concrete

1.1 INTRODUCTION

Cement and concrete are the most commonly used civil engineering materials in infrastructure construction structures due to the wide availability of raw materials, prominent adaptability, and low maintenance. With the development of modern engineering concepts and the projects' demands, great process has been obtained in the application of concrete technology, and traditional concrete is processed to the area of sustainability, high strength, and high performance. High-strength concrete is often considered as a relatively new material in the past time and has been gradually developed over the last several decades. In the 1950s, concrete with a compressive strength of 34 MPa was considered as high strength. In the 1960s, concrete with a compressive strength of 41–52 MPa was used commercially. In the early 1970s, concrete with a compressive strength of 62 MPa was produced (ACI 363R 1997). In the 1990s, the remarkable progress associated with the development of superplasticizers led to the development of high-performance concrete (HPC).

HPC is made with carefully selected high-quality ingredients and optimized mixture design, which is batched, mixed, placed, compacted, and cured to the highest industry standards. Typically, it has a low water-to-binder (w/b) ratio ranging from 0.24 to 0.45 and shows higher performance than that of conventional concrete. With the use of HPC, it is possible to obtain concrete with very good workability, high compressive strength of about 100 MPa, and favorable durability. HPC shows promising prospects in infrastructures. However, the presence of coarse aggregate becomes the weakest part in concrete and restricts its further performance enhancement. Advances in the techniques and understanding the behaviors of concrete on the microstructural level render people realize how to control microstructure and performance. This eventually led to the development of more advanced concrete, namely ultra-high performance concrete (UHPC).

UHPC is an ideal building material that can meet the requirements for strong, durable, and sustainable structures. Reactive powder concrete (RPC)

DOI: 10.1201/9781003203605-1

is the prototype of UHPC, which was developed and patented by Richard and Cheyrezy in Bouygues' laboratory in France in the early 1990s. It was named UHPC later, and has been recognized as one of the most innovative cement-based composite materials in recent years. The compressive strength of UHPC is typically around 150 MPa, and can reach up to 810 MPa. Its design and production are mainly based on the theory of the close packing of particles and can be achieved by: (1) elimination of coarse aggregate by using fine particles with size ranging from 0.02 to 300 μm; (2) use of superplasticizer to decrease w/b ratio to less than 0.24; (3) optimization of grain size distribution to densify the mixture; (4) use of postset heat treatment to improve the microstructure; and (5) addition of approximately 2% steel fibers, by volume of concrete, to improve strength and toughness.

UHPC shows good workability, high mechanical properties, and superior durability capable of ensuring a service life of 100 years or more, even under adverse environmental exposure. It is an excellent candidate to meet the lightweight, high rise, large span, and increased durability for civil engineering construction to ensure prolongated service life, reduced formwork, labor, and maintenance, lighter weight and thinner elements, fast construction speed, and lower carbon footprint. Until now, the applications of UHPC in Europe, North America, Australia, Asia, and New Zealand have been reported (Rebentrost et al. 2008; Schmidt et al. 2005; Graybeal 2008). The prestressed hybrid pedestrian bridge, which was completed in 1997 at Sherbrooke in Canada, was the first engineering structure application of UHPC (Resplendino 2004). In 1997 and 1998, UHPCs were cast in beams as the first industrial application. In 2001, the first UHPC road bridge was designed and constructed at Bourg-lès-Valence in France (Hajar 2004). However, numerous challenges are encountered before widespread implementation due to the lack of commonly accepted standards for testing methods, design guides for engineers, and quality control methods in manufacturing facilities (Ahlborn 2012).

1.2 ADVANCES IN CONCRETE MATERIALS

The commonly used ingredients for cement paste, mortar, and concrete include cement, coarse aggregate, sand, water, supplementary cementitious materials, chemical admixtures, and fibers. The birth of HPC and UHPC results from the development of chemical admixtures and processing techniques geared toward dense and homogenous microstructure of cement paste and high toughness and ductility. This part introduces the basic knowledge of main parts of constituents of concrete.

1.2.1 Cementitious materials

Cement is the most crucial constituent that sets and hardens to bind other materials together in the presence of water under a series of chemical reactions. It is a very fine powder with 60% to 70% particles, typically ranging between 3 and 30 μm. It is mainly made up of limestone (calcium oxide), sand or clay (silicon oxide), bauxite (aluminum oxide), iron ore (ferric oxide), and/or shells, chalk, marl, shale, and blast furnace slag. Cement has been used in various applications since the advent of human civilization, despite its difference from modern cement.

The Greeks and Romans produced the first calcium silicate cement. They discovered that volcanic ash, if finely ground and mixed with lime and water, can produce a hardened mortar. The underlying core to bind the other components is the corresponding gel-like product associated with the pozzolanic reaction (Newman and Choo 2003). Such reaction can happen when using supplementary cementitious materials, such as fly ash, silica fume, and metakaolin, contributing to performance enhancement. In 1793, John Smeaton discovered that certain impure limes containing appropriate reactive silica and alumina contents exhibited hydraulic properties. Later, John Smeaton used this material to rebuild the Eddystone Lighthouse in Cornwall in England. In 1824, Joseph Aspdin was credited with the invention of modern portland cement, which was soon in relatively high demand. The hydraulic potential of ground granulated blast furnace slag (GGBS) was first discovered in 1862 in Germany by Emil Langen.

Cement produced in the early 19th century did not have the same compound composition as modern portland cement. This is because the temperature achieved was not high enough to form the main constituent mineral of tricalcium silicate (C_3S). The introduction of the rotary kiln at the end of the 19th century enables a consistently high enough temperature of 1500°C, which results in a mixed and homogeneous product with C_3S. Since cement manufacturing is highly energy and emissions-intensive, people started to utilize coal fly ash and silica fume in concrete in the 1930s–1950s. Later, natural pozzolan, such as rice hush ash and metakaolin, is used for producing concrete. Because of the fine and/or glassy particle characteristics, hydraulic and/or pozzolanic activity of supplementary cementitious materials can increase social sustainability by replacing cement content and reducing CO_2 emissions. They can also improve fresh and hardened performance of concrete, including increased fluidity, enhanced strength, and reduced permeability.

The production of clinker of portland cement is energy-intensive and responsible for 5% of man-made CO_2 emissions. By reducing the clinker content with supplementary cementitious materials, large CO_2 savings can be achieved. However, these supplementary cementitious materials (SCMs) are gradually being depleted, and alternative sustainable materials are required. With the increasing rapid demand for cement, a new type of cement, limestone calcined clay cement (LC^3), is developed and gained significant attention these years. LC^3 is an appropriate solution for developing sustainable concrete. LC^3 is a ternary blended cement made using clinker, limestone, calcined clay, and gypsum. LC^3 can reduce CO_2 emissions by up to 40%, which is only about 0.3 kg of CO_2 per kg of calcined clay. Using calcined clay and lower kaolinite content, referred to as low-grade calcined clay, has remarkable advantages, including worldwide abundance, comparable early-age strength, capillary porosity refinement, and lower cost. The availability of the materials required to produce LC^3 and the achieved good performance make LC^3 a sustainable replacement for portland cement.

1.2.2 Chemical admixtures

Chemical admixtures are materials added to concrete, mortar, or grout before or during mixing. In addition to reducing the cost of concrete construction and overcoming certain emergencies during concrete operations, chemical admixtures are used to modify the fresh properties to ensure the quality of concrete during mixing, transporting, placing, and curing, and to improve the hardened properties to meet the requirements of specific applications.

In ancient Roman times, organic materials, such as milk, blood, and lard, were used as admixtures in lime-pozzolan mixtures to enhance plasticity (Mielenz 1984). The admixtures that could bring air entrainment are gas-performing agents, such as aluminum, zinc powder, hydrogen peroxide, or surface tension reducers like vinsol, resin, animal, and vegetable fats. In 1930, air-entraining agents were developed and added to concrete to introduce extremely small and closely spaced air bubbles. Hence, the workability and freezing-thawing resistance of concrete are greatly improved. Simultaneously, sulfonated naphthalene formaldehyde, lignosulfonates, and hydroxycarboxylic acid salts were used as plasticizers or retarders. These are typically conventional plasticizers with a low water-reducing range. Although high dosages of plasticizer can provide concrete with lower water demand or enough slump to meet specifications, the water-reducing ability that can be reached is limited. Furthermore, higher dosages can often lead to excessive retardation and slow down early-age strength development (Nmai et al. 1988).

A search for alternatives with a higher slump range led to the introduction of superplasticizers. In the mid-1990s, polycarboxylate-based high-range water reducer was introduced in North America, thus initiating a dramatic paradigm change in people's understanding of how to design and use highly workable concrete or high-performance concrete. It was rapidly accepted due to its flexibility, enhanced workability, workability retention with minimal set retardation, and excellent finishing characteristics. Concrete producers began to use polycarboxylate-based high-range water reducer to develop self-consolidating concrete and applied it to various products. Meantime, another class of chemical admixtures, i.e., viscosity modifying admixtures, has been commercialized to address the need to improve the water tolerance and segregation resistance of self-consolidating concrete and underwater concrete. In 1996, shrinkage-reducing admixtures were followed and helped address cracking issues associated with autogenous and drying shrinkage in HPC.

1.2.3 Fibers

The use of plain concrete as a structural material is limited to a certain extent due to deficiencies like brittleness, low tensile strength, poor resistance to impact and fatigue, as well as low ductility and durability. Incorporating fibers into concrete can prevent and control the initiation, propagation, or coalescence of cracks, thus leading to improved strength and ductility and enhanced fatigue, impact, and wear resistance. Historically, fibers were used in bricks and plaster in the form of straw and horsehair. In the early 1900s, asbestos fibers were used in concrete. Since health risks were found to be associated with asbestos, it is essential to find alternative reinforcements in concrete. In the 1950s, fiber-reinforced concrete became one of the topics of interest. By the 1970s, steel, glass, and certain synthetic fibers, such as polypropylene fibers, were used in concrete. By the 1980s, deformed steel fibers, microfiber, carbon fiber, Spectra, and Kevlar were employed in concrete.

Among these fibers, steel fiber is one of the most popular and widely used ones in both research and practice based on the following facts: (1) steel presents a good affinity with concrete, showing a comparable coefficient of thermal expansion; (2) greatly enhanced strength and toughness of concrete under static and dynamic loading; and (3) ease of use. Owing to these favorable characteristics, steel fiber reinforced concrete has been used at an increasing rate in various applications, including highway and airport pavements, earthquake-resistant and explosion-resistant structures, mine and tunnel linings, bridge repair, hydraulic structures, and rock-slope stabilization (Zhang et al. 2014). In UHPC, steel and carbon fibers are commonly used because of their superior tensile strength. Table 1.1 summarizes chronological advances in cementitious materials, chemical admixtures, and fibers.

1.3 HPC AND UHPC—PAST, PRESENT, AND FUTURE

Concrete developed before 1960 showed heaviness, low tensile-to-compressive strength ratio, poor volumetric stability, poor toughness, and low durability with a shortened lifespan. Increasing interest in the development and use of HPC and UHPC has arisen later. Aalborg Portland initiated research work to develop cement-based materials with ultra-high strength and durability around 1964. Concrete and Research Laboratory in Denmark started to investigate the possibility to produce soft-cast concrete with higher compressive strength. Concrete with a compressive strength in

Table 1.1 Chronological advances in cementitious materials, chemical admixtures, and fibers

Time	*Materials*	*Achievements*
Greek and Roman times	First calcium silicate cement	Mix volcanic ash with lime and water to produce a hardened mortar
Roman times	Retarders	Milk, blood, and lard, as well as organic materials such as molasses, eggs, and rice paste, were added to lime-pozzolan mixtures to enhance their plasticity
	Fibers	Straw and horsehair were used in bricks and plaster
1793	Hydraulic lime	Certain impure limes showed hydraulic properties underwater
1824	Portland cement	Patents in modern portland cement
1849	Reinforced concrete	A French gardener named Joseph Monier first invented the reinforced concrete
1862	GGBS	Hydraulic potential of ground granulated blast furnace slag was discovered
1873	Calcium chloride	Use of calcium chloride in concrete
1900s	Asbestos fiber	Asbestos fibers were used in concrete
1932	Plasticizer	Patents in sulfonated naphthalene formaldehyde plasticizer, but not available in commercial quantities
1930s		Lignosulfonates were used as a plasticizer
1930s		Hydroxycarboxylic acid salts were used as plasticizers and retarders
1930s	Waterproofers	Fatty acids, stearates, and oleates
1930s	Air-entraining agent	Air-entraining agents were developed
1930–1950 s	Coal fly ash	Fly ash, as a partial cementitious material, was used in research and dam application
1941	Air-entraining agent	Tallow and fatty acid soaps used for frost resistance
1947	Silica fume	Silica fume was first "obtained" in Norway and has been extensively investigated at Norwegian Institute of Technology
1963	Superplasticizer	Sulfonated melamine-formaldehyde patent and available
1970s	Fibers	Smooth steel fibers; glass fibers; some synthetic fibers
1980s	Fibers	Deformed steel fibers; microfiber, low-modulus synthetic fibers (polypropylene, nylon, etc.), high-performance polymer fibers, such as carbon, Spectra, Kevlar, etc.
1979	Corrosion-inhibiting admixture	The first corrosion-inhibiting admixture was introduced to help mitigate the impact of chloride salt (NaCl) attack on steel reinforcement
1980–1999	Superplasticizer	Polycarboxylate ether was developed and introduced as admixtures
1996	Shrinkage-reducing admixture	Addressing cracking issues associated with autogenous and drying shrinkage in HPC

the range of 60–80 MPa was developed at that time. Later, Powers from the Portland Cement Association developed a heavily compacted cement paste with a strength of about 280 MPa, indicating the possibility to achieve a much denser packing of cement and resulting in higher strength (1960). The development of superplasticizers in the 1970s led to the emergence of a series of high-performance cement-based materials. In 1972, Yudenfreund et al. (1972a and 1972b) developed cement paste with a compressive strength of 230 MPa at a water-to-cement ratio (w/c) of 0.2 with the use of vacuum mixing process and ultra-fine-ground cement. Roy et al. (1972) obtained a compressive strength of 510 MPa by applying a pressure of 50 MPa and a temperature of up to 250°C. Later in the 1980s, macro-defect-free (MDF) paste with compressive strength over 200 MPa and equivalent bending strength over 150 MPa was achieved by Birchall et al. (1981). MDF was formed under a pressure of 4–10 MPa and a temperature of 80–100°C with the use of water-soluble polymers. Bache (1981) introduced densified small particles (DSP) composed of cement, ultra-fine powders with an average particle size of 0.1–0.2 μm, superplasticizer, and water, based on the particle packing models. Different from the molding process of MDF, DSP is much easier to produce. However, there are some deficiencies in both MDF and DSP. For example, the polymers in MDF are easily hydrolyzed in a highly humid environment, resulting in decreased strength and increased brittleness.

In 1984, Lankard (1984) developed slurry infiltrated fiber concrete (SIFCON) with high compressive and flexural strengths of 238 and 38.5 MPa, respectively, as well as good ductility. A unique forming method of first placing steel fibers in the mold and then infiltrating the slurry into the network was employed. The steel fiber volume in SIFCON is up to 20%, making it quite expensive. In 1986, compact reinforced composite (CRC) was developed with a strong and brittle cementitious matrix, toughened with a high concentration of fine steel fibers, and further reinforced with a high content of larger steel bars (1987). In the 1990s, Haekman et al. (1992) used steel fiber mats to develop slurry-infiltrated mat concrete (SIMCON). Its construction method is similar to that of SIFCON, but the steel fiber volume was reduced from 20% to 4%–6%. SIFCON and SIMCON can be used for special concrete structures or members, such as antiearthquake structures. However, complicated procedures and high costs associated with special fiber-made networks hinder its wide application. In 1992, Li (1992) developed engineered cementitious composites (ECC) based on micromechanics using short metallic and/or nonmetallic fibers. ECC exhibits strain-hardening behavior with a tensile strain capacity greater than 3% and a multiple-cracking behavior with a maximum crack width of less than 100 μm. However, traditional ECC shows low compressive strength of 40–80 MPa, which is not a good option for the application in a harsh environment that needs to resist intense compression loads and severe corrosion. Later,

Table 1.2 Key developments of HPC and UHPC

Year	*Ref.*	*Name*	*Compressive strength*	*Characteristics*
1972	Yudenfreund et al.	-	230 MPa	Paste; vacuum mixing; low porosity; small specimens
1972	Roy et al.	-	510 MPa	Paste; high pressure and high heat; small specimens
1981	Birchall et al.	MDF	200 MPa	Paste; addition of polymer; compressive strength up to 150 MPa
1981	Bache, Young, Aitcin	DSP	120–250 MPa	Improved particle packing; use of silica fume; use of superplasticizers
1980s	Shah, Zia, Russel, Aitcin	HSC, HPC	80–120 MPa	Concrete with special additives and aggregates for structural applications; use of superplasticizers; normal curing; better durability
1984	Lankard, Naaman	SIFCON	Up to 210 MPa	Fine sand mortar with high volume fractions of steel fibers from 8% to 15%
1987	Bache	CRC	Up to 140 MPa	Concrete with high volume of steel fibers used with reinforcing bars
1987	Naaman	HPFRCC	Open range	Mortar and concrete with fibers lead to a strain-hardening response in tension
1992	Li and Wu	ECC	Up to 100 MPa	Mostly mortar with synthetic fibers; strain-hardening behavior in tension
1994	Richard, Cheyrezy	RPC	Up to 800 MPa	Paste and concrete; heat and pressure curing; particle packing
1994	De Larrard	UHPC	Over 150 MPa	Optimized material with dense particle packing and ultra-fine particles and ductile fibers
1998	Lafarge	DUCTAL	Up to 200 MPa	90°C heat curing for 3 d; steel fibers up to 6% (commercially available)
2011	Renade et al.	HSHDCC	Over 150 MPa	Mortar consists of 2% PE fibers; standard room curing for 1 d, and then standard water curing for 7 d, and 90°C hot water curing for 4 d followed by 90°C air curing for 2 d
Until now	Many researchers worldwide	UHPC	>120 MPa open range	Use of superplasticizer, steel fiber, and even coarse aggregate or nanoparticles

Source: Naaman 2012.

Richard and Cheyrezy (1994 and 1995) used fine and active components to develop RPC) through high-temperature curing. Its production emphasizes the use of high binder content up to 800–1200 kg/m^3, low w/b ratio of 0.2, use of fine aggregate, and superplasticizer. To increase the ductility and flexural strength, metallic fibers (e.g., steel fibers) are added. The RPC was classified into two grades based on compressive strength, i.e., RPC200 and RPC800. De Larrard (1994) then introduced the term UHPC. UHPC can obtain a compressive strength of over 120 MPa, a tensile strength of 5–15 MPa, and a bending strength of 25–60 MPa. Later, Renade et al. (2011) developed high strength, high-ductility concrete (HSHDC) with an average compressive strength of 160 MPa and an average tensile strain capacity of 3.5% in terms of the micromechanical-based design approach. Table 1.2 summarizes the key developments of HPC and UHPC from 1970 to date.

1.4 CHARACTERISTICS OF UHPC

UHPC is a newly developed concrete material that exhibits very high compressive strength, dependable tensile strength, and excellent durability properties. It is characterized by a low w/b ratio, small aggregate size, and the use of steel fibers and superplasticizers. ACI 239 (2018) defines UHPC as a new composite material that possesses a minimum specified compressive strength of 150 MPa with specified durability, tensile ductility, and toughness requirements; fibers are generally included to achieve specified requirements. On the other hand, ASTM C1856 (2017) addresses the UHPC specimens with a specified compressive strength of at least 120 MPa, with a nominal maximum size aggregate of less than 5 mm, and a flowability between 200 and 250 mm, for the purpose of determining the performance. The French Standard defines UHPC with a minimum compressive strength of 130 MPa when using Φ110 × 220 mm cylinders (French Standard Institute 2016 and 2017). The Chinese GB/T 31387 (2015) and Swiss prSIA 2052 (2014) believed that the compressive strengths of UHPC should be greater than 100 and 120 MPa, respectively, when using 100 × 100 × 100 mm cubes. Since materials and specimen size tested in each country vary, the minimum strength defined for UHPC is different. These definitions exclusively consider compressive strength to classify a concrete material because testing agencies primarily focus on structural applications.

In addition to the high compressive strength, UHPC also has extremely low porosity, high packing density, and superior durability. UHPC generally exhibits a uniform pore distribution and low porosity. Compared to ordinary concrete with a porosity of 9%–14% (Wu et al. 2017), the porosity of UHPC can be as low as 0.91%–1.32% (Li et al. 2019). Therefore, the resistance of UHPC to harmful gases, liquids, and chloride diffusion can be substantially improved. The water absorption coefficient of UHPC with a w/b ratio of 0.4 was 0.04 at 14 d, but decreased to 0.0025 at a w/b ratio of 0.17. In addition, the permeability coefficient of concrete before 28 d decreased with the prolongation of curing age. The corresponding value of UHPC after 98 d was 0.0005, which is one-third of that of conventional concrete (Tam et al. 2012). The chloride ion diffusion coefficient of UHPC ranges from 0.2×10^{-13} to 4.1×10^{-13} m^2/s, depending on the w/b ratio, curing regime, medium solution concentration, steel fiber volume, and testing age (Li et al. 2020). Due to the low w/c ratio, extremely dense structure, and low porosity, UHPC without fiber shows a relatively gentle decreasing trend in stress-crack opening curve during bar pullout test after initial cracking (Mechtcherine 2009), compared to normal-strength concrete and high-strength concrete. In the presence of steel fiber, the postcrack behavior of UHPC is improved significantly since fibers can transfer stress between matrix and fibers at rupture (Paipetis et al. 1999). As a result, energy dissipation is increased and crack propagation is limited (Naaman et al. 2003). UHPC also has high impact resistance and toughness. It can retain flexural response with minimal damage, due to its high strength and ductility. Under a high-velocity impact loading above 11.2 m/s, the dissipated energy of UHPC can reach up to 10,000 kJ/m^3, depending on the temperature (Ren et al. 2019).

The performances of UHPC are totally different from conventional concrete and HPC, due to the different compositions and mixture proportions. Experimental results have confirmed that the increment of the w/b ratio results in a reduction in the compressive strength of concrete. Figure 1.1 illustrates the classification of concrete type according to the compressive strength and the w/b ratio. Conventional concrete commonly consists of coarse and fine aggregates, water, and cementitious materials at a w/b ratio greater than 0.5. The strength of conventional concrete starts to develop after 7 d, ranging from 10 to 50 MPa. At 28 d, 75%–80% of the total strength can be attained. The slump value and setting time are 25–101 mm and 30–90 mins, respectively, depending on the moisture in the atmosphere, cement fineness, and content, etc. Conventional concrete is weak in tension and is not durable against severe conditions, such as freezing-thawing. HPC emerged due to the development of superplasticizers and in pursuit of long service life. HPC is produced by carefully selected high-quality ingredients and optimized mixture design at a lower w/b ratio of approximately 0.25–0.45. The amount of binder used in HPC manufacturing is generally greater than 600 kg/m^3 or even higher. Superplasticizer plays a significant role in making fluid and workable HPC. The composition of HPC is almost the same as that of conventional concrete, but its compressive strength can reach up to 80 MPa. UHPC is prepared with a low w/b ratio of 0.14–0.24, the use of steel fibers and superplasticizer, and/or absence of coarse aggregate.

Table 1.3 summarizes the typical mechanical properties and durability characteristics of conventional concrete, HPC, and UHPC. Compared to conventional concrete and HPC, UHPC can be developed with very high mechanical properties, superior durability, extended service life, and

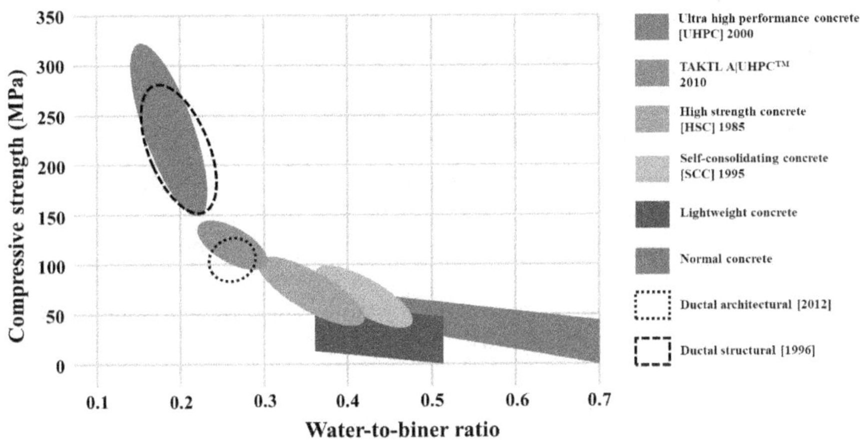

Figure 1.1 Classification of concrete type according to compressive strength and w/b ratio.

Source: Modified based on Bianchi and Gannon 2018.

Table 1.3 Mechanical properties and durability of conventional concrete, HPC, and UHPC

Parameter	*Conventional concrete*	*HPC*	*UHPC*
w/b ratio	~0.5	0.25–0.35	0.14–0.24
Binder content (kg/m^3)	≤450	400–700	800–1200
Chemical admixture	Superplasticizer not necessary	Superplasticizer necessary	Superplasticizer essential
Supplementary cementitious materials	Not necessary	Silica fume or fly ash	Silica fume or ultra-fine essential
Fibers	Beneficial	Beneficial	Essential
Compressive strength (MPa)	20–60	60–100	≥150
Flexural strength (MPa)	3–5	6–10	25–60
Tensile strength (MPa)	≤3	≤5	5–15
Elastic modulus (GPa)	20–30	30–40	40–60
Fracture energy (J/m^2)	30–100	120–500	20,000–40,000
Chloride-ion diffusion coefficient (steady state) ($\times 10^{-12}$ J/m^2)	1	0.6	0.02
Carbonation depth after 3 years (mm)	7	4	≤1.5
Salt-scaling resistance (g/m^2)	≤1500	150	20–50

reduced maintenance cost. Besides, hardened UHPC can have a considerable amount of unhydrated cement particles that can further hydrate, providing self-healing potential for cracked structures.

1.5 THE SCOPE OF THIS BOOK

In a word, the advances in the concrete industry associated with the development of HPC and UHPC mainly include the discoveries of superplasticizers, supplementary cementitious materials, fibers, and special processing procedures. UHPC is one of the most innovative cement composite materials developed in recent years, which is designed based on the theory of dense particle packing through carefully selected high-quality raw materials and special processing technologies. Compared to conventional concrete and HPC, UHPC is characterized by a low w/b ratio of 0.2, high compressive strength, high toughness, and superior durability. These superior properties make UHPC an excellent candidate to meet the requirements of lightweight, high rise, large span, and high durability for civil engineering construction with great potential applications in bridges, antiexplosion structures, thin-walled structures, architectural ornaments, marine structures, and rehabilitated and strengthening members, etc. Thus, UHPC can provide robust solutions for resilient, durable, and sustainable infrastructures.

This comprehensive book intends to summarize the recent progress in UHPC, from materials selection to mixture design, manufacture, fresh and hardened properties, and potential applications. It is dedicated to a group of users composed of universities and testing laboratories, building material companies and industries, material scientists and experts, building and infrastructure authorities, designers, and civil engineers. It aims at extending the new frontiers of concrete materials and structures and renders more strong, ductile, durable, sustainable, and environment-friendly infrastructures.

The book is divided into 13 chapters, including more than 1000 references. Chapter 1 is devoted to the historical evolution of advances in cement and concrete, and the differences in conventional concrete, HPC, and UHPC. Chapter 2 deals with the theoretical principles for designing UHPC and its production technology. Chapter 3 describes raw materials that can be used in UHPC. Chapter 4 focuses on the UHPC mixture design using traditional and/or advanced approaches. Chapter 5 elaborates on the fresh properties of UHPC. Chapter 6 introduces the setting and hardening process and microstructure development of UHPC. Chapter 7 subdivides the static mechanical properties of UHPC, including fiber-matrix bond, compressive, tensile, and flexural behaviors under static loadings. Chapter 8 deals with the dynamic mechanical properties of UHPC and its simulating methods. Chapter 9 describes the autogenous shrinkage and its mitigation techniques, drying shrinkage, as well as the creep of UHPC. Chapter 10 covers the durability aspects of UHPC to various degradation processes, including water permeability, chloride ion permeability, corrosion of steel reinforcement, carbonation, freezing-thawing resistance, chemical attack resistance, alkali-silica reaction, abrasion resistance, and fire resistance. Chapter 11 discusses the self-healing properties of UHPC associated with the continued hydration of unhydrated cement particles. Chapter 12 explores a new type of UHPC prepared using seawater and sea sand, i.e., seawater and sea-sand UHPC. Chapter 13 presents typical and potential applications and case studies of UHPC.

REFERENCES

ACI 239. (2018). Ultra-high-performance concrete: an emerging technology report (ACI 239R-18). ACI Committee, American Concrete Institute.

ACI 363R-92. (1997). State-of-the-art report on high-strength concrete. ACI Committee, American Concrete Institute.

Ahlborn, T.M., Steinberg, E.P. (2012). An overview of UHPC efforts through the North American Working Group, in: Proceedings of the 3rd International Symposium on UHPC and Nanotechnology for High Performance Construction Materials, Kassel University Press GmbH, Kassel, Germany, 43–49.

ASTM C1856/C1856M - 17 (2017). Standard practice for fabricating and testing specimens of ultra-high performance concrete. ASTM International. West Conshohocken, PA, USA.

Bache, H. H. (1981). Densified cement/ultrafine particle-based materials. Second International Conference on Superplasticizers in Concrete, Ottawa, ON, Canada, June 10–12.

Bache, H. H. (1987). Introduction to compact reinforced composite. *Nord. Concr. Res.* 6, 19–33.

Bianchi, G. Q., Gannon, K. (2018). Classification + reference standards for UHPC in architectural applications. The Fifth Best Conference Building Enclosure Science & Technology (BEST5), Philadelphia, USA.

Birchall, J. D., Howard, A. J., Kendall, K. (1981). Flexural strength and porosity of cements. *Nature* 289(5796), 388–390.

Charles, N., Schlagbaum, T., Violetta, B. (1988). A history of mid-range water-reducing admixtures. *Concr. Int.* 20(4), 45–50.

French Standard Institute. (2016). National addition to Eurocode 2-design of concrete structures: specific rules for ultra high performance fiber-reinforced concrete (UHPFRC): AFNORN F. P18-710. France: European Committee for Standardization.

French Standard Institute. (2017). Ultra-high performance fibre-reinforced concrete (UPHFRC)—specification, performance, production and conformity, NF P 18-470, France.

GB/T 31387-2015 (2015). *Reactive powder concrete (GB/T 31387-2015)*, China Architecture and Building Press, Beijing, China.

Graybeal, B. (2008). UHPC in the U.S. highway transportation system, in: Proceedings of the Second International Symposium on Ultra High Performance Concrete, Kassel University Press GmbH, Kassel, Germany, p. 11.

Hackman, L. E., Farrell, M. B., Dunham, O. O. (1992). Slurry infiltrated mat concrete (SIMCON). *Concr. Int.* 14(12), 53–56.

Hajar, Z., Lecointre, D., Simon, A., Petitjean, J. (2004). Design and construction of the world first ultra-high performance concrete road bridges, in: Proceeding of the International Symposium on Ultra High Performance Concrete, University of Kassel, Kassel, pp. 39–48. www.builtconstructions.in/OnlineMagazine/Builtconstructions/Pages/Terry-Harris-and-Dr.-ARA-A.-Jeknavorian-0380.aspx

Lankard, D. R., Newell, J. K. (1984). Preparation of highly reinforced steel fiber reinforced concrete composites. *ACI Spec. Publ.* 81, 287–306.

Larrard, F. D., Sedran, T. (1994). Optimization of ultra-high-performance concrete by the use of a packing model. *Cem. Concr. Res.* 24(6), 997–1009.

Li, V. C., Leung, C. K. (1992). Steady-state and multiple cracking of short random fiber composites. *J. Eng. Mech.* 118(11), 2246–2264.

Li, P.P., Yu, Q.L., Brouwers, H.J.H., Chen, W. (2019). Conceptual design and performance evaluation of two-stage ultra-low binder ultra-high performance concrete. *Cem. Concr. Res.* 125, 105858.

Li, J., Wu, Z., Shi, C., Yuan, Q., Zhang, Z. (2020). Durability of ultra-high performance concrete–A review. *Constr. Build. Mater.* 255, 119296.

Mechtcherine, V. (2009). Fracture mechanical behavior of concrete and the condition of its fracture surface. *Cem. Concr. Res.* 39, 620–628.

Mielenz, R.C. (1984). History of chemical admixtures for concrete. *Concr. Int.* 6(4), 40–53.

Naaman, A.E. (2003). Engineered steel fibers with optimal properties for reinforcement of cement composites. *J. Adv. Concr. Technol.* 1(3), 241–252.

Naaman A.E., Wille K. (2012). The path to ultra-high performance fiber reinforced concrete (UHP-FRC): five decades of progress. *Proceedings of Hipermat* 3–15.

Nmai, C., Tomita, R., Hondo, F., Buffenbarger, J. (1998). Shrinkage-reducing admixtures. *Concr. Int.* 20(4), 31–37.

Newman, J., Choo, B. S. (Eds.) (2003). *Advanced concrete technology 2: concrete properties*. Elsevier.

Paipetis A., C. Galiotis, Y.C. Liu, J.A. (1999). Nairn, Stress transfer from the matrix to the fibre in a fragmentation test: Raman experiments and analytical modeling. *J. Compos. Mater.* 33(4), 377–399.

Powers, T. C. (1960). Physical properties of cement paste. Proceedings of the Fourth International Symposium on the Chemistry of Cement, session V, Washington, USA; 577–613.

prSIA 2052. (2014). Bétons fibrés ultra-performant: matériaux, dimensionnement et exécution (UHPC: Material, dimensioning and construction). Swiss Society of Engineers and Architects.

Ranade, R., Stults, M.D., Li, V.C., Rushing, T. S., Ronth, J., Heard, W. F. (2011). Development of high strength high ductility concrete. 2nd Int. RILEM Conference of Strain Hardening Cement Composites, Rio de Janeiro, Brazil; 1–8.

Rebentrost, M., Wight, G. (2008). Experience and applications of ultra-high performance concrete in Asia, in: Proceedings of the 2nd International Symposium on Ultra-High Performance Concrete, Kassel University Press GmbH, Kassel, Germany, 19–30.

Ren L., Z. Fang, R. Zhong, Wang K. (2019). Experimental and numerical investigations of the seismic performance of UHPC Box Piers. *KSCE J. Civ. Eng.* 23(2), 597–607.

Resplendino J. (2004). First recommendations for ultra-high-performance concretes and examples of application, in: Proceeding of the International Symposium on Ultra High Performance Concrete, University of Kassel, Kassel, 79–90.

Richard, P., Cheyrezy, M. H. (1994). Reactive powder concretes with high ductility and 200–800 MPa compressive strength. *ACI Special Publication* 144, 507–518.

Richard, P., Cheyrezy, M. (1995). Composition of reactive powder concretes. *Cem. Concr. Res.* 25(7), 1501–1511.

Roy, D. M., Gouda, G. R., Bobrowsky, A. (1972). Very high strength cement pastes prepared by hot pressing and other high pressure techniques. *Cem. Concr. Res.* 2, 349–366.

Schmidt M., Fehling E. (2005). Ultra-high-performance concrete: research, development and application in Europe, in: 7th International Symposium on the Utilization of High-Strength- and High-Performance-Concrete, ACI Washington, SP.228-4, pp. 51–78.

Tam, C.M., Tam, V.W., Ng, K.M. (2012). Assessing drying shrinkage and water permeability of reactive powder concrete produced in Hong Kong. *Constr. Build. Mater.* 26(1), 79–89.

Wu, Z., Wong, H., Buenfeld N. (2017). Transport properties of concrete after drying wetting regimes to elucidate the effects of moisture content, hysteresis and microcracking. *Cem. Concr. Res.* 98, 136–154.

Yudenfreund, M., Odler, I., Brunauer, S. (1972a). Hardened portland cement pastes of low porosity. I. Materials and experimental methods. *Cem. Concr. Res.* 2(3), 313–330.

Yudenfreund, M., Skalny, J., Mikhail, R. S., Brunauer, S. (1972b). Hardened portland cement pastes of low porosity. II. Exploratory studies. Dimensional changes. *Cem. Concr. Res.* 2(3), 331–348.

Zhang, P., Zhao, Y. Z., Li, Q. F., Wang, P., Zhang, T. H. (2014). Flexural toughness of steel fiber reinforced high performance concrete containing nano-SiO_2 and fly ash. *Sci. World J.* 2014, 403743.

Chapter 2

Theoretical principles for design and production of UHPC

2.1 INTRODUCTION

UHPC mixtures are designed to achieve ultra-high strength, high toughness, and superior durability through carefully selecting high-quality constituents with the aid of scientific design theories. It is well known that the porosity of concrete can be decreased and the compressive strength and durability can be increased with the reduction of w/b ratio. The porosity of concrete is defined as the fraction of void volume over total volume. UHPC usually shows a w/b ratio ranging from 0.14 to 0.24 with a compressive strength over 120 or 150 MPa, while conventional concrete exhibits a w/b ratio ranging from 0.45 to 0.7 with a compressive strength varying from 10 to 50 MPa. The pores in concrete can be mainly categorized into three types, including gel pores formed in calcium-silicate-hydrate (C-S-H) gel, capillary pores formed in the cement matrix due to the anhydrate cement, and air voids due to entrained air during concrete mixing. The size of gel pores usually ranges from 0.0018 to 0.0025 μm, which are too small to have any significant effect on mechanical and transport properties of concrete. However, the loss of physical and chemical absorbed water in gel pores can lead to shrinkage of concrete. The capillary pores of concrete vary from 10 to 50 nm and can be up to 3–5 μm at a high w/b ratio, while the air voids have a normal range of 50–20 μm, and could be up to 3 mm. Capillary pores and air voids adversely affect the strength and durability of concrete, particularly when they are interconnected. The use of superplasticizer in UHPC results in low total porosity due to the low water demand. The addition of approximately 25% silica fume significantly decreases the capillary porosity and reduces the connectivity of pores. Besides, the absence of coarse aggregates in conventional UHPC further reduces the porosity since the aggregate-matrix interfacial transition zones are more porous than the cementitious matrix due to the wall and bleeding effects, which additionally induces some pores and accounts for partial porosity.

Increasing particle packing density is also a major contributor in achieving low porosity, improving flowability and durability, and reducing defects in

DOI: 10.1201/9781003203605-2

concrete, which plays a significant role in the design of UHPC. The particle packing density of a solid skeleton of concrete mixture can be defined as the volume of solids, including cement, aggregate, and/or supplementary cementitious materials, in a unit volume. It can also be expressed as unity minus porosity. Optimizing the particle packing density of concrete mixtures has several advantages for the fresh and hardened properties. Firstly, fine particles can help fill up the voids in the solid skeleton to render minimum space for water, thus reducing water demand. Secondly, a higher packing density leads to a smaller void ratio and eventually a lower cement amount. This concept has been fully used to produce UHPC to optimize the grading of particles by using the modified Andreasen and Andersen model. Besides, the use of fine quartz sand and high packing density render UHPC with a stronger skeleton, which can restrain the shrinkage and creep of concrete.

UHPC is produced by removing the coarse aggregate, replacing part of the cement with supplementary cementitious materials, adding superplasticizers, and incorporating steel fibers. Special mixing, casting, and curing practice are also needed to produce and handle UHPC. The basic principles involved in the design and production of UHPC can be summarized as follows: (1) increase in homogeneity by eliminating coarse aggregate; (2) increase in density by optimizing grain size distribution of particles to achieve maximum packing of particles; (3) reduction in water demand using superplasticizer; (4) improvement in microstructure by heat treatment; and (5) enhancement in ductility by adding steel fibers. By adhering to the first four principles, high compressive strength and superior durability can be achieved. With the addition of fibers, the tensile and flexural strengths and ductility of UHPC are improved.

This chapter describes the main theoretical principles for the production of UHPC, which involve reduction in porosity, improvement in microstructure, enhancement in homogeneity, and increase in toughness. The scientific basis for mixture design and common production technologies, including mixing, casting and placing, and curing, of UHPC are also discussed.

2.2 THEORETICAL PRINCIPLES FOR THE PRODUCTION OF UHPC

2.2.1 Reduction in porosity

Generally, the porosity and pore structure significantly influence the strength of hardened cement-based materials. The porosity–strength relationship laid the solid foundation for the subsequent research on HPC and UHPC. UHPC is designed based on the close particle packing theory. Thus, the maximum possible packing density of the particles in the system can be achieved. The pore size, distribution, shape, and volume play a decisive role in the performance of UHPC. From a technical perspective, the partial replacement of

Table 2.1 Strength and porosity relationships

Researchers	*Equation*	*Parameters*	*Type of materials*
Powers (1958)	$\sigma = S_0\ (1\text{-}P)^3$	σ is the compressive strength; P is the porosity; S_0 is the compressive strength at zero porosity; P_0 is the porosity at zero strength; n, k_r, k_H, k_s are empirical constants	Cement mortars
Balshin (1949)	$\sigma = S_0\ (1\text{-}P)^n$		Ceramics
Ryshkevitvh (1953)	$\sigma = S_0 \cdot e^{-krP}$		Sintered alumina and zirconia
Schiller (1971)	$\sigma = k_s \bullet \ln\left(\frac{P_{os}}{P}\right)$		Gypsum pastes
Hasselmann (1962)	$\sigma = S_0 - k_H P$		Polycrystalline refractory materials

cement and/or silica fume with other supplementary cementitious materials can further decrease the porosity and the pore connectivity through the filler effect. Furthermore, more C-S-H gels can be formed due to the pozzolanic reaction of silica fume, especially under heat treatment.

The relationships between porosity and compressive strength of cement-based materials are considered as an effective method to predict strength. There have been several equations reported from previous literature (Odler and Rößler 1985; Röβler and Odler 1985). Table 2.1 summarizes the typical relationships between strength and porosity that are suitable for cement-based materials. Although these semiempirical relationships were initially developed for ceramics and crystalline materials, the models were widely applied to concrete. Most other relationships are variations of one of the four typical models of Balshin (1949), Ryshkevitvh (1953), Hasselmann (1962), and Schiller (1971). Röβler and Odler (1985) evaluated the effects of porosity on the strength of cement paste and compared the data to the above-mentioned four strength-porosity models. It is found that for a rational porosity ranging from 5% to 30%, all the models were observed to provide similar values (Röβler and Odler 1985). The optimized constants in the models to best fit experimental data were also found in cement paste with a porosity ranging from 0.1 to 0.3, as presented in Figure 2.1. It is noteworthy that Ryshkevitvh's equation is particularly suitable for low porosity systems, while Schiller's Equation is appropriate for high porosity systems. A reduction in porosity results in a higher strength. On the other hand, given the uses of supplemental cementitious materials and nanoparticles in UHPC, the reliability of these equations would be decreased.

As mentioned earlier, most other relationships in literature are the variations of one of the four typical models presented in Table 2.1. For example, Shi et al. (2015) proposed a relationship between porosity and strength for high-strength concrete with a w/b ratio of 0.16 as follows:

$$\sigma = 245.92 \cdot e^{-0.0832P} \tag{2.1}$$

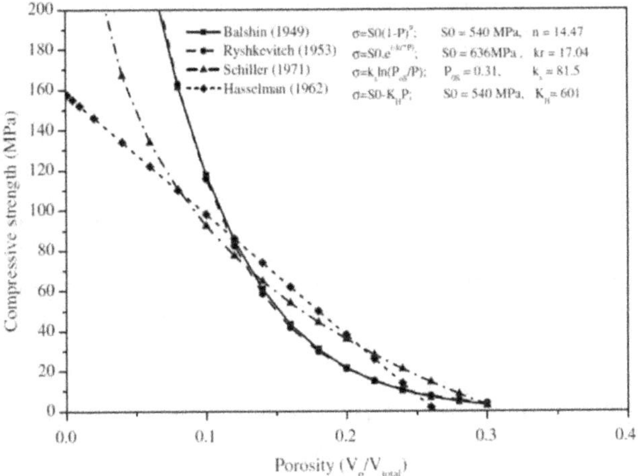

Figure 2.1 Four strength-porosity models with optimized constants.

Source: Rößler and Odler 1985.

where P is the porosity; and σ is the compressive strength of concrete with porosity P. Yu et al. (2015) used the model proposed by Chen et al. (2013) to determine the relationship between the water-permeable porosity and compressive strength of UHPC, as shown in Eq. (2.2):

$$\sigma = \sigma_0 \cdot \left[\left(\frac{P_c - P}{P_c} \right)^{1.85} \cdot \left(1 - P^{2/3}\right) \right]^{1/2} \tag{2.2}$$

where σ is the compressive strength of concrete with porosity of P; P_c is the percolation porosity at the failure threshold; and σ_0 is the compressive strength at zero porosity.

A reduction in the w/b ratio is regarded as the simplest way to reduce the porosity of concrete. The w/b ratio of UHPC usually ranges from 0.14 to 0.22. However, unhydrated particles in the UHPC matrix exist due to the low water content and thus the incomplete hydration of cementitious particles, despite a reduction in the porosity.

2.2.1.1 Close packing of particle materials

UHPC is a typical system of multiple components with a particle size ranging from 0.1 µm to 4.75 cm. The particles should be carefully selected to fill up the voids between larger and smaller particles, hence achieving a high packing density. At the same water content, higher packing density can improve the flowability of concrete by releasing excess water entrapped in the aggregate clusters (Ferraris et al. 2001). Therefore, cement paste can have more water for lubrication. Meanwhile, it can improve the strength of concrete under the same w/b ratio without compromising flowability (Fennis and Walraven 2011; Kwan et al. 2010).

The maximum particle packing density of concrete mixtures can be achieved by optimizing particle size distribution (PSD) and close packing models. Various close-packing models have been proposed, which can be classified into discrete and continuous models. Discrete models use idealized sets of specifically sized particles in creating packing models, which are represented by the Furnas model (Furnas 1931), Aim and Goff model (Aïm and Goff 1968), Toufar model (Toufar et al. 1976), etc. These models were established based on binary or ternary systems, and were unsuitable for the packing density calculation of concrete (Dewar 1986; Dewar 1999). Stovall et al. (1986) proposed a basic multimodal model, such as linear packing density, after considering interactions between the size and classes of the materials used. Afterward, de Larrardand Sedra (1994) improved the model by introducing virtual packing density. However, this model based on the maximum packing density was only attainable if the particles were placed one by one. The improvements of the linear packing density model resulted in the solid suspension model, which includes a virtual packing factor that accounts for the difference between ideal and random packing of particles. Later, a compaction index (K) was introduced to the compressive packing model (de Larrard and Sedran 2002). This index considers the difference between actual and virtual packing densities and the characteristics of particles. Meantime, the compressive packing model still uses the packing of the mono-sized particles to predict the packing of mixtures with different-sized particles. The Andreasen and Andersen particle packing model is another possibility for mixture design based on an optimal PSD of all particle materials used in the mixture.

2.2.1.2 Water reduction using superplasticizer

As mentioned above, the reduced porosity and increased strength of hardened cement-based materials are achieved by decreasing the w/b ratio. The addition of superplasticizer can significantly reduce the required w/b ratio for a given workability, thus effectively reducing the porosity and increasing the strength of concrete (Yoshioka et al. 2002; Gołaszewski and Szwabowski 2004). Superplasticizers with a water-reducing capacity of over 20% are often used in UHPC. Compared with conventional concrete and HPC, fresh UHPC is more viscous due to the high contents of fine particles and fibers. Thus, the selection and use of superplasticizers become more important for producing high-quality UHPC.

2.2.2 Improvement in microstructure

UHPC possesses a very dense and uniform microstructure due to the following fundamental effects: (i) close packing of solid particles; (ii) hydration and pozzolanic reactions of cementitious materials; and (iii) improvement in the interfacial transition zone between aggregates and bulk matrix. Generally, the microstructure of UHPC is mainly comprised of unhydrated cement clinker particles, quartz sand, steel fiber, and hydration products, such as C-S-H and calcium hydroxide (Sorelli et al. 2008). The low porosity of UHPC associated with low w/b ratio restricts the space available for the growth of $Ca(OH)_2$ crystals to a low content. The utilization of elevated temperature curing can accelerate the hydration process of cementitious materials, thus rendering more hydration products and denser microstructure. It is observed that no significant $Ca(OH)_2$ was detected by X-ray diffraction analysis (Reda et al. 1999). The C-S-H with high density in UHPC is characterized by higher stiffness and hardness values than the C-S-H in conventional concrete.

The interfacial transition zone, which is a 30–100 μm wide zone between aggregate and cement matrix, is considered to be the weakest link in concrete due to high porosity and $Ca(OH)_2$ content. It also accounts for part of porosity due to the cracks and pores associated with bleeding and the wall effects of large aggregates stacked on fine particles. The interfacial transition zone in conventional concrete and UHPC from scanning electron microscope (SEM) observations are shown in Figure 2.2. It can be seen that the interfacial transition zone in conventional portland cement concrete is relatively porous. By contrast, the interfacial transition zone observed in UHPC is as dense as the matrix due to the consumption of $Ca(OH)_2$ by pozzolanic reactions and a very low w/b ratio (Richard and Cheyrezy 1995; Chan and Chu 2004). This homogenous structure contributes to excellent performance of UHPC.

2.2.3 Enhancement in homogeneity

Aggregate used in conventional concrete usually has a higher elastic modulus than cement paste and acts as a skeleton. Nevertheless, shear and tensile stresses may induce microcracks at the interfacial transition zone due to the difference in thermal and mechanical properties between aggregate and matrix. These microcracks would result in stress concentration under loading, thus leading to cracking.

The crack size is proportional to the aggregate size. For the initially developed UHPC, coarse aggregate is eliminated instead of using quartz

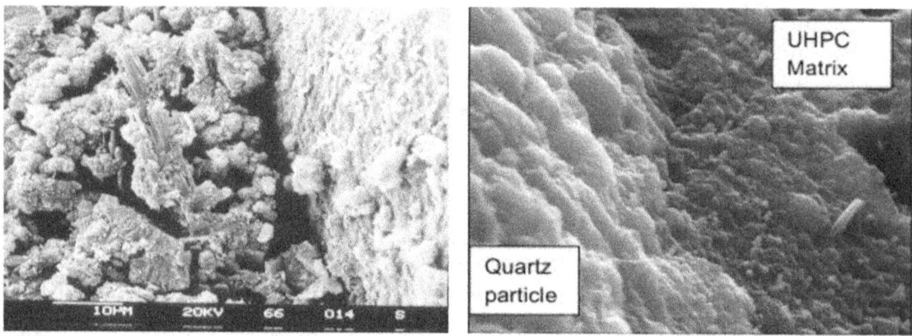

(a) Interfacial transition zone in conventional mortar (b) Interfacial transition zone in UHPC

Figure 2.2 Comparison of SEM observations of interfacial transition zone between conventional concrete and UHPC. (a) Interfacial transition zone in conventional mortar (Shi et al. 2015); (b) Interfacial transition zone in UHPC.

Source: Schmidt and Fehling 2007.

sand with a maximum diameter of 600 μm, which would considerably decrease the size of microcracks in UHPC (Richard and Cheyrezy 1995). Meanwhile, the aggregate size reduction also decreases the defect, thereby reducing the inhomogeneity of concrete. As discussed in Section 2.2.2, the interfacial transition zone in UHPC looks the same as the matrix, indicating the homogeneity of its microstructure.

2.2.4 Increase in toughness

Toughness is used to characterize the ability of a material to resist fracture by measuring the energy absorption capacity (Prabha et al. 2010). As a typical quasi-brittle material, plain concrete has low tensile strength, strain capacity, and fracture toughness. The incorporation of fibers into concrete can prevent and control the initiation, propagation, or coalescence of cracks. Defects, such as microcracks and/or air voids, intrinsically exist in concrete. Under loading, the whole system of fiber-reinforced concrete (FRC) sustains loads, and the microcracks propagate. With the increase in loading, the load is then transferred to the fibers through the fiber–matrix interface, hence the propagation of cracks is restrained. Macrocracks form and fibers are pulled out or ruptured with further loading.

Figure 2.3 illustrates tensile load-deflection curves of normal concrete (NC), FRC, and UHPC. It can be observed that UHPC shows much higher tensile strength and more ductile behavior than FRC. The tensile behavior can be distinguished into three phases, including linear elastic, strain hardening, and strain softening phases. UHPC exhibits linear elastic

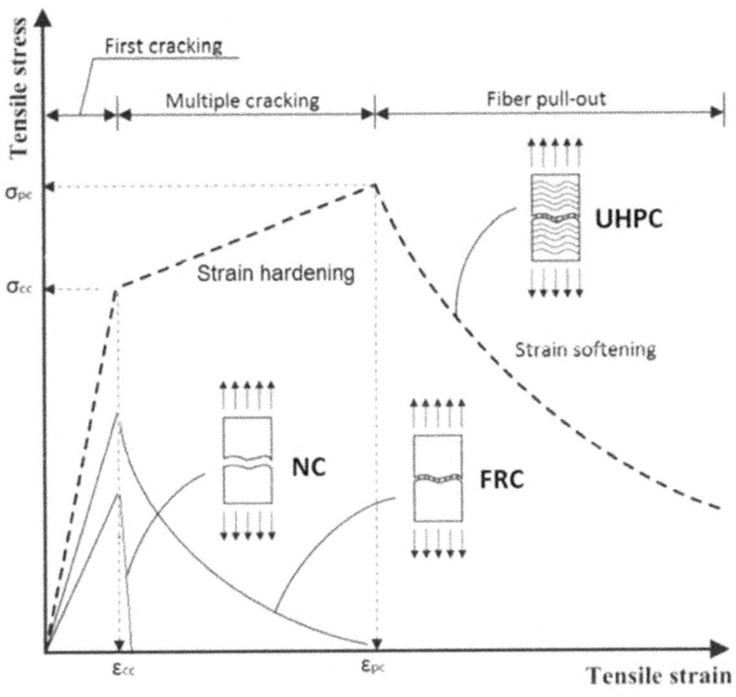

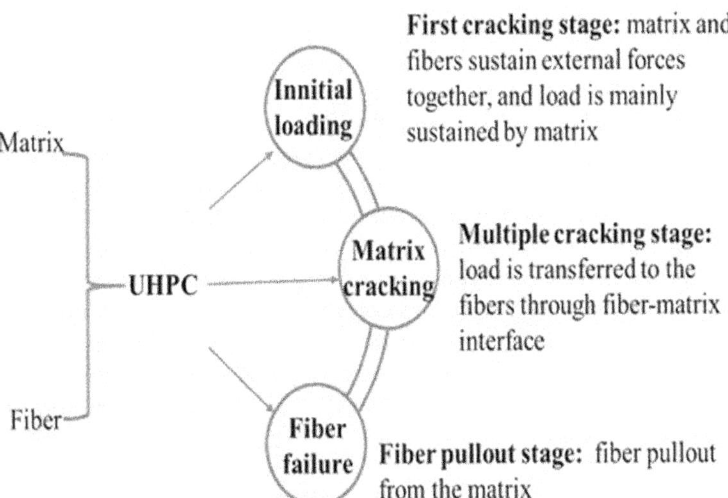

Figure 2.3 Tensile load-deflection curves of NC, FRC, and UHPC (Tran et al. 2014). (a) Tensile load-deflection curves of NC, FRC, and UHPC; (b) Various stages experienced during loading.

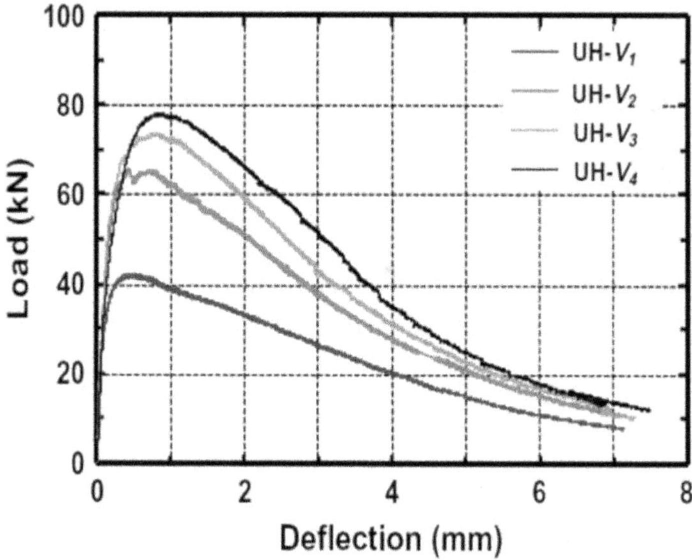

Figure 2.4 Flexural load-deflection curves of UHPC with various fiber contents.

Source: Yoo et al. 2013.

behavior below stress of approximately 70%–90% of the ultimate tensile strength. The increase in fiber content can further improve the tensile/flexural strength and toughness of UHPC.

Figure 2.4 shows flexural load-deflection curves of UHPC with 1%–2% steel fibers, by the volume of concrete. Steel fiber volume has a limited effect on the first cracking strength and first cracking deflection of UHPC, but shows a significant effect on the postcracking behavior. Higher steel fiber content can result in greater peak load and toughness. Deformed fiber, such as hooked and corrugated fibers, increases the flexural strength of UHPC by 20% to 40%, compared to that of straight fiber with the same fiber content (Wu et al. 2018). Longer fiber enhances the flexural strength and energy absorption capacity of UHPC, but exerts a limited effect on the first cracking properties (Yoo et al. 2016). The incorporation of 2.5% steel fibers can render UHPC with tensile strain-hardening behavior (Wille et al. 2011).

2.3 COMPOSITION DESIGN

The design of UHPC aims to achieve a dense cementitious matrix with good workability and excellent hardened performance through carefully

Table 2.2 Mixture proportions of UHPC in literature

Ref. Materials	*Graybeal (2006)*	*Azmee and Shafiq (2018)*	*Yu et al. (2014)*	*Ritter and Curbach (2015)*	*Mohammed (2015)*
				kg/m³	
Cement	712	1114	700	832	900
Silica fume	231	169	44	135	135
Fine sand	1020	1072	1055	-	1125
Micro sand	-	-	219	-	-
River sand	-	-	-	-	-
Ground quartz	211	-	175	207	-
Quartz sand	-	-	-	975	-
Steel fiber	156	234	-	192	160
Superplasticizer	30.7	40	46	30	54
W/C ratio	0.15	0.19	0.29	0.20	0.23
28d compressive strength (MPa)	-	-	149	174	194

selecting high-quality materials. The composition design should combine optimal proportions of all constituents to fulfill multiple requirements for fresh and hardened properties of concrete with a particular application. The raw materials used in UHPC usually include cementitious materials, quartz powder, quartz sand, superplasticizer, and fibers, as shown in Table 2.2 (Xue et al. 2020). Quartz sand usually has a particle size ranging from 150 to 600 μm, and it is dimensionally the largest granular material. Powder quartz possesses a PSD ranging from 0.1 to 100 μm and is generally considered as an inert filler. The fiber used in UHPC is often micro steel fibers with a length of 13 mm and a diameter of 0.2 mm. Optimization of raw materials and their mixture proportion results in an improved performance of UHPC. Detailed information on raw materials and their influence on the performance of UHPC will be described in Chapter 3.

2.3.1 Principles for selecting raw materials

The raw materials for UHPC must be selected based on availability and potential for high workability, strength, and durability, as well as low autogenous shrinkage. C_3S and C_2S contribute the most to the strength of cement-based materials, while C_3S has a remarkable effect on the volume stability of hardened paste. Cement with higher contents of C_3S and C_2S, lower C_3A and alkali contents, and greater Blaine surface area or lower fineness is recommended for use in UHPC. Replacing cement with supplementary cementitious materials can reduce the hydration heat and postpone the occurrence of its peak, improve workability, decrease early-age strength, increase later-age strength and durability, and reduce the cost of UHPC.

Therefore, supplementary cementitious material with high quality has the priority to be selected.

Aggregate takes up about 30%–40% of UHPC volume and greatly affects the volume stability. The quality of aggregates, such as chemical components, mineral compositions and structure, strength, density, thermal performance, particle size, shape, etc., has an essential effect on the technical performance and economic benefits of UHPC. For example, the soundness, size, and cleanliness of aggregate are important parameters affecting the strength and elastic modulus of UHPC. Aggregate gradation mainly influences aggregate and cement proportions and water demand of UHPC. Thus, aggregates used in UHPC are required to meet appropriate specifications and should be clean, strong, and durable in general. The high soundness, clean surface, good gradation, and low linear expansion coefficient can ensure high strength and low shrinkage of UHPC. Proper lightweight aggregate with saturated water can be employed to mitigate the autogenous shrinkage of UHPC, which is considered to be a major concern for this material.

Chemical admixtures are generally products used in relatively small quantities to improve the properties of fresh and hardened concrete. Superplasticizers or high-range water reducers are typically used in UHPC for water reduction and higher workability. The admixture coats the surfaces of the cementitious particles by electrostatic repulsion and hindrance, and

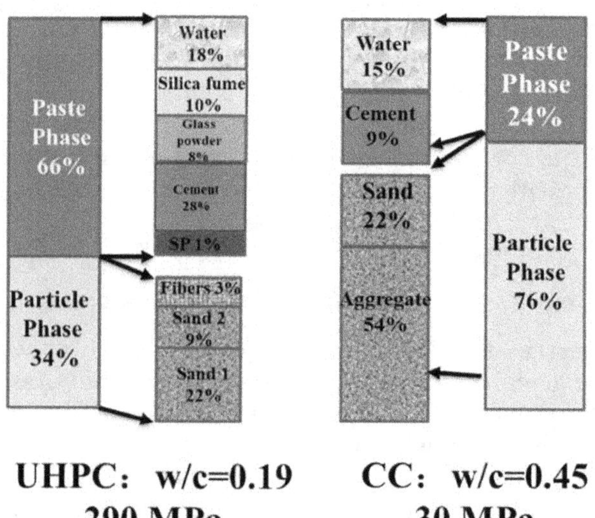

Figure 2.5 Typical volumetric proportions of conventional concrete and UHPC.

Source: Adapted based on Naaman and Wille 2012.

helps the particles stay dispersed in water. An appropriate superplasticizer dosage should be secured. Otherwise, incompatibility issues would occur between cementitious materials and admixture, leading to adverse effects on the setting and properties of UHPC.

As the nonfibrous UHPC matrix is very brittle, fiber is added to obtain elastic-plastic or strain-hardening behavior in tension. High-strength fibers, especially steel fiber, with great capability to improve ductility and to eliminate some of the mild steel reinforcements, are normally recommended for UHPC. In those structural building applications with required fire resistance, polypropylene fibers are often used in combination with steel fibers. The melting of polypropylene fibers at about 165°C can form pores and channels to release vapor and reduce the build-up of pore pressure, thus improving the fire resistance of UHPC.

2.3.2 Volumetric mixture proportions of UHPC

As depicted in Figure 2.5, the volumetric mixture proportion between conventional concrete and UHPC varies in cementitious materials, aggregates, and water contents. Conventional concrete with a w/b ratio of 0.45 contains about 60%–70% aggregate, 9% cement, and 15% water, by volume of concrete. The aggregate volume in UHPC significantly decreases to 31%, while the cement content increases to 28% in the presence of 10% silica fume and 5% glass powder. This greatly increases the viscosity and cost of UHPC.

2.3.3 Optimization of particle size distribution curves

Concrete properties can be improved by optimizing the aggregate grading, as first presented by Feret (1897). The optimized aggregate mixtures can be obtained through reaching the predefined grading curves. Grading models to optimize the PSD mainly include the Fuller and Thompson model (1907), the Andreasen and Andersen packing model (1930), the modified Andreasen and Andersen model or Funk and Dinger model, and the Rosin-Rammler (R-R) model (Mehdipour and Khayat 2018). The modified Andreasen and Andersen model believes that any real size distribution of particles must have a finite lower size limit. The modified Andreasen and Andersen curve considers the minimum particle size in the mixture as follows:

$$P(D_i) = \frac{D_i^q - D_{\min}^q}{D_{\max}^q - D_{\min}^q} \tag{2.3}$$

where $P(D_i)$ is the fraction of the total solids being smaller than size D_i (cumulative percent passing a sieve of size D_i); D_i is the particle size (μm);

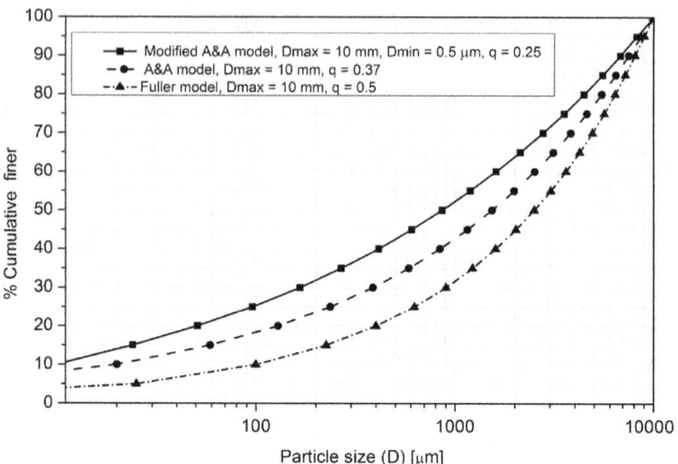

Figure 2.6 PSDs for Andreasen and Andersen, modified Andreasen and Andersen, and Fuller and Thompson models.

Source: Sohail et al. 2018.

D_{max} and D_{min} are the maximum and minimum particle sizes (μm), respectively; and q is the distribution modulus, which is in the range from 0 to 1.

Different types of concrete can be designed using Eq. (2.3) by applying different distribution modulus values of q, which determines the proportion between the fine and coarse particles in the mixture. Mathematically, q determines the curvature of this cumulative PSD. Higher values greater than 0.5 can lead to a coarse mixture, while lower q values less than 0.25 result in a concrete mixture rich in fine particles. Figure 2.6 illustrates PSDs for Andreasen and Andersen, modified Andreasen and Andersen, and Fuller and Thompson models with q values of 0.37, 0.25, and 0.5, respectively. The modified Andreasen and Andersen curve was used as a target function to optimize the composition of granular materials in UHPC with a high proportion of fine particles (Yu et al. 2014; Van Tuan et al. 2011c; Yu et al. 2015). Hunger (2010) recommended using a q value in the range of 0.22–0.25 for the design of self-consolidating concrete. Yu et al. (2014) employed a q value of 0.23 for the UHPC mixture design.

The proportions of each material in the mixture are adjusted until an optimum fit between the composed mixture and the target curves is reached, using an optimization algorithm based on the least squares method, as presented in Eq. (2.4).

$$RSS = \sum_{i=1}^{n} \left[P_{mix}(D_i) - P_{tar}(D_i)\right]^2 \rightarrow \min \quad (2.4)$$

where RSS is the sum of the squares of the residuals at defined particle size D_i; P_{mix} is the composed mixture; and P_{tar} is the target grading calculated from Eq. (2.3).

When the deviation between the target and the composed mixture curves, expressed by the sum of the squares of the residuals at defined particle sizes, is minimized, the composition of the concrete is treated as the best one. This mixture design method is often carried out through software, such as MATLAB®. The obtained mixture proportion of UHPC is verified by experiments.

2.4 PRODUCTION TECHNOLOGIES

2.4.1 Mixing

Due to the low water content with respect to fine content (<0.125 mm) and the high dosage of superplasticizer, the production of UHPC requires more mixing energy to homogenize the components. Thus, its mixing time is greater than that of conventional concrete, usually in the range of 7–18 minutes (Mazanec et al. 2010). This impedes continuous production processes and reduces the capacity of concrete plants.

The mixing efficiency and mixing performance are greatly affected by mixing time, mixing speed, temperature, mixing sequence, and way of addition (Dils et al. 2015). Almost all conventional concrete mixers can be used to mix UHPC (Russell et al. 2013). For the most efficient and consistent mixing of UHPC, high-shear mixers, especially counter-current pan mixers providing accelerated mixing times, can be used. These high-shear mixers efficiently disperse water and admixtures onto cement particles. However, the increased energy input combined with the eliminated coarse aggregate and low water content can generate heat during the mixing process. To ensure that the UHPC does not overheat during mixing, the use of high-energy mixer, lowering the temperature of constituents, and partially or fully replacing mixing water with ice, can be applied (Russell et al. 2013). Other mixers, such as mortar, horizontal shaft or pan mixers with different mixing speeds, can also be used.

For the mixing sequence, dry powder components are mixed first for 1–2 minutes to ensure they are fully blended. Water and superplasticizer are then introduced and mixed for approximately 5–10 minutes. Weighted fibers are added deliberately when the mortar matrix shows appropriate flowability to ensure uniform distribution through the matrix. The fiber introduction process is completed within 2 minutes. In the case of incorporating hybrid fibers with different sizes, microfibers are added first by hand into the mortar mixture, and then macro fibers are dispersed (Kim et al. 2011). SAMARIS research program in Europe recommended a mixing procedure for UHPC as follows: (i) cement and silica fume are put in a dry mixer (without prewetting) and mixed for 2 min; (ii) the mixing is stopped and fine quartz sand is added and mixed for 1 min; (iii) water and superplasticizer are added and mixed for about 4 min; and (iv) fibers are then added and mixed for about 1 min (Denarié 2006).

The use of external or internal energy, such as vacuum mixing and pressure before and during setting in sample preparation, is considered to be an effective way to reduce the pore volume and obtain the desired mechanical properties of UHPC (Dils et al. 2012, 2015). Vacuum mixing with three different pressure levels at vacuum of 100 mbar can influence the performance of UHPC. The compressive strength increased with the decrease of mixing pressure due to a decrease in the air content of concrete under low pressure (Dils et al. 2012). Vandenberg and Wille (2018) employed a novel mixing technology that combines the principles of reciprocating movement agitation and acoustic streaming micromixing zones called resonant acoustic mixing technology to prepare UHPC. Compared to a tabletop paddle mixer, the resonant acoustic mixing reduced the workability, but improved the mechanical properties. However, these sophisticated technologies are too costly to meet the demand of large-scale project engineering.

UHPC can be successfully mixed in the field using commonly available mortar mixers. Onsite mixing has special challenges that may go beyond plant mixing. For example, mixing and casting UHPC on a warm day leads to two complications: (1) a reduction in the slump flow because the excessive temperature compromises the effectiveness of the superplasticizer; and (2) the potential for evaporation of water during mixing and placement. To address the former, it is recommended that about 40% of the mixing water should be replaced with ice. Substantially hotter days will require greater ice quantities, which can be ascertained by trial and error. The objective of this action is to cool the mixture to less than 30°C to ensure the effectiveness of superplasticizer. The latter issue can only be resolved by speeding up the mixing and placing processes.

2.4.2 Casting and finishing

The casting of all UHPC specimens should be completed within appropriate duration, roughly 20 minutes, after mixing. Air bubbles are more easily entrapped in UHPC during mixing due to the lower w/b ratio and higher viscosity in a fresh state than conventional concrete. This can generate a negative effect on the strength of concrete. UHPC specimens can be cast on a vibrating table and are allowed to remain on the table for a short time, i.e., 30 s to 2 min, after filling to reduce air content entrapped during casting. It should be noted that over-vibration of fresh UHPC can lead to fiber settlement in the formwork and inadequate fiber distribution. When placing the self-leveling UHPC material into formworks, it is important to take advantage of its fluid characteristics.

The casting method significantly influences the reinforcing and toughening effects of fibers by affecting fiber dispersion and orientation, thus the mechanical properties of composite materials (Yoo and Banthia 2016; Kang et al. 2011). The casting sequence of UHPC should be planned to achieve preferential fiber dispersion and orientation. Molds are filled slowly to prevent entrapped air, and limited external vibration can be used to aid in air removal. The fresh UHPC can be placed either from one end to the other end of the mold or from the center. It is reported that UHPC placed from one end of the prismatic mold has better mechanical performance than that from the center (Yang et al. 2010). Yoo et al. (2014) reported that the beams cast in the edge had lower flexural behavior due to more fibers located in the middle of the beam specimens than those cast in the middle. Barnett et al. (2010) studied the effect of casting methods on flexural strength by pouring three round panel specimens in three different ways. It is found that fibers tended to align perpendicular to the flow of concrete. Panels poured from the center were found to have the highest strength. Kang et al. (2010) evaluated the fiber distribution characteristic of the flexural strength of UHPC under two different placing directions. The initial cracking and ultimate flexural strengths of UHPC placed parallel to the longitudinal direction of the mold were 5.5% and 61% greater than that placed transversely, respectively.

UHPC is often difficult to finish because of its sticky and viscous nature. High cementitious materials contents, high dosages of chemical admixtures, low water content, and air entrainment all contribute to the difficulty of finishing UHPC. Since fresh UHPC sticks to the trowels and other finishing equipments, finishing activities should be minimized. The finishing sequence should be modified from that used for conventional concrete.

2.4.3 Curing

Curing is the maintenance of satisfactory moisture content and temperature in a freshly placed cementitious mixture for a sufficient period to allow hydraulic reaction and/or pozzolanic reaction to occur so that the desired properties may develop. UHPC specimens are often demold after approximately 24 h of casting and then cured to achieve the required performance. The curing of UHPC is even more important than that of conventional concrete. Curing methods for UHPC vary for different circumstances and rely on two components, i.e., temperature and moisture. Maintaining an appropriate temperature within a certain duration is critical to achieve the desired rate for chemical reactions of cementitious materials, thus determining the microstructure and hardened properties of UHPC. In addition, given the low water content used in UHPC, UHPC shows high autogenous shrinkage, resulting in potential cracking and impaired hardened properties. Eliminating internal water loss by sealing the system or maintaining a high-humidity environment by compensating for additional water using internal curing agents, such as superabsorbent polymer, is also critical.

The curing of UHPC includes two phases: initial and second curing (Graybeal 2011). Given that UHPC tends to exhibit a dormant period before the initial setting, the initial curing phase consists of maintaining an appropriate temperature while precluding moisture loss until setting has occurred and rapid mechanical property growth occurs. The second curing phase may or may not include elevated temperature conditions and a high moisture environment, depending on whether accelerated attainment of particular material characteristics is desired.

Postset curing procedures, including standard room temperature curing, air curing, heat curing under atmospheric pressure, and autoclave curing, can be employed for UHPC in both laboratory and practical applications. Since UHPC contains a small amount of mixing water, surface drying may affect its hardening properties. It is important to cover the specimen surfaces with plastic sheets as quickly as possible after pouring to prevent moisture loss. Onsite or laboratory-manufactured specimens should be cured according to specifications, such as ASTM C31/C31M (2022) and ASTM C192/C192M (2019), respectively. The following section discusses the above curing regimes for UHPC.

2.4.3.1 Standard room temperature curing

Standard room temperature curing is the most common, economical, and environmentally friendly method in practice. The specimens are kept in a standard laboratory environment with a temperature of 20°C and a relative humidity greater than 95% or saturated limestone water from demolding until testing age. Normally, 28-d standard curing can render UHPC with relatively stable strength. If the curing age is reasonably prolonged, the compressive strength of UHPC can continually increase up to 200 MPa (Zhang et al. 2008). Moreover, UHPC with wet curing at 20°C for 91 d showed similar compressive strength to that under steam curing at 90°C (Koh et al. 2007). This indicates that adequate standard room temperature curing is comparable to heat curing with less carbon emission, which can decrease fabrication costs.

2.4.3.2 Steam curing under atmospheric pressure

Steam curing often starts at about 24 h after casting and keeps in a chamber of 60 or 90°C temperature for 24–48 h. This procedure often includes increasing steam and decreasing steam with preset changing rates. Compared to 28-d standard room temperature curing, an increased compressive strength of about 14–30 MPa was achieved by 24 h steam curing (Zhang et al. 2008). This is because elevated temperature curing can accelerate the hydration of cement and promote the pozzolanic effects of mineral admixtures. In addition, C-S-H and microstructure are changed under the elevated curing temperature. The pozzolanic effect is weak due to the formation of C-S-H with short-chain length when the curing temperature is 20°C. However, the pozzolanic effect becomes strong and more C-S-H with long-chain length is generated as the curing temperature increases to 90°C. When the curing temperature rises to 250°C, new crystals could be formed (Masse et al. 1993).

2.4.3.3 Autoclave curing

Autoclave curing is a technique in which concrete is cured with high pressure and temperature in a sealed vessel, which is suitable for HPC and UHPC. The main parameters in autoclave curing, including autoclave temperature, pressure, and time, have significant effects on the mechanical properties of UHPC. Zhang (2018) employed the following autoclave curing procedures for the preparation of UHPC. The autoclave was first evacuated for 0.5 h, then it was heated, and steam was pressed in. The temperature was increased to 190–200°C, and the pressure was increased to 1.2 MPa in 1 h. The set temperature can be high, up to 250°C in some research. Afterward, the temperature and pressure in the autoclave were kept stable for about 6 h. The temperature was then gradually decreased to room temperature, and the pressure was reduced to the atmospheric pressure in 2 h. It is found that the autoclave temperature, pressure, and period greatly influence the properties of UHPC. Yazıcı et al. (2013) investigated the effects of pressure-temperature level (1 MPa – 180°C, 2 MPa – 210°C, and 3 MPa – 235°C) and duration (0, 4, 6, 8, 10, 12, and 24 h) on properties of UHPC under autoclave curing with the heating rate of 1.1°C/m. It was found that there is a critical duration time for each pressure and temperature condition. Extended time or increased temperature and pressure can increase the crystallinity of binder, resulting in an increase in strength up to the maximum value and then a slight decrease in mechanical performance (Yazıcı et al. 2013).

Usually, the compressive strength of UHPC after autoclave curing was higher than those after standard room temperature curing and heat curing due to a significant change in the hydration of cementitious materials. It is reported that compressive strength higher than 200 MPa could be achieved for UHPC with 3% or 4% steel fibers at 8 h autoclave curing (Yang et al. 2009). The compressive strength of UHPC mixtures cured with 8 h high-pressure steam at 210°C was higher than the specimens cured in water at room temperature (Yazıcı 2007). Autoclave curing leads to the formation of tobermorite at temperatures below 200°C. Andxonotlite forms at temperatures of about 250°C in pure heat treatment process without steam curing (Cheyrezy et al. 1995). Massidda et al. (2001) studied the effects of autoclave curing at 180°C on the physical properties of reactive powder concrete (RPC). It is reported that specimens cured at 180°C for 3 h after 3 d precuring at ambient temperature reached 30 and 200 MPa of flexural and compressive strengths, respectively. It should be noted that α-C_2S from the hydration of C_3S and C_2S could be formed in the lack of SiO_2. The final hydration product of α-C_2S is a hydrogarnet phase, instead of C-S-H, C_3A, and C_3AF (Yazıcı et al. 2008).

2.5 SUMMARY

This chapter summarizes the theoretical design principles and production technologies of UHPC as follows:

(1) The main principles for the production of UHPC are reduction in porosity, improvement in microstructure, enhancement in homogeneity, and increase in toughness. Raw materials, preparation techniques, and curing regimes significantly influence the properties of UHPC.
(2) The raw materials for UHPC must be selected based on availability and potential for high workability, strength, and durability, as well as low autogenous shrinkage. Appropriate alternative materials are recommended to replace certain constituents of UHPC to reduce production costs and improve the homogeneity and properties of UHPC.
(3) Due to the low water content with respect to high content of fine particles and superplasticizer dosage, the production of UHPC requires more mixing energy to homogenize the components. Thus, its mixing times are greater than that of conventional concrete. Self-leveling or dry-casting formulations are possible, depending on the casting technique and performance requirements. Because of the sticky and viscous nature of UHPC, its finishing process is more challenging.
(4) UHPC specimens are often demold after approximately 24 h of casting and then cured to achieve performance requirements. Postset curing procedures, including standard room temperature curing, air curing, steam curing under atmospheric pressure, and autoclave curing regimes, can be employed for UHPC in both laboratory and practical applications.

REFERENCES

Aïm, R. B., P. L. Goff. (1968). Effet de paroidans les empilementsdésordonnés de sphères et application à laporosité de mélanges binaires. *Powder Technol.* 1(5), 281–290.

Aïtcin, P. C. (1998). *High performance concrete*. CRC press.

Aldea, C., F. Young, Wang, K., Shah, S. P. (2000). Effects of curing conditions on properties of concrete using slag replacement. *Cem. Concr. Res.* 30(3), 465–472.

Andreasen, A. H. M., Andersen, J. (1930). Über die BeziehungzwischenKornabstufung und Zwischenraum in ProduktenauslosenKörnern (miteinigenExperimenten). *Colloid Polym. Sci.* 50(3), 217–228 (in German).

ASTM C31/C31M (2022). Standard Practice for Making and Curing Concrete Test Specimens in the Field. ASTM International: West Conshohocken, PA, USA.

ASTM C192/C192M (2019). Standard Practice for Making and Curing Concrete Test Specimens in the Laboratory. ASTM International: West Conshohocken, PA, USA.

Azmee, N.M., Shafiq, N. (2018). Ultra-high performance concrete: from fundamental to applications, Case Stud. *Constr. Mater*. 9, e00197.

Bai, J, Wang, M., Shi, C., Sha S., Xiang, S, Zhou B. (2020). Design and synthesis of viscosity-reducing polycarboxylic acid water-reducing agent and its role in cement-silica fume system with low water-binder ratio. *Mater. Rep*. 34(06), 176–183 (in Chinese).

Balshin, M.Y. (1949). Relation of mechanical properties of powder metals and their porosity and the ultimate properties of porous metal-ceramic materials. *Dokl. Akad. Nauk. SSSR* 67(5), 831–834.

Barnett, S. J., Lataste, J., Parry, T., Millard, S. G., Soutsos, M. N. (2010). Assessment of fibre orientation in ultra high performance fibre reinforced concrete and its effect on flexural strength. *Mater. Struct.* 43(7), 1009–1023.

Chan, Y., S. Chu (2004). Effect of silica fume on steel fiber bond characteristics in reactive powder concrete. *Cem. Concr. Res.* 34(7), 1167–1172.

Chen, X., S. Wu, Zhou, J. (2013). Influence of porosity on compressive and tensile strength of cement mortar. *Constr. Build. Mater.* 40, 869–874.

Cheyrezy, M., Maret, V., Frouin, L. (1995). Microstructural analysis of RPC (Reactive Powder Concrete). *Cem. Concr. Res.* 25, 1491–500.

Collepardi, S., L. Coppola, Troli, R., Collepardi, M. (1997). Mechanical properties of modified reactive powder concrete. *ACI Special Publications* 173, 1–22.

Courtial, M. and M. N. de Noirfontaine, Dunstetter, F., Signes-Frehel, M., Mounanga, P., Cherkaoui, K., Khelidj, A. (2013). Effect of polycarboxylate and crushed quartz in UHPC: Microstructural investigation. *Constr. Build. Mater.* 44, 699–705.

de Larrard, F., T. Sedran (1994). Optimization of ultra-high-performance concrete by the use of a packing model. *Cem. Concr. Res.* 24(6), 997–1009.

de Larrard, F., T. Sedran (2002). Mixture-proportioning of high-performance concrete. *Cem. Concr. Res.* 32(11), 1699–1704.

Denarié, E. (2006). SAMARIS D25b-Guidance for the use of UHPFRC for rehabilitation of concrete highway structures, MCS/EU.

Dewar, J. D. (1986). Ready-mixed concrete mix design. *Munic. Eng.* 3, 35–43.

Dewar, J. (1999). *Computer modelling of concrete mixtures*. CRC Press.

Dils, J., G. De Schutter, Boel, V. (2012). Influence of mixing procedure and mixer type on fresh and hardened properties of concrete: a review. *Mater. Struct.* 45(11), 1673–1683.

Dils, J., G. De Schutter, Boel, V., Braem, E. (2012). Influence of vacuum mixing on the mechanical properties of UHPC. 3rd International Symposium on UHPC and Nanotechnology for High Performance Construction Materials: Ultra-High

Performance Concrete and Nanotechnology in Construction (HIPERMAT-2012), Kassel University Press GmbH.

Dils, J., Boel V., De Schutter, G. (2015). Vacuum mixing technology to improve the mechanical properties of ultra-high performance concrete. *Mater. Struct.* 48 (11), 3485–3501.

Dupont, D., Vandewalle L. (2005). Distribution of steel fibres in rectangular sections. *Cem. Concr. Compos.* 27(3), 391–398.

El-Tawil, S., Tai, Y. S., Belcher II, J. A., Rogers, D. (2020). Open-Recipe Ultra-High-Performance Concrete. *Concr. Int.* 42(6), 33–38.

Fennis, S., Walraven, J. (2011). Design of ecological concrete by particle packing optimization. Tese (doutorado), Technische Universiteit Delft, Delft, Netherlands, 256.

Feret, R. (1897). Etudes sur la constitution intime des mortiershydrauliques. Bulletin de la Socitd'Encouragement pour l'Industrie Nationale 2, 1591–1625 (in French).

Ferrara, L., Ozyurt, N., Di Prisco, M. (2011). High mechanical performance of fibre reinforced cementitious composites: the role of "casting-flow induced" fibre orientation. *Mater. Struct.* 44(1), 109–128.

Ferraris, C. F., Obla, K. H., Hill, R. (2001). The influence of mineral admixtures on the rheology of cement paste and concrete. *Cem. Concr. Res.* 31(2), 245–255.

Fuller, W., Thompson, S. E. (1907). The laws of proportioning concrete. *Trans. Am. Soc. Civ. Eng.* 59(2), 67–143.

Funk, J. E., Dinger, D. R. (1994). Introduction to predictive process control. Predictive process control of crowded particulate suspensions, Springer, New York, 1–16.

Furnas, C.C. (1931). Grading Aggregates – I. – Mathematical Relations for Beds of Broken Solids of Maximum Density. *Ind. Eng. Chem.* 23(9), 1052–1058.

Gołaszewski, J., Szwabowski J. (2004). Influence of superplasticizers on rheological behaviour of fresh cement mortars. *Cem. Concr. Res.* 34(2), 235–248.

Graybeal. B. (2006). Material property characterization of ultra-high performance concrete, Report No. FHWA-HRT-06-103, FHWA, U.S. Department of Transportation.

Graybeal, B. (2011). Ultra-High Performance Concrete, *TechNote*, FHWA-HRT-11-038, Federal Highway Administration, McLean, VA.

Hasselman, D. P. H. (1962). On the porosity dependence of the elastic moduli of polycrystalline refractory materials. *J. Am. Ceram. Soc.* 45(9), 452–453.

Hunger, M. (2010). An integral design concept for ecological self-compacting concrete, PhD thesis, Eindhoven University of Technology, Eindhoven, the Netherland, 2010.

Kang, S. T., Lee, B. Y. Kim, J. K., Kim, Y.Y. (2011). The effect of fibre distribution characteristics on the flexural strength of steel fibre-reinforced ultra high strength concrete. *Constr. Build. Mater.* 25(5), 2450–2457.

Kang, S., Lee, Y., Park, Y. D., Kim, J. K. (2010). Tensile fracture properties of an ultra high performance fiber reinforced concrete (UHPFRC) with steel fiber. *Compos. Struct.* 92(1), 61–71.

Kim, D. J., Park, S. H., Ryu, G. S., Koh, K. T. (2011). Comparative flexural behavior of hybrid ultra high performance fiber reinforced concrete with different macro fibers. *Construction and Building Materials* 25(11), 4144–4155.

Koh, K. T., Park, J. J., Ryu, G. S., Kang, S. T. (2007). Effect of the compressive strength of ultra-high strength steel fiber reinforced cementitious composites on curing method. *KSCE J. Civ. Environ. Eng. Res.* 27(3A), 427–432.

Kwan, A. K. H., Chen, J. J., Fung, W. W. S., Li, L. G. (2010). Improving packing density of powder in cement paste for production of high-performance concrete. *Adv. Mater. Res.* 168–170, 1640–1647.

Le Thanh Ham, K. S., Ludwig H. (2012). Synergistic effect of rice husk ash and fly ash on properties of self-compacting high performance concrete. Ultra-High Performance Concrete and Nanotechnology in Construction. Proceedings of Hipermat 2012. 3rd International Symposium on UHPC and Nanotechnology for High Performance Construction Materials, Kassel University Press GmbH.

Masse, S., Zanni, H., Lecourtier, J., Roussel, J. C., Rivereau, A. (1993). ^{29}Si solid state NMR study of tricalcium silicate and cement hydration at high temperature. *Cem. Concr. Res.* 23(5), 1169–1177.

Massidda, L., Sanna, U., Cocco, E., Meloni, P. (2001). High pressure steam curing of reactive-powder mortars. *Spec. Publ.* 200, 447–464.

Mazanec, O., Lowke, D., Schießl, P. (2010). Mixing of high performance concrete: effect of concrete composition and mixing intensity on mixing time. *Mater. Struct.* 43(3), 357–365.

Mehdipour, I., Khayat, K. H. (2018). Understanding the role of particle packing characteristics in rheo-physical properties of cementitious suspensions: A literature review. *Constr. Build. Mater.* 161, 340–353.

Mohammed, H. (2015). Mechanical properties of ultra high strength reinforced concrete, Master's thesis, University of Akron, Akron, OH, USA.

Monosi, S., Pignoloni, G., Collepardi, S., Troli, R., Collepardi, M. (2000). Modified reactive powder concrete with artificial aggregates. *Spec. Publ. 195,* 447–460.

Naaman, A. E., Wille, K. (2012). The path to ultra-high performance fiber reinforced concrete (UHP-FRC): five decades of progress. *Proceedings of Hipermat*, 3–15.

Nehdi, M. and S. Mindess, Aïtcin, P. C. (1996). Optimization of high strength limestone filler cement mortars. *Cem. Concr. Res.* 26(6), 883–893.

Odler, I., Rößler M. (1985). Investigations on the relationship between porosity, structure and strength of hydrated Portland cement pastes. II. Effect of pore structure and degree of hydration. *Cem. Concr. Res.* 15(3), 401–410.

Park, S. H., Kim, D. J., Ryu, G. S., Koh, K. T. (2012). Tensile behavior of ultra high performance hybrid fiber reinforced concrete. *Cem Concr. Compos.* 34(2), 172–184.

Peng, Y., Hu, S., Ding, Q. (2010). Preparation of reactive powder concrete using fly ash and steel slag powder. *J. Wuhan Univ. Technol., Mater. Sci. Ed.* 25(2), 349–354.

Powers, T. C. (1958). Structure and physical properties of hardened Portland cement paste. *J. Am. Ceram. Soc.* 41(1), 1–6.

Prabha, S. L., Dattatreya, J. K., Neelamegam, M., Seshagirirao, M. V. (2010). Study on stress-strain properties of reactive powder concrete under uniaxial compression. *Int. J. Eng. Sci. Technol.* 2(11), 6408–6416.

Qian, C. X., Stroeven P. (2000). Development of hybrid polypropylene-steel fibre-reinforced concrete. *Cem. Concr. Res.* 30(1): 63–69.

Reda, M. M., Shrive, N. G., Gillott J. E. (1999). Microstructural investigation of innovative UHPC. *Cem. Concr. Res.* 29(3), 323–329.

Richard, P., Cheyrezy M. (1995). Composition of reactive powder concretes. *Cem. Concr. Res.* 25(7), 1501–1511.

Ritter, R., Curbach, M. (2015), Material behavior of ultra-high-strength concrete under multiaxial stress states. *ACI Mater. J.* 112 (5), 641–652.

Rößler, M., Odler I. (1985). Investigations on the relationship between porosity, structure and strength of hydrated Pportland cement pastes I. Effect of porosity. *Cem. Concr. Res.* 15, (2): 320–330.

Russell, H. G., Graybeal, B. A., Russell, H. G. (2013). Ultra-high performance concrete: A state-of-the-art report for the bridge community (No. FHWA-HRT-13-060). United States. Federal Highway Administration. Office of Infrastructure Research and Development.

Ryshkevitch, R. (1953). Compression strength of porous sintered alumina and zirconia. *J. Am. Ceram. Soc.* 36(2), 65–68.

Sanchez, F., Sobolev K. (2010). Nanotechnology in concrete – A review. *Constr. Build. Mater.* 24(11), 2060–2071.

Schiller, K. K. (1971). Strength of porous materials. *Cem. Concr. Res.* 1(4), 419–422.

Shhmidt, M., Fehling, E. (2007). Grundlagen der Betontechnologie von Hochund Ultra Hochleistungsbeton und Anwendung von UHPC imBruckenbau. Ultra High Performance Concrete – 10 Years of Research and Development at the University of Kassel, Kassel, Germany, 70–81.

Shi, C., Wu, Z., Xiao, J., Wang, D., Huang, Z., Fang, Z. (2015). A review on ultra high performance concrete: Part I. Raw materials and mixture design. *Constr. Build. Mater.* 101, 741–751.

Shi, C., D. Wang, Wu, L., Wu, Z. (2015). The hydration and microstructure of ultra high-strength concrete with cement–silica fume–slag binder. *Cem. Concr. Compos.* 61, 44–52.

Shi, C., Hu S. (2003). Cementitious properties of ladle slag fines under autoclave curing conditions. *Cem. Concr. Res.* 33(11), 1851–1856.

Shi, Y., Matsui, I., Feng, N. (2002). Effect of compound mineral powders on workability and rheological property of HPC. *Cem. Concr. Res.* 32(1), 71–78.

Sobolev, K., Gutiérrez M. F. (2005). How nanotechnology can change the concrete world. *Am. Ceram. Soc. Bull.* 84(10), 14.

Sohail, M. G., Wang, B., Jain, A., Kahraman, R., Ozerkan, N. G., Gencturk, B., Belarbi, A. (2018). Advancements in concrete mix designs: High-performance and ultrahigh-performance concretes from 1970 to 2016. *J. Mater. Civ. Eng.* 30(3), 04017310.

Soliman, N. A., Tagnit-Hamou A. (2017). Partial substitution of silica fume with fine glass powder in UHPC: Filling the micro gap. *Constr. Build. Mater.* 139, 374–383.

Soliman, N. A., Tagnit-Hamou A. (2017). Using glass sand as an alternative for quartz sand in UHPC. *Constr. Build. Mater.* 145, 243–252.

Sorelli, L., Constantinides, G., Ulm, F. J., Toutlemonde, F. (2008). The nanomechanical signature of ultra high performance concrete by statistical nanoindentation techniques. *Cement and Concrete Research* 38(12), 1447–1456.

Stovall, T., de Larrard, F., Buil, M. (1986). Linear packing density model of grain mixtures. *Powder Technol.* 48(1), 1–12.

Svermova, L., Sonebi, M., Bartos, P. J. (2003). Influence of mix proportions on rheology of cement grouts containing limestone powder. *Cem. Concr. Compos.* 25(7), 737–749.

Tafraoui, A., Escadeillas, G., Lebaili, S., Vidal, T. (2009). Metakaolin in the formulation of UHPC. *Constr. Build. Mater.* 23(2), 669–674.

Toufar, W., Born M., Klose E. (1976). Contribution of optimisation of components of different density in polydispersed particles systems. *Freiberger Booklet* A558, 29–44.

Tran N. T., Kim D. J. (2014). High strain rate effects on direct tensile behavior of high performance fiber reinforced cementitious composites. *Cem. Concr. Compos.* 45, 186–200.

Tue, N. V., Ma, J., Orgass, M. (2008). Influence of addition method of superplasticizer on the properties of fresh UHPC. Proceedings of the 2nd International Symposium on Ultra-High Performance Concrete, Kassel, Germany.

Van Tuan, N., Ye, G., Van Breugel, K., Copuroglu, O. (2011a). Hydration and microstructure of ultra high performance concrete incorporating rice husk ash. *Cem. Concr. Res.* 41 (11), 1104–1111.

Van Tuan, N., Ye, G., van Breugel, K., Fraaij, A. L. A., Bui, D. D.(2011c). The study of using rice husk ash to produce ultra high performanceconcreteperformance concrete. *Constr. Build. Mater.* 25, 2030–2035.

Van Tuan, N., Ye, G., Van Breugel, K., Fraaij, A. L., Dai Bui, D. (2011b). The study of using rice husk ash to produce ultra high performance concrete. *Constr. Build. Mater.* 25 (4), 2030–2035.

Van, V., Ludwig H. (2012). Proportioning optimization of UHPC containing rice husk ash and ground granulated blast-furnace slag. Proceedings of the 3rd International Symposium on UHPC and Nanotechnology for High Performance Construction Materials, Kassel, Germany.

Vandenberg, A., Wille, K. (2018). Evaluation of resonance acoustic mixing technology using ultra high performance concrete. *Constr. Build. Mater.* 164, 716–730.

Wille, K., Kim, D. J., Naaman, A. E. (2011). Strain-hardening UHP-FRC with low fiber contents. *Mater. Struct.* 44(3), 583–598.

Wu, Z., Shi, C., He, W., Wu, L. (2016). Effects of steel fiber content and shape on mechanical properties of ultra high performance concrete. *Constr. Build. Maer.* 103, 8–14.

Wu, Z., Shi, C., He, W. (2017a). Comparative study on flexural properties of ultra-high performance concrete with supplementary cementitious materials under different curing regimes. *Constr. Build. Mater.* 136, 307–313.

Wu, Z., Shi, C., He, W., Wang, D. (2017b). Static and dynamic compressive properties of ultra-high performance concrete (UHPC) with hybrid steel fiber reinforcements. *Cem. Concr. Compos.* 79, 148–157.

Wu, Z., Khayat, K. H., Shi, C. (2018). How do fiber shape and matrix composition affect fiber pullout behavior and flexural properties of UHPC?. *Cem. Concr. Compos.* 90, 193–201.

Xue, J., Briseghella B., Huang, F., Nuti, C., Tabatabai, H., Chen, B. (2020). Review of ultra-high performance concrete and its application in bridge engineering. *Constr. Build. Mater.* 260, 119844.

Yang, I. H., Joh, C., B. S. Kim (2010). Structural behavior of ultra high performance concrete beams subjected to bending. *Eng. Struct.* 32(11), 3478–3487.

Yang, S. L., Millard, S. G., Soutsos, M. N., Barnett, S. J., Le, T. T. (2009). Influence of aggregate and curing regime on the mechanical properties of ultra-high performance fibre reinforced concrete (UHPFRC). *Constr. Build. Mater.* 23(6), 2291–2298.

Yang, Y., Ingham J. M. (2000). *Manufacturing reactive powder concrete using common New Zealand materials*. Department of Civil and Resource Engineering, University of Auckland, Auckland, New Zealand.

Yanni, V. Y. G. (2009). *Multi-scale investigation of tensile creep of ultra-high performance concrete for bridge applications*. Georgia Institute of Technology, Atlanta, USA.

Yazıcı, H. (2007). The effect of curing conditions on compressive strength of ultra high strength concrete with high volume mineral admixtures. *Build. Environ.* 42(5), 2083–2089.

Yazıcı, H., Yiğiter, H., Karabulut, A. Ş., Baradan, B. (2008). Utilization of fly ash and ground granulated blast furnace slag as an alternative silica source in reactive powder concrete. *Fuel* 87(12), 2401–2407.

Yazıcı, H., Yardımcı, M. Y., Yiğiter, H., Aydın, S., Türkel, S. (2010). Mechanical properties of reactive powder concrete containing high volumes of ground granulated blast furnace slag. *Cem. Concr. Compos.* 32(8), 639–648.

Yoo, D. Y., Lee, J. H., Yoon, Y. S. (2013). Effect of fiber content on mechanical and fracture properties of ultra high performance fiber reinforced cementitious composites. *Compos. Struct.* 106, 742–753.

Yoo, D., Banthia N. (2016). Mechanical properties of ultra-high-performance fiber-reinforced concrete: A review. *Cem. Concr. Compos.* 73, 267–280.

Yoo, D., Kang, S., Yoon, Y. S. (2014). Effect of fiber length and placement method on flexural behavior, tension-softening curve, and fiber distribution characteristics of UHPFRC. *Constr. Build. Mater.* 64, 67–81.

Yoo, D. Y., Kang, S. T., Yoon, Y. S. (2016). Enhancing the flexural performance of ultra-high-performance concrete using long steel fibers. *Compos. Struct.* 147, 220–230.

Yoshioka, K., Tazawa, E., Kawai, K., Enohata, T. (2002). Adsorption characteristics of superplasticizers on cement component minerals. *Cem. Concr. Res.* 32(10), 1507–1513.

Yu, R., Spiesz, P., Brouwers, H. J. H. (2014). Mix design and properties assessment of ultra-high performance fibre reinforced concrete (UHPFRC). *Cem. Concr. Res.* 56, 29–39.

Yu, R., Spiesz, P., Brouwers, H. J. H. (2015). Development of an eco-friendly ultra-high performance concrete (UHPC) with efficient cement and mineral admixtures uses. *Cem. Concr. Compos.* 55, 383–394.

Zhang, Y.S., Sun, W., Liu, S.L., Jiao, C.J., Liu, J.Z. (2008). Preparation of C200 green reactive powder concrete and its static–dynamic behaviors. *Cem. Concr. Compos.* 30(9), 831–838.

Zhang, H., Ji, T., Zeng, X., Yang, Z., Lin, X., Liang, Y. (2018). Mechanical behavior of ultra-high performance concrete (UHPC) using recycled fine aggregate cured under different conditions and the mechanism based on integrated microstructural parameters. *Constr. Build. Mater.* 192, 489–507.

Zhao, S., Fan, J., Sun, W. (2014). Utilization of iron ore tailings as fine aggregate in ultra-high performance concrete. *Constr. Build. Mater.* 50, 540–548.

Zhu, W., Gibbs, J. C. (2005). Use of different limestone and chalk powders in self-compacting concrete. *Cem. Concr. Res.* 35(8), 1457–1462.

Chapter 3

Raw materials

3.1 INTRODUCTION

The initially developed commercial UHPC consists of 700–1100 kg/m^3 cement, 160–500 kg/m^3 silica fume, 700–1300 kg/m^3 quartz sand, 109–209 kg/m^3 water, 140–470 kg/m^3 steel fibers, and 13–44 kg/m^3 superplasticizer (Graybeal 2016; Williams 2009). In some recipes, 211–295 kg/m^3 quartz power with sizes ranging from 0.1 to 100 μm is used as filler to enhance the packing density of UHPC. Finely grounded quartz with sizes ranging from 150 to 600 μm is dimensionally the largest granular material. The large particle followed is cement with an average diameter of approximately 15 μm and crushed quartz powder with an average diameter of 10 μm. Silica fume with an average diameter of 0.2 μm is the smallest particle if no nanoparticles are included. It can fill the interstitial voids between cement and quartz powder. Thus, with an optimized mixture proportion of granular particles, the maximum packing density of UHPC can be achieved to render low porosity and high strength. Due to the high initial materials cost and CO_2 emission, strategies have been employed to develop cost-effective UHPC without compromising mechanical properties. These include: (1) use of supplementary cementation materials and fillers to reduce cement and/or silica fume contents; (2) replacing finely grounded quartz sand using river sand and/or other types of aggregates; and (3) decreasing steel fiber volume through fiber hybridization. Since the quality of those alternative materials is closely related to the source, region, production technology, etc., their uses in UHPC can sometimes lead to unfavorable effects, such as incompatibility issues between superplasticizer and cementitious materials, as well as decreased compressive strength. Therefore, the raw material should be carefully selected and examined before its use to ensure high mechanical properties and sustainability of UHPC.

This chapter presents the available raw materials, including cement, supplementary cementitious materials, aggregate, nanoparticles, mixing water, chemical admixture, fibers, and internal curing agents, used for the preparation of UHPC. The characteristics of each raw material and their

DOI: 10.1201/9781003203605-3

effects on the workability and hardened properties of UHPC are briefly discussed.

3.2 CEMENT

Portland cement mainly consists of finely ground clinker and gypsum. The main compositions of cement clinker include C_3S, C_2S, C_3A, and C_4AF. The hydration product of C-S-H forms from the chemical reactions of C_2S and C_3S with water and is the main contributor to concrete strength. Cement with C_2S and C_3S content greater than 65% is preferred for developing UHPC (Wille et al. 2011). The hydration products of C_3A do not contribute to the strength of concrete, but they can bind superplasticizers and result in less superplasticizer dosage to affect workability. The C_3A and gypsum contents in cement significantly influence the properties of concrete, especially workability. It is suggested that the C_3A content of cement used for UHPC should be less than 8% (Cheyrezy et al. 1995).

Alkali content also influences the properties of UHPC. The increase in alkali content can reduce the compressive strength of concrete at a later age by depressing the Ca^{2+} released from gypsum. It can increase the porosity of the microstructure and result in the generation of low-strength alkali-containing C-S-H gel. An increase in the alkali content of cement also increases the potential of the alkali-silica reaction in UHPC. The equivalent alkali content (Na_2O_{eq}) should be limited to 0.9% to restrain the alkali-silica reaction expansion.

Cement content in UHPC is generally about 700–1000 kg/m^3, and the cement used should have a low alkali content, a low to medium fineness, and a low C_3A content (Talebinejad et al. 2004). Hence, CEM I Portland cement of strength classes 42.5 R and 52.5 R according to EN 197 and Type III and white cement in terms of ASTM C150, as well as P·I 42.5 and P·II 52.5 cements following the Chinese Standard GB175-2007, are preferred for UHPC. These types of cement are advantageous because of their high fast setting time and high strength potential.

3.3 SUPPLEMENTARY CEMENTATION MATERIALS

3.3.1 Silica fume

Silica fume is an industrial byproduct from ferrosilicium alloys production. It has an average particle diameter of 0.1–0.3 μm and a specific surface area of 13,000–30,000 m^2/kg, as shown in Figure 3.1. It is a spherical, amorphous, and highly reactive pozzolan with SiO_2 content usually over 90%. The SiO_2 content in silica fume varies with the alloy type produced. For example, 98% silicon metal produces silica fume with 87%–98% SiO_2

content, whereas the 50% ferrosilicon alloy produces silica fume with 74%–84% SiO_2 content (ACI-Committee-234 2006).

The silica fume content in UHPC mixtures is generally about 50–250 kg/m^3, which is 5%–25%, by mass of cementitious materials (Long et al. 2002; Wu et al. 2019a). Due to its small particle size and smoothness, the addition of silica fume content under a critical content, typically 5%–15%, can improve the particle packing density of UHPC. It can release the entrapped water among particles to lubricate the paste and contribute to workability. In the meantime, silica fume can react with $Ca(OH)_2$ to form dense C-S-H, thus enhancing the interfacial transition zone between aggregate and paste. Moreover, silica fume can change the average Ca/Si ratio in the C-S-H phase from about 1.7 to 1.2, which is beneficial for limiting steel corrosion, especially in the presence of alkaline ions. However, when the silica fume content exceeds the critical value, the workability of UHPC will be greatly reduced (Wu et al. 2016). This is due to the more required liquid to coat their surface associated with ultra-high specific surface area, which is approximately 100 times greater than that of ordinary Portland cement. The optimal silica fume content is highly dependent on the w/c ratio and characteristics of other materials used. Lower silica fume content is required for a lower w/c ratio (Van Der Putten et al. 2016). It is worth noting that individual silica fume particles can easily agglomerate to form 1 to 100 μm particles. Thus, superplasticizers used in UHPC must be capable of breaking up the agglomerated silica fume particles during the mixing process.

The variable carbon content in silica fume can result in changeable water and superplasticizer required to affect the fluidity of UHPC. Silica fume with low carbon content is preferred to achieve good workability. Additionally, high unburnt coal in silica fume may result in dark surface color of concrete. It is shown that pulverized slag and fly ash, metakaolin, limestone powder, rice husk ash, etc., can be used to replace silica fume in UHPC (Shi et al. 2015; Yazıcı et al. 2010; Yazıcı et al. 2008). The combined use of silica fume with light-colored fly ash or slag can balance the color of UHPC.

3.3.2 Fly ash

Fly ash is a byproduct of burning pulverized coal in electric power generating plants. It is composed of mainly spherical particles with sizes ranging from less than 0.5 to about 300 μm (Figure 3.1(b)). The average specific gravity of fly ash is estimated at around 2.2, with a standard deviation of about 0.3. The Brunauer, Emmett, and Teller (BET) specific surface area may range from less than 500 to more than 10,000 m^2/kg. Fly ash is one of the most commonly used supplementary cementitious materials to prepare UHPC, and its optimal content ranges from 10% to 60%, depending on the chemical composition of fly ash and UHPC mixture proportion.

Figure 3.1 SEM images of silica fume and fly ash (Xi et al. 2022; Chen and Zhu 2018). (a) Silica fume. (b) Fly ash.

The addition of fly ash in UHPC can improve workability due to its spherical shape and glassy structure, which provides balling effect to reduce the friction between particles. However, the ignition loss of fly ash significantly affects the workability of concrete by influencing the tendency of absorbed superplasticizers. The use of fly ash in UHPC can retard cement hydration (Park et al. 2021). This is because the fly ash surface acts like a calcium sink. As Ettringite phases preferentially form on the surface of fly ash, the calcium in the solution is absorbed by fly ash. This depresses the Ca^{2+} concentration in solution during the first 6 h of hydration and postpones a Ca-rich surface layer on the clinker minerals. Therefore, the $Ca(OH)_2$ and C-S-H nucleation and crystallization are delayed, and the cement hydration is simultaneously retarded.

UHPC shows high potential cracking associated with high autogenous shrinkage. Fly ash can mitigate the shrinkage deformation of UHPC by releasing the entrapped water associated with its spherical shape and glassy structure, thus lowering the rapid drop of internal relative humidity. Moreover, unreacted fly ash can act as micro-aggregates in UHPC to restrain the shrinkage deformation (Yang et al. 2019). Although fly ash depresses the development of mechanical properties of UHPC at an early age, the pozzolanic reaction of fly ash at later ages can be fully exerted to deplete calcium hydroxide and form more supplemental C-S-H gel, thus improving the mechanical properties and durability (Ferdosian et al. 2017). Fly ash is usually combined with slag, silica fume, steel slag powder, etc., as a binary, ternary, or quaternary system. The compressive strength of UHPC containing high slag and fly ash contents was over 200 MPa (Peng et al. 2010; Yazıcı et al. 2009). However, the elasticity modulus of UHPC decreased if the fly ash replacement with cement was greater than 30% (Yazıcı et al. 2009).

3.3.3 Ground granulated blast-furnace slag

Ground granulated blast-furnace slag (GGBS) is a glassy material from the byproduct of blast furnace iron-making. GGBS with fewer impurities has particle sizes of 31–50 μm (Ganesh and Murthy 2019). It mainly contains calcium silicoaluminate with high reactivity. Replacing cement with 20%–60% GGBS can effectively enhance the properties of UHPC under different curing conditions. Firstly, the use of GGBS can significantly reduce the required superplasticizer content and reach the target slump-flow value of UHPCs (Yalçınkaya and Çopuroğlu 2020). Due to its angular shape and relatively low absorbed superplasticizer content, the flowability of UHPC made with GGBS is lower than that with the same fly ash content. The replacement of cement with GGBS can retard the hydration of cement, resulting in reduced chemical shrinkage and self-desiccation. However, it should be noted that the effect of GGBS on the autogenous shrinkage of UHPC is controversial in the literature. The addition of 8% and 20% GGBS effectively decreased the autogenous shrinkage of UHPC (Ghafari et al. 2012). Tazawa et al. (1995) reported that cement pastes with 70% slag with a Blaine fineness of 400 m^2/kg showed increased autogenous shrinkage. However, no increase in autogenous shrinkage was observed in paste made with slag at a Blaine fineness of 300 m^2/kg. The enhancement of slag activity associated with higher fineness is probably responsible for the more refined pore structure and greater chemical shrinkage, which induces faster and

Figure 3.2 SEM image of GGBS in UHPC cured for 90 d (QZ: quartz grain, C: unreacted cement, S: GGBS particle).

Source: Yalçınkaya and Çopuroğlu 2020.

greater self-desiccation, thus resulting in greater autogenous shrinkage (Yang et al. 2019).

The hydraulic and pozzolanic reactivity of GGBS can promote the formation of C-S-H gel in UHPC, improving the aggregate-mortar interfacial transition zone (Figure 3.2), mechanical properties, and durability of UHPC (Ganesh and Murthy 2019). Yu et al. (2015) indicated that UHPC prepared with GGBS had superior mechanical properties at 91 d compared to that with fly ash. It should be noted that there is still a large quantity of nonhydrated cement and GGBS particles in UHPC after 90-d curing. Heat curing can further enhance the mechanical properties of UHPC. Yazıcı (2010) used 20%, 40%, and 60% GGBS to prepare UHPC and obtained compressive strength over 250 MPa after autoclaving.

3.3.4 Metakaolin

Metakaolin, with particle sizes ranging from 0.5 to 20 μm, is a highly reactive pozzolan made by calcining natural clay. Table 3.1 presents the physical and chemical properties of metakaolin. Its micromorphology characteristic is shown in Figure 3.3. Metakaolin consists of amorphous silica and alumina, and its structure is in long order or hexagonal layers. Metakaolin is a very reactive pozzolan and performs a similar reaction to silica fume (Tafraoui et al. 2009). Its pozzolanic reactivity is mainly controlled by its calcining temperature (Ma 2008). The use of metakaolin may delay the hydration of cementitious components in UHPC and act as an

Table 3.1 Physical and chemical properties of metakaolin

Compound (%)								Specific gravity	Specific surface area (m^2/g)
CaO	SiO_2	Al_2O_3	MgO	Fe_2O_3	K_2O	SO_3	Na_2O		
1.15	58.10	35.14	0.20	1.21	1.05	0.03	0.07	2.50	15–30

Source: Tafraoui et al. 2009.

Figure 3.3 SEM image of metakaolin.

Source: Scherb et al. 2020.

internal curing agent, causing lower autogenous shrinkage (Norhasri et al. 2019). Staquet and Espion (2004) used metakaolin to substitute 33% silica fume and observed a reduction of around 50% in 6-d autogenous shrinkage of UHPC.

Metakaolin powder can refine pore structure and improve the early strength and durability of UHPC (Ma 2008; Tafraoui et al. 2009; Tafraoui et al. 2016). The compressive strength of UHPC specimens decreases with the increasing metakaolin content (Ma 2008; Tafraoui et al. 2009). However, the substitution of silica fume with metakaolin leads to a negligible increase in flexural strength (Mo et al. 2020). Its availability and low price can favor its use in manufacturing UHPC. Meantime, its white color can be an esthetic advantage.

3.3.5 Rice husk ash

Rice husk ash is an agricultural waste obtained by burning rice husk. The physical and chemical composition of rice husk ash strongly depends on the source and processing variables during the grinding process, as shown in Table 3.2. When the husk is completely incinerated in controlled conditions, 90%–96% amorphous silica may exist. The average particle size of rice husk ash is between cement and silica fume particles. Usually, it ranges from 5 to 10 μm with a very high specific surface area due to its porous structure (even more than 250 m^2/g), as shown in Figure 3.4. Typical rice husk ash content, ranging from 5% to 30%, by the mass of binder, is often employed in UHPC, which can be used to replace cement, silica fume, and quartz powder. A higher amount of rice husk ash requires more superplasticizer and water contents to maintain suitable desired workability because of its porous structure and high specific surface area. When rice husk ash is used instead of silica fume or quartz powder, the workability of UHPC tends to decrease (Kang et al. 2019; Huang et al. 2017). The combination of appropriate rice husk ash content with silica fume, GGBS, and fly ash can improve the workability of UHPC (Rößler et al. 2014).

Rice husk ash shows various benefits in densifying microstructure, improving mechanical properties, and lowering autogenous shrinkage of UHPC. UHPC containing rice husk ash possesses superior or comparable compressive strength compared to that made with silica fume (Rößler et al. 2014; Mosaberpanah and Umar 2020). The 10% replacement ratio of cement with rice husk ash increases the 3 and 28 d compressive strength by 10.6% and 8.8%, respectively (Van aTuan et al. 2011; Van bTuan et al.

Table 3.2 Variation of physical and chemical properties due to different grinding times

Physical and chemical properties	*Raw rice husk ash*	*Grinding time*		
		5 min	*30 min*	*120 min*
SiO_2 (%)	81.4	82.90	82.90	83.00
Al_2O_3 (%)	0.26	0.28	0.29	0.34
Fe_2O_3 (%)	0.93	1.69	1.83	1.83
CaO (%)	2.42	2.35	2.29	2.16
MgO (%)	1.02	0.98	0.93	1.07
SO_3 (%)	1.47	1.56	1.34	1.58
Na_2O (%)	0.18	0.11	0.06	0.01
K_2O (%)	6.79	6.83	6.82	6.83
Loss on ignition	3.13	3.14	3.14	3.13
Particle mean size (mm)	-	8.61	5.45	6.95
Pore volume (ml/g)	0.073	0.023	0.053	0.031

Source: Mosaberpanah and Umar 2020.

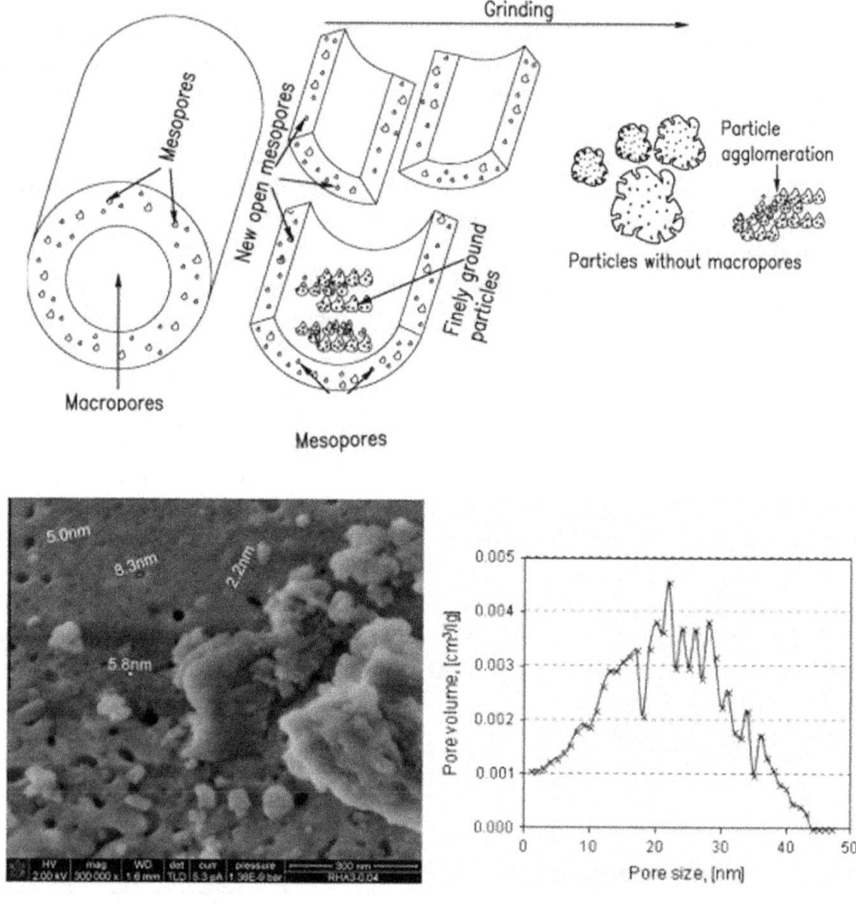

Figure 3.4 Porous surface structure and pore size distribution of ground rice husk ash. (a) Morphological characteristics of rice husk ash particles during grinding (Van et al. 2013). (b) Porous surface structure and pore size distribution of ground rice husk ash (Van et al. 2014).

2011). The compressive strength of UHPC incorporating rice husk ash with a mean size between 3.6 and 9 μm could reach over 150 MPa under normal curing regime (Le Thanh Ham and Ludwig 2012). Ye (2012) found that the 28-d autogenous shrinkage values of UHPC were reduced by around 40% and 95% when rice husk ash contents were increased by 10% and 20%, respectively. The underlying mechanisms in mitigating shrinkage can be summarized as enhancement in hydration degree (Van aTuan et al. 2011), reduction in the drop rate of internal relative humidity (Rößler et al.

2014), and increase in pores larger than 50 nm (Kang et al. 2019). The porous structure of rice husk ash acts as an internal curing agent by storing moisture reservoirs to reduce autogenous shrinkage of UHPC as detailed in Section 3.9.

3.4 LIMESTONE POWDER

The main constituent of limestone used in concrete is calcite, while dolomite and quartz are also identified as minor constituents by X-ray diffraction analysis, as shown in Table 3.3 (Tsivilis et al. 2003). Its density is approximately 2600–2800 kg/m^3 with a specific gravity ranging from 2 to 3. The particle size of limestone powder is smaller than that of cement. Since the w/b ratio of UHPC is around 0.2, the hydration degree of cement at 28 d is usually less than 35%, with a majority of unhydrated cement particles (Korpa et al. 2009). Thus, the unhydrated cement in UHPC can be replaced by inner filler, such as limestone powder, to maximize the packing density and reduce materials cost without sacrificing the properties.

Limestone powder is usually used to replace cement in UHPC at a replacement ratio of 5% to 60% with a d_{50} of 8–20 μm (Figure 3.5). The use of limestone powder can improve the workability of UHPC because of the repulsion between OH^- group located on the Ca^{2+} surface and its lower water absorption (Shi et al. 2002; Wang et al. 2012; Zhu and Gibbs 2005). It can also accelerate cement hydration because of the nucleation and dilution effects (Huang et al. 2017). The dilution effect of limestone powder can increase the effective w/c ratio of UHPC. Thus, more free water is available for cement hydration and to maintain the relatively high internal relative humidity. Therefore, proper use of limestone powder in UHPC can improve microstructure, increase strength, and decrease autogenous shrinkage (Svermova et al. 2003; Zhu and Gibbs 2005). Moreover, the combination of limestone powder and supplementary cementitious materials, such as metakaolin, can promote the generation of C-S-H gel (Mo et al. 2020; Nehdi et al. 1996). It should be noted that an excessive limestone powder amount can lead to an

Table 3.3 Physical and chemical properties of limestone powder

Compound (%)										*Specific gravity*	*Specific surface area (m^2/g)*
CaO	*SiO_2*	*Al_2O_3*	*MgO*	*Fe_2O_3*	*K_2O*	*SO_3*	*Na_2O*	*ZrO_2*	*LOI (950°C)*		
56.90	0.07	-	0.13	0.02	-	0.05	0.07	0.03	42.73	2.70	0.80

Source: Tafraoui et al. 2009.

Figure 3.5 SEM observation of limestone powder.

Source: Souza et al. 2020.

increase in porosity and a decrease in strength because of insufficient reactive materials and the corresponding generated C-S-H content in UHPC.

3.5 AGGREGATE

Aggregate is the most inexpensive constituent and accounts for a large volume of concrete. It plays a vital role in the highly packed particle skeleton of UHPC, thus directly affecting the mechanical properties and durability and restraining shrinkage deformation. The concrete failure originates from weak interfacial transition zones between cementitious matrix and aggregates, as well as micropores or microcracks. The decrease in aggregate size can help decrease interfacial transition zone amounts and enhance the packing density to render more homogeneous and robust concrete. In the initial UHPC mixture developed by Richard and Cheyrezy, finely grounded quartz with a size from 150 to 600 μm was used. Since the cost of fine quartz is high, alternative aggregates are used to replace it.In some recent studies, coarse aggregate with a size up to 20 mm is even used in UHPC to decrease the volume of cementitious paste, reduce viscosity, increase elastic modulus, and restrain autogenous shrinkage (Ma et al. 2004; Li et al. 2018).

3.5.1 Quartz sand

Table 3.4 shows the chemical composition and physical properties of quartz sand, river sand, and other raw materials. The main mineralogical

Table 3.4 Chemical composition and physical properties of quartz sand, river sand, and other raw materials

Properties	*Identification*	*Quartz sand*	*River sand*	*Silica fume*	*Cement*
Chemical composition (%)	SiO_2	99.43	77.43	98.86	22.80
	Fe_2O_3	0.0255	4.23	0.14	4.43
	Al_2O_3	0.18	15.71	0.12	4.01
	CaO	0.11	3.60	0.57	65.32
	MgO	0.01	-	0.20	2.07
	Na_2O	-	0.04	0.35	0.07
	K_2O	0.03	0.05	0.21	0.56
Physical properties	Packing density (kg/m^3)	1536.31	1498.09	-	-
	Mean particle diameter d_{50} (μm)	300	300	-	-
	Maximum particle diameter d_{max} (μm)	600	600	-	-
	Crushing values (%)	94.72	94.44	-	-

Source: Jiao et al. 2020.

composition of quartz sand is silicon dioxide. The quartz sand used in UHPC accounts for 40%–65% of the total volume. An optimal sand-to-cement ratio of 1.4 was found for UHPC with a quartz particle size of 0.8 mm (Wille et al. 2012). The use of quartz sand can reduce the cement demand and improve the flowability of UHPC. However, its production process always creates environmental risks, such as dust, noise, and soil pollution. Besides, its high cost and difficulty in meeting the optimal gradation requirement of quartz sand restrict its wide usage in UHPC. Other candidates, including river sand, masonry sand, limestone sand, lightweight sand, and even recycled fine aggregate, can be used as replacements for refined quartz sand (Yang et al. 2009; Yang and Ingham 2000).

3.5.2 River sand

River sand costs about 1/15 of the cost of silica sand, which significantly reduces the cost of UHPC. Its main mineralogical composition is silicon dioxide, and it is the most widely used fine aggregate in UHPC (Wang et al. 2017; Yu et al. 2014a). Its specific gravity and fineness modulus are 2.66 and 2.8 g/cm^3, respectively (Qi et al. 2018). River sand with particle size distributions in the ranges of 0–1.25, 0–2.36, or 0–4.75 mm is often used as fine aggregate in UHPC (Liu et al. 2019; Wang et al. 2019). It is believed that the river sand can produce UHPC with similar strength and ductility compared to fine quartz sand with the similar particle size curve (Yang et al. 2009). Under 20°C standard curing, the 28-d compressive strength of nonfibrous UHPC matrix with 100% river sand can reach up to 120 MPa, and the strength of UHPC with 2% steel fibers achieves 140 MPa (Wu et al. 2019a).

3.5.3 Lightweight sand

Lightweight sand has a low-particle relative density, lower strength, and elastic modulus because of its cellular pore system compared to normal sand. Several kinds of lightweight aggregate, including cold bonded and sintered fly ash lightweight aggregate, expanded clay and slate, oil-palm-boiler clinker, oil palm shell, and other lightweight aggregates, such as scoria and tuff expanded shale, can be used to manufacture lightweight sand. Due to its porous structure, the presaturated lightweight aggregate has high water absorption and desorption ability. For example, the water absorption value of normal sand is 0.14%, while the value for lightweight sand after soaking in water for 24 h is 17.6%. The absorbed water can be held during concrete mixing and then released when the internal relative humidity is dropped to a certain value. This can enhance the flowability and promote the hydration of cement to result in refined microstructure, thus improving mechanical properties and reducing autogenous shrinkage of UHPC. An optimal lightweight sand content, in the range of 15%–25%, by the volume of total aggregate, is often preferred (Liu et al. 2019; Meng and Khayat 2017). Beyond this, an adverse effect on the performance of UHPC would be observed due to the porous structure of the excessive content of lightweight sand.

3.5.4 Sea sand

In addition to manufactured sands, sea sands are being mined and used in large quantities in coastal areas because of the shortage of river sand resources. However, natural sea sands contain a large number of chloride ions, and their use in reinforced concrete can cause steel bars or fibers corrosion. Therefore, its chloride ions need to be removed before sea sands can be used in concrete, which significantly increases costs. Since the particle size of sea sand is typically between quartz sand and river sand, it can be successfully used in producing UHPC (Teng et al. 2019; Huang et al. 2021). Besides, the detrimental effects of salt ions in sea sand are expected to be much smaller for UHPC than for normal concrete due to its dense structure.

3.5.5 Recycled waste materials as aggregate

Given the environmental and economic impacts derived from quartz sand and river sand, recycling waste materials, such as construction waste, rock dust, and iron ore tailings, are used in UHPC as alternatives.

3.5.5.1 Recycled aggregate

Recycled aggregate produced from waste concrete has been used to produce UHPC (Qian et al. 2020). Compared with natural river sand, recycled aggregate contains a large amount of old cement matrix. The use of recycled aggregate has an essential influence on the development of hydration and microstructure of UHPC. On the one hand, recycled aggregate can reduce the water content required for hydration due to its high water absorption capability. This is harmful to workability and delays the hydration kinetics. On the other hand, a large amount of old cement matrix existing in recycled aggregate can lead to more interfacial transition zones. Hence, 25% recycled aggregate is suitable for producing green UHPC with dense microstructure (Zhang et al. 2018; Yu et al. 2020).

3.5.5.2 Waste glass sand

Waste glass sand is a recycling waste material from landfills and can be used to prepare UHPC. The main mineralogical composition of waste glass sand is silicon dioxide, but its content is lower than that of river sand and quartz sand. Waste glass sand has a smooth surface than quartz sand and a higher crush value of about 96.71% than quartz sand and river sand (Figure 3.6). The use of waste glass sand with a proper amount of about 50% can improve the flowability and compressive strength of UHPC (Jiao et al. 2020). However, the use of waste glass sand without any special treatment of washing or sieving can reduce the flowability, compressive strength, flexural strength, and fracture energy of UHPC due to its sharper and more angular grain shapes compared to that of river sand and quartz sand, as illustrated in Figure 3.6 (Yang et al. 2009). Therefore, it is suggested to

Figure 3.6 Photomicrograph of quartz sand (left) and glass sand (right).

Source: Soliman and Tagnit-Hamou 2017.

do some special treatment, including washing, sieving, and grinding, before producing UHPC.

3.5.5.3 *Recycled rock dust*

Recycled rock dust can be obtained from the stone-crushing and polishing industry (Upadhyaya et al. 2019) and can also be generated from recycling marble tops, terrazzo, and stone scraps. The rock dust employed to replace quartz sand and river sand in UHPC supports the sustainable development of the construction industry (Yang et al. 2020). The main minerals of rock dust are calcite and dolomite, and the surface of rock dust is rough. The rock dust has greater apparent density, crush value index, and water absorption capacity than quartz sand. The fine particle content (<150 μm) in rock sand is about 36.5% in total, greater than that of quartz sand. The added rock dust has a limited effect on the flowability and hydration process of UHPC. However, the rough surface of rock dust can guarantee a better connection between the aggregate and cementitious paste, thus improving the compressive strength of UHPC (Yang et al. 2020).

3.5.5.4 *Iron ore tailings*

Iron ore tailings are a common type of hazardous solid waste and have caused severe environmental problems. It shows the potential to be used as a fine aggregate because they are relatively inert. The particle size of tailings is significantly greater than cement and smaller than river sand (Zhao et al. 2014). The use of iron ore tailings as fine aggregate in UHPC can mitigate environmental deterioration and reduce the production cost of UHPC (Zhao et al. 2014). In addition to quartz, the iron ore tailings also contain other minerals, such as albite, diopside, clinochlore, calcite magnesium, meixnerite, and magnesium ferrous oxide. When the replacement level of iron ore tailings is less than 40%, it has a limited influence on the workability and compressive strength of UHPC. However, 100% replacement of natural aggregate by the iron ore tailings significantly decreases the workability and compressive strength of UHPC.

3.5.6 Crushed basalt as coarse aggregate

The utilization of coarse aggregate in UHPC can decrease the volume of cementitious paste, reduce material cost, and mitigate shrinkage. Crushed basalt with maximized size of up to 20 mm is a common coarse aggregate used in concrete. It has an apparent density of 2800–2930 kg/m^3 and contains much less silicon dioxide and more aluminum oxide than river sand and quartz sand (Van Der Putten et al. 2016). UHPC containing coarse aggregate is easier to be fluidized and homogenized, thus reaching high workability and comparable compressive strength (Ma et al. 2004). The workability of UHPC with coarse aggregate mainly depends on the content used. However, the introduction of the interfacial transition zones between coarse aggregate and matrix can impair the strength and durability of UHPC to some extent. Besides, the coarse aggregate can disturb the dispersion of fibers, which eventually affects the mechanical properties of UHPC.

3.6 NANOPARTICLES

Nanomaterials refer to ultrafine materials with particle sizes in the range of 1–100 nm, much smaller than those of other materials used in concrete (Figure 3.7). The particle size is smaller than ordinary powders, but larger than atomic clusters. They can be divided into four groups, including zero-dimensional (0D), one-dimensional (1D), two-dimensional (2D), and three-dimensional (3D) nanoparticles. The 0D nanoparticles include nanoclusters and nanocrystals, such as nano-SiO_2, nano-TiO_2, nano-ZnO, and nano-$CaCO_3$, nano-Al_2O_3. Nanotubes, nanofibers, nanorods, and nanowires belong to 2D nanomaterials. The 3D nanomaterials include fibrous, multilayer, and polycrystalline materials, in which the 0D, 1D, and 2D structural elements are in close contact.

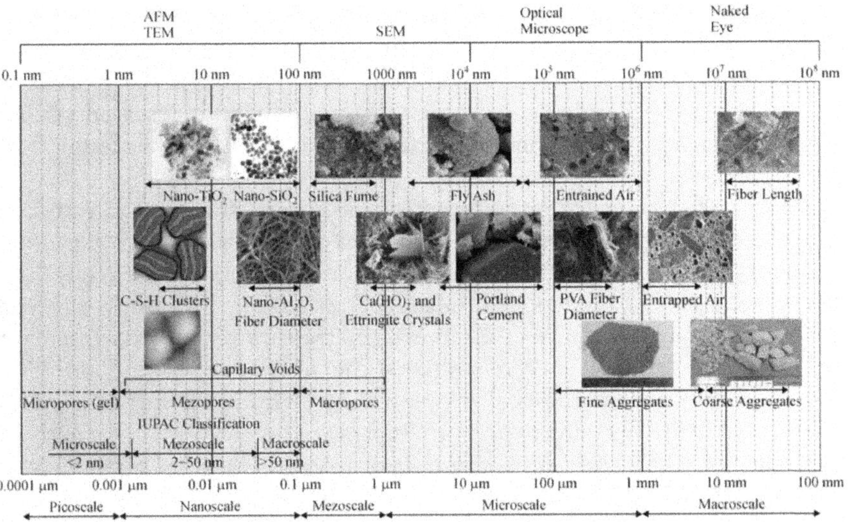

Figure 3.7 Scale range of several materials used in concrete fabrication.

Source: Flores-Vivian et al. 2017.

Various nanomaterials, such as nano-SiO_2, nano-Al_2O_3, nano-$CaCO_3$, grapheme oxide, carbon nanotube, and nanocarbon fiber, have been employed in UHPC since the breakthrough of nanotechnology in the field of construction. Several underlying mechanisms are often involved in enhancing the performance of cement-based materials associated with nanoparticles: (1) filling effect—fill the spaces and pores between the coarse and fine particles to render close packing of the solid system; (2) nucleation effect—act as nucleating sites for cement phases, further promoting cement hydration to form more hydration products; (3) chemical effect—react with $Ca(OH)_2$ crystals and/or C_3A to produce C-S-H gel and reduce the quantity and size of $Ca(OH)_2$, especially for nano-SiO_2 and nano-$CaCO_3$; and (4) reinforcing effect—play a bridging role to restrict the formation and propagation of cracks, especially for nanotubes, thus improving the mechanical properties of concrete.

3.6.1 Nano-SiO_2

Nano-SiO_2 is a white cubic powder composed of silicon oxide particles ranging from 10 to 30 nm with a density of 2.4 g/cm^3. The commonly used nano-SiO_2 materials include pyrogenic nano-SiO_2 (fumed nanosilica), precipitated nano-SiO_2, nano-SiO_2 gels, and nano-SiO_2 hydrosols (Madani et al. 2012). The addition of nano-SiO_2 has a significant effect on the workability of UHPC. Due to the ball bearing associated with its round shape and filling effect, the fluidity of UHPC can be enhanced. On the other hand, nano-SiO_2 can lead to decreased fluidity of UHPC due to its large specific surface area (Wu et al. 2016). When the cement content is below 500 kg/m^3, the replacement of cement with nano-SiO_2 decreases the workability (Yu et al. 2014b). Besides, the decreased flowability and increased viscosity can also increase the air content because of more difficult discharge in the mixing process.

Because of the nucleation and pozzolanic effects, the inclusion of nano-SiO_2 can accelerate the heat of hydration and form more hydration products to density the microstructure. The small pores with diameters from 0.3 to 40 nm are significantly reduced (Ghafari et al. 2014). It can also efficiently improve the microstructure of the interfacial transition zone between matrix and aggregate and/or fiber to enhance the fiber-matrix bond properties and eventually the mechanical properties of UHPC (Ghafari et al. 2015; Wu et al. 2016). The effects of 1% nanosilica content on the strength and durability of UHPC are similar to that of using 10% silica fume (Gesoglu 2016). It should be noted that a threshold limit of nano-SiO_2 exists because of dispersion issues, beyond which the mechanical properties of UHPC can be impaired. This optimal content is controversial in literature and depends on many factors, such as w/b ratio, mixture proportion, size, and type of nano-SiO_2.

Figure 3.8 Morphology of nano-$CaCO_3$ particles.

Source: Yang et al. 2020.

3.6.2 Nano-$CaCO_3$

Nano-$CaCO_3$ is a cubic or hexagonal white powder material with typical diameters of 10–100 nm and a specific surface area of 30–60 m^2/g (Figure 3.8). Its density is approximately 2.93 g/cm^3, and it is a low-activity mineral micropowder material rich in natural resources with a cost of about one-tenth of nano-SiO_2. Nano-$CaCO_3$ can accelerate the hydration of cement due to its nucleation and dilution effects, densifying the microstructure and performance of UHPC. The porosity and the critical pore size are decreased with the increase of nano-$CaCO_3$ content to an appropriate content of up to 3% (Ding et al. 2020). Besides, the fiber-matrix bond properties, mechanical strength, including compressive, flexural, and tensile strengths, and elastic modulus, gradually increase. Because the volume of hydration products formed is lower than that of the master materials, the use of nano-$CaCO_3$ can increase the autogenous shrinkage of UHPC (Li et al. 2016). It should be noted that suitable nano-$CaCO_3$ content should be secured to avoid the agglomeration of nano-$CaCO_3$ and the resulting adverse effect.

3.6.3 Nano-TiO_2

Nano-TiO_2 is characterized by good stability, low cost, and photocatalysis. When used as a pigment, it is called titanium white. TiO_2 occurs in nature as well-known minerals rutile, anatase, and brookite (Maravelaki-Kalaitzaki et al. 2013). It is mainly derived from ilmenite ore, the most widespread form of TiO_2-bearing ore worldwide. Rutile is the next most abundant and contains around 98% titanium dioxide in the ore. The metastable anatase and brookite phases convert after being heated above temperatures of 600–800°C.

Typically, the addition of TiO_2 in UHPC has a great effect on self-cleaning ability and contributes to the application of green material in construction (Maravelaki-Kalaitzaki et al. 2013). The self-cleaning effect of TiO_2 has been utilized in buildings and paving material, presented by the construction of Jubilee Church in Rome, Italy (Sanchez and Sobolev 2010). Another benefit of nano-TiO_2 is the increase in strength of UHPC. The use of 2.32% nano-TiO_2 in UHPC increased the 28-d compressive and flexural strengths by 18% and 47%, respectively (Li et al. 2017). In addition, nano-TiO_2 can reduce the drying shrinkage rate of UHPC and improve its chloride penetration resistance, frost resistance, and carbonation resistance (Gu et al. 2018). This is attributed to the fact that nano-TiO_2 can act as a glass layer or pigment outside and inside the microstructure of UHPC. The layer reacts to the hydration gel during mixing and acts as a protective layer, endowing the self-cleaning ability of the UHPC surface. Besides, nano-TiO_2 forms a fiber-reinforced system in UHPC, which can be seen as mimicking the glass fiber effect. Refining and tailoring the hydration gel with fiber contributes to strength enhancement and results in more durable UHPC (Chen and Poon 2009).

Nano-TiO_2 can cause safety and health issues to human beings, including cytotoxicity, phytotoxicity, and phototoxicity through organelle dysfunction. Dusty and small TiO_2 particles can lead to environmental effects, inflammatory effects, and even dangerously cancer in factory workers during packing and production (Pacheco-Torgal and Jalali 2011). Therefore, care during the handling and mixing processes related to nano-TiO_2 must be taken seriously.

3.6.4 Graphene oxide

Graphene oxide is a carbon material with a two-dimensional honeycomb crystal structure composed of monolayer carbon atom SP2. The ideal structure of graphene is a planar hexagonal lattice with each carbon atom attached to its three adjacent carbon atoms in a firm covalent bond. This unique structure also presents excellent mechanical properties, with an elastic modulus of 1000 GPa and tensile strength of 130 GPa, making it a promising material for producing UHPC. Graphene oxide is the product of strong oxidation of graphene and is one of the essential derivatives of graphene (Schniepp et al. 2006). After oxidation, graphene oxide is linked to many oxygen-containing functional groups, such as hydrocarbon and carboxyl, in its two-dimensional plane structures. These hydrophilic oxygen-containing functional groups can give graphene oxide better dispersion and enable graphene oxide to deeply participate in the chemical reaction process through active functional groups (He et al. 1998). However, graphene oxide is less active and insoluble in water, making it difficult to apply directly to cement-based materials.

The mechanism underlying the performance enhancement of UHPC associated with graphene oxide is quite different from fiber (Kim et al. 2019; Ragalwar et al. 2020; Wu et al. 2020). With graphene oxide content increasing from 0 to 0.05%, the compressive and flexural strengths of UHPC initially increase to reach the maximum values at a critical content of 0.02% and then decrease (Wu et al. 2020). Using molecular dynamics modeling, the C-S-H gel layer in UHPC with graphene oxide contains more calcium ions and hydroxide than ordinary concrete with graphene oxide (Wan and Zhang 2020). Calcium ions combining with graphene oxide oxygen-containing functional groups and water molecules can form hydrogen bonds between the functional groups and C-S-H gel. These two actions enhance the bond between C-S-H and graphene oxide and improve the ductility and tensile strength of the microstructure.

3.6.5 Carbon nanotubes

Carbon nanotubes belong to one-dimensional nanocarbon materials with a length-to-diameter ratio of up to 132,000,000: 1, significantly greater than that of other materials. It can be viewed as a piece of graphene rolled up, with a hollow interior and a cylinder-cap-like seal at both ends. According to the number of layers, carbon nanotubes can be divided into single-walled carbon nanotubes and multiwalled carbon nanotubes (Jiang et al. 2018). These cylindrical carbon molecules have unusual thermal conductivity and mechanical and electrical properties, making it possible to prepare intelligent concrete (Quercia et al. 2012).

Carbon nanotubes can introduce piezoresistivity and make UHPC a self-sensing material due to the improved conductivity and pressure sensitivity, which helps evaluate the strain and time response of UHPC and establish the model. The effect of carbon nanotubes on the performance of UHPC mainly depends on its content. You et al. (2017) investigated the electrical and self-sensing capacities of UHPC with 2% microsteel fibers and 0.1%, 0.3%, and 0.5% multiwalled carbon nanotubes. The carbon nanotubes have little influence on the fluidity when the content increases from 0.1% to 0.3%. Besides, the electrical resistivity of UHPC is substantially reduced by incorporating carbon nanotubes, while the compressive strength of UHPC is decreased. However, it decreases the fluidity obviously at high contents due to poor dispersion (You et al. 2017). At present, several methods, including ultrasonic dispersion (Shah et al. 2010), surface modification (Li et al. 2005), and composite method (Luo et al. 2011), are proven to be effective in improving the dispersion of carbon nanotubes. Yoo et al. (2018) studied the feasibility of achieving self-strain and damage sensing in UHPC by incorporating 2% microsteel fibers and 0.5% multiwalled carbon nanotubes. It is indicated that the hybridization of steel fibers and carbon nanotubes provided a significant improvement in tensile performance, including strength and postpeak ductility, compared to the single use of carbon nanotubes.

3.6.6 Nanocarbon fiber

Nanocarbon fiber is a novel nanocarbon material formed by rolling a layered graphite sheet. It has a relatively low cost and a great production advantage compared with carbon nanotube. Carbon nanotube has a tensile strength of 2.7 GPa, an elastic modulus of 400 GPa, and a diameter between 0.5 and 100 nm (Shi et al. 2019). It can be divided into plate, fishbone, and tube types, depending on the stacking way of grapheme (Shi et al. 2019). In addition to excellent mechanical properties, nanocarbon fiber also has high electrical conductivity and biocompatibility (Yang et al. 2020; He et al. 2020). Similar to nanocarbon fiber, the dispersion of nanocarbon fiber is influenced by surface energy, high length-diameter ratio, and van der Waals force, causing them to intertwine and be difficult to disperse in cement-based materials. Its dispersion in cement-based materials can also be enhanced by utilizing ultrasonic treatment and surface modification.

Incorporating nanocarbon fiber into UHPC has little effect on its fluidity due to its larger diameter, making its mixed air content lower and microstructure denser. In addition, the use of nanocarbon fiber can reduce the self-shrinkage and significantly improve the tensile and flexural strengths of UHPC because of its fiber-bridging effect (Lim et al. 2019). However, it has little influence on compressive strength and overall compactness of UHPC (Meng et al. 2016).

3.7 MIXING WATER

Water is an essential gradient in UHPC. The quantity of water determines how much cement can be hydrated and controls fresh and hardened properties in concrete, including workability, compressive strength, and durability. In conventional concrete with a w/b ratio of 0.4 to 0.5, partially water hydrates with cement, while the excess water would develop pores to increase the permeability and reduce the strength of concrete. In UHPC, the water amount is far less than enough for the complete hydration of cement used. Thus, using high-range water-reducer is essential to help maintain workability without the need for excess water.

The quality of water plays a vital role in affecting the performance of concrete. Impurities and contaminations, such as oils, acids, salts, or other components in water, may interfere with cement setting and be harmful to the quality of concrete. Water suitable for making conventional concrete can be used to prepare UHPC. The most commonly used one is drinkable water, also known as potable water. Various existing and new water sources are available, which may be suitable for the complete or partial replacement of potable water for concrete fabrication. These include reclaimed water, groundwater, treated water from sewers, streams, rivers, and ready-mix concrete plant water.

It should be noted that the impurities in the water should be limited to avoid affecting the workability, setting, strength, and durability without polluting the concrete surface. In arid regions and coastal areas, the shortage and scarcity of freshwater address the challenges of searching for alternatives. Seawater with a high concentration of chloride ions cannot be used in reinforced concrete due to the corrosion of steel bars. However, due to the high density and low porosity, it may be difficult for chloride ions to migrate into UHPC, and the contents of water and oxygen in UHPC may be very low. Therefore, unpurified seawater provides technical feasibility for preparing UHPC, which is approved by some researchers in the laboratory (Teng et al. 2019; Huang et al. 2021). Detailed information regarding UHPC made with seawater and sea sand will be discussed in Chapter 12.

3.8 CHEMICAL ADMIXTURE

3.8.1 Superplasticizer

Superplasticizer is a critical component in UHPC, which plays a significant role in reducing water content, maintaining suitable flowability, and influencing the dispersing ability of fine particles, especially silica fume. Polycarboxylate-based superplasticizer is the most effective one for UHPC because it can induce a steric hindrance between binder particles to release entrapped water among particles and achieve good compatibility with different binder materials (Schröfl et al. 2012; Ferrari et al. 2010; Hanehara and Yamada 1999). Methacrylic acid ester (MPEG) and allylether (APEG)-based polycarboxylate superplasticizers were shown to be effective dispersants for UHPC (Meng et al. 2017; Plank et al. 2009). The blends of these two superplasticizer types can reduce the superplasticizer demand by around 50% (Plank et al. 2009). This is because the MPEG superplasticizer can better disperse cement, whereas the APEG superplasticizer fluidizes silica fume particularly well. The behavior is explained by the preferential adsorption of APEG superplasticizer on silica fume, while MPEG superplasticizer exhibits a more balanced affinity to cement and silica (Figure 3.9). Silica fume has a much higher surface area than that of cement. Therefore, the successful dispersion of silica fume is critical for the high fluidity of UHPC (Plank et al. 2009). The incorporation of APEG superplasticizer can

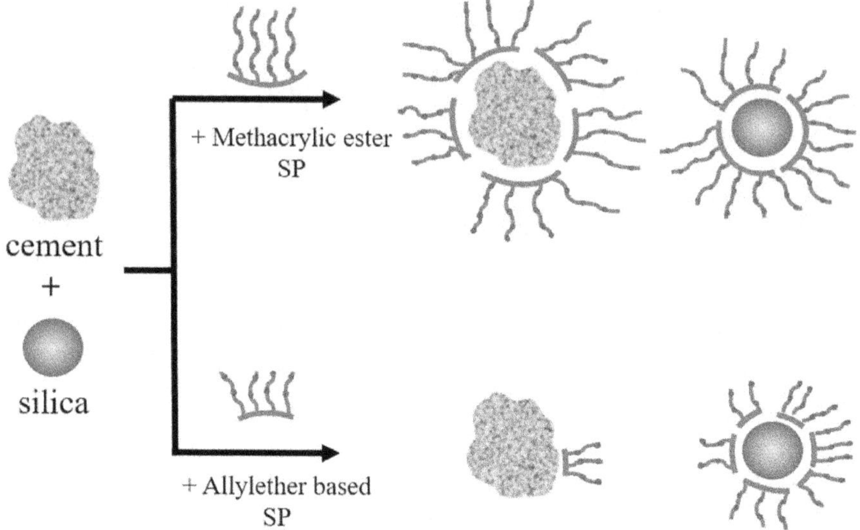

Figure 3.9 Schematic representation of different surface coverages of cement and silica fume by methacrylic acid ester-based superplasticizer and allylether-based superplasticizer.

Source: Adapted from Schröfl et al. 2012.

predominantly cover the surface of silica fume and consequently disperse it. Therefore, combining two superplasticizers can reduce the dosage required to achieve the same fluidity.

The length of the side chain of superplasticizer mainly influences the retardation time and the workability is mainly controlled by the side chain density (Feylessoufi et al. 1997). Hirschi and Wombacher (2008) investigated eight types of polycarboxylate-based superplasticizers on the fresh and hardened properties of UHPC. It was found that mixtures with long side-chain lengths showed higher early strength than those with medium side-chain lengths.

The adsorbed superplasticizer mainly determines the fluid-retaining ability of UHPC and it will not increase after obtaining the saturation dosages of superplasticizer due to the complete coverage of particles. Li et al. (2017) indicated that the chemical structure and its adsorption ability on particles mainly affect the dispersing ability of polycarboxylate-based superplasticizer. An exponential relationship between the flowability of pastes and superplasticizer dosage was observed. Depending on binder type and content and superplasticizer type, the most effective superplasticizer dosage varies. Courtial et al. (2013) studied the effect of polycarboxylate superplasticizer dosages in the range of 0.5%–2.0% on the microstructure of UHPC. It was found that the belite phase content significantly decreased when the polycarboxylate superplasticizer dosage varied from 1.8% to 1.2%. The increasing superplasticizer dosage has little effect on the packing density and water film thickness of the paste. However, UHPC usually demonstrates very high viscosity due to the low w/b ratio. Furthermore, high polycarboxylate superplasticizer dosage with very high water reduction capability increases the space between the binder particles, thus increasing the viscosity because of the high concentration of the unadsorbed superplasticizer in the interstitial solution (Khayat et al. 2019). It is necessary to design specific polycarboxylate superplasticizer with high water reduction capability but low viscosity that targets for UHPC. The addition procedure of superplasticizer also influences the workability of fresh UHPC (Tue et al. 2008). Stepwise addition remarkably enhances the dispersing effect of superplasticizer and increases the flowability of UHPC, compared with the one-time direct addition.

3.8.2 Viscosity modifying admixture

Viscosity modifying admixture, known as antiwashout admixtures, is a water-soluble polymer able to increase the viscosity and cohesion of cement-based materials (Khayat 1998). The enhancement of the liquid-phase viscosity is essential in flowable systems to reduce the separation rate of material constituents and improve homogeneity. Viscosity modifying admixture can be used along with superplasticizer to improve the rheological properties of UHPC, thus regulating the dispersion and orientation of fibers to enhance the mechanical properties (Meng and Khayat 2017).

Polysaccharide gum from microbial source, such as welan gum, can enhance the rebuilt-up kinetics at rest, thus increasing yield stress (Bouras et al. 2012). This effect can be more significant at low shear rates (Ghio et al. 1994). It can give shear thinning behavior at low shear rates and shear thickening behavior at high shear rates. The shear-thinning behavior is attributed to the disentanglement and alignment of admixture molecules, while the increase of repulsive interparticle forces results in shear thickening behavior (Bouras et al. 2012). Sonebi (2006) compared the rheological properties of cement grout made with welan gum and diutan gum. It is reported that the increase in gum content at a given superplasticizer dosage can significantly increase the rheological properties. For a given dosage of viscosity modifying admixture, diutan gum showed a higher viscosity than welan gum. This is because of a higher molecular weight and longer side chains of diutan gum, thus leading to greater entanglement and intertwining.

Meng and Khayat (2017) reported that the increase of high molecular ethylenoxide viscosity modifying admixture gradually enhanced the plastic viscosity of UHPC with targeted minislump flow from 270 to 290 mm. On the contrary, Boulekbache et al. (2010) found that increasing the plastic viscosity reduced the mechanical properties by introducing more air voids. Therefore, the optimal plastic viscosity needs to be determined to ensure uniform fiber distribution and minimized flaws to enhance the tensile/flexural performance (Teng et al. 2020).

3.8.3 Air-detraining admixture

Since fresh UHPC mixture is very viscous, a large amount of air bubbles would be entrained during mixing and casting processes to affect mechanical properties and quality of surface appearance adversely. These air bubbles can be eliminated by applying about 1–3 minutes of vibration during casting. Some studies used air-detraining admixture in self-consolidating UHPC to counteract the air entrainment caused by water reducers (Meng et al. 2017; Meng and Khayat 2017). The air-detraining admixtures can destabilize air in fresh UHPC, allowing it to escape during mixing, thus reducing the air content and increasing strength. It should be noted that the compatibility between air-detraining agent and other materials in UHPC should be carefully checked.

3.8.4 Incompatibility issues

Incompatibility of concrete refers to the adverse interaction between acceptable materials that result in unexpected or unacceptable performance. Common problems include early stiffing, flash setting, false setting, rapid slump loss, unstable air entrainment, retarded strength gain, and unordinary cracking. Incompatibility issues can easily occur in UHPC compared to conventional concrete. Such issues may include binder-driven incompatibility, incompatibility related to supplementary cementitious materials, binder-superplasticizer incompatibility, and chemical admixture–chemical admixture interaction.

Incompatibility issue between superplasticizer and cementitious materials is often accounted for UHPC. For example, an insufficient superplasticizer dosage would make UHPC compaction difficult and lead to high porosity. However, excess superplasticizer dosage can cause delayed setting and fiber segregation, resulting in nonuniform mixing or higher porosity. Schröfl et al. (2012) investigated the interactions between binders and methacrylic acid ester-based or allylether-based polycarboxylate ether superplasticizers in UHPC. They found that the methacrylate-based polycarboxylate ether superplasticizers strongly interacted with cement, whereas the allylether-based polycarboxylate ether superplasticizers were more effective with silica fume. The combination of the two superplasticizers showed a significantly better dispersion than that using a single type. Li et al. (2017) showed that a polycarboxylate-based superplasticizer with rapid absorption to cement particles is incompatible with the microsand in UHPC. Van Tuan et al.

(2011b) indicated that UHPC with limestone microfiller demonstrated a good fluidity compared to metakaolin, pulverized fly ash, and siliceous microfiller at the same content superplasticizer. When cement replacement by silica fume was increased from 30% to 40%, the superplasticizer dosage increased dramatically from 1.43% to 2.38%.

3.9 FIBERS

Concrete is an inherently brittle material with a relatively low tensile strength as compared to its compressive strength. Although the compressive strength of UHPC is over 120 MPa, its tensile capacity is only about 1/10 to 1/20 of its compressive strength. Fiber is the most essential ingredient for UHPC due to economic, strength, ductility, and durability reasons. Fiber can inhibit and control the formation of intrinsic cracking caused both in the plastic and hardened stage of concrete, thus restraining shrinkage and improving the strength and toughness of UHPC. Fiber is incorporated into UHPC, usually by volume of concrete, in mono or hybrid ways.

3.9.1 Mono fiber

The fiber used in concrete includes steel fiber, synthetic fiber, natural fiber, etc. Table 3.5 shows the typical properties of the commonly used fibers in concrete. Among all available fibers in the market, high-strength microsteel fiber is commonly employed for structural applications in UHPC. These fibers are generally used in UHPC to enhance toughness and postcrack load-carrying capacity. Such steel fiber usually has a diameter of 0.009–0.5 mm and a length of 6–60 mm (Figure 3.10). The steel fiber volume ranges from 1% to 3%, or higher. Steel fibers can be processed into different shapes, such as hooked, corrugated, and twisted fibers, to enhance the fiber-matrix bond and mechanical properties of UHPC by providing mechanical anchorage through the deformed sections (Wu et al. 2018; Kang et al. 2016; Wu et al. 2019).

The mechanical properties of UHPC are influenced by fiber type, content, geometry, length, dispersion and orientation, and fiber-matrix bond. The increase in steel fiber volume can make UHPC more ductile (Kang et al. 2010). In addition, the incorporation of steel fibers has a significant effect in restraining shrinkage of UHPC, depending on steel fiber volume and shape (Wu et al. 2019). Compared to the reference mixture without any fiber, the autogenous and drying shrinkage values of UHPC were decreased by 50%–58% and 25%–36%, respectively (Wu et al. 2019). However, it should be noted that the increased amount of fibers can lead to fiber cluster issues during mixing, thus exerting an unfavorable influence on workability and mechanical properties of UHPC (Dupont and Vandewalle 2005).

Table 3.5 Typical properties of fibers

Types	*Diameter (μm)*	*Relative density (g/cm³)*	*Tensile strength (MPa)*	*Young modulus (GPa)*	*Elongation (%)*
-	*250–1000*	*7.8*	*280–2800*	*200–250*	*0.5–40*
Acrylic	5–17	1.18	800–950	16–23	9–11
Polyester	10–80	1.38	735–1200	6–18	11–15
Polyethylene	800–1000	0.96	200–300	5–6	3–4
Polypropylene	20–70	0.91	300–770	3.5–11	15–25
Nylon	23	1.16	900–960	4.2–5.2	18–20
Polyvinyl Alcohol	1.30	1.30	600–2500	5–50	6–17
Aramid	10–12	1.44	2500–3100	60–120	2.1–4.5
Glass	10–16	2.74	1400–2500	70–80	2.5–3.5
Asbestos	≤0.5	2.75	500–980	84–140	0.3–0.6
Rock wool	2.7	2.7	490–770	70–119	0.6
Wood	25–400	1.40	50–1000	14–40	-
Carbon	7–18	1.75	1800–4000	200–480	1.2–1.6

Microsynthetic fibers are generally used for the protection and mitigation of plastic shrinkage cracking in UHPC. Most fiber types are manufactured from polypropylene, polyethylene, polyester, nylon, and other synthetic materials, such as carbon, aramid, and acrylics. These fiber types are generally incorporated at low volume, ranging from 0.03% to 0.2%, by volume of concrete. Cellulose fibers, manufactured from processed wood pulp products, are used in UHPC to control and mitigate shrinkage. These fibers are efficient in restraining shrinkage, but their improvement in strength and toughness of UHPC is limited. Thus, these types of fiber are often blended with steel fibers in UHPC to achieve multiple functions.

3.9.2 Hybrid fibers

A cost analysis of raw materials of UHPC with 2% steel fibers shows that the fiber cost takes up to approximately 60%–80%. This ratio can be much higher if either deformed steel fibers or a higher fiber amount is used. To minimize the steel fiber content and reduce material cost, fiber blending is proposed to prepare UHPC. Such fiber blending usually includes the use of different shapes, sizes, and types of fiber. For nonstructural applications to control plastic shrinkage of UHPC, polypropylene fibers are often combined with steel fibers to achieve a certain degree of fire resistance. For architectural applications with either no or very low structural requirements, probably polyvinyl alcohol (PVA) and stainless-steel fibers can be used.

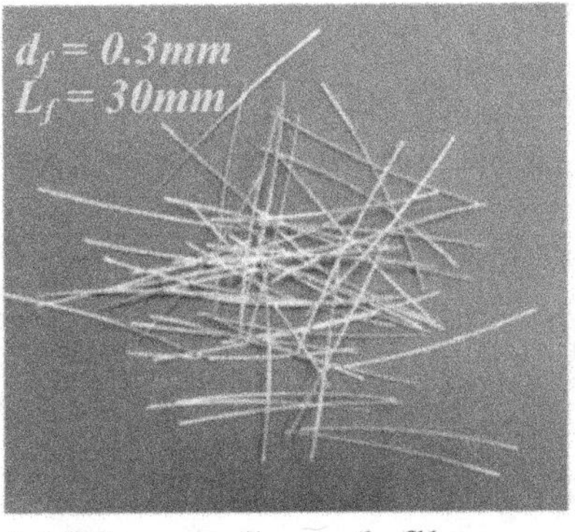

(a) Macro – Smooth fiber

(b) Macro – Hooked fiber A

(c) Macro – Hooked fiber B

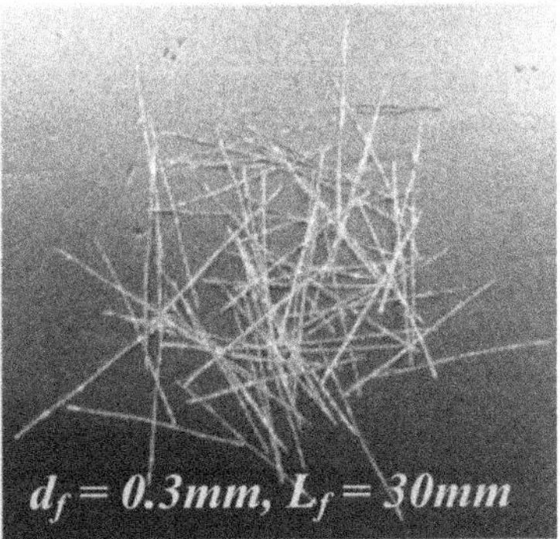

(d) Macro – Twisted fiber

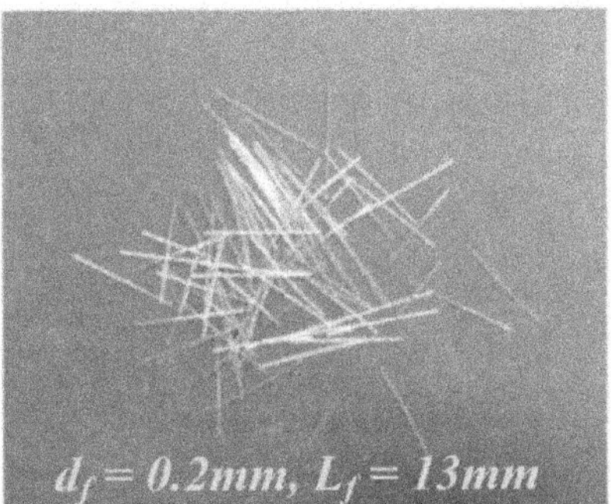

(e) Micro – Smooth fiber

Figure 3.10 Steel fibers commonly used in UHPC.

Source: Kim et al. 2011.

Compared with steel fibers, the use of polypropylene and glass fibers might decrease the strength due to the low strength and the weak interfacial transition zone between synthetic fibers and the matrix (Hannawi et al. 2016). Park et al. (2012) found that UHPC incorporated with 1.0% macrosteel fibers and more than 0.5% microsteel fibers showed tensile strain hardening behavior. Kim et al. (2018) studied the effects of hybridization of micro- and macrofibers on the flexural performance of UHPC. They revealed that the deflection capacity (d_{MOR}) and energy absorption capacity (T_{MOR}) of UHPC with 1.0% micro and 1.0% macrofibers were 45.4%–75.9% and 48.7%–67.9%, respectively, greater than those with 2.0% microfibers.

3.10 INTERNAL CURING AGENTS

UHPC with a low w/b ratio can undergo considerable self-desiccation during hydration, resulting in relatively high autogenous shrinkage. However, traditional curing techniques, such as watering and covering burlap, cannot allow significant penetration of external water into these materials because of the dense microstructure and low permeability of UHPC, as shown in Figure 3.11. Internal curing effectively mitigates the autogenous shrinkage of UHPC since it can gradually release water to compensate for the drop in internal relative humidity (Liu et al. 2017; Ma et al. 2019). General requirements for internal curing materials that can be used in cement-based materials include thermodynamic availability and kinetic availability. The thermodynamic availability requires the water to have an activity close to one. In other words, an equilibrium relative humidity close to 100% is needed. Kinetic availability refers

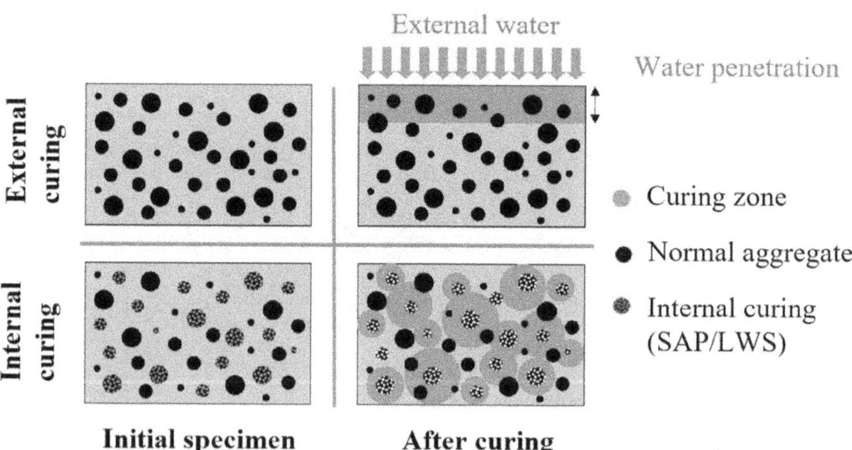

Figure 3.11 Schematic illustration of the mechanism of internal curing used in cement-based materials.

Table 3.6 Main properties of porous superfine powers

Type	*Rice husk ash*	*Acid-leaching residue of boron mud*	*Bottom ash*
Density (g/cm^3)	2.19	-	1.4–2.2
Blaine (BET) SSA (m^2/g)	50–150	50.2	-
Mean particle size (μm)	3–10	3.89	100–600
Pore size	2–50 nm	-	30 nm to 1 μm
Absorption (ml/g)	0.180	0.1–0.2	-
Content of silica	90%–96%	58.3%	-
References	Bui (2001); Sensale et al. (2008); Tuan et al. (2011); Van et al. (2014)	Huang and Zhang (2015)	Wyrzykowski et al. (2016)

to the transport of water from the reservoir to all parts of the self-desiccating cementitious material (Jensen and Lura 2006).

Lightweight expanded clay, pumice, expanded shale, zeolite, perlite, cenospheres, superabsorbent polymer, recycled aggregate, wood-derived powders and fibers, biolightweight aggregate, and Lechuguilla natural fiber can be used as internal curing agents. Among these materials, superabsorbent polymer and lightweight aggregates are the primary types used in UHPC (Liu 2019). Recently, the internal curing functions of other materials, such as rice husk ash, coal bottom ash, bentonite clay, and cellulose fiber, have been found. Table 3.6 summarizes the main properties of porous superfine powers. Due to the abundant supplies, superabsorbent polymers is more inexpensive than other candidate materials. Compared with lightweight aggregate, superabsorbent polymer presents superior absorption capacity and can be directly used in a dry state during the mixing process (Jensen and Lura 2006). The use of superabsorbent polymer permits the accessible design of the shape and size of formed inclusions (Jensen and Lura 2006). However, the voids or holes of superabsorbent polymer residues after water release might reduce mechanical properties and structural loading capacity of concrete (Lv et al. 2015). This is not found in concrete when using lightweight aggregate (Babcock and Taylor 2015; Byard and Schindler 2010). Ground rice husk ash and coal bottom ash have porous structures with nanosize pores (Van et al. 2013). The following section describes the characteristics of these internal curing materials.

3.10.1 Superabsorbent polymer

Superabsorbent polymers are cross-linked polyelectrolytes, which start to swell upon contact with water or aqueous solutions and form hydrogel

(Mechtcherine and Reinhardt 2012). They can be divided into acrylamide/acrylic acid copolymer and polyacrylic acid (Mechtcherine and Reinhardt 2012). The main driving force for the swelling of superabsorbent polymers is the osmotic pressure, and the pressure is proportional to the concentration of ions in the aqueous solution (Mechtcherine and Reinhardt 2012). The superabsorbent polymer has water absorption of up to 5000 times greater than its mass but is only 50 g/g in salt solution (Jensen and Hansen 2001). Because of the fast absorption rate of superabsorbent polymer and avoidance of "gel blocking" (Mechtcherine and Reinhardt 2012), it is necessary to blend superabsorbent polymer with cement and supplementary cementitious materials before adding water. Gel blocking is a special property of very fine superabsorbent polymers with a particle size of less than 100 μm. If the superabsorbent polymer contacts water in pure form, little absorption occurs at the surface, and the slightly swollen particles stick together. The superabsorbent polymer dosage in concrete is calculated according to the need for internal curing water. Based on Powers model (1948), Jensen et al. (2001) proposed the effective water-to-cement ratio $[(w/c)_{effective}]$ for internal curing that is necessary to secure complete hydration of cement as follows:

(1) when $w/c \leq 0.36$, $(w/c)_{effective} = 0.18 \times (w/c)$
(2) when $0.36 \leq w/c \leq 0.42$; $(w/c)_{effective} = 0.42 - (w/c)$.

It is also necessary to consider the composition, dosage, type, particle size, and the water-saturated state of superabsorbent polymer since they influence the resulting internal curing effect.

3.10.2 Lightweight aggregate

Lightweight aggregate is often used to replace aggregate or sand in cement-based materials. The internal curing effects of lightweight aggregate are affected by three key factors (Babcock and Taylor 2015): (1) water volume needed from the lightweight aggregate; (2) desorption properties of lightweight aggregate; and (3) lightweight aggregate distribution within the mixture. Firstly, the water volume needed or the internal curing water amount is determined by two measurable elements, including chemical shrinkage and autogenous shrinkage (Henkensiefken et al. 2009). The water absorption ability of lightweight aggregate is related to the continuous inner porosity and particle size. The porosity and pore size of lightweight aggregate should be appropriate to ensure adequate internal curing. Jensen and Lura (2006) thought that pure water held in a porous material had an equilibrium relative humidity of 99%, when the meniscus diameter was 100 nm. Thus, pores above approximately 100 nm help store internal curing water, while the water in pores smaller than 100 nm cannot be released in a timely manner to the hardened cement paste. Most water is absorbed in nanosized pores of zeolite aggregates and retained at low relative humidity levels (Ghourchian et al. 2013). As a result, zeolite aggregates provided insufficient internal curing capacity (Ghourchian et al. 2013).

The pore size of UHPC is mainly between 2 and 3 nm, with the most probable pore diameter of 2.0 nm, and it is in the range of 3.75–100 nm when it is cured between 150 and 200°C (Wang et al. 2015). Therefore, lightweight aggregate with nanosized pores, such as zeolite, can also be applied for internal curing (Zhang et al. 2017). On the other hand, if the pores in the lightweight aggregate are too large, the water retention capacity of the lightweight aggregate would be poor. Lightweight aggregate shows a high absorption rate and can release water at high relative humidity (Ghourchian et al. 2013). Lightweight aggregate has been successfully employed to partially replace conventional coarse aggregate and sand in UHPC (Hwang et al. 2012). The lightweight aggregate content needs to be sufficient to ensure uniform distribution in concrete (Akcay and Tasdemir 2010).

3.10.3 Other internal curing materials

In HPC, the water-absorbing capacity of rice husk ash is 17.15 l/m^3, according to Van et al. (2014). In addition, the pore size in rice husk ash particles, i.e., from 2 to 50 nm in diameter, is much smaller than the superabsorbent polymer particles of 450 μm and the size of lightweight aggregates of about 10–20 μm (Mechtcherine et al. 2009; Zhutovsky et al. 2004). According to the Kelvin equation, the pore size of rice husk ash in the range of 2–50 nm corresponds to a relative humidity change from about 75% to 98% (Lura et al. 2003). This means that when the internal relative humidity in UHPC drops below 98%, the absorbed water in these mesopores will be released to compensate for the self-desiccation during hydration. It postpones and slows the internal relative humidity reduction in the rice husk ash-blended matrices. Thus, rice husk ash also has the internal curing function to reduce self-desiccation, hence mitigating the autogenous shrinkage of UHPC. In addition, the rice husk ash content and particle size greatly influence the autogenous shrinkage of UHPC.

Cellulose fibers are hollow, and their surface can be easily modified with a more hydrophilic character or attached functional groups (Faruk et al. 2012). A high hydroxyl group content in cellulose increases its water absorption capacity (Alvarez et al. 2003). This allows them to absorb and retain water, as shown in Figure 3.12. Water existing in the structure of cellulose fibers usually includes free water in the lumen with diameters ranging from 10 to 200 μm, small cell wall pores, and bound water in the cell wall with a size range of 40–100 nm (mainly hydrogen bonds) (Topgaard and Soderman 2002). Both types of water can migrate to the surrounding

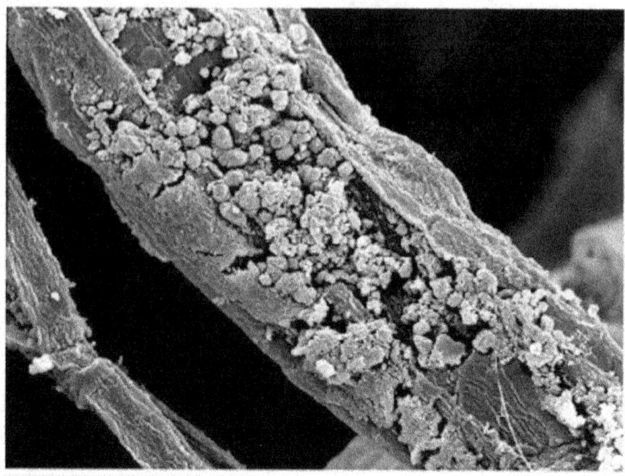

Figure 3.12 SEM micrograph of lumen in cellulose fibers.

Source: Ardanuy et al. 2015.

cement paste and contribute to reducing the autogenous shrinkage of cement-based materials (Jongvisuttisun et al. 2018). The amount of internal curing water released depends on the size, lumen structure, such as open, closed, and collapsed, and the capacity to bind water. The last one is closely related to its chemical compositions, i.e., cellulose, hemicellulose, and lignin contents (Wu et al. 2009). Thinner wall thickness and larger lumen diameter contribute to more water stored inside the lumen of cellulose fibers, but a thick cell wall can further reduce the autogenous shrinkage beyond 25 h of hydration because it will not release all water too early (Jongvisuttisun et al. 2018). Compared with lightweight aggregate and superabsorbent polymer, the obvious advantage of cellulose fibers is to additionally provide resistance to shrinkage-induced cracking due to an increase in tensile capacity. Besides, compared to superabsorbent polymer, cellulose fibers are not vulnerable to volume change during water migration. Although cellulose fibers have internal curing capabilities, the adverse effect of cellulose fibers on workability may make them unsuitable for an exclusive application due to dispersion issues.

Perforated cenospheres with particle sizes ranging from several microns to 400 μm are manufactured from hollow fly ash particles (Hanif et al. 2017). As shown in Figure 3.13(a), cenosphere particles have an aluminosilicate shell with high stiffness, strength, and thickness of several microns (Liu et al. 2017). The shell of cenospheres is covered by a glass-crystalline nanosize

Figure 3.13 SEM observations of cenospheres (Liu et al. 2017). (a) Cenospheres with impermeable shell. (b) Porous shell of the cenosphere.

film. After removing this thin film using chemical etching, perforating holes can be produced on the shell of the cenosphere, allowing internal curing water to penetrate and store in the inner pore of the cenosphere, as seen in Figure 3.13(b) (Chen et al. 2018). Although cenospheres have particle sizes like powder, they are not included as cementitious materials in the mixtures. Instead, they can be added to replace the same volume of sand (Chen et al. 2019; Liu et al. 2017).

3.11 SUMMARY

This chapter discusses the available constituents and their characteristics for the production of UHPC. Typical ingredients, including cement, supplementary cementitious materials, limestone powder, aggregates, nanoparticles, water, chemical admixtures, fibers, and internal curing agents, are described. The use of supplementary cementitious materials can reduce cement and/or silica fume contents and alter the hydration, microstructure, and properties of UHPC. River sand, lightweight sand, and recycled waste materials can replace quartz sand without sacrificing the mechanical properties of UHPC. The addition of nanoparticles can densify the microstructure and improve the mechanical properties of UHPC due to the filling, pozzolanic, nucleation, and reinforcing effects. The chemical admixtures can significantly adjust the workability, and the addition of internal curing agents can mitigate the autogenous shrinkage of UHPC. An appropriate selection of high-quality raw materials with optimal content and without incompatibility issues should be considered to prepare UHPC with excellent properties.

REFERENCES

ACI Committee 234. (2006). Guide for the Use of Silica Fume in Concrete, ACI–American Concrete Institute–Committee, Detroit, USA, 2006, pp. 1–63.

Akcay, B., Tasdemir, M. A. (2010). Effects of distribution of lightweight aggregates on internal curing of concrete. *Cem. Concr. Compos.* 32(8), 611–616.

Alvarez, V. A., Ruscekaite, R. A., Vazquez, A. (2003). Mechanical properties and water absorption behavior of composites made from a biodegradable matrix and alkaline-treated sisal fibers. *J. Compos. Mater.* 37(17), 1575–1588.

Ardanuy, M., Claramunt, J., Toledo, R. D. (2015). Cellulosic fiber reinforced cement-based composites: A review of recent research. *Constr. Build. Mater.* 79, 115–128.

Arora, A., Yao, Y., Mobasher, B., Neithalath, N. (2019). Fundamental insights into the compressive and flexural response of binder-and aggregate-optimized ultra-high performance concrete (UHPC). *Cem. Concr. Compos.* 98, 1–13.

Babcock, A. E., Taylor, P. (2015). Impacts of internal curing on concrete properties.

Bai, P., Sharratt, P., Yeo, T. Y., Bu, J. (2014). A facile route to preparation of high purity nanoporous silica from acid-leached residue of serpentine. *J. Nanosci. Nanotechno.* 14(9), 6915–6922.

Boulekbache, B., Hamrat, M., Chemrouk, M., Amziane, S. (2010). Flowability of fibre-reinforced concrete and its effect on the mechanical properties of the material. *Constr. Build. Mater.* 24(9), 1664–1671.

Bouras, R., Kaci, A., Chaouche, M. (2012). Influence of viscosity modifying admixtures on the rheological behavior of cement and mortar pastes. *Korea-Aust. Rheol J.* 24(1), 35–44.

Bui, D. D. (2001). *Rice husk ash a mineral admixture for high performance concrete.* Delft University of Technology, Delft, the Netherlands.

Byard, B. E., Schindler, A. K. (2010). *Cracking tendency of lightweight concrete.* Highway Research Center, Auburn University, Auburn, AL.

Chen, J., Poon, C. S. (2009). Photocatalytic construction and building materials: From fundamentals to applications. *Build. Environ.* 44(9), 1899–1906.

Chen, P. Y., Wang, J. L., Liu, F. J., Qian, X., Xu, Y., Li, J. (2018). Converting hollow fly ash into admixture carrier for concrete. *Constr. Build. Mater.* 159, 431–439.

Chen, P. Y., Wang, J. L., Wang, L., Xu, Y. (2019). Perforated cenospheres: A reactive internal curing agent for alkali activated slag mortars. *Cem. Concr. Compos.* 104, 103351.

Chen, W., Zhu, Z. (2018). Utilization of fly ash to enhance ground waste concrete-based geopolymer. Advances in Materials Science and Engineering, 2018, 1–10.

Cheyrezy, M., Maret, V., Frouin, L. (1995). Microstructural analysis of RPC (Reactive Powder Concrete). *Cem. Concr. Res.* 25(7), 1491–1500.

Courtial, M., de Noirfontaine, M. N., Dunstetter, F., Signes-Frehel, M., Mounanga, P., Cherkaoui, K., Khelidj, A. (2013). Effect of polycarboxylate and crushed quartz in UHPC: Microstructural investigation. *Constr. Build. Mater.* 44, 699–705.

Ding, Y., Liu, J. P., Bai, Y. L. (2020). Linkage of multi-scale performances of nano-$CaCO_3$ modified ultra-high performance engineered cementitious composites (UHP-ECC). *Constr. Build. Mater.* 234, 117418.

Dupont, D., Vandewalle, L. (2005). Distribution of steel fibres in rectangular sections. *Cem. Concr. Compos.* 27(3), 391–398.
Faruk, O., Bledzki, A. K., Fink, H. P., Sain, M. (2012). Biocomposites reinforced with natural fibers: 2000–2010. *Prog. Polym. Sci.* 37(11), 1552–1596.
Ferdosian, I., Camões, A., Ribeiro, M. (2017). High-volume fly ash paste for developing ultra-high performance concrete (UHPC). *Ciência & Tecnologia dos Materiais* 29(1), e157–e161.
Ferrari, L., Kaufmann, J., Winnefeld, F., Plank, J. (2010). Interaction of cement model systems with superplasticizers investigated by atomic force microscopy, zeta potential, and adsorption measurements. *J. Colloid. Interf. Sci.* 347(1), 15–24.
Feylessoufi, A., Crespin, M., Dion, P., Bergaya, F., Van Damme, H., Richard, P. (1997). Controlled rate thermal treatment of reactive powder concretes. *Adv. Cem. Based Mater.* 6(1), 21–27.
Flores-Vivian, I., Pradoto, R. G., Moini, M., Kozhukhova, M., Potapov, V., Sobolev, K. (2017). The effect of SiO_2 nanoparticles derived from hydrothermal solutions on the performance of Portland cement based materials. *Front. Struct. Civ. Eng.* 11(4), 436–445.
Ganesh, P., Murthy, A. R. (2019). Tensile behaviour and durability aspects of sustainable ultra-high performance concrete incorporated with GGBS as cementitious material. *Constr. Build. Mater.* 197, 667–680.
Gesoglu, M., Güneyisi, E., Sabah Asaad, D., Fakhraddin Muhyaddin, G., 2016. Properties of low binder ultra-high performance cementitious composites: comparison of nanosilica and microsilica. *Constr. Build. Mater.* 102, 706–713.
Ghafari, E., Arezoumandi, M., Costa, H., Julio, E. (2015). Influence of nano-silica addition on durability of UHPC. *Constr. Build. Mater.* 94, 181–188.
Ghafari E, Costa H, Julio E, Portugal A, Duraes L. (2014). The effect of nanosilica addition on flowability, strength and transport properties of ultra high performance concrete. *Mater. Des.* 59, 1–9.
Ghafari, E., Costa, H., Júlio, E., Portugal, A., Durães, L. Enhanced durability of ultra high performance concrete by incorporating supplementary cementitious materials (2012). Proc., The 2nd international conference microdurability, Delft, Netherland, 86–94.
Ghio, V. A., Monteiro, P. J., Demsetz, L. A. (1994). The rheology of fresh cement paste containing polysaccharide gums. *Cem. Concr. Res.* 24(2), 243–249.
Ghourchian, S., Wyrzykowski, M., Lura, P., Shekarchi, M., Ahmadi, B. (2013). An investigation on the use of zeolite aggregates for internal curing of concrete. *Constr. Build. Mater.* 40, 135–144.
Graybeal, B. A. (2006). Material property characterization of ultra-high performance concrete. Mclean, Virginia, United States: Federal Highway Administration. Office of Infrastructure Research and Development.
Gu C., Wang Q., Liu J., Sun W. (2018). The effect of nano-TiO_2 on the durability of ultra-high performance concrete with and without flexural load. *Ceram-Silikáty.* 62(4), 374–381.
Hanehara, S., Yamada, K. (1999). Interaction between cement and chemical admixture from the point of cement hydration, absorption behaviour of admixture, and paste rheology. *Cem. Concr. Res.* 29(8), 1159–1165.

Hanif, A., Lu, Z. Y., Li, Z. J. (2017). Utilization of fly ash cenosphere as lightweight filler in cement-based composites—A review. *Constr. Build. Mater.* 144, 373–384.

Hannawi, K., Bian, H., Prince-Agbodjan, W., Raghavan, B. (2016). Effect of different types of fibers on the microstructure and the mechanical behavior of ultra-high performance fiber-reinforced concretes. *Compos. Part B-Eng.* 86, 214–220.

Helmy, S., Hoa, S. V. (2014). Tensile fatigue behavior of tapered glass fiber reinforced epoxy composites containing nanoclay. *Compos. Sci. Technol.* 102, 10–19.

Henkensiefken, R., Bentz, D., Nantung, T., Weiss, J. (2009). Volume change and cracking in internally cured mixtures made with saturated lightweight aggregate under sealed and unsealed conditions. *Cem. Concr. Compos.* 31(7), 427–437.

He, H., Klinowski, J., Forster, M., Lerf, A. (1998) A new structural model for graphite oxide. *Chem. Phys.* Lett. 287(1–2), 53–56.

He, H., Cheng, M. X., Liang, Y. T., Zhu, H. X., Sun, Y. P., Dong, D., Wang, S. F. (2020). Intelligent cellulose nanofibers with excellent biocompatibility enable sustained antibacterial and drug release via a pH-responsive mechanism. *J. Agric. Food. Chem.* 68(11), 3518–3527.

Hirschi, T., Wombacher, F. (2008). Influence of different superplasticizers on UHPC. Proc., Proceedings of the Second International Symposium on Ultra High Performance Concrete, Kassel University Press, Kassel, 77–84.

Huang, W., Kazemi-Kamyab, H., Sun, W., Scrivener, K. (2017). Effect of cement substitution by limestone on the hydration and microstructural development of ultra-high performance concrete (UHPC). *Cem. Concr. Compos.* 77, 86–101.

Huang, H., Gao, X., Wang, H., Ye, H. (2017). Influence of rice husk ash on strength and permeability of ultra-high performance concrete. *Constr. Build. Mater.* 149, 621–628.

Huang, Z., Zhang, C. (2015). Effect of porous superfine powder on the performance of ultra-high performance concrete. *Bull. Chines. Ceram. Soc.* 7, 22–27.

Huang, B. T., Wang, Y. T., Wu, J. Q., Yu, J., Dai, J. G., Leung, C. K. (2021). Effect of fiber content on mechanical performance and cracking characteristics of ultra-high-performance seawater sea-sand concrete (UHP-SSC). *Advan. Struct. Eng.* 24(6), 1182–1195.

Hwang, S. D., Khayat, K. H., Youssef, D. (2012). Effect of moist curing and use of lightweight sand on characteristics of high-performance concrete. *Mater. Struct.* 46(1–2), 35–46.

Jensen, O. M., Hansen, P. F. (2001). Water-entrained cement-based materials I. Principles and theoretical background. *Cem. Concr. Res.* 31(4), 647–654.

Jensen, O. M., Lura, P. (2006). Techniques and materials for internal water curing of concrete. *Mater. Struct.* 39(9), 817–825.

Jiang, Y., Song, H., Xu, R. (2018). Research on the dispersion of carbon nanotubes by ultrasonic oscillation, surfactant and centrifugation respectively and fiscal policies for its industrial development. *Ultrason Sonochem* 48, 30–38.

Jiao, Y., Zhang, Y., Guo, M., Zhang, L., Ning, H., Liu, S. (2020). Mechanical and fracture properties of ultra-high performance concrete (UHPC) containing waste glass sand as partial replacement material. *J. Clean. Prod.* 277, 123501.

Jongvisuttisun, P., Leisen, J., Kurtis, K. E. (2018). Key mechanisms controlling internal curing performance of natural fibers. *Cem. Concr. Res.* 107, 206–220.

Kang, S. H., Hong, S. G., Moon, J. (2019). The use of rice husk ash as reactive filler in ultra-high performance concrete. *Cem. Concr. Res.* 115, 389–400.

Kang, S. T., Choi, J. I., Koh, K. T., Lee, K. S., Lee, B. Y. (2016). Hybrid effects of steel fiber and microfiber on the tensile behavior of ultra-high performance concrete. *Compos. Struct.* 145, 37–42.

Kang, S. T., Lee, Y., Park, Y. D., Kim, J. K. (2010). Tensile fracture properties of an Ultra High Performance Fiber Reinforced Concrete (UHPFRC) with steel fiber. *Compos. Struct.* 92(1), 61–71.

Khayat, K. H. (1998). Viscosity-enhancing admixtures for cement-based materials—An overview. *Cem. Concr. Compos.* 20(2–3), 171–188.

Khayat, K. H., Meng, W. N., Vallurupalli, K., Teng, L. (2019). Rheological properties of ultra-high-performance concrete—An overview. *Cem. Concr.* Res. 124, 105828.

Kim, D. J., Park, S. H., Ryu, G. S., Koh, K. T. (2011). Comparative flexural behavior of hybrid ultra high performance fiber reinforced concrete with different macro fibers. *Construct. Build. Mater.* 25(11), 4144–4155.

Kim, H. K., Jang, J. G., Choi, Y. C., Lee, H. K. (2014). Improved chloride resistance of high-strength concrete amended with coal bottom ash for internal curing. *Constr. Build. Mater.* 71, 334–343.

Kim, J., Yoo, D. (2019). Effects of fiber shape and distance on the pullout behavior of steel fibers embedded in ultra-high-performance concrete. *Cem. Concr. Compos.* 103, 213–223.

Kim, M. J., Yoo, D. Y., Kim, S., Shin, M., Banthia, N. (2018). Effects of fiber geometry and cryogenic condition on mechanical properties of ultra-high-performance fiber-reinforced concrete. *Cem. Concr. Res.* 107, 30–40.

Konstagdoutos, M. S., Metaxa, Z. S., Shah, S. P. (2010). Highly dispersed carbon nanotube reinforced cement based materials. *Cem. Concr. Res.* 40(7), 1052–1059.

Korpa, A., Kowald, T., Trettin, R. (2009). Phase development in normal and ultra high performance cementitious systems by quantitative X-ray analysis and thermoanalytical methods. *Cem. Concr. Res.* 39(2), 69–76.

Le Thanh Ham, K. S., Ludwig, H. M. (2012). Synergistic effect of rice husk ash and fly ash on properties of self-compacting high performance concrete. Proc., Ultra-High Performance Concrete and Nanotechnology in Construction. Proceedings of Hipermat. 3rd International Symposium on UHPC and Nanotechnology for High Performance Construction Materials, Kassel University Press GmbH, 187.

Li, G., Wang, P. M., Zhao, X.(2005). Mechanical behavior and microstructure of cement composites incorporating surface treated multi-walled carbon nanotubes. *Carbon.* 43(6), 1239–1245.

Li, P. P., Yu, Q. L., Brouwers, H. J. H. (2017). Effect of PCE-type superplasticizer on early-age behaviour of ultra-high performance concrete (UHPC). *Construct. Build. Mater.* 153, 740–750.

Li, P., Yu, Q., Brouwers, H. (2018). Effect of coarse basalt aggregates on the properties of Ultra-high Performance Concrete (UHPC). *Construct. Build. Mater.* 170, 649–659.

Li, W., Huang, Z., Zu, T., Shi, C., Duan, W. H., Shah, S. P. (2016). Influence of nanolimestone on the hydration, mechanical strength, and autogenous shrinkage of ultrahigh-performance concrete. *J. Mater. Civ. Eng.* 28(1), 04015068.

Li, Z., Han, B., Yu, X., Dong, S., Zhang, L., Dong, X., Ou, J. (2017). Effect of nano-titanium dioxide on mechanical and electrical properties and microstructure of reactive powder concrete. *Mater. Res. Express* 4(9), 095008.

Lim, J. L., Raman, S. N., Safiuddin, M., Zain, M. F. M., Hamid, R. (2019). Autogenous shrinkage, microstructure, and strength of ultra-high performance concrete incorporating carbon nanofibers. *Mater.* 12(2), 320.

Liu, F. J., Wang, J. L., Qian, X., Hollingsworth, J. (2017). Internal curing of high performance concrete using cenospheres. *Cem. Concr. Res.* 95, 39–46.

Liu, J. (2019). Internal curing of SAP and FLWA in ultra-high performance concrete. *Doctoral thesis*. Hunan University.

Liu, J. H., Shi, C. J., Farzadnia, N., Ma, X. W. (2019). Effects of pretreated fine lightweight aggregate on shrinkage and pore structure of ultra-high strength concrete. *Constr. Build. Mater.* 204, 276–287.

Liu, J. H., Shi, C. J., Ma, X. W., Khayat, K. H., Zhang, J., Wang, D. H. (2017). An overview on the effect of internal curing on shrinkage of high performance cement-based materials. *Constr. Build. Mater.* 146, 702–712.

Liu, K., Yu, R., Shui, Z., Li, X., Ling, X., He, W., Yi, S., Wu, S. (2019). Effects of pumice-based porous material on hydration characteristics and persistent shrinkage of ultra-high performance concrete (UHPC). *Mater.* 12(1), 11.

Long, G. C., Wang, X. Y., Xie, Y. J. (2002). Very-high-performance concrete with ultrafine powders. *Cem. Concr. Res.* 32(4), 601–605.

Luo, J., Duan, Z., Xian, G., Li, Q., Zhao, T. J. (2011). Fabrication and fracture toughness properties of carbon nanotube-reinforced cement composite. *Eur. Phys. J. Appl. Phys.* 53(3), 30402.

Lura, P., Bentz, D. P., Lange, D. A., Kovler, K., Bentur, A. (2004). Pumice aggregates for internal water curing. Proc., Proceedings of International RILEM Symposium on Concrete Science and Engineering: A Tribute to Arnon Bentur, RILEM Publications SARL Evanston, Paris, France, 137–151.

Lura, P., Jensen, O. M., van Breugel, K. (2003). Autogenous shrinkage in high-performance cement paste: An evaluation of basic mechanisms. *Cem. Concr. Res.* 33(2), 223–232.

Lv, Y., Huang, H., Ye, G., De Schutter, G. (2015). Autogenous shrinkage of zeolite cement pastes with low water-binder ratio. Proc., 14th International Congress on the Chemistry of Cement (ICCC 2015), Beijing, China, 1–8.

Ma, A. (2008). Experimental study on reactive powder concrete with environment-protecting and energy-saving properties. *Doctoral thesis*. Beijing Jiaotong University (in Chinese), Beijing, China.

Ma, J., Orgass, M., Dehn, F., Schmidt, D., Tue, N. (2004). Comparative investigations on ultra-high performance concrete with and without coarse aggregates. Proc., International Symposium on Ultra High Performance Concrete, Kassel, Germany, 205–212.

Ma, X. W., Liu, J. H., Shi, C. J. (2019). A review on the use of LWA as an internal curing agent of high performance cement-based materials. *Constr. Build. Mater.* 218, 385–393.

Madani, H., Bagheri, A., Parhizkar, T. (2012). The pozzolanic reactivity of monodispersed nanosilica hydrosols and their influence on the hydration characteristics of Portland cement. *Cem. Concr. Res.* 42(12), 1563–1570.

Maravelaki-Kalaitzaki, P., Agioutantis, Z., Lionakis, E., Stavroulaki, M., Perdikatsis, V. (2013). Physico-chemical and mechanical characterization of hydraulic mortars containing nano-titania for restoration applications. *Cem. Concr. Compos.* 36, 33–41.

Markovic I. (2006). High-performance hybrid-fiber concrete: development and utilisation. *Ph.D. Thesis.* Delft University of Technology, Delft, Netherlands.

Mechtcherine, V., Dudziak, L., Hempel, S. (2009). Mitigating early age shrinkage of ultra-high performance concrete by using super absorbent polymers (SAP). Proceedings of the Creep, Shrinkage, and Durability Mechanics of Concrete and Concrete Structures, Mitigating, Japan, 847–853.

Mechtcherine, V., Reinhardt, H. W. (2012). *Application of super absorbent polymers (SAP) in concrete construction: state-of-the-art report prepared by Technical Committee 225.* Springer Science & Business Media.

Meng, W., Khayat, K. H. (2016). Mechanical properties of ultra-high-performance concrete enhanced with graphite nanoplatelets and carbon nanofibers. *Compos. Part B: Eng.* 107, 113–122.

Meng, W., Khayat, K. H. (2018). Effect of hybrid fibers on fresh properties, mechanical properties, and autogenous shrinkage of cost-effective UHPC. *J. Mater. Civ. Eng.* 30(4), 04018030.

Meng, W., Valipour, M., Khayat, K. H. (2017). Optimization and performance of cost-effective ultra-high performance concrete. *Mater. Struct.* 50, 1–16.

Meng, W. N., Khayat, K. (2017). Effects of saturated lightweight sand content on key characteristics of ultra-high-performance concrete. *Cem. Concr. Res.* 101, 46–54.

Meng, W. N., Khayat, K. H. (2017). Improving flexural performance of ultra-high-performance concrete by rheology control of suspending mortar. *Compos. Part B-Eng.* 117, 26–34.

Mo, Z., Gao, X., Su, A. (2020). Mechanical performances and microstructures of metakaolin contained UHPC matrix under steam curing conditions. *Constr. Build. Mater.* 268, 121112.

Mosaberpanah, M., Umar, S. (2020). Utilizing rice husk ash as supplement to cementitious materials on performance of ultra high performance concrete: A review. *Mater. Today Sustain.* 7, 100030.

Nehdi, M., Mindess, S., Aïtcin, P. C. (1996). Optimization of high strength limestone filler cement mortars. *Cem. Concr. Res.* 26(6), 883–893.

Norhasri, M.S.M., Hamidah, M.S., Fadzil, A.M. (2019). Inclusion of nano metaclayed as additive in ultra-high performance concrete (UHPC). *Constr. Build. Mater.* 201, 590–598.

Pacheco-Torgal, F., Jalali, S. (2011). Nanotechnology: Advantages and drawbacks in the field of construction and building materials. *Constr. Build. Mater.* 25(2), 582–590.

Park, S. H., Kim, D. J., Ryu, G. S., Koh, K. T. (2012). Tensile behavior of ultra high performance hybrid fiber reinforced concrete. *Cem. Concr. Compos.* 34(2), 172–184.

Park, S., Wu, S., Liu, Z., Pyo, S. (2021). The role of supplementary cementitious materials (SCMs) in ultra high performance concrete (UHPC): a review. *Mater.* 14(6), 1472.

Peng, Y., Hu, S., Ding, Q. (2010). Preparation of reactive powder concrete using fly ash and steel slag powder. *Journal of Wuhan University of Technology-Mater. Sci. Ed.* 25(2), 349–354.

Plank, J., Schroefl, C., Gruber, M., Lesti, M., Sieber, R. (2009). Effectiveness of polycarboxylate superplasticizers in ultra-high strength concrete: the importance of PCE compatibility with silica fume. *J. Adv. Concr. Technol.* 7(1), 5–12.

Powers, T., Brownyard, T. (1948). Studies of the physical properties of hardened cement paste. *J. Proc.* 22, 101–132.

Qi, J. A., Wu, Z. M., Ma, Z. J., Wang, J. Q. (2018). Pullout behavior of straight and hooked-end steel fibers in UHPC matrix with various embedded angles. *Constr. Build. Mater.* 191, 764–774.

Qian, D., Yu, R., Shui, Z., Sun, Y., Jiang, C., Zhou, F., Ding, M., Tong, X., He, Y. (2020). A novel development of green ultra-high performance concrete (UHPC) based on appropriate application of recycled cementitious material. *J. Clean Prod.* 261, 121231.

Quercia, G., Hüsken, G., Brouwers, H. (2012). Water demand of amorphous nano silica and its impact on the workability of cement paste. *Cem. Concr. Res.* 42(2), 344–357.

Ragalwar, K., Heard, W. F., Williams, B. A., Kumar, D., Ranade, R. (2020). On enhancing the mechanical behavior of ultra-high performance concrete through multi-scale fiber reinforcement. *Cem. Concr. Compos.* 105, 103422.

Rößler, C., Bui, D. D., Ludwig, H. M. (2014). Rice husk ash as both pozzolanic admixture and internal curing agent in ultra-high performance concrete. *Cem. Concr. Compos.* 53, 270–278.

Sanchez, F., Sobolev, K. (2010). Nanotechnology in concrete—A review. *Constr. Build. Mater.* 24(11), 2060–2071.

Scherb, S., Köberl, M., Beuntner, N., Thienel, K. C., Neubauer, J. (2020). Reactivity of metakaolin in alkaline environment: correlation of results from dissolution experiments with XRD quantifications. *Mater.* 13(10), 2214.

Schniepp, H. C., Li J., Mcallister, M. J., Sai H., Herreraalonso, M., Adamson, D. H., Aksay, I. A. (2006). Functionalized single graphene sheets derived from splitting graphite oxide. J. *Phys. Chem. B.* 110(17), 8535–8539.

Schröfl, C., Gruber, M., Plank, J. (2012). Preferential adsorption of polycarboxylate superplasticizers on cement and silica fume in ultra-high performance concrete (UHPC). *Cem. Concr. Res.* 42(11), 1401–1408.

Sensale, G. R. D., Ribeiro, A. B., Gonçalves, A. (2008). Effects of RHA on autogenous shrinkage of Portland cement pastes. *Cem. Concr. Compos.* 30(10), 892–897.

Shah, S. P., Konsta-Gdoutos, M. S., Metaxa, Z. S. (2010). Exploration of fracture characteristics, nanoscale properties and nanostructure of cementitious matrices with carbon nanotubes and carbon nanofibers. In *Proceedings of the 7th international conference on fracture mechanics of concrete and concrete structures*. Seoul, South Korea, 9–12.

Shi, C. J., Wu, Z. M., Xiao, J. F., Wang, D. H., Huang, Z. Y., Fang, Z. (2015). A review on ultra high performance concrete: Part I. Raw materials and mixture design. *Constr. Build. Mater.* 101, 741–751.

Shi, Y. X., Matsui, I., Feng, N. Q. (2002). Effect of compound mineral powders on workability and rheological property of HPC. *Cem. Concr. Res.* 32(1), 71–78.

Shi, T., Li, Z., Guo, J., Gong, H., Gu, C. (2019). Research progress on CNTs/CNFs-modified cement-based composites—a review. *Constr. Build. Mater.* 202, 290–307.

Sobolev, K., FI, H. R., Torres-Martinez, L. Nanomaterials and nanotechnology for high-composites. SP-254, 8, 93–120.

Sobolev, K., Ferrada-Gutiérrez, M. (2005). How nanotechnology can change the concrete world: part 2. *Am. Ceram. Soc. Bull.* 11, 16–19.

Soliman, N. A., Tagnit-Hamou, A. (2017). Using glass sand as an alternative for quartz sand in UHPC. *Constr. Build. Mater.* 145, 243–252.

Sonebi, M. (2006). Rheological properties of grouts with viscosity modifying agents as diutan gum and welan gum incorporating pulverised fly ash. *Cem. Concr. Res.* 36(9), 1609–1618.

Souza, A. T., Barbosa, T. F., Riccio, L. A., dos Santos, W. J. (2020). Effect of limestone powder substitution on mechanical properties and durability of slender precast components of structural mortar. *J. Mater. Res. Technol.* 9(1), 847–856.

Staquet, S., Espion, B. (2004). Early-age autogenous shrinkage of UHPC incorporating very fine fly ash or metakaolin in replacement of silica fume. International Symposium on UHPC, Kassel, Germany, 587–599.

Svermova, L., Sonebi, M., Bartos, P. J. M. (2003). Influence of mix proportions on rheology of cement grouts containing limestone powder. *Cem. Concr. Compos.* 25(7), 737–749.

Tafraoui, A., Escadeillas, G., Lebaili, S., Vidal, T. (2009). Metakaolin in the formulation of UHPC. *Constr. Build. Mater.* 23(2), 669–674.

Tafraoui, A., Escadeillas, G., Vidal, T. (2016). Durability of the ultra high performances concrete containing metakaolin. *Constr. Build. Mater.* 112, 980–987.

Talebinejad, I., Bassam, S. A., Iranmanesh, A., Shekarchizadeh, M. Optimizing mix proportions of normal weight reactive powder concrete with strengths of 200–350 MPa. Proc., Proceedings of the International Symposium on UHPC, Kassel, Germany, 133–141.

Tazawa, E., Miyazawa, S. (1995). Influence of cement and admixture on autogenous shrinkage of cement paste. *Cem. Concr. Res.*25(2), 281–287.

Teng, J.G., Xiang, Y., Yu, T., Fang, Z. (2019). Development and mechanical behaviour of ultra-high-performance seawater sea-sand concrete. *Adv. Struct. Eng.* 22(14), 310–3120.

Teng, L., Meng, W., Khayat, K. H. (2020). Rheology control of ultra-high-performance concrete made with different fiber contents. *Cem. Concr. Res.* 138, 106222.

Topgaard, D., Soderman, O. (2002). Changes of cellulose fiber wall structure during drying investigated using NMR self-diffusion and relaxation experiments. *Cellulose* 9(2), 139–147.

Tsivilis, S., Tsantilas, J., Kakali, G., Chaniotakis, E., Sakellariou, A. (2003). The permeability of Portland limestone cement concrete. *Cem. Concr. Res.* 33(9), 1465–1471.

Tuan, N. V., Ye, G., Breugel, K. V., Copuroglu, O. (2011). Hydration and microstructure of ultra high performance concrete incorporating rice husk ash. *Cem. Concr. Res.* 41(11), 1104–1111.

Tuan, N. V., Ye, G., van Breugel, K. (2010). Internal Curing of Ultra High Performance Concrete by Using Rice Husk Ash. *International Rilem Conference on Material Science (Matsci)* Iii, 77, 265–274.

Tue, N. V., Ma, J., Orgass, M. (2008). Influence of addition method of superplasticizer on the properties of fresh UHPC. Proc., Proceedings of the 2nd International Symposium on Ultra-High Performance Concrete, Kassel, Germany, 93–100.

Upadhyaya, S., Nanda, B., Panigrahi, R. (2019). Experimental analysis on partial replacement of fine aggregate by granite dust in concrete. In Sustainable Construction and Building Materials: Select Proceedings of ICSCBM, 335–344.

Van Der Putten, J., Lesage, K., De Schutter, G. (2016). *Influence of the particle shape on the packing density and pumpability of UHPC.* Ultra-High Performance Concrete and High Performance Construction Materials (HIPERMAT). 27, 1–8.

Van Tuan, N., Ye, G., van Breugel, K., Copuroglu, O. (2011a). Hydration and microstructure of ultra high performance concrete incorporating rice husk ash. *Cem. Concr. Res.* 41(11), 1104–1111.

Van Tuan, N., Ye, G., Van Breugel, K., Fraaij, A. L., Dai Bui, D. (2011b). The study of using rice husk ash to produce ultra high performance concrete. *Constr. Build. Mater.* 25(4), 2030–2035.

Van, V. T. A., Rossler, C., Bui, D. D., Ludwig, H. M. (2013). Mesoporous structure and pozzolanic reactivity of rice husk ash in cementitious system. *Constr. Build. Mater.* 43, 208–216.

Van, V. T. A., Rossler, C., Bui, D. D., Ludwig, H. M. (2014). Rice husk ash as both pozzolanic admixture and internal curing agent in ultra-high performance concrete. *Cem. Concr. Compos.* 53, 270–278.

Wan, H., Zhang, Y. (2020). Interfacial bonding between graphene oxide and calcium silicate hydrate gel of ultra-high performance concrete. *Mater. Struct.* 53(2), 1–12.

Wang, C., Yang, C. H., Liu, F., Wan, C. J., Pu, X. C. (2012). Preparation of ultra-high performance concrete with common technology and materials. *Cem. Concr. Compos.* 34(4), 538–544.

Wang, D. H., Shi, C. J., Wu, Z. M., Xiao, J. F., Huang, Z. Y., Fang, Z. (2015). A review on ultra high performance concrete: Part II. Hydration, microstructure and properties. *Constr. Build. Mater.* 96, 368–377.

Wang, X. P., Yu, R., Song, Q. L., Shui, Z. H., Liu, Z., Wu, S., Hou, D. S. (2019). Optimized design of ultra-high performance concrete (UHPC) with a high wet packing density. *Cem. Concr. Res.* 126, 105921.

Wang, X., Yu, R., Shui, Z., Song, Q., Zhang, Z. (2017). Mix design and characteristics evaluation of an eco-friendly ultra-high performance concrete incorporating recycled coral based materials. *J. Clean Prod.* 165, 70–80.

Weber, S., Reinhardt, H. W. (1997). A new generation of high performance concrete: concrete with autogenous curing. *Adv. Cem. Based Mater.* 6(2), 59–68.

Wille, K., Naaman, A. E., El-Tawil, S., Parra-Montesinos, G. J. (2012). Ultra-high performance concrete and fiber reinforced concrete: achieving strength and ductility without heat curing. *Mater. Struct.* 45(3), 309–324.

Wille, K., Naaman, A. E., Parra-Montesinos, G. J. (2011). Ultra-high performance concrete with compressive strength exceeding 150 MPa (22 ksi): a simpler way. *ACI Mater. J.* 108(1), 46–54.

Williams, E.M., Graham, S.S., Reed, P.A., Rushing, T. S. (2009). *Laboratory characterization of Cor-Tuf concrete with and without steel fibers.* Engineer Research and Development Center.

Wu, N., Hubbe, M. A., Rojas, O. J., Park, S. (2009). Permeation of polyelectrolytes and other solutes into the pore spaces of water-swollen cellulose: a review. *Bioresources* 4(3), 1222–1262.

Wu, Z., Shi, C., Khayat, K. H. (2016). Influence of silica fume content on microstructure development and bond to steel fiber in ultra-high strength cement-based materials (UHSC). *Cem. Concr. Compos.* 71, 97–109.

Wu, Z., Shi, C., He, W., Wang, D. (2017). Static and dynamic compressive properties of ultra-high performance concrete (UHPC) with hybrid steel fiber reinforcements. *Cem. Concr. Compos.* 79, 148–157.

Wu, Z., Khayat, K. H., Shi, C. (2019a). Changes in rheology and mechanical properties of ultra-high performance concrete with silica fume content. *Cem. Concr. Res.* 123, 105786.

Wu, Z. M., Shi, C. J., Khayat, K. H. (2019b). Investigation of mechanical properties and shrinkage of ultra-high performance concrete: Influence of steel fiber content and shape. *Compos. Part B-Eng.* 174, 107021.

Wu, Z. M., Shi, C. J., Khayat, K. H., Wan, S. (2016). Effects of different nanomaterials on hardening and performance of ultra-high strength concrete (UHSC). *Cem. Concr. Compos.* 70, 24–34.

Wu, Y-Y, Zhang, J, Liu, C., Zheng, Z., Lambert, P. (2020). Effect of graphene oxide nanosheets on physical properties of ultra-high-performance concrete with high volume supplementary cementitious materials. *Mater.* 13(8), 1929.

Wyrzykowski, M., Ghourchian, S., Sinthupinyo, S., Chitvoranund, N., Chintana, T., Lura, P. (2016). Internal curing of high performance mortars with bottom ash. *Cem. Concr. Compos.* 71, 1–9.

Xi, J., Liu, J., Yang, K., Zhang, S., Han, F., Sha, J., Zheng, X. (2022). Role of silica fume on hydration and strength development of ultra-high performance concrete. *Constr. Build. Mater.* 338, 127600.

Yalçınkaya, Ç., Çopuroğlu, O. (2020). Hydration heat, strength and microstructure characteristics of UHPC containing blast furnace slag. *J. Build. Eng.* 34, 101915.

Yang, C., Ren, J., Zheng, M., Zhang, M., Zhong, Z., Liu, R., Yang, X. (2020). High-level N/P co-doped Sn-carbon nanofibers with ultrahigh pseudocapacitance for high-energy lithium-ion and sodium-ion capacitors. *Electrochimica Acta* 359, 136898.

Yang, H., Li, W., Che, Y. (2020). 3D printing cementitious materials containing nano-$CaCO_3$: workability, strength, and microstructure. *Front. Mater.* 7, 260.

Yang, L., Shi, C. J., Wu, Z. M. (2019). Mitigation techniques for autogenous shrinkage of ultra-high-performance concrete—A review. *Compos. Part B-Eng.* 178, 107456.

Yang, R., Yu, R., Shui, Z., Gao, X., Xiao, X., Fan, D., Chen, Z., Cai, J., Li, X., He, Y. (2020). Feasibility analysis of treating recycled rock dust as an environmentally friendly alternative material in ultra-high performance concrete (UHPC). *J. Clean Prod.* 258, 120673.

Yang, S. L., Millard, S. G., Soutsos, M. N., Barnett, S. J., Le, T. T. (2009). Influence of aggregate and curing regime on the mechanical properties of ultra-high performance fibre reinforced concrete (UHPFRC). *Constr. Build. Mater.* 23(6), 2291–2298.

Yang, Y., Ingham, J. M. (2000). *Manufacturing reactive powder concrete using common New Zealand materials.* Department of Civil and Resource Engineering, University of Auckland.

Yazıcı, H., Yardımcı, M. Y., Aydın, S., Karabulut, A. Ş. (2009). Mechanical properties of reactive powder concrete containing mineral admixtures under different curing regimes. *Constr. Build. Mater.* 23(3), 1223–1231.

Yazıcı, H., Yardımcı, M. Y., Yiğiter, H., Aydın, S., Türkel, S. (2010). Mechanical properties of reactive powder concrete containing high volumes of ground granulated blast furnace slag. *Cem. Concr. Compos.* 32(8), 639–648.

Yazıcı, H., Yiğiter, H., Karabulut, A. Ş., and Baradan, B. (2008). Utilization of fly ash and ground granulated blast furnace slag as an alternative silica source in reactive powder concrete. *Fuel* 87(12), 2401–2407.

Ye, G. (2012). Mitigation of autogenous shrinkage of ultra-high performance concrete by rice husk ash. *J. Chin. Ceram. Soc.* 40(2), 212–216 (In Chinese).

Yoo, D. Y., Banthia, N. (2016). Mechanical properties of ultra-high-performance fiber-reinforced concrete: A review. *Cem. Concr. Compos.* 73, 267–280.

Yoo, D. Y., Kim, S., Lee, S. H. (2018). Self-sensing capability of ultra-high-performance concrete containing steel fibers and carbon nanotubes under tension. *Sens. Actuators, A* 276, 125–136.

You, I., Yoo, D. Y., Kim, S., Kim, M. J., Zi, G. (2017). Electrical and self-sensing properties of ultra-high-performance fiber-reinforced concrete with carbon nanotubes. *Sensors* 17(11), 2481.

Yu R., Spiesz P. R. P., Brouwers HJ. Effect of nano-silica on the hydration and microstructure development of ultra-high performance concrete (UHPC) with a low binder amount. (2014b). *Constr. Build. Mater.* 65, 140–150.

Yu, L., Wu, R. (2020). Using graphene oxide to improve the properties of ultra-high-performance concrete with fine recycled aggregate. *Constr. Build. Mater.* 259, 120657.

Yu, R., Spiesz, P., Brouwers, H. J. H. (2014a). Mix design and properties assessment of Ultra-High Performance Fibre Reinforced Concrete (UHPFRC). *Cem. Concr. Res.* 56, 29–39.

Yu, R., Spiesz, P., Brouwers, H. J. H. (2015). Development of an eco-friendly ultra-high performance concrete (UHPC) with efficient cement and mineral admixtures uses. *Cem. Concr. Compos.* 55, 383–394.

Zhang, D., Zhou, C. H., Lin, C. X., Tong, D. S., Yu, W. H. (2010). Synthesis of clay minerals. *Appl. Clay Sci.* 50(1), 1–11.

Zhang, H., Ji, T., Zeng, X., Yang, Z., Lin, X., Liang, Y. (2018). Mechanical behavior of ultra-high performance concrete (UHPC) using recycled fine aggregate cured under different conditions and the mechanism based on integrated microstructural parameters. *Constr. Build. Mater.* 192, 489–507.

Zhang, J., Wang, Q., Zhang, J. J. (2017). Shrinkage of internal cured high strength engineered cementitious composite with pre-wetted sand-like zeolite. *Constr. Build. Mater.* 134, 664–672.

Zhao, S., Fan, J., Sun, W. (2014). Utilization of iron ore tailings as fine aggregate in ultra-high performance concrete. *Constr. Build. Mater.* 50, 540–548.

Zhu, W. Z., Gibbs, J. C. (2005). Use of different limestone and chalk powders in self-compacting concrete. *Cem. Concr. Res.* 35(8), 1457–1462.

Zhutovsky, S., Kovler, K., Bentur, A. (2004). Influence of cement paste matrix properties on the autogenous curing of high-performance concrete. *Cem. Concr. Compos.* 26(5), 499–507.

Chapter 4

Mixture design

4.1 INTRODUCTION

Concrete mixture design is the process to obtain the optimal combination of cement, aggregate, water, mineral admixtures, and chemical admixtures with desired workability, strength, durability, and economy according to given specifications. A highly efficient mixture design should have a good balance among these four aspects. Most countries around the world have formulated specifications for the mixture design of conventional concrete. The workability of concrete is usually decided by the construction engineer based on the site conditions, while the strength is specified by the structural designer. The durability requirements are usually covered in terms of minimum and maximum w/b ratios and cement content. Maximizing the aggregate amount in concrete with the use of appropriate cement paste to bind the particles together is a general philosophy of mixture design. A continuous gradation of aggregate particles is often employed to achieve the closest packing of the solid skeleton. This aspect is treated loosely for the mixture design of conventional concrete. However, close packing of solid particles is critical to the design of UHPC to achieve low porosity, dense matrix, high mechanical properties, and superior durability.

UHPC mixture design usually includes the following contents: (1) determine the type and composition of cementitious materials; (2) preliminarily select w/b ratio; (3) determine aggregate type, binder-to-sand ratio, and fiber volume; and (4) adjust the UHPC mixture proportion according to the required properties. The key to UHPC mixture design is to produce a tightly packed structure by optimizing the particle size distribution of solid particles based on the close particle packing theory. This requires strictly selected materials to fill the pores between larger particles with smaller ones to achieve a possible maximum packing density. The modified Anderson and Andreasen model is often used to obtain the maximum particle packing density of UHPC by achieving the target mixture curve in terms of particle size and content of each material. Normally, this model considers only the solid particles in the system and assumes that the particles are

DOI: 10.1201/9781003203605-4

spherical. In the presence of elongated fibers and liquids, such as water and superplasticizer, the real packing state of UHPC can be disrupted, resulting in changes in microstructure and performance. Therefore, finding a good UHPC mixture to achieve the closest possible system with suitable workability and hardened properties is not an easy task for concrete engineers and scientists. There always exist conflicts and contradictions when optimizing UHPC proportions to meet the overall performances. For example, the increase in compressive strength of UHPC is not that closely related to the w/b ratio and the fundamental parameters governing the properties of conventional concrete (Aïtcin 1998). Thus, conventional mixture design methods for conventional concrete and high-performance concrete based on empirical parameters cannot directly deal with these complexities of UHPC without appropriate modification. Therefore, the mixture design for UHPC is more complicated and can be considered a multiobjective optimization process.

This chapter focuses on four categories of mixture design principles, which include close packing method based on either dry or wet packing, rheology–based mixture design method, statistical design method, and artificial neural network–based method. The basic design principles, procedures, features, and advantages and disadvantages of these methods are described and compared. The influences of key raw materials and interaction among superplasticizer, aggregates, fibers, and other components on the design efficiency and performance of UHPC are discussed. This chapter can provide a scientific basis to produce UHPC with required workability, mechanical properties, durability, and economic efficiency.

4.2 CLOSE PACKING MODEL–BASED METHOD

Various models have been applied to the UHPC mixture design, such as Furnas model, Powers model, linear packing density model, Fuller and Thomson model, Andreasen and Andersen packing model, and the modified Andreasen and Andersen packing model or Funk and Dinger model. The interaction between particles and the packing density is often emphasized in these models.

The interaction force between particles can determine the overall properties and behavior as particle size reaches the micron and nanometer scales, but it can be ignored for larger particles (Kjeldsen et al. 2004). Studies have proved that UHPC with a densely packed skeleton can be designed using a particle packing model (Yu et al. 2015; Van Der Putten et al. 2016; Wang et al. 2019; Hou et al. 2021; Shi et al. 2021). Nevertheless, the existing particle packing models neglect the effect of liquid phase and its interaction with solid particles, especially fines, of mixture. Therefore, separating the packing and interaction between particles in dry and wet states is indispensable.

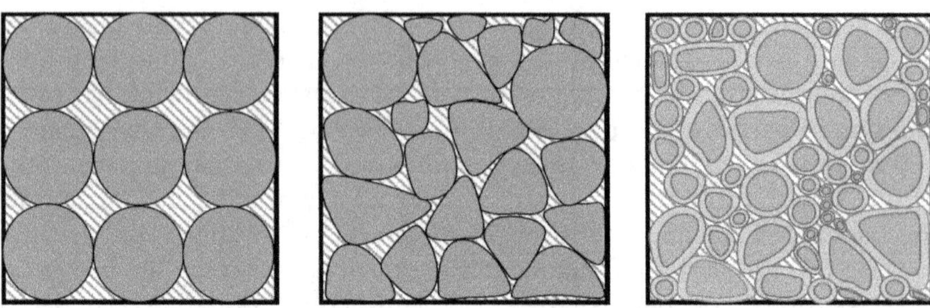

Figure 4.1 Aggregate packing diagram (Wong and Kwan 2005). (a) Concrete composes of single-sized aggregate. (b) Multisized aggregate reduces cement paste. (c) Aggregate surface coating.

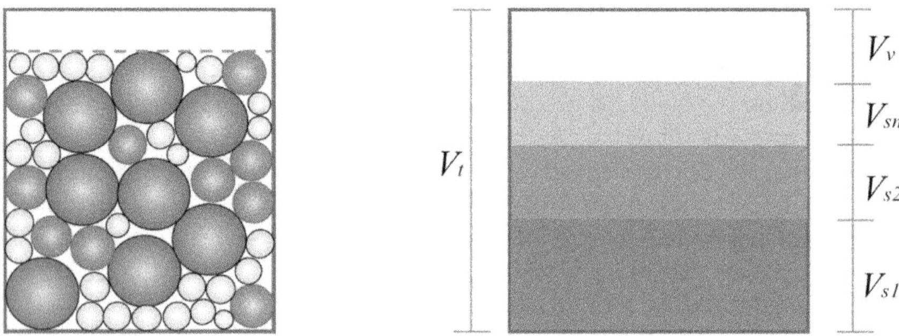

Figure 4.2 Packing density diagram of a granular mixture system. (a) Normal particle packing system. (b) Actual volume occupied by each component.

UHPC with a close particle packing structure is achieved by applying a particle packing model considering particle shape, content, grading, and size. The voids in cement-based materials with a monoaggregate are primarily introduced by insufficient filling between aggregates, as shown in Figure. 4.1(a). The addition of smaller sizes of aggregate reduces void content due to the packing of variable sizes of particles. Besides, multiparticle aggregate leads to extra paste on the aggregate surface at constant binder content, which forms a film with a certain thickness that improves concrete flowability, as shown in Figure 4.1.

Dry packing density refers to the proportion of solid volume to the total volume of system, as shown in Eq. (4.1). Figure 4.2 illustrates the packing density diagram of a granular mixture system. Decreases in binder content and porosity in UHPC can be achieved with a better dry packing density of solid materials. Improvement in packing density of solid skeleton is crucial for UHPC production. In mixtures containing aggregate with multiple particle sizes, the assumption that at least one aggregate is filled is indispensable. This can ensure that the mixture is compacted.

$$\textit{dry packing density } (\phi) = \frac{V_s}{V_t} \tag{4.1}$$

where ϕ is the packing density (unitless); V_s is the volume of each solid (ml); and V_t is the sum of solids and pore volume (mL).

The value of dry packing density fluctuates due to the external compaction energy exerted and agglomeration among fine particles under a dry state. When cementitious materials contact water, agglomeration easily occurs because of the high surface energy associated with the high specific surface area of fine particles. Besides, the further dried water molecules onto particles form hydrogen bonds that bridge each other, which trigger capillaries to shrink and further promote particle agglomeration (Yao et al. 2002). Unlike dry packing density, wet packing density considers the effects of water and superplasticizer on the dispersion of fine particles and can be calculated according to Eqs. (4.2–4.3) (Wong and Kwan 2008; Teng et al. 2020). Detailed information regarding the measurement of wet packing of particles and its effect on the performance of cement-based material will be explained in Section 4.2.1.

$$V_c = \frac{M_{\max}}{\rho_w u_w + \rho_{SP} u_{SP} + \rho_\alpha R_\alpha + \rho_\beta R_\beta + \rho_\gamma R_\gamma} \tag{4.2}$$

$$\textit{wet packing density } (\phi) = \frac{V_C}{V} \tag{4.3}$$

where ϕ is the packing density (unitless); u_w and u_{SP} represent the ratio of water and superplasticizer volumes to solid volume, respectively (unitless); ρ_w and ρ_{SP} are the densities of water and superplasticizer, respectively (kg/m^3); V is the mold volume (ml); V_C is the total solid volume (ml); ρ_α, ρ_β, and ρ_γ are the solid densities of components α, β, and γ, respectively (kg/m^3); and R_α, R_β, and R_γ represent the volumetric ratios of components α, β, and γ to the total volume of materials in the system, respectively (unitless).

The accuracy of the close packing model is governed by the interaction between particles, such as the wall effect, loosening effect, and wedging effect, as shown in Figure 4.3 (De Larrard 1999; de Larrard and Sedran 2002). The wall effect suggests that large particles can prevent the dense packing of small particles. When the volume of small particles exceeds pore

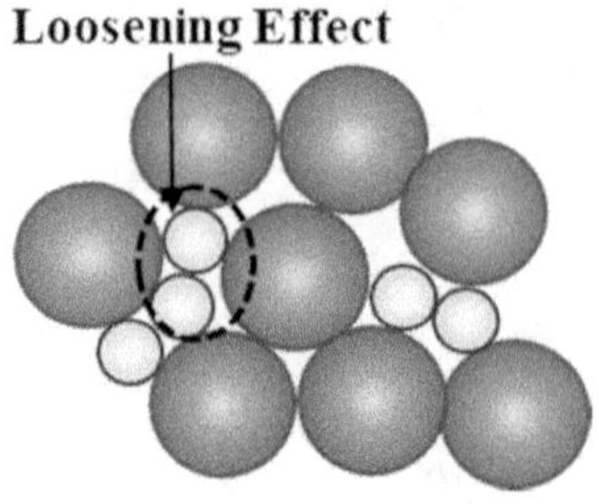

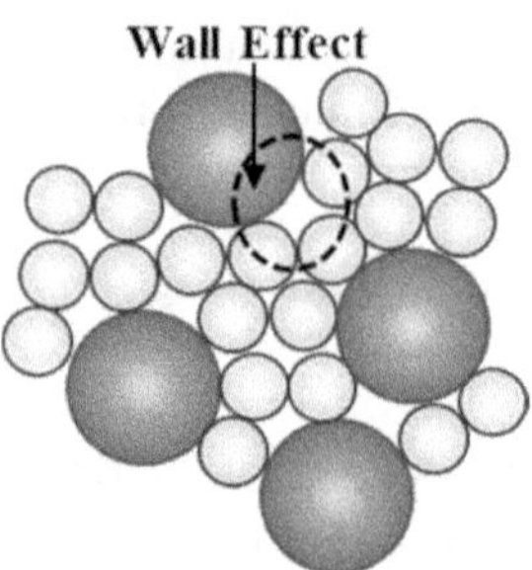

Figure 4.3 Schematic diagrams of particle loosening and wall effects (adapted from Mehdipour and Khayat 2018). (a) Coarse particles are mixed with single fine particles. (b) Coarse particles wrapped in fine particles.

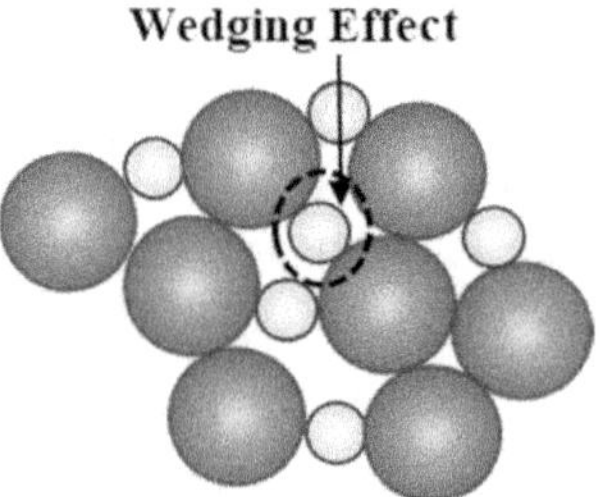

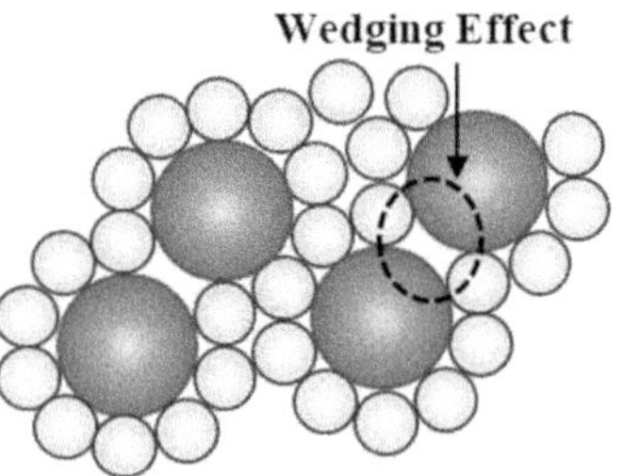

Figure 4.4 Illustration of particle wedging effect.

Source: Adapted from Kwan et al. 2013.

volume between large particles, the loosening effect occurs. A new particle interaction called the wedging effect was identified by Kwan et al. (2013). Like the loosening and wall effects, the wedging effect can also reduce the packing density. The wedging effect presents two main forms, as shown in Figure 4.4. One is that some isolated fine particles could be entrapped in the gaps among the coarse particles, thereby wedging the coarse particles apart. The other is that the fine particle layer wrapping on the surface of coarse particles is disconnected somewhere, causing voids to be formed there and the apparent wedging of the fine particles against the coarse particles (Chan and Kwan 2014). The geometric interactions between particles should be considered during the mixture design of UHPC.

4.2.1 Close packing models based on dry packing density

Various particle packing models have been proposed to estimate the packing density of granular materials in a dry state with the consideration of structure and interaction effects, such as filling, wedging, loosening, and wall effects (Kwan et al. 2013; Mehdipour and Khayat 2018). These models include Furnas (1931), Powers (1968), and Aim and Le Goff (1968) models for binary mixture, and Toufar and modified Toufar models for the ternary mixture (Goltermann et al. 1997). The proposed linear packing density model and solid suspension model in discrete models are suitable for multicomponent mixtures (Stovall et al. 1986; Puri et al. 1992; De Larrard and Sedran 1994; Dewar 1999; Pedersen 1999). In 1907, the grading curve of the maximum density of particles was proposed by Fuller and Thompson (1907), which laid the foundation of the continuous model.

In terms of classification, the close packing models can be either discrete or continuous. The use of a single particle packing model always brings general mathematical problems in the prediction of particle-graded concrete. When the particle size distribution curve of mixture gets closed to the close packing model, the mixture proportion can be considered the target one.

4.2.1.1 Discrete models

(1) Binary mixture model

The discrete model based on specific particle size assumes that the volume of each class of particles can be fully compressed. The binary mixture model divides particles into coarse and fine particles. The Furnas model (Furnas 1931) assumes that particles do not interfere with each other and the ratio of continuous size particles is constant. The total volume and packing density of solid with different particles are derived by binary mixture model, as shown in Eqs. (4.4–4.6), respectively.

$$Z = \frac{1}{1+V} \tag{4.4}$$

$$S = \frac{1+V+V^2+V^3}{1+V} \tag{4.5}$$

$$\phi = \frac{G_a}{1-\dfrac{y(S-Z)}{S}} \tag{4.6}$$

where Z_{is} the proportion of absolute volume of large particles in maximum packing density (unitless); V_i is the volume of void (mL); S is the absolute volume of solid when solid particle is infinite (mL); y is the ratio of volume reduction after mixing for a specific set of examples to volume reduction and mixing of infinitesimal particles (unitless); G_a is the apparent specific gravity of particles (unitless); and ϕ is dry packing density (unitless).

Based on Eqs. (4.4–4.5), Eq. (4.6) assumes that the particles are evenly distributed; the size of the first layer of particles is quite large, while other particles are very small. This is different from the actual situation; that is, the particles should be infinitesimal. Since concrete is a multicomponent mixture, Furnas revised the model and found when it exceeds the normal range of engineering practice, the maximum density batch classification curve can well predict the actual density of the mixture. In this model, the whole mixture system is divided into several binary mixture combinations, which are composed of iterative packing.

Powers (1968) considered the wall effect and loosening effect of particle packing in the study of aggregate mixture. Under specific combinations, the porosity of binary mixture system was minimized. Aim and Le Goff (1968) used a simple geometric model to observe the excess porosity in a layer of spherical grains in contact with a smooth plane. The model is used to explain the function of porosity of binary mixture. The wall effect interference coefficient is introduced to modify the effect on porosity, as shown in Eqs. (4.7) and (4.8). It is found that the wall edge interference particle packing only affects the first layer of particles in contact with the wall, and the porosity at the edge of wall is the largest. The limit value of porosity appears far away from the wall and has nothing to do with the shape of the container. Besides, the action position of the wall effect is limited to d/2.

$$w = \frac{\pi d^3}{6} \tag{4.7}$$

$$Nw = V_0\left(1 - \zeta_0\right) = V\left(1 - \zeta\right) \tag{4.8}$$

where w is the volume of a single particle (mm^3); N is the number of particles (unitless); V_0 is the apparent volume of mixture (mm^3); V is the volume disturbed by wall effect (mm^3); and ζ is the wall effect interference coefficient (unitless).

Based on the premise that diameter difference between coarse and fine particles is infinite, the ideal binary packing density is expressed as Eq. (4.9) according to the Furnas model. When the difference between coarse and fine particles decreases, the actual packing density is lower than that predicted by Fuller model. According to the size ratio of aggregate, the modified packing density is proposed as Eq. (4.10) (Zheng et al. 1995).

$$PE_{\max} = PE_c + (1 - PE_c) PE_f \tag{4.9}$$

$$PE_{\text{mix}} = PE_c + (1 - PE_c) PE_f F_1 (X_f) F_2 (R) \tag{4.10}$$

where PE_c is the filling efficiency of coarse aggregate; PE_f represents the filling efficiency of fine aggregate; F_1 is the volume fraction of fine particles; F_2 is the size ratio of two coarse and fine particles (unitless); and R is the average particle size ratio (unitless).

Since concrete is usually a multicomponent mixture and cement paste composition contains a variety of supplementary cementitious materials, the binary mixture model can be extended to ternary mixture and multicomponent mixture.

(2) Ternary mixture model

Assuming that the whole particle packing system is composed of large and small particles, Toufar model extends binary mixture to ternary mixture. It means that any two components constitute a binary system, and the packing of multiple binary systems forms the whole system. Three unrealistic are assumed in this model: (1) spherical aggregate; (2) monoaggregate size; and (3) different fine and coarse aggregate sizes (Shah et al. 2020).

Toufar model was first modified and changed parameter design in 1993 (Roy et al. 1993). The packing density is affected by particle distributions, shapes, and manner of packing if surface forces are ignored. According to the fact that the diameter of fine particles is much smaller than that of coarse particles, the packing density can be divided into the following two cases according to the volume fraction of particles. The particle sizes are assumed to be $d_1 \le d_2$, the volume fractions of fine and coarse particles are r_1 and r_2, respectively, and the maximum packing densities are ϕ_1 and ϕ_2. In the case of high small particle content and $r_1 \ge r_2$, the total volume of solids in the mixture is v, as shown in Eq. (4.11).

$$v = \frac{r_1}{\phi_1} + r_2 \tag{4.11}$$

The packing density of the mixture is:

$$\phi_{\min} = \frac{1}{v} = \frac{1}{\left(\frac{r_1}{\phi_1 + r_2} \right)} \tag{4.12}$$

When the volume fraction of coarse particles is large, the total volume packing density of mixture can be given by Eqs. (4.13–4.14).

$$v = \frac{r_2}{\phi_2} \tag{4.13}$$

$$\phi_{\min} = \frac{\phi_2}{d_2} \tag{4.14}$$

Goltermann et al. (1997) modified Toufar model, introduced the concept of "Eigen Packing", and compared the packing density of mixtures by Aim and modified Toufar model. It was found that the packing density calculated by the modified Toufar model was close to the experimental value. At the same time, based on the modified Toufar model, the theory of particle packing density contour was established to study the ternary packing of particles. The modified model used linear interpolation and packing degree to calculate the Eigen diameter of particles and the final packing density of the multisize mixture was obtained after repeated iterations.

(3) Multicomponent mixture model

For multicomponent mixtures, a linear packing density model was proposed in 1986 (Stovall et al. 1986), which can be expressed as a particle solid volume function to the derived relationship between continuous distribution of particles. This lays a foundation for a follow-up study on the close packing of particles. The linear packing density model for grain mixtures is based on the concepts of packing density and maximum paste thickness. According to the linear packing density model, fine sand is selected to fill concrete to optimize the pores and thus improve compressive strength. The linear packing density model is suitable for angular or circular particles, as follows in Eqs. (4.15–4.17) (de Larrard et al. 1994).

$$c(t) = \frac{\alpha(t)}{1 - \int_d^t y(x) f(x/t)\,dx - [1 - \alpha(t)] \int_t^D y(x) g(t/x)\,dx} \tag{4.15}$$

$$f(z) = 0.7(1 - z) + 0.3(1 - z)^{12} \tag{4.16}$$

$$g(z) = (1 - z)^{1.3} \tag{4.17}$$

where c is the packing density (unitless); t is the grain size (μm); d is the minimum grain sizes (μm); D is the maximum grain sizes (μm); $f(z)$ is the function of loose effect; and $g(z)$ represents the function of wall effect.

The linear packing density model shows good efficiency in concrete mixture design. But owing to its linear characteristics, the packing density of particles appears turning point at the maximum value, and it is necessary to optimize the model to eliminate this defect. The particle size and shape affect the packing effect and porosity. Therefore, the concept of specific volume variation can be used to describe the particle packing model.

The shape of particles affects the porosity of concrete due to the inhomogeneity of particles. Some previous packing models can predict the porosity but have some limitations. In this context, Yu et al. (1996) established an improved spherical particle linear packing model based on earlier studies on nonspherical particles to predict the porosity of nonspherical particles. Assuming that system consists of n particles of equal density, under certain packing conditions, the volume of the system can be shown in Eq. (4.18).

$$V = f\left(V_s, X_1, X_2, \ldots X_n, d_{V1}, d_{V2}, \ldots d_{Vn}, \psi_1, \psi_2, \ldots \psi_n\right) \tag{4.18}$$

where V is the volume of the system; V_s is the function of specific volume; d_{vi} is the volume diameter; and X_i is the volume fraction of sphere.

After extending spherical particle to nonspherical particle, the above formula is improved, as shown in Eqs. (4.19–4.23).

$$f(r) = (1-r)^{3.3} + 2.8r(1-r)^{2.7} \tag{4.19}$$

$$g(r) = (1-r)^{2.0} + 0.4r(1-r)^{3.7} \tag{4.20}$$

$$\overline{V}_{Sj} = V_j\left[1-f(r)\right] \tag{4.21}$$

$$\overline{V}_{Lj} = \left[V_j - \left(V_j - 1\right)g(r)\right] \tag{4.22}$$

$$V_i^T = \sum_{j=1}^{i-1}\left[V_j - \left(V_j - 1\right)g(r)\right]X_j + V_iX_i + \sum_{j=i+1}^{n} V_j\left[1-f(r)\right]X_j \tag{4.23}$$

where $f(r)$ is the interaction functions between components i; $g(i)$ is the interaction functions between components j; and r is the size ratio between components I and j (small to large) (μm).

It is worth noting that when particle size is relatively large, some defects can be found in the above formula. When the initial porosity of each component is significantly different, the model extending from binary mixture to multicomponent mixture is challenging to satisfy the required accuracy. Nevertheless, various mathematical models that can provide a basis for accurate porosity estimation of system contribute to better understanding of the effect of particle packing and size on particle packing, but the linear packing density model still needs to be further improved. The solid suspension model considers random particle packing, like suspension, which can be used to produce fluid mortar. The value of packing density tends to be higher if particles are assumed to accumulate as a suspension with viscosity in the solid suspension model.

The compressible packing model was originally improved by the linear packing density model and is a relatively complete model at present. It can be used to predict the optimal combination of multiple mixtures from three parameters. It includes packing density of monosize classes, size distribution of mixture, compaction energy, and compaction coefficient (de Larrard and Sedran 2002). Besides, in the case of an unknown compaction coefficient, the proportion of components in the mixture is uncertain. The optimum combination of UHPC mixtures can be predicted when the compaction coefficient is determined (Van Der Putten et al. 2016). When *n* kinds of particles are mixed, *n* virtual packing density equations will be generated. The real packing density is the minimum value of *n*. The compaction index is a premise to calculate packing density. It refers to the proximity between actual packing conditions and ideal conditions. It can be expressed by Eq. (4.24) (de Larrard and Sedran 2002).

$$K = \sum_{i=1}^{n} K_i = \sum_{i=1}^{n} \frac{\frac{\Phi_i}{\Phi_i^*}}{1 - \frac{\Phi_i}{\Phi_i^*}} \tag{4.24}$$

where K is the compaction index (unitless) and Φ_i / Φ_i^* is the filling ratio; Φ_i is the volume of the compacted mixture (unitless); and Φ_i^* is the maximum value of Φ_i, while packing density refers to the volume fraction of solids in total volume of system (unitless).

For concrete with high filler content, it is necessary to select an appropriate filler to control the water demand of UHPC, and an appropriate mixture proportion design method must be used. The generation of a compressible packing model increases the packing density of particles, optimizes pores in structure, reduces water demand, and improves concrete strength. The compressible packing model considers the wall effect and loosening effect, introducing virtual and actual packing density. However, this model only considers the packing of particles under dry conditions and neglects the effects of water and superplasticizer.

To consider the optimal particle size distribution of all particles in reality, Yu et al. (2014) used the modified Andreasen and Andersen particle packing model to design UHPC by using limestone and quartz powder as fillers, as shown in Eq. (4.25).

$$P(D) = \frac{D^q - D^q_{\min}}{D^q_{\max} - D^q_{\min}} \tag{4.25}$$

where *P(D)* is the fraction of the total solids with a size smaller than *D*; D_{min} and D_{max} are the minimum and maximum particle sizes, respectively; and q is the distribution modulus representing the proportion between fine particles and coarse particles, which is determined to be 0.23.

The proportions of each material in the mixture are adjusted until an optimum fit between the composed mixture and the target curves is reached, using an optimization algorithm based on the least-squares method, as presented in Equation. (4.26).

$$RRS = \sum_{i=1}^{n} \left(P_{mix}\left(D_i^{i+1}\right) - P_{tar}\left(D_i^{i+1}\right)\right)^2 \tag{4.26}$$

where P_{tar} is the target grading calculated from Eq. (4.25).

However, both the first Furnas model and subsequent compressible packing model have common defects. That is, only dry particles are considered in the mixture system, which is quite different from the actual situation of concrete. The development of discrete model is shown in Table 4.1.

4.2.1.2 Continuous models

Concerning the continuous methods, it assumes the particle size distribution is continuous and is expressed in densification with a small particle system. It means that all possible particle sizes exist in the system. Fuller and Thompson (1907) proposed a grading curve of the maximum density of particles called Fuller curve, as shown in Eq. (4.27). The Fuller model modified by Shakhmenko and Birsh (1998) was applied to concrete mixture design, as shown in Equation. (4.28).

$$CPFT = \left(\frac{d}{D}\right)^n 100 \tag{4.27}$$

$$CPFT = T_n\left(d_i - d_0\right)^n \tag{4.28}$$

Table 4.1 Summarization of discrete model for particle packing

Year	Models	*Packing system*			*Effect on particle packing*		Main features	References
Year	*Models*	*Binary*	*Ternary*	*Multicomponent*	*Wall effect*	*Loosening effect*	*Main features*	*References*
1929	Furnas model	✓					Assumed that particles are independent	Furnas (1931)
1967	Aim and Goff model	✓			✓		Proposed coefficient of wall effect	Aim and Le Goff (1968)
1969	Powers model	✓			✓	✓	Considered wall and loosening effects	Powers (1968)
1977	Toufar model		✓		✓		Divided ternary system into multiple binary systems	Goltermann et al. (1997)
1986	Linear packing density model			✓	✓	✓	Derived packing density when particles are continuously distributed	Stovall et al. (1986); Pedersen (1999)
1994	Solid suspension model			✓	✓	✓	Proposed virtual packing density and is widely used	De Larrard and Sedran (1994)
1999	Compressible packing model			✓	✓	✓		

Source: Adapted from Mangulkar and Jamkar 2013.

where *CPFT* is the percentage of cumulative volume (unitless); n is the degree of curve equation and its value is 0.45 in Eq. (4.28); and T is a coefficient related to the maximum diameter of aggregate (unitless).

Andreassen then investigated the particle size distribution of particle packing by using a continuous method (Mangulkar and Jamkar 2013). Assuming that the smallest particle is infinitesimal, an ideal packing equation was proposed. However, this model is still empirical. Since the actual particle size is limited, Dinger and Funk modified the model considering the minimum particle size as shown in Eq. (4.29) (Mangulkar and Jamkar 2013).

$$CPFT = \left(\frac{(d - d_0)}{(D - d_0)} \right)^q \times 100 \tag{4.29}$$

where d is the particle size (μm); d_0 is the minimum particle size of distribution (μm); D is the maximum particle size (μm); and q is the distribution coefficient (unitless). The value of q is related to the workability of concrete, and the index decreases with the increase of fine particles. For high-performance concrete and ordinary concrete, the value of q is ranging from 0.25 to 0.3. For self-compacting concrete, $q < 0.23$.

Rosin-Rammler-Bennett equation described the characteristic diameter of concrete particle size, as shown in Eq. (4.30).

$$R = \exp\left(\frac{-D}{D'} \right)^n \tag{4.30}$$

where R is the percentage of residual fraction (unitless); D is the diameter (μm); D' is the characteristic diameter (μm); and n is the distribution index (unitless).

Wang (1999) used the above Rosin-Rammler-Bennett equation to deduce the relationship between particle size distribution and packing density. This equation is suitable for the following two conditions: (1) the cement particles are spherical, and particle size does not affect the degree of hydration; (2) the packing density of the same particle size is a fixed value.

The relationship between packing density and particle size distribution is as follows:

$$\phi = \inf_{0 \le t < \infty} \left[\frac{\phi_0}{1 - (1 - \phi_0) \int_t^\infty g(t,x) \Phi(x) dx - \int_t^\infty f(t,x) \Phi(x) dx} \right] \tag{4.31}$$

$$g(t,x)=(1-t/x)^{1.6} \tag{4.32}$$

$$f(t,x)=(1-x/t)^{3.1}-3.1\frac{x}{t}(1-x/t)^{2.9} \tag{4.33}$$

where ϕ_o is the packing density of one size particle (unitless); and x is the diameter of particle (μm).

The continuous models provide a good theoretical basis for experiments. Hoang et al. (2016) proposed a UHPC mixture design method based on the progressive optimization of particle packing density. In this method, four kinds of self-compacting UHPC with the largest grain size of 1, 2.5, 4, and 8 mm, respectively, were prepared using cement, quartz powder, silica fume, and mixed aggregate. The compressive strength is more than 190 MPa. Besides, a particle packing rule is proposed, which assumes that particles in the group (i+1) are always thinner than those in the group (i). A combination with the largest packing density is obtained by continuing to connect with group (i+2) and repeating this process. Figure 4.5 shows the flow chart of the mixture design. To determine the packing density of powder with water, the water demand measurement method was adopted (Marquardt et al. 2002). The powder's packing density can be calculated using Eq. (4.34). The results showed that the optimal stacking density

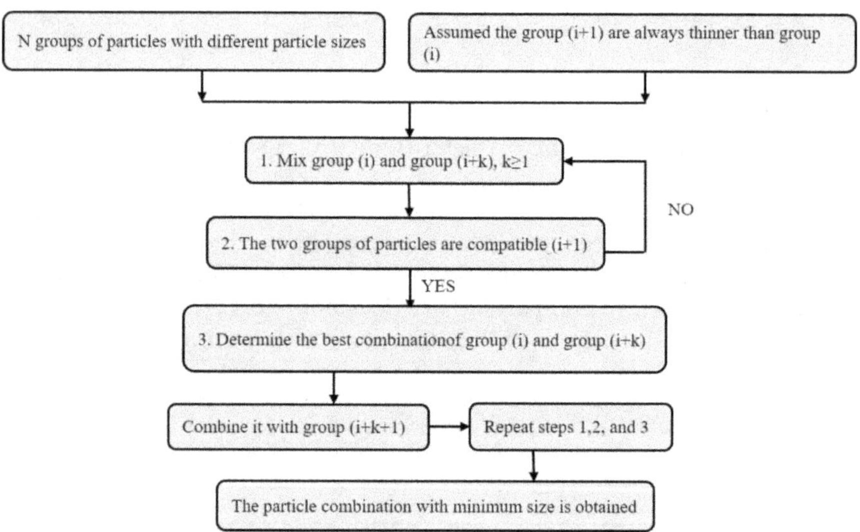

Figure 4.5 A new mixture design method proposed by Hoang et al. (2016).

obtained by this method is reliable because it can consider the properties of various materials and their compatibility.

$$\text{Packing density} = \frac{V_P}{V_P + V_L} \tag{4.34}$$

where V_P is the volume of powder; and V_L is the total volume of liquid contained in mixture.

The mixture design method for UHPC considering dry packing density is established under dry conditions without taking into consideration the role of mixed solutions, such as water and superplasticizer. Besides, the dry packing of particles is greatly governed by the degree of compaction and particle interaction related to particle size, content, and shape. Therefore, particle agglomeration might occur and influence the measurement accuracy of dry packing density, which affects the performance of UHPC. Therefore, the wet close packing method considering the influence of the liquid phase is proposed (Wong and Kwan 2008; Li and Kwan 2014; Li et al. 2019).

4.2.2 Close packing models based on wet packing density

4.2.2.1 Superplasticizers and water film thickness

The wet packing density of particles refers to the sum of the particles' mass and water in voids per unit volume in a wet-mixed state. The maximum wet packing density is directly related to particle characteristics, superplasticizer, the thickness of water film onto solid particles, degree of particle dispersion, and agglomeration. Due to the filling effect of fine particles, such as silica fume, the water entrapped in voids can be released, increasing the wet packing density of UHPC mixture. However, pores will be formed around the agglomerated silica fume with large size due to the loosening effect, reducing the wet packing density of the system, as shown in Figure 4.6.

The effect of superplasticizer on the packing of particles can be summarized as dispersion and can increase the wet packing density. The effect depends on the superplasticizer type and concentration. Results showed that the wet packing density of paste mixed with polycarboxylate-based superplasticizer is higher than that of naphthalene-based superplasticizer (Li and Kwan 2015). At a low superplasticizer dosage, the wet packing density of paste is linearly increased with the superplasticizer dosage (Liu et al. 2017). Superplasticizers can improve the wet packing density of particle, which is mainly attributed to their dispersion effect including electrostatic repulsion and steric hindrance. The dispersion capability of superplasticizer is

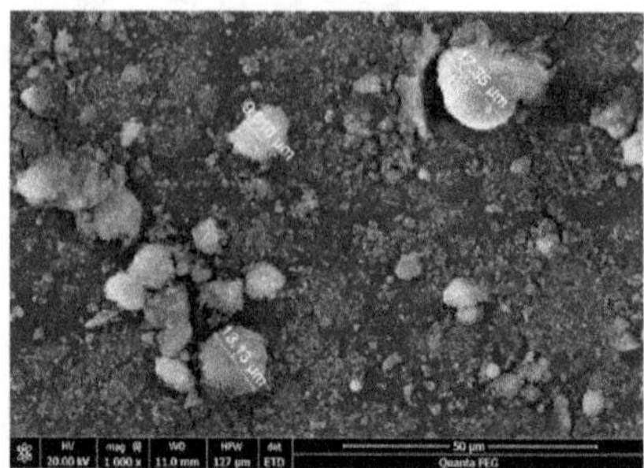

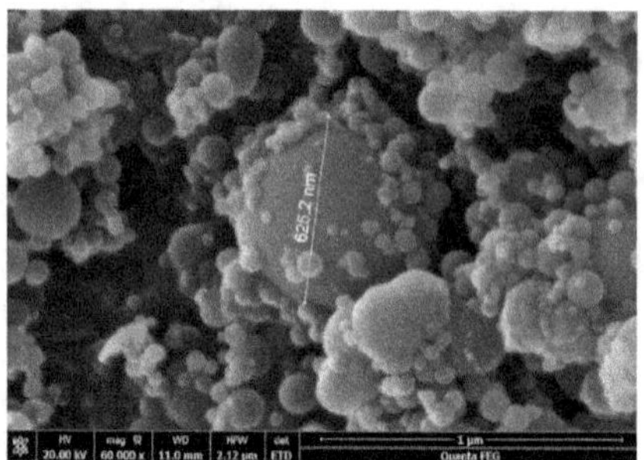

Figure 4.6 Scanning electron microscope images of size and shape of agglomerated and original silica fume (Wang et al. 2019). (a) Size and shape of silica fume agglomerates. (b) Original silica fume size.

controlled by molecular weight, side-chain density, side-chain length of the polymers, and the superplasticizer amount adsorbed on the surface of particles (Yamada et al. 2000). Particles flocculate with each other due to the surface charge of the particles at the initial stage of hydration (Chandra and Björnström 2002; Liu et al. 2017). The anions in the superplasticizer that are adsorbed onto cement surfaces neutralize the charge, generate electrostatic repulsion, and weaken the van der Waals force between particles (Gelardi and Flatt 2016; Wang et al. 2019; Sha et al. 2020). Steric hindrance arises from the presence of polymers adsorbed onto particle surfaces, and the side-chain length of polymers has a dominant effect on steric resistance (Gelardi and Flatt 2016).

It is shown that numerous irregular flocculation structures with large sizes exist in cement suspension without a superplasticizer, as shown in Figure 4.7(a). However, the flocculation size decreases in the presence of superplasticizer (Figure 4.7(b)) (Zhang et al. 2018). The water wrapped in the cement particles is released, increasing the wet packing density (Kwan and Fung 2012; Liu et al. 2017). Part of water released inside the particles after mutual repulsion is used to render thicker water film onto particles, as shown in Figure 4.8. This directly increases the flowability of UHPC (Daimon and Roy 1978).

The effect of water on particle agglomeration is complex and is classified into two main aspects: the formation of chemical bonds and liquid bridges. The chemical bond is primarily governed by a hydrogen bond that is formed by the combination of OH^- and H^+ in water with ions on the surface of particles. Besides, the liquid bridge appears in water adsorbed on the particle surface and in the capillary, promoting the agglomeration of particles (Liu et al. 2020). The particle agglomeration in water leads to a decrease in wet packing density, which is opposite to that of superplasticizer. In addition

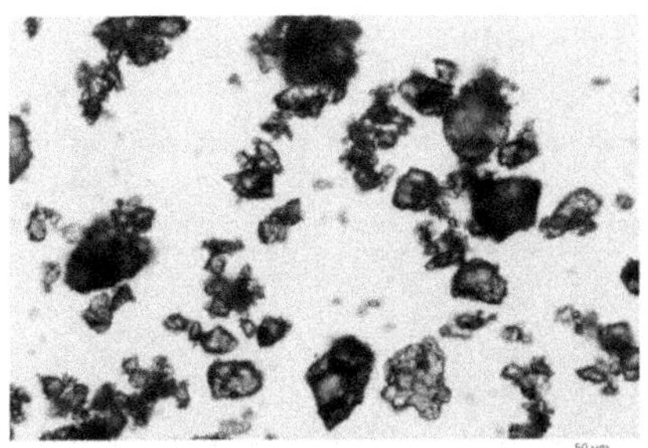

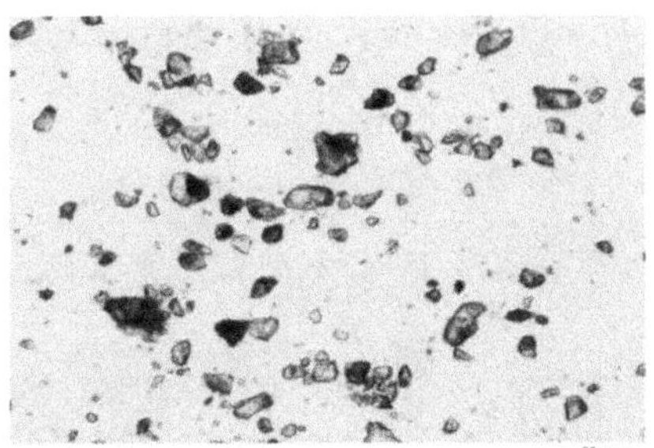

Figure 4.7 Optical microscope images of cement suspension with and without superplasticizer (Zhang et al. 2018). (a) Cement suspension without superplasticizer. (b) Cement suspension with superplasticizer.

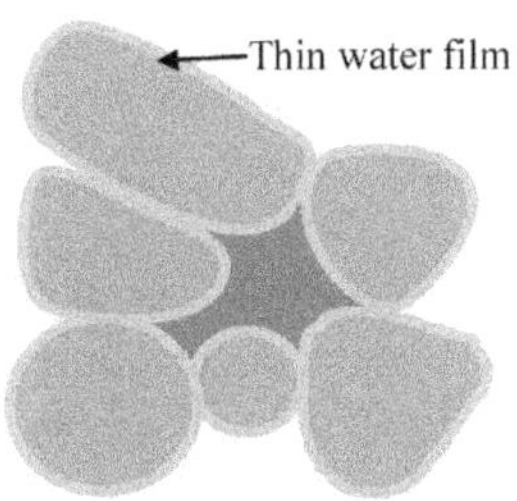

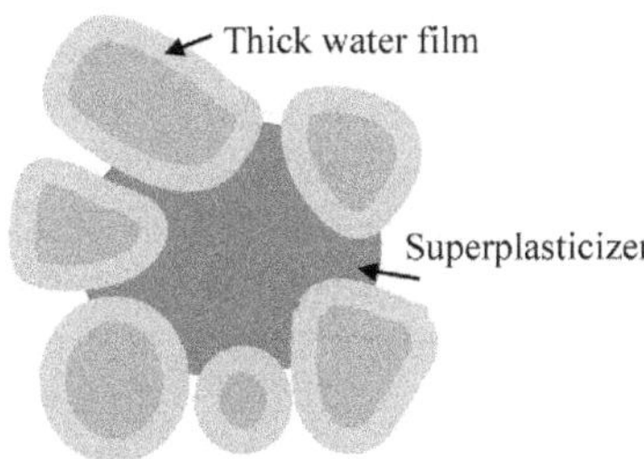

Figure 4.8 Comparison of dispersion of particles with and without superplasticizer. (a) Particle agglomeration and thin water film on particles. (b) Thick water film on particles with superplasticizer.

to filling pores, ultrafine particles also affect the packing density of the mixture through hydration reaction. The gradual diffusion of ultrafine particles in water provides nucleation sites and then promotes hydration, leading to densified C-S-H gels with needle-like morphologies (Gruskovnjak et al. 2008; Bullard et al. 2011; Lee et al. 2018; Pedrosa et al. 2020). Therefore, the particles in water exhibit a multicomponent, multisize, and multiform complex rather than the original form.

Water film thickness is an indicator used to evaluate wet packing density. It is defined as the ratio of excess water ratio to the specific surface area of the total solid, as shown in Eqs. (4.35–4.37) (Li and Kwan 2015). Some studies showed that the increase in particle content and decrease in water consumption reduced the water film thickness and flowability of cement paste (Ye et al. 2017; Qiu et al. 2020). Therefore, the relationship between water film thickness, flowability, and particle wet packing density is indispensable for evaluating the particles' wet packing.

$$u_w' = u_w - u_{\min} \tag{4.35}$$

$$A_S = A_\alpha R_\alpha + A_\beta R_\beta + A_\gamma R_\gamma + \cdots\cdots \tag{4.36}$$

$$WFT = \frac{u_w'}{A_S} \tag{4.37}$$

where u_w' is the ratio of excess water to the volume of solids in paste (unitless); A_s is the specific surface area of all solid components (m^2/g); A_α, A_β, and A_γ are the specific surface area values of solid α, β, and γ, respectively (m^2/g); and R_α, R_β, and R_γ are the volume ratios of solid α, β, and γ, respectively (unitless) (Li and Kwan 2015).

The range of cementitious materials and the optimal water consumption in UHPC are determined by this method using the theory of extra paste thickness and the maximum wet packing density (Figure 4.9). This method considers the role of liquid phase in concrete mixture design.

Theoretically, UHPC with a relatively low binder content (about 650 kg/m^3) can be produced using the close packing model, and its compressive strength reaches 140–165 MPa at 28 d (Van Der Putten et al. 2016; Wang et al. 2019; Shi et al. 2021). However, the wet close packing of particles is unstable in the presence of a large number of fine granular particles, such as silica fume. The interaction of silica fume particles causes the loose structure, and the loose effect increases as the particle size decreases. Meanwhile, the loose effect is more evident in a mixture with pure cement particles (Yu et al. 1997; Yu et al. 2014). Therefore, the commonly used particle packing model shows a significant error in predicting the packing state of mixture containing nanoparticles. The features and shortcomings of dry and wet close packing models are summarized in Table 4.2.

4.2.2.2 The maximum solid content

Packing density is usually measured under dry conditions but dried aggregated particles are susceptible to the compaction degree due to the interaction forces between solid particles. Many researchers explored the wet density testing of paste after adding water and superplasticizers, but no unified standard has been formulated (Wong et al. 2008; Fung et al. 2009; Kwan et al. 2012; Li and Kwan 2014).

The solid concentration is determined by measuring apparent density, packing density instead of consistency, and water consumption of mixture.

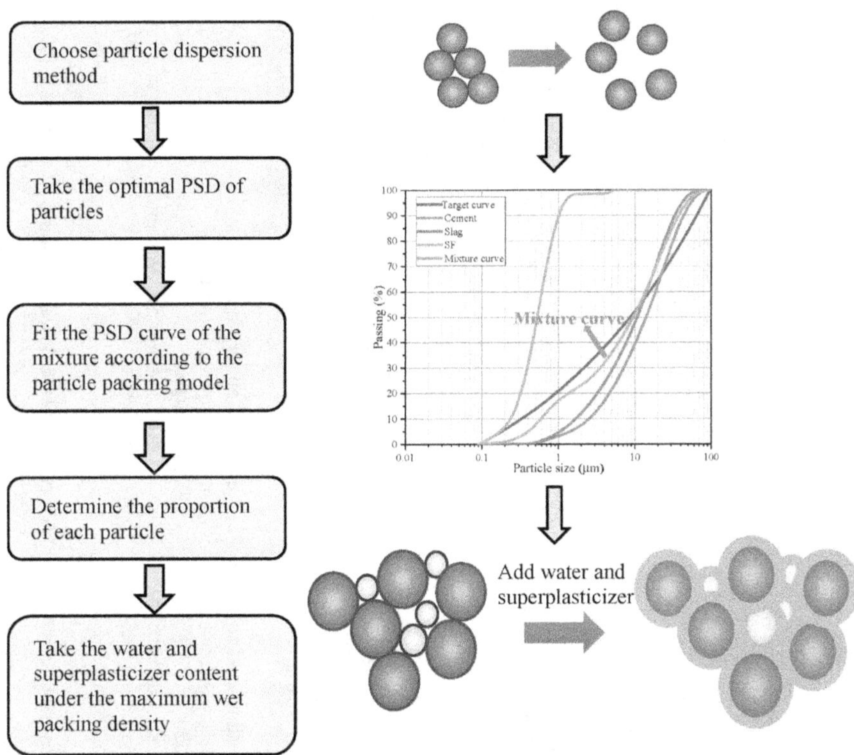

Figure 4.9 Mixture design based on the compact wet packing of particles.

The choice of water consumption as a measure of minimum porosity is based on the ideal assumption that there is no air in cement paste. The solid concentration can be calculated as follows using Eqs. (4.38–4.39) (Wong and Kwan 2008).

$$\phi = \frac{V_S}{V} \tag{4.38}$$

$$V_S = \frac{M}{\rho_w u_w + \rho_\alpha R_\alpha + \rho_\beta R_\beta + \rho_\gamma R_\gamma} \tag{4.39}$$

where M is the mass of paste (g); ρ_w, ρ_α, ρ_β, and ρ_γ are density values of water, solid α, solid β, and solid γ, respectively; u_w is the water ratio (unitless); and R_α, R_β, and R_γ are the volumetric ratios of solids α, β, and γ (unitless), respectively.

Table 4.2 Summarization of close packing model-based mixture design method of UHPC

Model			*Features*	*Shortcomings*	*References*
Based on dry packing density	Discrete models	Furnas model	Belongs to binary model; Puts forward the basic theory of particle packing.	Only suitable for mixtures with significant particle size differences.	Furnas (1931)
		Powers model	Considered the wall effect and loosening effect; proposed algorithm of minimum porosity.	Only considered packing of binary mixtures	Powers (1968)
		Aim and Goff model	Modified the packing density of binary mixture; Mitigated negative effect caused by wall effect.	Not considered more components	Aim and Le Goff (1968)
		Modified Toufar model	Proposed the Eigen packing; Established contour of packing density of ternary particles.	Unreasonable hypothesis of model	Roy et al. (1993); Goltermann et al. (1997)
		Linear packing density model	Belongs to the model of multiparticle packing density; Derived continuous distribution of grain size.	Suitable for spherical particles with deviation from the actual situation	Stovall et al. (1986)
		Compressible packing model	Introduced virtual and actual packing density; Corrected wall effect and porosity effect.	The particle size is specific; The packing density is governed by compaction process	De Larrard and Sedran (1994)
	Continuous models	Fuller Thomson model	Proposed maximum density grading curve.	Not suitable for mixture design of concrete	Fuller and Thompson (1907)
		Andreassen model	The continuous method is used to study particle size distribution; Proposed an ideal packing equation.	Still based on empirical model; Unreasonable assumption for smallest particle	
Considering wet packing density		Kwan model	Measured the wet packing density, i.e., solid concentration of cementitious materials and fine aggregate.	Rely on experimental data without theoretical basis	Kwan and Wong (2007); Wong and Kwan (2008)
		Extra paste thickness model	Considered the influence of liquid phase of concrete.	Affected by paste flowability and particle packing	Shi et al. (2018)

The voids ratio u can be determined using Eq. (4.40).

$$u = \frac{V - V_S}{V_S}, \qquad u = \frac{1 - \phi}{\phi} \tag{4.40}$$

However, it is uncertain whether the pore water in the particle gap in the saturated paste is saturated, and a large error exists in measuring the porosity of paste with water consumption.

4.2.2.3 Testing method for wet packing density

The role of solid phase and liquid relative packing models needs to be considered in particle packing. However, no unified standard for the wet packing density of cementitious materials exists (Kwan et al. 2012). Kwan and Wong (2007), Wong and Kwan (2008), Fung et al. (2009), and Li and Kwan (2014) extended the wet packing density of aggregate to cementitious materials and concrete mixtures and pointed out that the wet packing density of fine aggregates or concrete was higher than that in the dry state, as shown in Figure 4.10. The involved testing method for wet packing density of cementitious materials can be classified into the following four steps: (1) mix particles with water; (2) fill mixed sample into a designated container; (3) determine whether compaction is required according to experimental requirements; and (4) calculate the wet packing density, as shown in Figure 4.11.

On the one hand, since the state of water in particles is in unstable and unobvious saturated states, it is difficult to judge whether the liquid is just

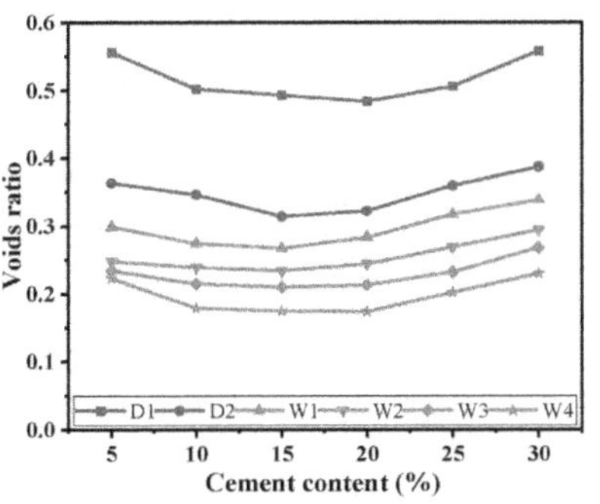

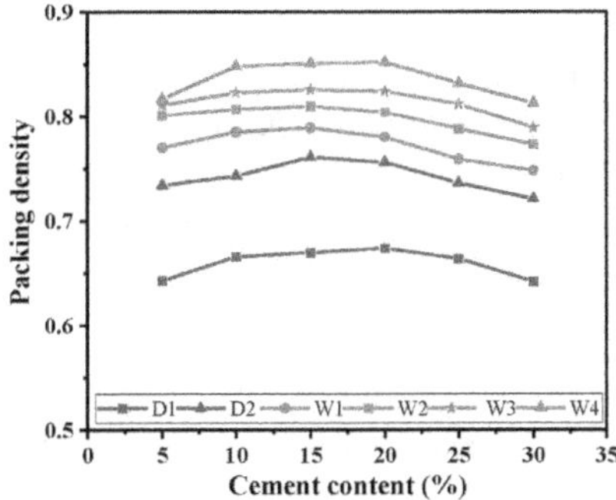

Figure 4.10 Voids ratio and packing density of concrete mixture with various cement contents under dry and wet conditions (Li and Kwan 2014). (a) Voids ratio under dry and wet conditions; (b) Packing density under dry and wet conditions.

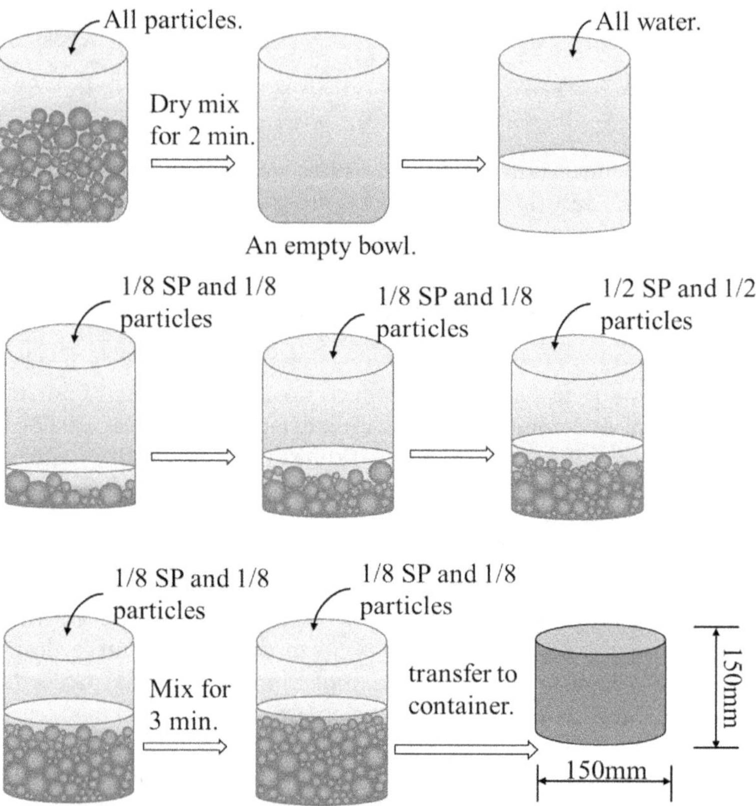

Figure 4.11 Testing procedures of wet packing density of cement-based materials.

Source: Adapted from Fung et al. 2009.

enough to fill up the voids. On the other hand, the air content in paste is often neglected. Compared with the method mentioned above, Kwan's method of calculating wet packing density using solid concentration shows the advantages of direct measurement of solid concentration with negligible experimental error and consideration of residual air content in cement paste.

4.2.2.4 Experimental examples

Based on the compressible packing model, Van Der Putten et al. (2016) selected round and angular aggregate and three different types of silica fume as fillers. The laser light scattering method and a dynamic light scattering method were used to determine the particle size distribution of silica fume, and particles were dispersed entirely in an ultrasonic bath in advance. Besides, the packing density of cement paste is compacted and is calculated by Eq. (4.41).

The compacted paste was obtained after cement particles were compacted, and a small amount of water was needed to fill the gap in cement matrix. Based on the existence of water film on the paste surface, it is reasonable to use the following formula to calculate the packing density of particles.

$$\text{packing density} = \frac{1000}{1000 + \rho_{\text{average}} \cdot \dfrac{m_{water}}{m_{powder}}} \tag{4.41}$$

where m_{water} and m_{powder} are the mass values of water and powder, respectively.

In this compaction test, smaller particles will migrate between larger particles, resulting in the segregation of cement paste. The adverse effects of segregation can be ignored if mixture reaches the optimum packing density. The results showed that van der Waals force between silica fume particles is higher than that of cement due to the finer silica fume size.

The D-optimal method is an effective method to transform concrete mixture design into mathematical models by statistics. Wang et al. (2019) first established the D-optimal method combined with the packing density model to determine the functional relationship between raw material and the packing density of filler. The maximum filling density was then used as a goal to find out the optimal content of superplasticizer, cement, fly ash, and sand, thus obtaining UHPC with high packing density. In this method, the relationship between mixture proportions and density of filler was determined using the secondary saturation D-optimal design, and the effect of superplasticizer on density is mainly considered. This method can simultaneously consider the effects of mixture of solid phase and liquid phase, which are more reliable than the previous models. The model based on the wet packing density can fully reflect the internal characteristics of mixture. This method determines the relationship between mixture proportion and packing density by statistical analysis, and the mixture proportion with maximum packing density was determined by contour plot.

In the design of concrete mixture proportion, water in cement paste can fill voids among particles, and water film will be formed on the surface of solid particles to increase flowability. Li and Kwan (2013) calculated the wet packing density of fine aggregate and cementitious material based on water and paste film thicknesses. Shi et al. (2018) determined the amount of cementitious material based on the surplus paste thickness theory, as shown in Eqs. (4.42–4.43).

$$V_{paste} = t_{paste} g\left(M_S A_{S,C} + M_G A_{G,C}\right) + V_{pore} \tag{4.42}$$

$$M = \frac{V_{paste}\rho}{\left(1 + m_w/m_c\right)} \tag{4.43}$$

where V_{paste} is the total volume of cement paste (m^3); V_{pore} is the volume of paste filling voids of aggregate particles (m^3); t_{paste} is the thickness of surplus paste (m); M_G and M_S are the amounts of coarse and fine aggregate, respectively (kg); $A_{S,C}$ and $A_{G,C}$ are the specific surface areas of coarse and fine aggregate after correction, respectively (m^2/kg); and m_w/m_c is the water-to-cement ratio. This method deduces cementitious materials content from the theory of surplus paste thickness, which considers the role of the liquid phase in concrete mixture design.

4.3 MIXTURE DESIGN METHOD BASED ON RHEOLOGICAL PROPERTIES OF PASTE

The rheology-based mixture design method uses the relationship between constituents and rheological properties of paste to determine the appropriate mixture composition and proportion of UHPC with target rheology and hardened performance. The rheological properties of UHPC refer to the deformation and flow properties under the action of external force. Yield stress and plastic viscosity are two main rheological parameters of UHPC. Yield stress is the maximum stress that hinders the plastic deformation of UHPC, and viscosity reflects the flow characteristics of UHPC and determines the dispersion and orientation of solid phases in a fresh UHPC matrix. UHPC with proper rheological properties can lead to uniform fiber distribution and satisfactory mechanical strength and toughness (Meng and Khayat 2017; Wang et al. 2017). Factors affecting the rheological properties of UHPC include particle characteristics (Khayat et al. 2019), fiber content (Kuder et al. 2007), the wet packing density of skeleton, admixture, water content (Vikan et al. 2007; Chen and Kwan 2012), and external environment humidity and temperature (Khayat et al. 2019). The influence of each component on the rheological properties of cement-based materials is shown in Figure 4.12.

Figure 4.13 demonstrates the process of mixture design method proposed by Meng et al. (2016), which aimed at optimizing the rheological properties and compressive strength of UHPC. After rheological properties adjustment and radar chart analysis, binary, ternary, and quaternary cementitious

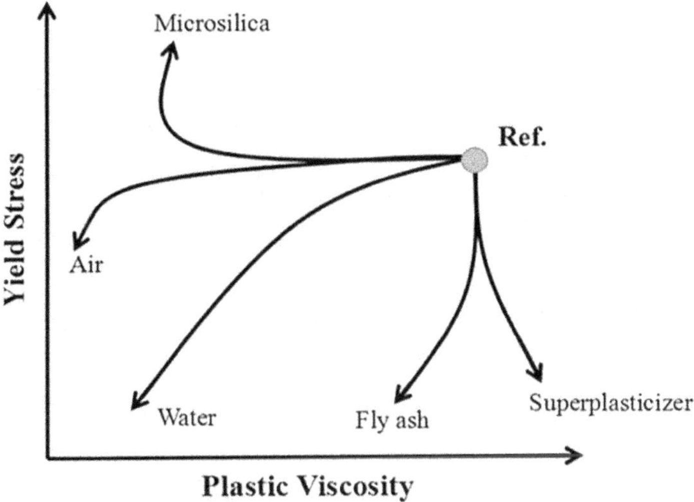

Figure 4.12 Effects of ingredients on the rheology of concrete.

Source: Jiao et al. 2017.

systems were selected with commonly used content ranges. The developed UHPC mixtures were self-consolidating and stable with 28-d compressive strength of 120–125 MPa under standard curing and 178 MPa under heat curing.

4.3.1 Effects of particle characteristics on rheological properties of UHPC

Before the acceleration period of cement hydration, the effect of particles on the rheological properties of UHPC originates from the surface morphology, volume fraction, gradation, and size of particles (Jiao et al. 2017). Before the acceleration period of cement hydration, the effect of particles on the rheological properties of UHPC originates from the surface morphology, volume fraction, gradation, and size of particles (Jiao et al. 2017). Particles with an angular shape and rough surface induce an increase in contact points between particles, significantly increasing the yield stress and viscosity due to the increased frictional resistance between particles (Hu and Wang 2011). It was found that the yield stress and viscosity increased with the increase of volume fraction of coarse particles. In contrast, the incorporation of fine particles reduced the plastic viscosity of UHPC (Hu and Wang 2011). This can be explained by the introduced higher solid specific surface

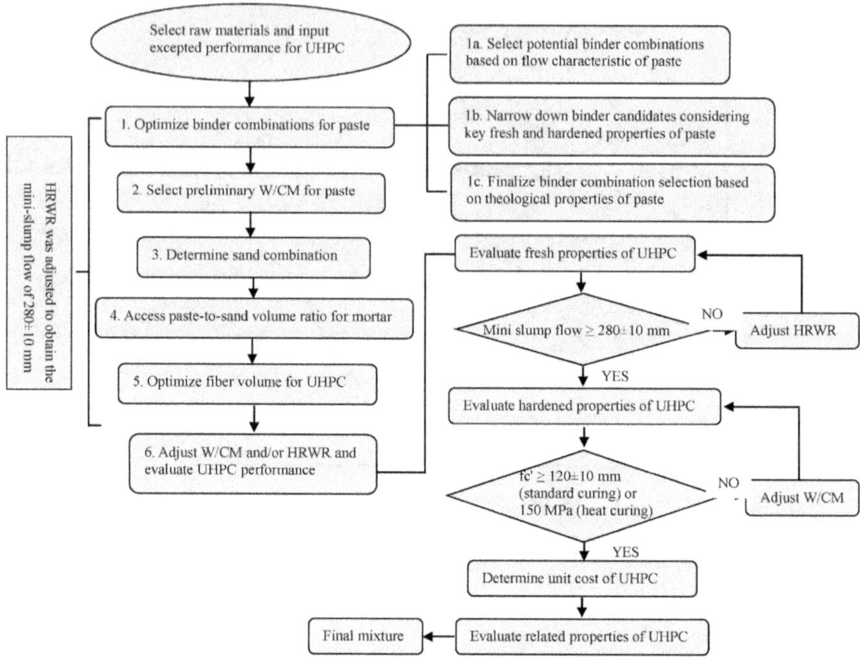

Figure 4.13 Mixture design of UHPC based on rheological properties and compressive strength of UHPC.

Source: Meng et al. 2016.

area and water demand at increasing fine particle content, leading to a thicker water film between the particles and reduced contact and friction. In addition, the free water content increased due to the filling effect associated with fine particles, which enhances the flowability of UHPC (Lee et al. 2003; Bentz et al. 2012). However, the increase in the total surface area of the solid introduced by fine particles may result in higher viscosity and lower flowability if the paste amount to coat the surface of aggregate is not enough. The particle size distribution and size of particles are critical to the rheological properties of UHPC. Particles with better particle size distribution can achieve a denser system to reduce the viscosity of paste and improve its flowability (Lee et al. 2003). The findings were also confirmed by Bentz et al. (2012).

A large amount of supplementary cementitious materials in UHPC can optimize the pore structure of UHPC and make the interface transition zone denser through its physical filling of submicron and nanometer particles, as well as the pozzolanic reaction. The reactivity and particle morphology of supplementary cementitious materials are the main factors affecting the mixture design of UHPC. Supplementary cementitious material with a smooth surface, such as fly ash and silica fume, significantly improves the flowability of UHPC by reducing the friction between particles. However, fresh UHPC containing silica fume has higher consistency than that of UHPC incorporating fly ash. This is caused by the higher specific surface area and the adsorption of a large amount of water by hydrogen bond on the surface of silica fume (Khatri et al. 1995; Artelt and Garcia 2008; Kwan and Chen 2013). Considering the flowability and compressive strength of UHPC, the content of silica fume is usually less than 15% (Mazloom et al. 2004). The addition of 20% limestone powder can produce UHPC with significantly improved 28-d compressive strength and slump of 240 mm by Nehdi et al. (1998) and Wang et al. (2012).

Arora et al. (2018) combined packing and rheological principles to select UHPC mixture proportions with slump as a rheological parameter. Before this method was used, all cement substitutes are dry mixed. The UHPC mixture was chosen by considering the close packing of structure and rheological parameters, and UHPC mixture design meeting standard was selected after screening all mixture combinations. The decrease in surface contact between particles leads to a decrease in plastic viscosity. Compared with cement paste, the yield stress and plastic viscosity of binary mixture paste containing fly ash or slag were lower.

Considering the difference in water consumption for hydration of particles of different sizes, Mehdipour and Khayat (2017) used a particle packing model to evaluate the effects of superplasticizer and mineral admixtures on the performance of cement paste. The results showed that water consumption per unit area of cement paste decreased at a higher replacement ratio of fine particles, while flowability increased. Therefore, to ensure the fresh performance of UHPC, particle size and content should be reasonably selected and its effects on rheological properties should be considered in UHPC mixture design.

4.3.2 Effect of fiber on rheological properties of concrete

Fiber is an essential part of UHPC to obtain high tensile strength and toughness (Mindess and Skalny 1991). The strengthening and toughening effects of fibers mainly depend on the fiber dispersion and orientation throughout the system and the resulting bond properties between fibers and matrix (Ozyurt et al. 2007; Wu et al. 2018). The fiber commonly used in UHPC can be divided into rigid and flexible fibers according to Young modulus, fiber shape, and mixture consistency (Martinie et al. 2010). The addition of fibers can greatly influence particle packing and finally the rheological properties of UHPC. For fiber with a small enough aspect ratio, when the fibers are altered from disorderly to densely, its packing density

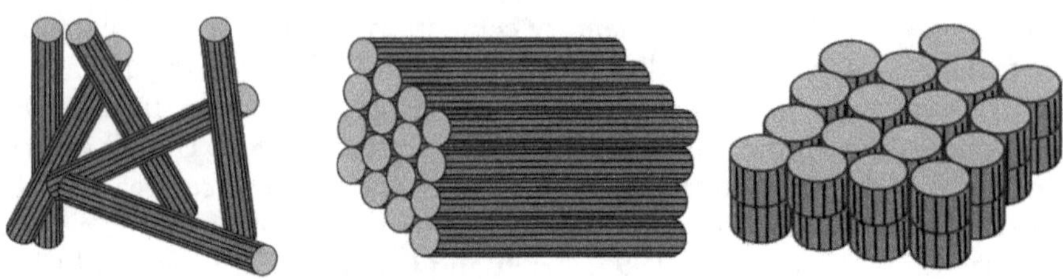

Figure 4.14 Different packing states of rigid fiber. (a) Random packing of fibers. (b) Most densely packed fibers. (c) Close to the packing of spherical particles.

Source: Adapted from Martinie et al. 2010.

is close to that of the particles, as shown in Figure 4.14 (Martinie et al. 2010). Rigid fiber can partially obstruct the packing of particles by pushing them away, while flexible fiber can increase the packing density by filling the space between particles (Khayat et al. 2019).

Fiber distribution in UHPC is closely related to the fresh performance of concrete. Concrete with higher cohesion has higher yield stress and plastic viscosity, reducing the segregation of steel fiber in UHPC and promoting its stability (Ozyurt et al. 2007). Since the depth of fiber distribution mainly depends on the rheological parameters of fresh UHPC, especially yield stress, appropriate rheological parameters are necessary conditions for uniform distribution of steel fibers in UHPC. The adverse effects of fibers on the flowability of UHPC include three main aspects: (1) fibers with high specific surface area absorb water between particles as lubricants, thus increasing friction between particles; (2) uniform distribution of fibers and network structure formed by mutual winding can reduce the flowability of UHPC; and (3) surface of fibers is not completely covered with enough cement paste (Jiang et al. 2018). On the other hand, the interlocking of steel fibers and coarse aggregate in UHPC significantly reduces its workability. The coarse aggregate with a large particle size cannot be completely wrapped by steel fibers, and the bond between aggregate and fibers would be weakened. Appropriate steel fiber length and reasonable aggregate gradation should be selected (Li et al. 2018). Studies showed that the maximum particle size of coarse aggregate is recommended to be smaller than 25 mm, and the steel fiber length should be between two and five times the maximum size of coarse aggregate (Li et al. 2021).

The fiber aspect ratio and volume in UHPC significantly affect the performance of UHPC. The probability of heterogeneous distribution and flocculation of fibers in UHPC increases with the increase in fiber aspect ratio (Su et al. 2016). To avoid excessive fiber content, less fiber is added to UHPC at an increased fiber aspect ratio (Marar et al. 2017). Research indicated

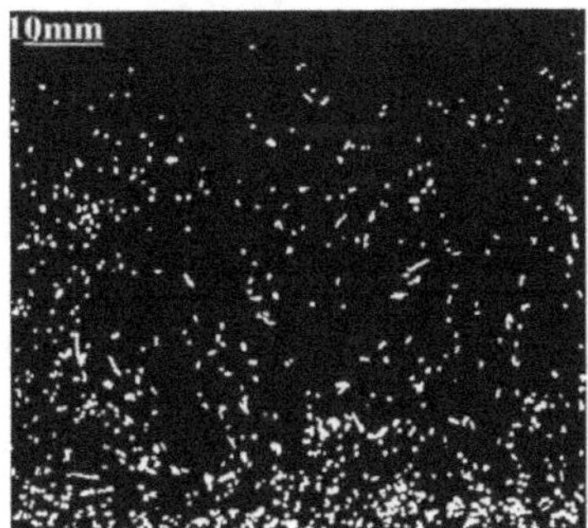

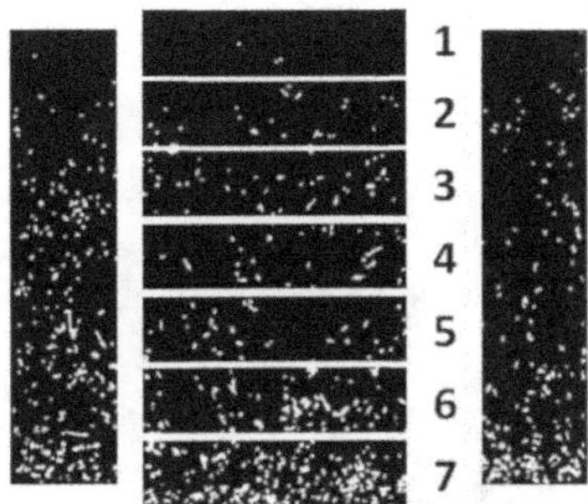

Figure 4.15 Fiber distribution image collected by OLYCIA m3-DSX software (Wang et al. 2017). (a) Fiber distribution image. (b) Regions image.

that on the premise of uniform fiber distribution, the flocculation of fibers makes flow under UHPC's weight impossible when the critical fiber content is surpassed (Grünewald and Walraven 2001). Generally, the fiber aspect ratio used in UHPC varies between 50 and 100 (Yazıcı et al. 2007), and the fiber volume is 0.5%–2.5% to make UHPC with good workability. In the case of the same fiber volume, short fiber can lead to a better compressive strength of UHPC. This may be because longer fibers are prone to agglomerate, causing UHPC difficulties in vibration and flow (Soulioti et al. 2011).

Fiber sinks naturally in concrete by gravity and accumulates in the middle and lower sections of specimens, as shown in Figure 4.15 (Wang et al. 2017). Wang et al. (2018) observed noticeable sedimentation of 2% fibers after mixing 1% and 1.6% superplasticizer in UHPC. However, steel fibers are uniformly distributed in UHPC with 1% superplasticizer, which exhibits high yield stress, as shown in Figure 4.16 (Wang et al. 2018).

Wang et al. (2017) selected four w/b ratios and superplasticizer dosages to adjust the rheological properties of cement mortar and controlled fibers distribution in UHPC. Poor dispersion of fibers was observed in the case of low w/b ratio and high fiber content. Besides, the effect of superplasticizer on the dispersion of steel fibers was more significant under a higher w/b ratio. Figure 4.17 indicates that the development trend of the distribution coefficients of UHPC specimens with different fiber contents are similar, and decreased rheological parameters can aggravate the nonuniform dispersion of fibers in a certain range. Better stability was observed when the fiber content was 3%, which may be due to the greater mechanical interlocking from more fibers. The optimal distribution of steel fibers occurs when the rheological properties of paste are moderate, which is also confirmed by other studies (Yao et al. 2002; Teng et al. 2020). The plastic viscosity of UHPC mortar under the same dispersion and orientation coefficient is higher when

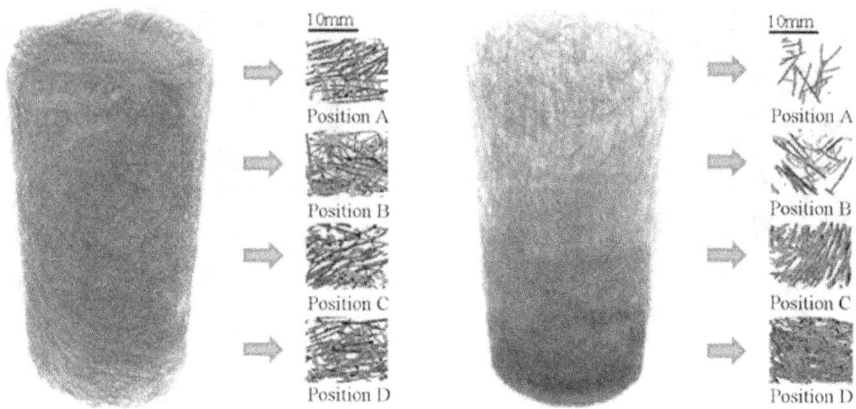

Figure 4.16 Distribution of steel fibers in UHPC cylinders.

Source: Wang et al. 2018.

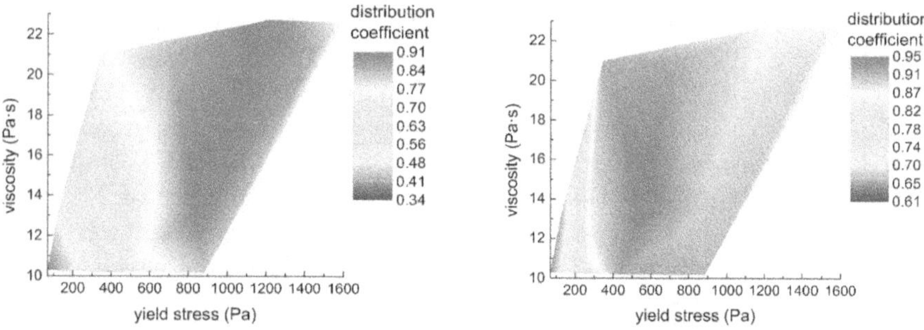

Figure 4.17 Changes in viscosity and yield stress values of cement mortar with various distribution coefficients at 1% and 3% fiber contents (Wang et al. 2017). (a) 1% fibers; (b) 3% fibers.

fiber content is increased from 1% to 3%, and the change in fiber distribution is limited after exceeding the optimal value, as shown in Figure 4.18. It also can be shown that for UHPC with higher fiber content, the optimal distribution of fibers requires a higher viscosity of the paste.

This method takes the rheological property of cement paste as an indicator to control the distribution of steel fiber in UHPC. After continuously adjusting the w/b ratio and superplasticizer, mixture proportions can meet the rheological properties of UHPC and make steel fiber uniformly distributed. However, the evaluation of UHPC mixture using rheological properties

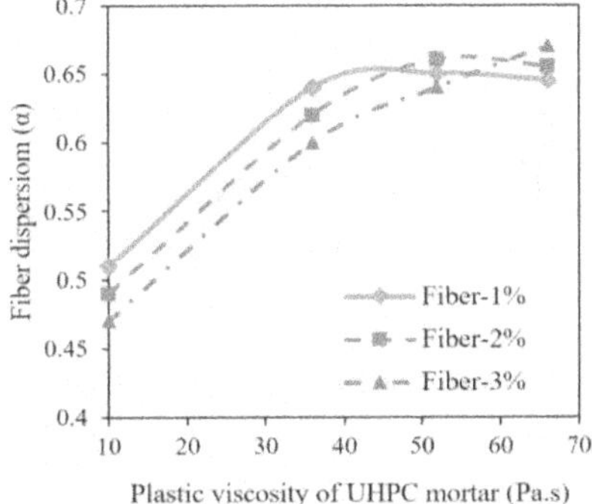

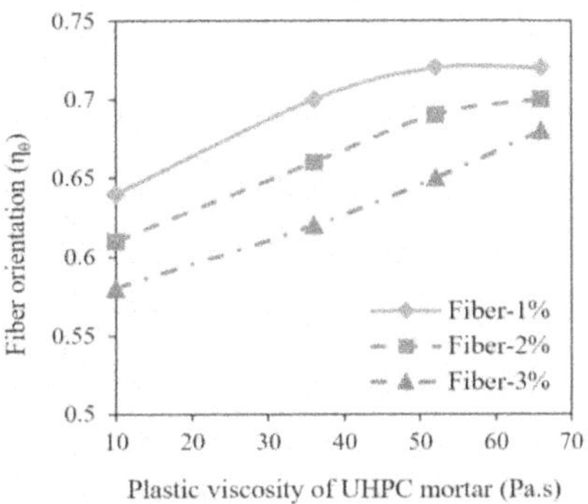

Figure 4.18 Changes in fiber dispersion and orientation coefficients of UHPC with plastic viscosity at various fiber contents (Teng et al. 2020). (a) Fiber distribution vs. plastic viscosity. (b) Fiber orientation coefficient vs. plastic viscosity.

alone is often one-sided, and it requires many repeated experiments. In addition, the rheological test data is very sensitive to variables, such as the external environment and testing parameters.

4.4 STATISTICAL MIXTURE DESIGN METHOD

The statistical analysis model is used to generate decision variables and objective data in the mixture proportion of UHPC (Barrett 2009). It is an effective experimental design method by varying mixture proportions to explore how variables affect the performance of mixture (Eriksson et al. 1998). It involves designing parameters, such as w/b ratio, contents of cement, mineral and chemical admixtures, and aggregate. An appropriate statistical model can be obtained within the reasonable parameter ranges to reflect the mathematical relationship between mixture proportion and fresh and hardened performance of UHPC.

4.4.1 The response surface method

The response surface method is based on statistical mathematics. The purpose of testing or experiment can be achieved by inputting the response generated by several variables. The relationship between response and variable and the interaction between variables are used to establish an analytical model (Nemeth 2003). The response surface analysis mainly includes three steps: (1) experimental design and data collection; (2) establishment of a numerical model to verify the accuracy of response surface; and (3) adjustment of the proportion of each component to get expected

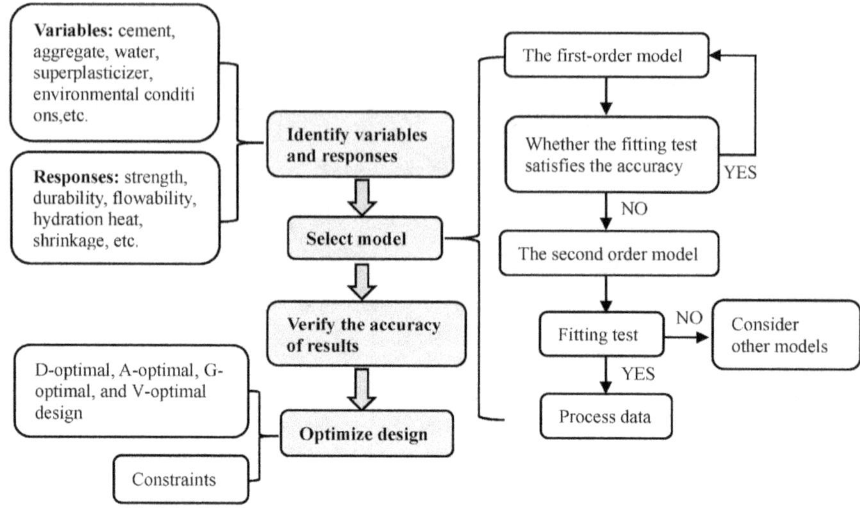

Figure 4.19 Design steps of response surface method.

Source: Adapted from Li et al. 2021.

response values (Ghafari et al. 2014). The ultimate purpose of the response surface method is to determine the optimal operating conditions and areas meeting operating specifications. In cement-based materials, the use of response surface method can satisfy nonlinear factors, be used to consider the relationship between variables that affect each other, and ultimately provide an ideal approximation for optimizing mixture proportion (Li et al. 2021). The design steps of response surface method are shown in Figure 4.19. As shown in Figure 4.20, the flowability and compressive strength model was proposed through the central composite design in response surface method. An environment-friendly UHPC was designed with cement and silica fume contents of 640 and 56.3 kg/m^3, respectively (Ferdosian and Camões 2017). Results showed that silica fume had the greatest effect on flowability reduction, and 2.5% steel fibers were the maximum content to render UHPC with optimal slump flow value (Ghafari et al. 2014).

The relationship between response and independent variables is unknown based on the response surface method in most problems. Therefore, the first step of the response surface method is to find a suitable approximation between response and independent variables. If the response is well modeled by a linear function of the independent variables, then the approximating function is the first-order model, as shown in Eq. (4.44). To satisfy

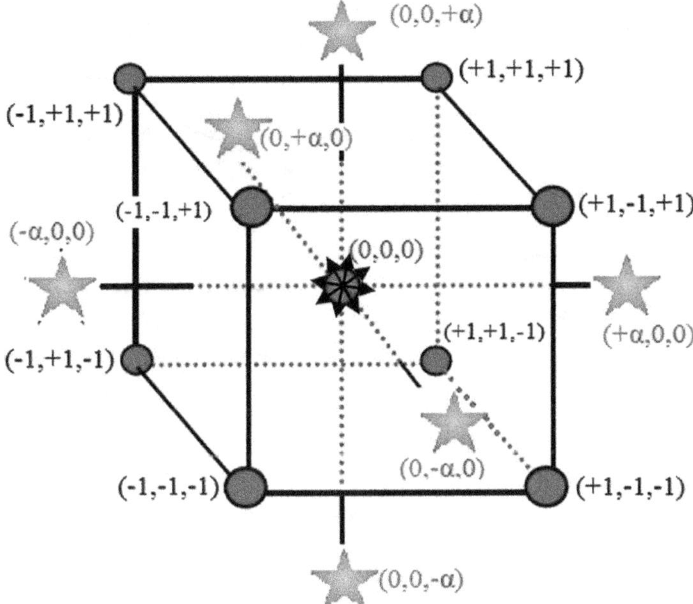

Figure 4.20 Geometric illustration of the central composite design.

Source: Ferdosian and Camões 2017.

the experiment data is relatively close to the optimal value, the second-order model is usually required to predict the response, as shown in Eq. (4.45).

$$y = \beta_0 + \beta_1 x_1 + \beta_2 x_2 + \cdots + \beta_k x_k + \varepsilon \tag{4.44}$$

$$y = \beta_0 + \sum_{i=1}^{k} \beta_i x_i + \sum_{i=1}^{k} \beta_{ii} x_i^2 + \sum\sum_{i<j} \beta_{ij} x_i x_j + \varepsilon \tag{4.45}$$

where Y is the response variable (the unit depends on the response); β_i, β_{ii}, and β_{ij} are the influence coefficients of factors (unitless); x_i and x_j are variables; and ε is the error of response y (unitless).

For environmental-friendly UHPC, the response surface method can optimize the proportion of raw materials and adjust the production cost of UHPC. Ferdosian and Camões (2017) established a second-order polynomial model with a central composite design of the response surface method by selecting three types of raw materials in UHPC as variables. The environmentally

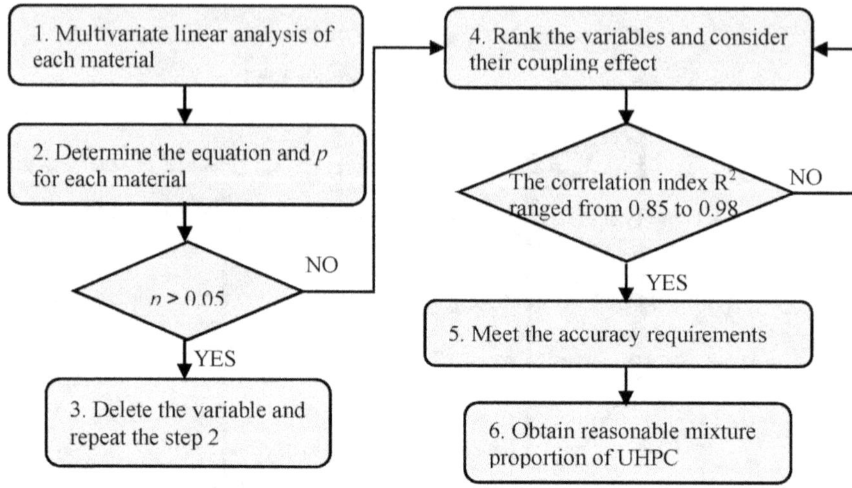

Figure 4.21 Statistical analysis method considering lightweight sand.

Source: Produced based on Meng et al. 2018.

friendly UHPC with cement consumption of less than 650 kg/m^3 was successfully designed. The mixture proportion of UHPC designed by this method meets the fresh and hardened performance of UHPC and reduces cement content. Lightweight sand is a common porous material to reduce the autogenous shrinkage of UHPC. Meng et al. (2018) used a statistical method to analyze the effect of replacement ratio of natural sand with lightweight sand, w/c ratio, and amount of substituted cementitious materials. The equation and probability value (p) of each material were analyzed by lightweight sand, and the coupled effect between parameters was expressed using significance degree. The procedures of the statistical analysis method considering lightweight sand are illustrated in Figure 4.21. Contrary to the assumption that most statistical analysis methods are based on independent variables, the components in UHPC usually affect each other. Therefore, the accuracy of the prediction results of statistical analysis methods is limited.

Statistical analysis can also be used to select binders, such as cement and silica fume. Ghafari et al. (2015) demonstrated the relationship between the different contents of each component and the properties of UHPC using the response surface method. This method can directly be used to compare and select the appropriate type and content of binder, as shown in Figure 4.22. It was demonstrated that cement has a dominant role in determining the strength of UHPC, and the increase in quartz content enhances the flowability of paste. Therefore, a practical reference for the selection

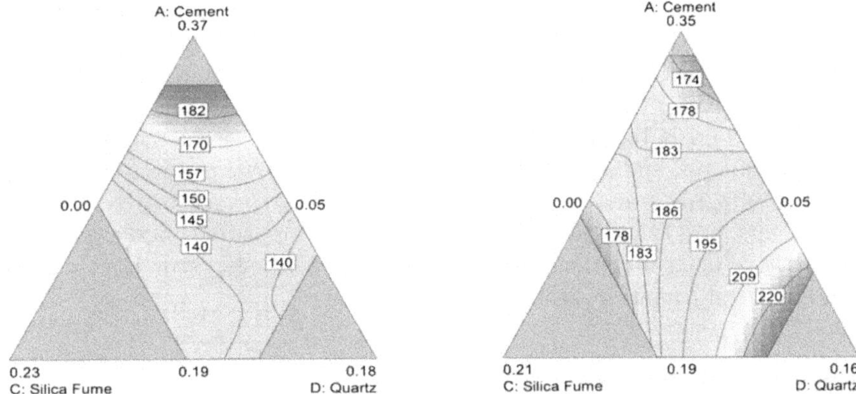

Figure 4.22 Contour plots of compressive strength and slump flow of UHPC with different types and contents of cementitious materials (Ghafari et al. 2015). (a) Compressive strength (MPa). (b) Slump flow (mm).

of type and content binder can be provided by contour plots. The model established in this method reflects the relationship between variables and compressive strength and predicts the performance of UHPC. Soliman and Tagnit-Hamou (2017) used a statistical model to optimize the mixture design based on the particle packing model considering the compatibility of cement and superplasticizer and developed an environmentally friendly UHPC.

Ghafari et al. (2014) combined the regression model and response surface method to predict the properties of UHPC. Three variables combined with the second-order polynomial model of Eq. (4.46) and an iterative model are used to check the accuracy of predicted results. The accuracy of model was tested by analysis of variance after eliminating the secondary items. When multiple components affect each other, the simplex centroid design method can be used for mixture design. Shi et al. (2015) assumed that the composition of cementitious materials includes cement, fly ash, and ground granulated blast-furnace slag. The model was designed using Eq. (4.46). Seven groups of different contents of cementitious materials were selected to draw the contour plots of the ternary system. It is worth noting that when the coefficient of a specific component is positive, the component will promote the response.

$$Y = \beta_1 x_1 + \beta_2 x_2 + \beta_3 x_3 + \beta_{12} x_1 x_2 + \beta_{13} x_1 x_3 + \beta_{123} x_1 x_2 x_3 \quad (4.46)$$

where Y is the response (the unit depends on the response); β_i is the coefficient (unitless); and x_1, x_2, and x_3 are the three components, respectively (unitless). Within the range of variables, the optimal mixture proportion scheme is based on a large number of well-performing data, which greatly reduces the number of experiments and labor. However, the response and accuracy of this method governed by its variables limit its application in UHPC.

This method is based on the interaction of design parameters, such as cement, mineral, and aggregate contents, as well as w/c ratio of UHPC. An optimal numerical model can be obtained based on the reasonable range of parameters, and finally it provides an optimal proportion of concrete mixture used to determine the effect of each component of mixture on the fresh and hardened properties of UHPC.

4.4.2 D-optimal design method

When the experimental design area is irregular, the workload of traditional experiments is huge. Besides first-order or second-order models, there is also a D-optimal design method, which minimizes the joint confidence region of regression coefficients (de Aguiar et al. 1995).

To overcome that the previous packing model cannot predict the performance of fresh UHPC, a statistical mixture design was proposed to optimize the relative proportion of mixture and determine the effective experimental area clearly (Ghafari et al. 2015). This method selected water, cement, sand, superplasticizer, fiber, silica powder, and quartzite as parameters. In this statistical method, the sum of components' proportions must be equal to 1. The ordinary quadratic polynomial is used to establish the model, as shown in Eq. (4.46), and irregular experimental area was treated with a computer-aided optimization design method. The seven-step mixture design and D-optimal technology were proposed to develop the design method, as shown in Figure 4.23. One advantage of D-optimal design is that it can consider any number of experimental points with any shape. Each variable and its variation range were input into model, and the accuracy of model was checked by experimental design. The optimal value obtained by this method can be used to ensure compressive strength and flowability of UHPC under low cement content.

$$E(Y) = \eta = \sum_{i=1}^{q} \beta_i x_i + \sum_{i<j}^{q} \sum^{q} \beta_{ij} x_i x_j \tag{4.46}$$

where β_i and β_{ij} are the regression coefficients calculated from experimental data by multiple regression (unitless) and x_i and x_j are the levels of independent variables (unitless).

After the relationship between variables and response was established, all variables were free to change, and the model can be gradually optimized.

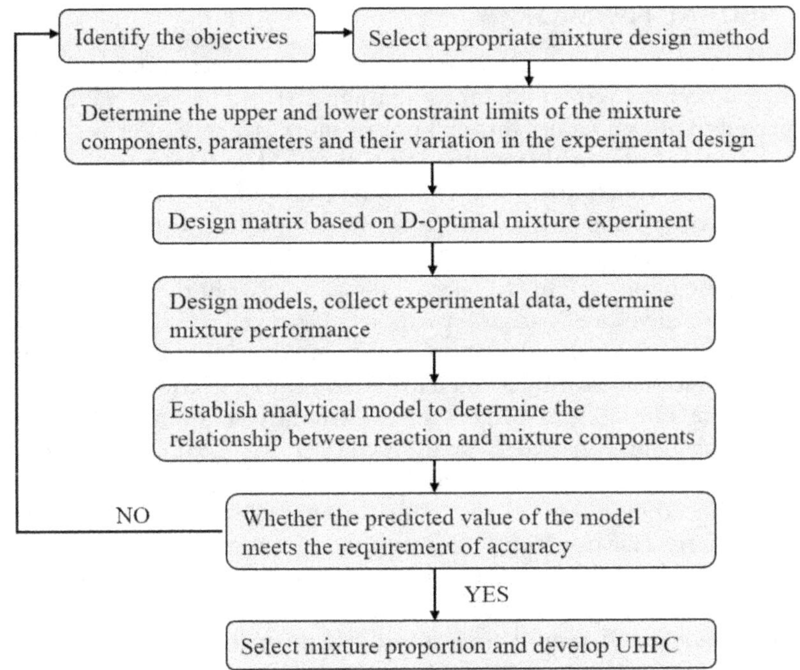

Figure 4.23 Seven-step mixture design method proposed by Ghafari et al. (2015).

The predicted value from model was compared with the experimental value, and standard deviation value was used as an evaluation index. Results showed that the predicted value differs little from the experimental value, and the accuracy meets the requirement. The standard deviation formula is shown in Eq. (4.47). The model established in this method can reflect the relationship between variables and compressive strength and predict UHPC performance based on determining components.

$$\text{Standard deviation}(\%) = \frac{\text{Experimental-Model}}{\text{Experimental}} \times 100\% \tag{4.47}$$

The statistical design method discusses the effect of a wide range of mixture design parameters on properties of UHPC and gradually optimizes and predicts concrete mixture proportion by analyzing experimental data. It can also minimize the number of experiments required and reduce production costs when ensuring that UHPC has a high flowability and compressive strength. However, the accuracy of the statistical design method is greatly affected by the complexity of concrete composition.

4.5 MIXTURE DESIGN METHOD BASED ON ARTIFICIAL NEURAL NETWORKS

Artificial neural networks are mathematical techniques originating from the biological nervous system (Shanmuganathan 2016). The input values can be connected to output values through one or more layers of intermediate. The output signal is obtained from the input signal of related neurons, weight, and activation function. In the UHPC mixture design using the artificial neural networks model, the input layer nodes are UHPC mixture decision variables, and the output layer is the objective (DeRousseau et al. 2018). By adjusting the input weights to narrow the gap between the experiment and input values, the model's accuracy can be guaranteed.

The backpropagation algorithm widely used in neural network modeling was first proposed by Rumelhart et al. (1985). It compares the errors between the output and target values. Ghafari et al. (2012) proposed an artificial neural network analysis model for UHPC based on the backpropagation neural network and Levenberg Marquardt algorithm. The typical artificial neural networks model structure consists of input, weight, function, activation function, and output, as shown in Figure 4.24. The calculation idea of backpropagation neural network is divided into the following two points: (1) error of neural network depends on input and measured value, and overall error is reduced through spreading error back, and (2) weight and error are readjusted until the output value is closest to the experimental value. The backpropagation neural network is simple, but the data convergence is slow due to an unknown mapping relationship. Moreover, the convergence speed of neural networks changes with the adjustment of weight (Chandwani et al. 2015). The weights and their

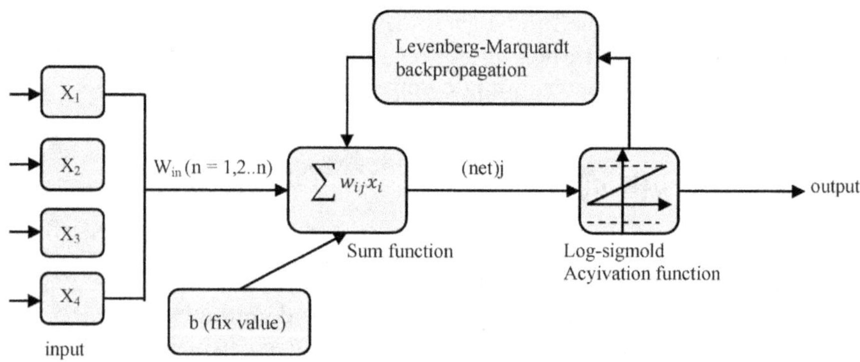

Figure 4.24 Typical structure diagram of artificial neural networks model.

Source: Ghafari et al. 2012.

sums in the backpropagation neural network model can be calculated by Eq. (4.48).

$$net_j = \sum_{i=1}^{n} w_{ij} x_i + b \tag{4.48}$$

where W_{in} represents the weight of each parameter (unitless) and net_j is the weighted sum of input components (unitless).

The artificial neural networks model can be applied to predict the fresh and hardened performance of UHPC, such as slump, filling performance, and compressive strength. Zhang and Zhao (2017) used an artificial neural networks model with 78 groups of experimental data to predict the compressive strength of UHPC. Root mean square error, mean absolute percentage error, and variance are used to evaluate the error. In the experiment, different combinations of fly ash and silica fume instead of cement and influence of particle size on compressive strength of UHPC were mainly considered. After the replacement of cement with fly ash and silica fume, cement equivalent C_{eq} shall be calculated, as shown in Eq. (4.49). The compressive strength was predicted by a transit function through Eq. (4.50).

$$C_{eq} = C + FA \times 0.4 + SF \times 2.0 \tag{4.49}$$

$$f_{cu(j)} = \text{tansig}\ \frac{e^{x} - e^{-x}}{e^{x} + e^{-x}} \tag{4.50}$$

Similarly, Abuodeh et al. (2019) input data into an artificial neural networks model based on a sequential feature selection algorithm and got a minimum error value after iteration. The results showed that with the increase of fly ash and silica fume contents, the compressive strength of UHPC increases significantly. It can be considered that an established artificial neural networks model can accurately predict the compressive strength of UHPC, but there was no theoretical mathematical model. Therefore, the artificial neural networks model was used to detect and analyze input parameters and reduce the number of parameters. However, it should be noted that the accuracy of the previous model is limited due to its single parameter that only considers the proportion of steel fiber.

Qu et al. (2018) collected 162 compressive strength and 102 flexural strength data to establish two artificial neural network models to predict the bending and compressive strengths of UHPC. About 80% of data was used to train neural networks, and the remaining was used to test model training results. The comparison between the predicted results of model and experimental values showed that they were close, as shown in Figure 4.25.

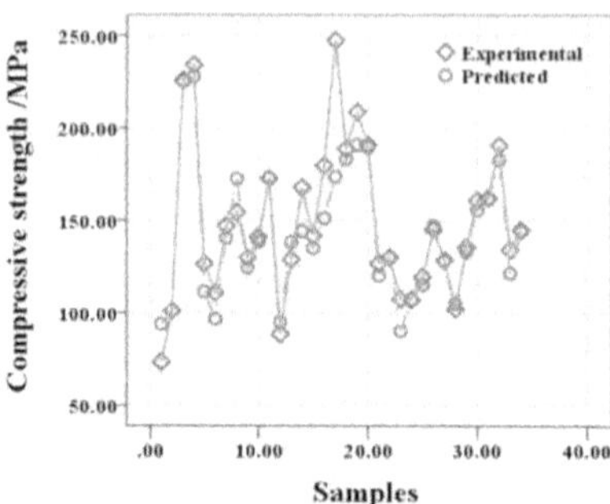

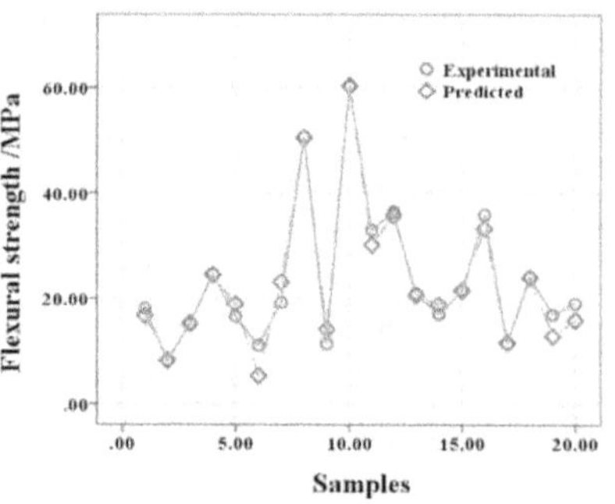

Figure 4.25 Comparison between model prediction results and experimental values (Qu et al. 2018). (a) Compressive strength. (b) Flexural strength.

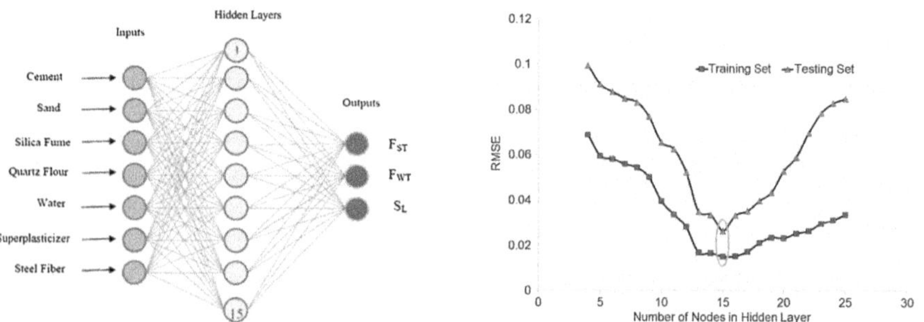

Figure 4.26 Structure of artificial neural network model and testing and training values through RMSE (Ghafari et al. 2015). (a) Structure of artificial neural network model. (b) Testing and training values through root mean squared error (RMSE).

Ghafari et al. (2015) selected seven different input values to establish the artificial neural networks model, as shown in Figure 4.26(a). The results showed that this artificial neural network model can efficiently predict concrete performance, as shown in Figure 4.26(b). The same conclusion was reported by Fan et al. (2020). The artificial neural networks model takes the consistency of fresh UHPC into account and can be used to predict the strength of UHPC. However, artificial nerve requires many experiments to train each neuron. Moreover, overfitting may occur, which causes neural networks to be unable to derive new data. The artificial neural networks model is still based on a lot of data and lacks theoretical derivation.

Although predicted results of the artificial neural network model are in good agreement with experimental values, this method can provide a theoretical basis for further study of the artificial neural network model.

However, parameters of this method were still not comprehensive enough. In terms of UHPC, only fiber length, diameter, and volume fraction can be considered. Therefore, the type, strength, and shape of fiber need to be further studied. Besides, artificial neurons require many experiments to train each neuron, and overfitting may occur, which may cause neural networks to be unable to generalize to new data.

4.6 CHARACTERISTICS OF MIXTURE DESIGN METHODS OF UHPC

According to the above discussion, the advantages and disadvantages of the existing design methods of UHPC are listed in Table 4.3. The design method based on particle packing is a widely used method at present. However, most of the models do not consider the effect of liquid phase on the

Table 4.3 Advantages and disadvantages of mixture design methods

Methods		*Advantages*	*Disadvantages*
Close packing model	Based on dry packing density	Increased packing density of particles; Reduced porosity, and improved strength of UHPC	Lower dry packing density without considering the interaction between solid and liquid phases of water and superplasticizer; Lower strength of UHPC
	Based on wet packing density	More precise and actual packing density value considering the effects of water and superplasticizer on solid particles	Decreased wet packing density due to agglomeration of fine particles
Mixture design method based on rheological properties of paste		Adjusted optimal rheology to secure relatively uniform dispersion of the solid skeleton, especially fiber, and enhanced hardened performance of UHPC	Lacked theoretical model for predicting the fresh performance of UHPC mixed with fiber
Statistical mixture design method		Established relationship between mixture proportioning and performance of UHPC; Reduced number of experiments	Reduced model accuracy due to interaction between components
Mixture design method based on artificial neural network model		Predicted consistency and strength of fresh UHPC; Reduced number of experiments	Required a large amount of data to train the model; Occurred data overfit

dispersion of solid particles, which results in a difference between the prediction results and the actual situation. The rheology-based mixture design method for UHPC needs to consider the effects of various components in UHPC, including aggregate, fiber, and superplasticizer, on the yield stress and viscosity of UHPC. A lot of trial tests are required to reach the optimal rheological properties by adjusting the mixture proportions. The statistical analysis method can reduce the number of trials, but the accuracy is affected by the independent variables and the selected model. The artificial neural network model is derived from the biological nervous system, but its application is limited due to overfitting.

4.7 SUMMARY

This chapter describes the basic design principles, procedures, features, and advantages and disadvantages of four mixture design methods of UHPC. The following conclusions can be drawn:

(1) The close packing method derived from particle continuous packing is by far the most widely used mixture design method for UHPC. It enables UHPC to achieve a densely compacted structure with significantly reduced porosity and enhanced strength of UHPC. Nevertheless, most of the particle packing models only consider the packing density of the solid skeleton without reflecting the interaction of liquids, such as water and superplasticizer, with solid particles. Thus, the measured packing density is lower than that under the actual wet state, leading to an inaccurately predicted strength of UHPC.
(2) Rheology-based mixture design method can produce UHPC with satisfactory flowability, which can ensure relatively uniform dispersion and orientation of strengthening materials, such as fiber and aggregate, to render better performance of UHPC. The rheology of UHPC can be adjusted by changing the mixture proportion, including superplasticizer dosage, particle size distribution, fibers type and content, and w/b ratio. The determination of water and superplasticizer demands should consider the particle packing of UHPC and evaluate the flowability of paste. Repeated experiments are required to determine the optimal rheological properties of UHPC mixed with fibers, and the rheological test data is sensitive to variables, such as the external environment and testing parameters.
(3) The statistical mixture design method quantitatively analyzes the relationship between the mixture proportions and the performance of UHPC and simplifies the scheme required to optimize mixture design. However, much data needs to be provided in the early stage to optimize and verify the reliability of the method.

(4) The artificial neural network can reduce the trial tests, enhance the cost-effectiveness of UHPC, and provide accurate predictions by continuously varying the input values and their weight. It is simpler and more direct than the traditional statistical method through linear and non-linear programming. However, overfitting may occur locally, and data cannot be extended due to the lack of data to train each neuron.

REFERENCES

Abuodeh, O., Abdalla, J. A., and Hawileh, R. A. (2019). Prediction of compressive strength of ultra-high performance concrete using SFS and ANN. 2019 8th International Conference on Modeling Simulation and Applied Optimization (ICMSAO), Manama, Bahrain, 1–5.

Aim, R. B., Le Goff, P. (1968). Effet de paroi dans les empilements désordonnés de sphères et application à la porosité de mélanges binaires. *Powder Technol.* 1(5), 281–290.

Aïtcin, P. C. (1998). *High performance concrete.* CRC press.

Arora, A., Aguayo, M., Hansen, H., Castro, C., Federspiel, E., Mobasher, B., Neithalath, N. (2018). Microstructural packing- and rheology-based binder selection and characterization for Ultra-high Performance Concrete (UHPC). *Cem. Concr. Res.* 103, 179–190.

Artelt, C., Garcia, E. (2008). Impact of superplasticizer concentration and of ultra-fine particles on the rheological behaviour of dense mortar suspensions. *Cem. Concr. Res.* 38(5), 633–642.

Barrett, P. (2009). A Review of "Statistical Models: Theory and Practice": DA Freedman. New York: Cambridge University Press, 2005.

Bentz, D. P., Ferraris, C. F., Galler, M. A., Hansen, A. S., Guynn, J. M. (2012). Influence of particle size distributions on yield stress and viscosity of cement–fly ash pastes. *Cem. Concr. Res.* 42(2), 404–409.

Bullard, J. W., Jennings, H. M., Livingston, R. A., Nonat, A., Scherer, G. W., Schweitzer, J. S., Scrivener, K. L., Thomas, J. J. (2011). Mechanisms of cement hydration. *Cem. Concr. Res.* 41(12), 1208–1223.

Chan, K. W., Kwan, A. K. H. (2014). Evaluation of particle packing models by comparing with published test results. *Particuology* 16, 108–115.

Chandra, S., Björnström, J. (2002). Influence of cement and superplasticizers type and dosage on the fluidity of cement mortars—Part I. *Cem. Concr. Res.* 32(10), 1605–1611.

Chandwani, V., Agrawal, V., Nagar, R. (2015). Modeling slump of ready mix concrete using genetic algorithms assisted training of Artificial Neural Networks. *Expert. Syst. Appl.* 42(2), 885–893.

Chen, J. J., Kwan, A. K. H. (2012). Superfine cement for improving packing density, rheology and strength of cement paste. *Cem. Concr. Compos.* 34(1), 1–10.

Daimon, M., Roy, D. M. (1978). Rheological properties of cement mixes: I. Methods, preliminary experiments, and adsorption studies. *Cem. Concr. Res.* 8(6), 753–764.

de Aguiar, P. F., Bourguignon, B., Khots, M. S., Massart, D. L., Phan-Than-Luu, R. (1995). D-optimal designs. *Chemom. Intell. Lab. Syst.* 30(2), 199–210.

De Larrard, F. (1999). *Concrete mixture proportioning: a scientific approach.* CRC Press.

De Larrard, F., Sedran, T. (1994). Optimization of ultra-high-performance concrete by the use of a packing model. *Cem. Concr. Res.* 24(6), 997–1009.

de Larrard, F., Sedran, T. (2002). Mixture-proportioning of high-performance concrete. *Cem. Concr. Res.* 32(11), 1699–1704.

de Larrard, F., Sedran, T. J. C.(1994). Optimization of ultra-high-performance concrete by the use of a packing model. *Cem. Concr. Res.* 24(6), 997–1009.

DeRousseau, M. A., Kasprzyk, J. R., Srubar, W. V. (2018). Computational design optimization of concrete mixtures: A review. *Cem. Concr. Res.* 109, 42–53.

Dewar, J. (1999). *Computer modelling of concrete mixtures.* CRC Press.

Eriksson, L., Johansson, E., Wikström, C. (1998). Mixture design-design generation, PLS analysis, and model usage. *Chemom. Intell. Lab. Syst.* 43(1–2), 1–24.

Fan, D., Yu, R., Shui, Z., Wu, C., Wang, J., Su, Q. (2020). A novel approach for developing a green ultra-high performance concrete (UHPC) with advanced particles packing meso-structure. *Constr. Build. Mater.* 265, 120339.

Ferdosian, I., Camões, A. (2017). Eco-efficient ultra-high performance concrete development by means of response surface methodology. *Cem. Concr. Compos.* 84, 146–156.

Fuller, W. B., Thompson, S. E. (1907). The laws of proportioning concrete. *Trans. Am. Soc. Civ. Eng.* 59(2), 67–143.

Fung, W. W. S., Kwan, A. K. H., Wong, H. H. C. (2009). Wet packing of crushed rock fine aggregate. *Mater. Struct.* **42**(5), 631–643.

Fung, W., Kwan, A., Wong, H. J. M. (2009). Wet packing of crushed rock fine aggregate. *Mater. Struct.* 42, 631–643.

Furnas, C. (1931). Grading aggregates-I.-Mathematical relations for beds of broken solids of maximum density. *Ind. Eng. Chem.* 23(9), 1052–1058.

Gelardi, G., Flatt, R. J. (2016). Working mechanisms of water reducers and superplasticizers. *In Science and technology of concrete admixtures.* 257–278.

Ghafari, E., Bandarabadi, M., Costa, H., Júlio, E. (2012). Design of UHPC using artificial neural networks. *Brittle Matrix Compos.* 10, 61–69.

Ghafari, E., Bandarabadi, M., Costa, H., Júlio, E. (2015). Prediction of fresh and hardened state properties of UHPC: Comparative study of statistical mixture design and an artificial neural network model. *J. Mater. Civ. Eng.* 27(11), 04015017.

Ghafari, E., Costa, H., Júlio, E. (2014). RSM-based model to predict the performance of self-compacting UHPC reinforced with hybrid steel micro-fibers. *Constr. Build. Mater.* 66, 375–383.

Ghafari, E., Costa, H., Júlio, E. (2015). Statistical mixture design approach for eco-efficient UHPC. *Cem. Concr. Compos.* 55, 17–25.

Ghafari, E., Ghahari, S. A., Costa, H., Júlio, E., Portugal, A., Durães, L. (2016). Effect of supplementary cementitious materials on autogenous shrinkage of ultra-high performance concrete. *Constr. Build. Mater.* 127, 43–48.

Goltermann, P., Johansen, V., Palbol, L. (1997). Packing of aggregates: An alternative tool to determine the optimal aggregate mix. *ACI Mater. J.* 94(5), 435–443.

Grünewald, S., Walraven, J. C. (2001). Parameter-study on the influence of steel fibers and coarse aggregate content on the fresh properties of self-compacting concrete. *Cem. Concr. Res.* 31(12), 1793–1798.

Gruskovnjak, A., Lothenbach, B., Winnefeld, F., Figi, R., Ko, S. C., Adler, M., Mäder, U. (2008). Hydration mechanisms of super sulphated slag cement. *Cem. Concr. Res.* 38(7), 983–992.

Hoang, K. H., Hadl, P., Nguyen, V. (2016). A new mix design method for UHPC based on stepwise optimization of particle packing density. First International Interactive Symposium on UHPC; Iowa State University Digital Press: Ames, USA, 1–8.

Hou, D., Wu, D., Wang, X., Gao, S., Yu, R., Li, M., Wang, P., Wang, Y. (2021). Sustainable use of red mud in ultra-high performance concrete (UHPC): design and performance evaluation. *Cem. Concr. Compos.* 115, 103862.

Hu, J., and Wang, K. (2011). Effect of coarse aggregate characteristics on concrete rheology. *Constr. Build. Mater.* 25(3), 1196–1204.

Huang, H., Gao, X., Li, L., Wang, H. (2018). Improvement effect of steel fiber orientation control on mechanical performance of UHPC. *Constr. Build. Mater.* 188, 709–721.

Jiang, S., Shan, B., Ouyang, J., Zhang, W., Yu, X., Li, P., Han, B. (2018). Rheological properties of cementitious composites with nano/fiber fillers. *Constr. Build. Mater.* 158, 786–800.

Jiao, D., Shi, C., Yuan, Q., An, X., Liu, Y., Li, H. (2017). Effect of constituents on rheological properties of fresh concrete-A review. *Cem. Concr. Compos.* 83, 146–159.

Khatri, R. P., Sirivivatnanon, V., Gross, W. (1995). Effect of different supplementary cementitious materials on mechanical properties of high performance concrete. *Cem. Concr. Res.* 25(1), 209–220.

Khayat, K. H., Meng, W., Vallurupalli, K., Teng, L. (2019). Rheological properties of ultra-high-performance concrete - An overview. *Cem. Concr. Res.* 124, 105828.

Kjeldsen, A. M., Bergström, L., Geiker, M. (2004). Centrifugal consolidation of MgO-suspensions-The Influence of Superplastisizers on Particle Packing. Proceedings, Nordic Rheology Conference, Iceland.

Kuder, K. G., Ozyurt, N., Mu, E. B., Shah, S. P. (2007). Rheology of fiber-reinforced cementitious materials. *Cem. Concr. Res.* 37(2), 191–199.

Kwan, A. K. H., Chan, K. W., Wong, V. (2013). A 3-parameter particle packing model incorporating the wedging effect. *Powder Technol.* 237, 172–179.

Kwan, A. K. H., Chen, J. J. (2013). Adding fly ash microsphere to improve packing density, flowability and strength of cement paste. *Powder Technol.* 234, 19–25.

Kwan, A. K. H., Fung, W. W. S. (2012). Roles of water film thickness and SP dosage in rheology and cohesiveness of mortar. *Cem. Concr. Compos.* 34(2), 121–130.

Kwan, A. K. H., Li, L. G., Fung, W. W. S. (2012). Wet packing of blended fine and coarse aggregate. *Mater. Struct.* 45(6), 817–828.

Kwan, A. K. H., Wong, H. H. C. (2007). Packing density of cementitious materials: part 2-packing and flow of OPC+PFA+CSF. *Mater. Struct.* 41(4), 773.

Lee, N. K., Koh, K. T., Kim, M. O., Ryu, G. S. (2018). Uncovering the role of micro silica in hydration of ultra-high performance concrete (UHPC). *Cem. Concr. Res.* 104, 68–79.

Lee, S. H., Kim, H. J., Sakai, E., Daimon, M. (2003). Effect of particle size distribution of fly ash–cement system on the fluidity of cement pastes. *Cem. Concr. Res.* 33(5), 763–768.

Li, L. G., Kwan, A. K. H. (2013). Concrete mix design based on water film thickness and paste film thickness. *Cem. Concr. Compos.* 39, 33–42.

Li, L. G., Kwan, A. K. H. (2014). Packing density of concrete mix under dry and wet conditions. *Powd. Technol.* 253, 514–521.

Li, L. G., Kwan, A. K. H. (2015). Effects of superplasticizer type on packing density, water film thickness and flowability of cementitious paste. *Constr. Build. Mater.* 86, 113–119.

Li, L. G., Zhuo, H. X., Zhu, J., Kwan, A. K. H. (2019). Packing density of mortar containing polypropylene, carbon or basalt fibres under dry and wet conditions. *Powder Technol.* 342, 433–440.

Li, L., Kwan, A. J. P. t. (2014). Packing density of concrete mix under dry and wet conditions. *Powd. Technol.* 253, 514–521.

Li, P. P., Cao, Y. Y. Y., Sluijsmans, M. J. C., Brouwers, H. J. H., Yu, Q. (2021). Synergistic effect of steel fibres and coarse aggregates on impact properties of ultra-high performance fibre reinforced concrete. *Cem. Concr. Compos.* 115, 103866.

Li, P. P., Yu, Q. L., Brouwers, H. J. H. (2018). Effect of coarse basalt aggregates on the properties of Ultra-high Performance Concrete (UHPC). *Constr. Build. Mater.* 170, 649–659.

Li, Z., Lu, D., Gao, X. (2021). Optimization of mixture proportions by statistical experimental design using response surface method—A review. *J. Build. Eng.* 36, 102101.

Liu, C., Zhang, C., Zhou, C., Mi, Y., Wang, Z. (2020). Effects of the solidification of capillary bridges on the interaction forces between hydrate particles. *Energy Fuels.* 34(4), 4525–4533.

Liu, J., Wang, K., Zhang, Q., Han, F., Sha, J., Liu, J. (2017). Influence of superplasticizer dosage on the viscosity of cement paste with low water-binder ratio. *Constr. Build. Mater.* 149, 359–366.

Mangulkar, M., Jamkar, S. J. C. P. (2013). Review of particle packing theories used for concrete mix proportioning. *Contributory Papers* 141.

Marar, K., Eren, Ö., Roughani, H. (2017). The influence of amount and aspect ratio of fibers on shear behaviour of steel fiber reinforced concrete. *KSCE. J. Civ. Eng.* 21(4), 1393–1399.

Marquardt, I., Vala, J., Diederichs, U. (2002). Concrete technology-Determination of the composition of self-compacting concretes on the basis of the water requirements of the constituent materials. *Betonwerk und Fertigteiltechnik* 68(11), 22–31.

Martinie, L., Rossi, P., Roussel, N. (2010). Rheology of fiber reinforced cementitious materials: classification and prediction. *Cem. Concr. Res.* 40(2), 226–234.

Mazloom, M., Ramezanianpour, A. A., Brooks, J. J. (2004). Effect of silica fume on mechanical properties of high-strength concrete. *Cem. Concr. Compos.* 26(4), 347–357.

Mehdipour, I., Khayat, K. H. (2017). Effect of particle-size distribution and specific surface area of different binder systems on packing density and flow characteristics of cement paste. *Cem. Concr. Compos.* 78, 120–131.

Mehdipour, I., Khayat, K. H. (2018). Understanding the role of particle packing characteristics in rheophysical properties of cementitious suspensions: A literature review. *Constr. Build. Mater.* 161, 340–353.

Meng, W., Khayat, K. H. (2017). Improving flexural performance of ultra-high-performance concrete by rheology control of suspending mortar. *Compos. Part-B. Eng.* 117, 26–34.

Meng, W., Samaranayake, V., Khayat, K. H. (2018). Factorial design and optimization of ultra-high-performance concrete with lightweight sand. *ACI Mater. J.* 115(1), 129–138.

Meng, W., Valipour, M., Khayat, K. H. (2016). Optimization and performance of cost-effective ultra-high performance concrete. *Mater. Struct.* 50(1), 1–16.

Mindess, S., Skalny, J. (1991). Fiber-reinforced cementitious materials. *J. Mater. Civ. Eng.* 5(3), 405–483.

Nehdi, M., Mindess, S., Aïtcin, P. C. (1998). Rheology of high-performance concrete: Effect of ultrafine particles. *Cem. Concr. Res.* 28(5), 687–697.

Nemeth, M. A. (2003). Response surface methodology: Process and product optimization using designed experiments. *J. Qual. Technol.* 35(4), 428–429.

Ozyurt, N., Mason, T. O., Shah, S. P. (2007). Correlation of fiber dispersion, rheology and mechanical performance of FRCs. *Cem. Concr. Compos.* 29(2), 70–79.

Pedersen, E. (1999). Packing calcuations applied for concrete mix design. Proceedings of the International Conference Held at the University of Dundee, Scotland, 1999, Thomas Telford.

Pedrosa, H. C., Reales, O. M., Reis, V. D., Paiva, M. d. D., Fairbairn, E. M. R. (2020). Hydration of Portland cement accelerated by C-S-H seeds at different temperatures. *Cem. Concr. Res.* 129, 105978.

Powers, T. C. (1968). *Properties of Fresh Concrete.* John Wiley & Sons, Inc. New York, USA.

Puri, B., Brown, B., McKee, H., Treasaden, I. (1992). *Manual of ready-mixed concrete.* CRC Press.

Qiu, J., Guo, Z., Yang, L., Jiang, H., Zhao, Y. (2020). Effects of packing density and water film thickness on the fluidity behaviour of cemented paste backfill. *Powder Technol.* 359, 27–35.

Qu, D., Cai, X., Chang, W. (2018). Evaluating the effects of steel fibers on mechanical properties of ultra-high performance concrete using artificial neural networks. *Appl. Sci.* 8(7), 1120.

Roy, D. M., Scheetz, B. E., Silsbee, M. R. (1993). Processing of optimized cements and concretes via particle packing. *MRS Bull.* 18(3), 45–49.

Rumelhart, D. E., Hinton, G. E., Williams, R. J. (1985). Learning internal representations by error propagation. In: *Parallel Distributed Processing: Explorations in the Microstructure of Cognition*, Cambridge, MA: MIT Press, vol. 1, pp. 318–362.

Sha, S., Wang, M., Shi, C., Xiao, Y. (2020). Influence of the structures of polycarboxylate superplasticizer on its performance in cement-based materials-A review. *Constr. Build. Mater.* 233, 117257.

Shah, S., Sharma, R., Paswan, S., Bhandari, A., Arukala, S. R., Pancharathi, R. K. (2020). Prioritizing the aggregate source based on particle packing density using modified Toufar Model and MCDM methods. *Adv.Sustainable Constr. Mater.* 171–182.

Shakhmenko, G., Birsh, J. (1998). Concrete mix design and optimization. 2nd Int. PhD Symposium in Civil Engineering. Budapest, Hungary.

Shanmuganathan, S. (2016). Artificial neural network modelling: An introduction. 1–14. Springer International Publishing.

Shi, C., Wang, D., An, X., Jiao, D., Li, H. (2018). Method for mixture design of concrete with multiple performance requirements. *J. Chin. Ceram. Soc.* 46(02), 230–238.

Shi, C., Wang, D., Wu, L., Wu, Z. (2015). The hydration and microstructure of ultra high-strength concrete with cement–silica fume–slag binder. *Cem. Concr. Compos.* 61, 44–52.

Shi, C., Wu, Z., Lv, K., Wu, L. (2015). A review on mixture design methods for self-compacting concrete. *Constr. Build. Mater.* 84, 387–398.

Shi, C., Wu, Z., Xiao, J., Wang, D., Huang, Z., Fang, Z. (2015). A review on ultra high performance concrete: Part I. Raw materials and mixture design. *Constr. Build. Mater.* 101, 741–751.

Shi, Y., Long, G., Zen, X., Xie, Y., Shang, T. (2021). Design of binder system of eco-efficient UHPC based on physical packing and chemical effect optimization. *Constr. Build. Mater.* 274, 121382.

Soliman, N. A., Tagnit-Hamou, A. (2017). Using particle packing and statistical approach to optimize eco-efficient ultra-high-performance concrete. *ACI Mater. J.* 114(6), 847–858.

Soulioti, D. V., Barkoula, N. M., Paipetis, A., Matikas, T. E. (2011). Effects of fibre geometry and volume fraction on the flexural behaviour of steel-fibre reinforced concrete. *Strain.* 47, e535–e541.

Stovall, T., de Larrard, F., Buil, M. (1986). Linear packing density model of grain mixtures. *Powder Technol.* 48(1), 1–12.

Su, Y., Li, J., Wu, C., Wu, P., Li, Z.X. (2016). Effects of steel fibres on dynamic strength of UHPC. *Constr. Build. Mater.* 114, 708–718.

Teng, L., Meng, W., Khayat, K. H. (2020). Rheology control of ultra-high-performance concrete made with different fiber contents. *Cem. Concr. Res.* 138, 106222.

Teng, L., Zhu, J., Khayat, K. H., Liu, J. (2020). Effect of welan gum and nanoclay on thixotropy of UHPC. *Cem. Concr. Res.* 138, 106238.

Van Der Putten, J., Dils, J., Minne, P., Boel, V., De Schutter, G. (2016). Determination of packing profiles for the verification of the compressible packing model in case of UHPC pastes. *Mater. Struct.* 50(2), 118.

Vikan, H., Justnes, H., Winnefeld, F., Figi, R. (2007). Correlating cement characteristics with rheology of paste. *Cem. Concr. Res.* 37(11), 1502–1511.

Wang, C. Z. N. (1999). The theoretic analysis of the influence of the particle size distribution of cement system on the property of cement. *Cem. Concr. Res.* 29(11), 1721–1726.

Wang, C., Yang, C., Liu, F., Wan, C., Pu, X. (2012). Preparation of ultra-high performance concrete with common technology and materials. *Cem. Concr. Compos.* 34(4), 538–544.

Wang, R., Gao, X., Huang, H., Han, G. (2017). Influence of rheological properties of cement mortar on steel fiber distribution in UHPC. *Constr. Build. Mater.* 144, 65–73.

Wang, R., Gao, X., Zhang, J., Han, G. (2018). Spatial distribution of steel fibers and air bubbles in UHPC cylinder determined by X-ray CT method. *Constr. Build. Mater.* 160, 39–47.

Wang, X., Huang, J., Ma, B., Dai, S., Jiang, Q., Tan, H. (2019). Effect of mixing sequence of calcium ion and polycarboxylate superplasticizer on dispersion of a low grade silica fume in cement-based materials. *Constr. Build. Mater.* 195, 537–546.

Wang, X., Yu, R., Song, Q., Shui, Z., Liu, Z., Wu, S., Hou, D. (2019). Optimized design of ultra-high performance concrete (UHPC) with a high wet packing density. *Cem. Concr. Res.* 126, 105921.

Wong, H. H. C., Kwan, A. K. H. (2008). Packing density of cementitious materials: part 1—measurement using a wet packing method. *Mater. Struct.* 41(4), 689–701.

Wong, H. H., Kwan, A. K. J. M., structures (2008). Packing density of cementitious materials: part 1—measurement using a wet packing method. *Mater. Struct.* 41, 689–701.

Wong, H., Kwan, A. (2005). Packing density: a key concept for mix design of high performance concrete. Proceedings of the Materials Science and Technology in Engineering Conference. HKIE Materials Division, Hong Kong, 1–15.

Wu, Z., Khayat, K. H., Shi, C. (2018). How do fiber shape and matrix composition affect fiber pullout behavior and flexural properties of UHPC? *Cem. Concr. Compos.* 90, 193–201.

Yamada, K., Takahashi, T., Hanehara, S., Matsuhisa, M. (2000). Effects of the chemical structure on the properties of polycarboxylate-type superplasticizer. *Cem. Concr. Res.* 30(2), 197–207.

Yang, L., Shi, C., Wu, Z. (2019). Mitigation techniques for autogenous shrinkage of ultra-high-performance concrete—A review. *Compos. B. Eng.* 178, 107456.

Yao, W., Guangsheng, G., Fei, W., Jun, W. (2002). Fluidization and agglomerate structure of SiO_2 nanoparticles. *Powder Technol.* 124(1–2), 152–159.

Yazıcı, Ş., İnan, G., Tabak, V. (2007). Effect of aspect ratio and volume fraction of steel fiber on the mechanical properties of SFRC. *Constr. Build. Mater.* 21(6), 1250–1253.

Ye, H., Gao, X., Wang, R., Wang, H. (2017). Relationship among particle characteristic, water film thickness and flowability of fresh paste containing different mineral admixtures. *Constr. Build. Mater.* 153, 193–201.

Yoo, D., Kang, S. T., Yoon, Y. S. (2016). Enhancing the flexural performance of ultra-high-performance concrete using long steel fibers. *Compos. Struct.* 147, 220–230.

Yu, A., Bridgwater, J., Burbidge, A. (1997). On the modelling of the packing of fine particles. *Powder Technol.* 92(3), 185–194.

Yu, A., Zou, R. P., Standish, N. (1996). Modifying the linear packing model for predicting the porosity of nonspherical particle mixtures. *Ind. Eng. Chem. Res.* 35(10), 3730–3741.

Yu, R., Spiesz, P., Brouwers, H. J. H. (2014). Effect of nano-silica on the hydration and microstructure development of Ultra-High Performance Concrete (UHPC) with a low binder amount. *Constr. Build. Mater.* 65, 140–150.

Yu, R., Spiesz, P., Brouwers, H. J. H. (2014). Mix design and properties assessment of Ultra-High Performance Fibre Reinforced Concrete (UHPFRC). *Cem. Concr. Res.* 56, 29–39.

Yu, R., Spiesz, P., Brouwers, H. J. H. (2015). Development of an eco-friendly Ultra-High Performance Concrete (UHPC) with efficient cement and mineral admixtures uses. *Cem. Concr. Compos.* 55, 383–394.

Zhang, J., Zhao, Y. (2017). Prediction of Compressive Strength of Ultra-High Performance Concrete (UHPC) Containing Supplementary Cementitious Materials. 2017 International Conference on Smart Grid and Electrical Automation (ICSGEA), Changsha, China, 522–525.

Zhang, Y., Luo, X., Kong, X., Wang, F., Gao, L. (2018). Rheological properties and microstructure of fresh cement pastes with varied dispersion media and superplasticizers. *Powder Technol.* 330, 219–227.

Zheng, J., Carlson, W. B., Reed, J. S. (1995). The packing density of binary powder mixtures. J. *Eur. Ceram. Soc.* 15(5), 479–483.

Chapter 5

Fresh properties

5.1 INTRODUCTION

Fresh properties refer to the concrete possessing original workability, enough to make it be placed and consolidated using the intended methods (ASTM C125 2018). The fresh properties are of crucial importance to ensure the quality and mechanical properties of hardened concrete. The properties of fresh concrete include workability, temperature, air content, and setting time. Workability is a vital property of concrete, which indicates the ease of mixing, placing, compacting, and finishing freshly mixed concrete with minimum effort and without much bleeding and segregation. The workability of concrete mainly contains two aspects: consistency and cohesiveness. Consistency is an indicator representing the mobility and flowability of concrete in its fresh state, which can be measured using the slump test, compacting factor test, Vebe test, and flow table test. The initial flow and slump retention capabilities of concrete are related to the rheological properties of cement paste, the internal frictions between aggregate particles, and the external interactions between concrete and other surfaces. The other aspect of workability is cohesiveness, sometimes called stability. It identifies whether a concrete mixture is plastic, harsh, or sticky and qualitatively describes the ability of fresh mixture to hold all ingredients together without segregation. The science of rheology has been widely used to address the concepts of flowability and stability of fresh concrete mixture using quantitative fundamental physical terms, such as yield stress and viscosity. Thus, workability is a complex system that involves different parameters, such as flowability, mobility, stability, pumpability, compactability, and resistance to segregation and bleeding.

The components of UHPC are complex, containing a high binder content in conjunction with a variety of mineral admixtures and superplasticizers. Fresh UHPC is required to have high fluidity and low fluidity loss over time to meet the technological requirements of centralized mixing, transportation, pouring, and finishing of concrete. The superplasticizer must have a strong dispersing effect on cement slurries and a high water reduction rate

DOI: 10.1201/9781003203605-5

greater than 30%. Fresh UHPC should remain homogeneous and not show fiber segregation or a solid fraction of the constituents after taking into account the placing methods. The workability of UHPC is commonly verified with a flow table test as specified by ASTM C1437 (2020), the Chinese standard of GB2419 (2005), and the French standard of NF P 18-470 (2016). According to the French standard, the consistency of UHPC in fresh state is normally subject to a specification associated with a target value (NF P 18-470 2016). Otherwise, the consistency belongs to one of the following three categories: UHPC, which may be self-compacting, viscous UHPC, and UHPC exhibiting a flow threshold. Self-compacting UHPC is capable of being placed during pouring without vibration or the use of a mechanical flow aid. The viscous UHPC can be placed without vibration but requires the use of a mechanical flow aid. The third category exhibits a flow threshold, representing the fresh mixture generally able to flow under the effect of dynamic shearing but whose free surface at rest may keep sloping.

The workability of UHPC is affected by multiple factors, such as characteristics of raw materials (Hallal et al. 2010; Schröfl et al. 2012), dispersion, and retention capabilities of superplasticizers (Felekoğlu and Sarıkahya 2008; Shin et al. 2008). Supplementary cementitious materials, such as silica fume, fly ash, and rice husk ash, can enhance the mechanical properties and durability of UHPC (Kang et al. 2019; Madandoust et al. 2011; Nadeem et al. 2013) but have a side negative on the workability, depending on their water adsorption and chemical reactivities (Juenger and Siddique 2015; Shetty 2006). The shape, size, texture, gradation, specific gravity, soundness, and moisture content of aggregate are key parameters affecting the fresh and hardened properties (Collepardi et al. 1997; Madandoust et al. 2011; Yu et al. 2014). Binders and aggregates of different sizes can be used to achieve the maximum packing density. Such a grade system increases the ball-bearing effect among particles and decreases the requirement of cement paste for the lubricating surface of particle, thus improving the workability of fresh concrete (Rossi 2001). Chemical admixtures, such as water-reducing agents, viscosity-modifying admixtures, and air-entraining agents, can also be added to counteract the negative effect caused by the low w/b ratio and the use of fibers to maintain desired workability (Rossi 2001).

This chapter describes the fresh properties of UHPC, including the mini-slump flow, slump-retention ability (slump flow life), setting time, temperature, air content, and rheological properties. The effects of different factors, including characteristics of cementitious materials, chemical admixture contents and types, fibers, nanomaterials, and superabsorbent polymers, on the workability are discussed. It can provide a relationship between the characteristics or content of used materials and fresh properties of UHPC.

5.2 MINI-SLUMP FLOW

The mini-slump flow is considered a standard among field-friendly methods to quantify the consistency of concrete due to its importance in construction workability (ASTM C143/C143M 2012; Zingg et al. 2008) and provides information on the filling ability of mortar (De Schutter et al. 2008). The mini-slump flow test can be also performed to evaluate the consistency and unconfined horizontal flow potential of nonfibrous UHPC matrix. Wille et al. (2011) recommended employing a flow diameter spread limit of 200–350 mm for dense UHPC without fibers, according to ASTM C230 (2021). The influence of constituents on the slump flow of UHPC is described as follows.

5.2.1 Cementitious materials and fillers

Cement is a crucial factor affecting the mechanical properties and durability of UHPC. To obtain excellent workability of UHPC, cement with a low alkali content (K_2O and Na_2O) is recommended (Flatt and Houst 2001; Sakai et al. 2008). Nevertheless, some alkalis are in the form of highly soluble Na_2SO_4 and K_2SO_4. The sulfate ions can consume C_3A molecules in competition with the superplasticizers, leading to a better slump value (Flatt and Houst 2001; Zingg et al. 2009).

Dils et al. (2013) studied the effects of six types of cement on the slump flow of UHPC. The chemical and compound composition of those types of cement and their influence on the slump flow of UHPC are shown in Table 5.1 and Figure 5.1, respectively. The specific surface of C_3A and C_3A/SO_4^{2-} ratio of cement significantly affects the slump flow (Spiratos et al. 2003). Higher C_3A content and reactivity (corresponding to a higher fineness of cement particles) promote the formation of more ettringite, indicating a higher specific surface and more superplasticizer consumption. On the other hand, greater water content is consumed, which causes an increased solid volume and a decreased volume of free pore solution. Consequently, UHPC made with C1 cement containing the highest C_3A-specific surface exhibits a lower slump flow. The competitive consumption of anchor points on the C_3A between sulfate ions and superplasticizer molecules is stronger with the amount of SO_4^{2-}. This means that the adsorption of superplasticizer on the surface of C_3A results in a decrease in initial slump flow.

Generally, if more superplasticizer is available in the pore solution for dispersion at a later time, the UHPC will present a longer slump life (Mazanec and Schiessl 2008). The presence of alkalis changes C_3A into a more reactive form, affecting its reactions with sulfate and superplasticizer (Spiratos et al. 2006). A reduction in the workability of UHPC is the result of high alkali content of C5 cement. As a result, flowable UHPC can be prepared using cement with a good combination of low C_3A-specific surface, an adequate SO_3^-, and acceptable alkalis K_2O and Na_2O amount.

Table 5.1 Chemical and compound composition of cement used for UHPC

Oxides	*C1*	*C2*	*C3*	*C4*	*C5*	*C6*
SiO_2	21.4	21.35	21.48	20.90	21.03	21.24
Al_2O_3	3.60	3.58	3.61	3.64	4.06	3.87
Fe_2O_3	4.20	4.09	4.20	5.19	5.06	5.18
CaO	63.2	63.25	63.37	63.68	63.86	64.11
MgO	2.00	1.77	1.64	0.77	1.06	0.95
SO_3	2.90	2.64	2.54	3.03	1.75	2.30
Na_2O	0.15	0.17	0.17	0.17	0.25	0.18
K_2O	0.54	0.50	0.56	0.63	0.59	0.52
Blaine (cm^2/g)	5600	3490	4392	3975	3300	4300
D50 (μm)	9.90	18.59	13.08	10.40	16.27	11.27
C_2S	19.10	17.74	18.55	14.88	14.68	16.07
C_3S	56.12	57.74	57.16	59.82	60.58	59.53
$C_2S + C_3S$	75.22	75.48	75.71	74.70	75.26	75.60
C_3A	2.44	2.57	2.47	0.87	2.21	1.50
C_4AF	12.80	12.40	12.80	15.80	15.40	15.70
C_3A-spec. surf (cm^2/g C_3A)	136.75	89.86	108.42	34.78	72.85	64.56

Source: Dils et al. 2013.

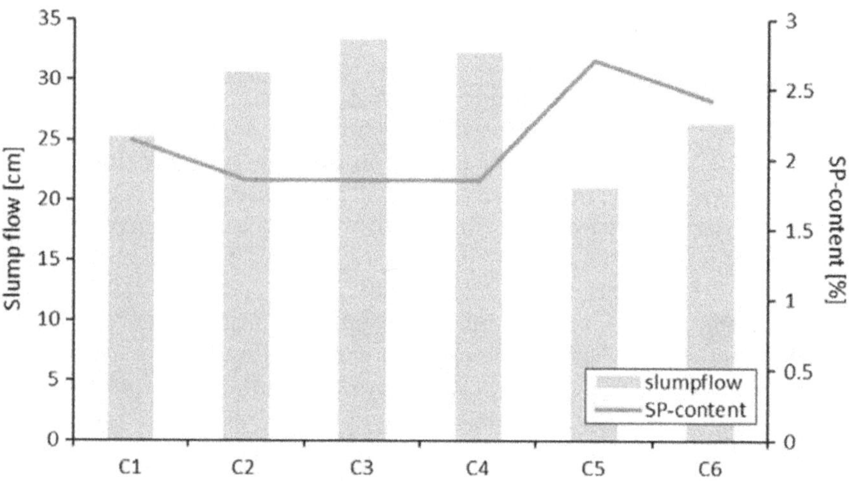

Figure 5.1 Effect of cement type on slump flow of UHPC.

Source: Dils et al. 2013.

Superfine cement has much higher fineness than ordinary Portland cement. Its addition decreases the voids ratio and increases the packing density from 0.64% to 0.68% when the replacement of ordinary cement by superfine cement achieves 20%. This indicates that the amount of water needed to fill in voids decreases and free water amount increases. Therefore, the spread diameter of paste at a w/b ratio of 0.22 increases from 113.5 to 163.0 mm (Chen and Kwan 2012).

Silica fume is an essential constituent for UHPC, which affects the overall properties due to its high amorphous silica content and the ability to enhance the packing density of the matrix. It consumes calcium hydroxide to form C-S-H gel, which can improve the mechanical properties of concrete (Yan et al. 1999). Moreover, it can eliminate the large preferentially oriented calcium hydroxide crystals and improve homogeneity, resulting in the enhancement in the interfacial transition zone between aggregate and fiber/matrix (Wang et al. 2009). Silica fume particles can provide a "ball-bearing" effect and release the water entrapped between the coarse and fine particles, thus increasing the workability (Rao 2003; Vikan et al. 2007). However, some publications showed that the carbon content present in silica fume causes a workability loss (Shi et al. 2015a; Yazıcı et al. 2010). The effect of silica fume on workability relates to its content and characteristics (Aldahdooh et al. 2013; Duval and Kadri 1998). Van Tuan et al. (2011) found that 10%–20% silica fume led to the optimal workability of UHPC. The typical silica fume content used in UHPC mixture is approximately 20%–30%, by mass of cementitious materials (Dong et al. 2018; Wille et al. 2014). This substitution can improve the fiber-matrix bond as well as compressive and flexural strengths of UHPC (Chan and Chu 2004; Wu et al. 2016). Low silica fume content of 5%–15% can be also found in previous studies, given the specific material characteristics and mixture proportion (Arora et al. 2018; Meng et al. 2016). The mini-slump flow of UHPC matrix with 2% superplasticizer increases from 113 to 175 mm when silica fume content increases from 0% to 15% but decreases with further increasing content (Wu et al. 2016). The matrix with 30% silica fume has no flowability, and the optimal silica fume content is found to be 15%–25%, as shown in Figure 5.2(a). Superplasticizer dosage is also an essential factor affecting the workability of UHPC. At a target mini-slump flow of 280 ± 10 mm, the superplasticizer demand of UHPC decreases initially and then increases with a further increase in silica fume content, regardless of fiber content, as illustrated in Figure 5.2(b) (Wu et al. 2019). The lubrication effect of fine silica fume favors the release of the water entrapped among particles, hence increasing slump flow. However, high silica fume content harms the workability due to its high specific surface areas (Shi et al. 2015a).

The limited output and the high cost constrain the application of silica fume in the modern construction industry. In this context, rice husk ash, ground granulated blast-furnace slag (GGBS), limestone powder, etc., are

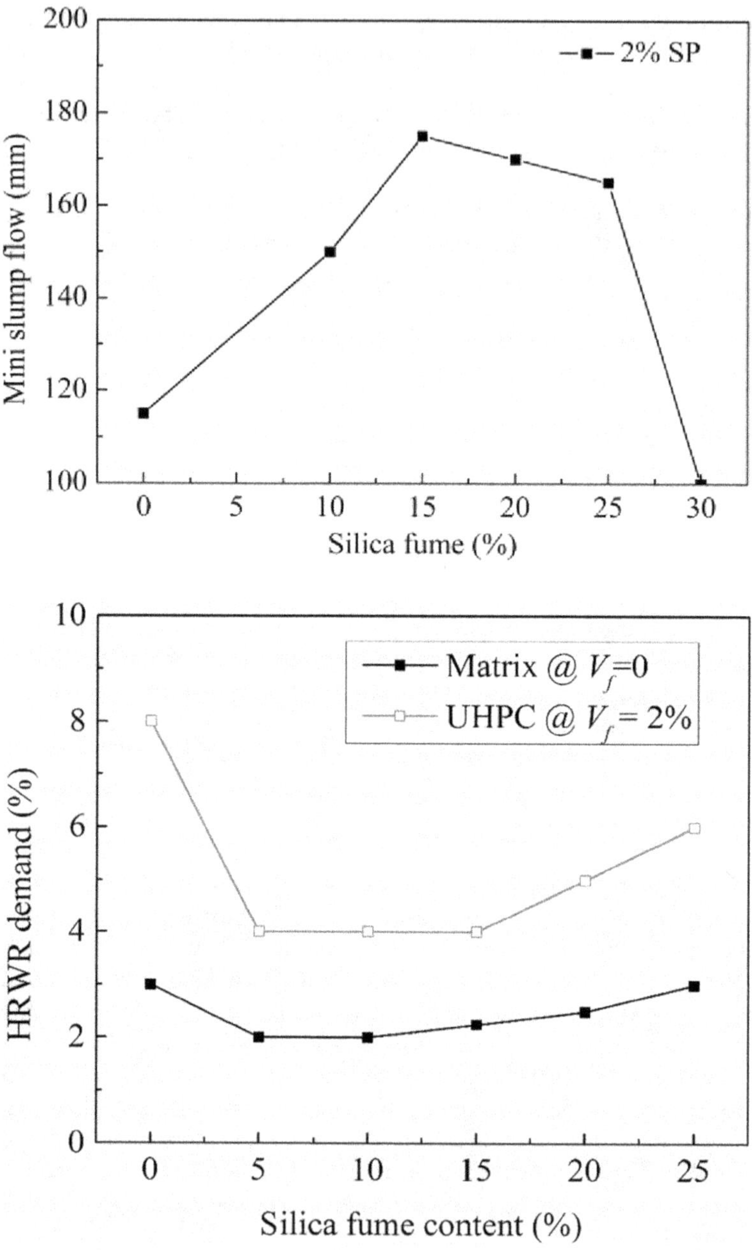

Figure 5.2 Effects of silica fume content on mini-slump flow and superplasticizer demand of nonfibrous UHPC matrix and UHPC (Wu et al. 2019; Wu et al. 2016): (a) Mini-slump flow; (b) Superplasticizer demand at a target mini-slump flow.

used to replace silica fume in UHPC (Yazıcı et al. 2010; Yazıcı et al. 2008). Shi et al. (2015a) found that the silica fume and slag exhibited positive synergistic effects on workability. The flowability of ultra-high-strength concrete (UHSC) with more than 10% slag increased from 110 to 195 mm with the silica fume increasing from 0% to 22%. However, it showed negative effects on the flowability at 22% silica fume. In comparison, the flowability of UHSC with more than 40% slag was less than 140 mm. Silica fume can absorb a certain amount of superplasticizer on its surface due to its fine particle size (Khayat and Aïtcin 1993). Slag also requires more water to achieve the same flowability due to its larger specific surface area than that of cement. As a result, the control of the content of cementitious materials is significant for UHSC to achieve its desired workability.

At the same spread flow of UHPC mixtures, the amount of superplasticizer increases with an increased replacement of rice husk ash. A blended cement with 30% rice husk ash and 10% silica fume is considered the maximum replacement amount for good flowability (Van Tuan et al. 2011). The massive cement replacement by limestone filler can break the workability barrier and solve the problem of incompatibility between cement and superplasticizers (Denarié and Houst 2011). Figure 5.3 illustrates the effects of limestone power and GGBS contents on flowability of UHPC. The slump flow of UHPC firstly increases and then decreases as limestone powder content increases up to 40% (Figure 5.3(a)). It then achieves a higher slump of 240 mm with 20% replacement of limestone power. In comparison, the mixture containing 20% GGBS content exhibits similar flowability to control samples, while excessive addition is not beneficial to the improvement of workability, as illustrated in Figure 5.3(b) (Wang et al. 2012). A high replacement of 15%–45% GGBS is incorporated for preparing reactive power concrete to optimize the hydration of the binder (Yazıcı et al. 2010).

In addition, the flowability of UHPC is affected by replacing cement with other fillers or additives. For example, Bornemann and Schmidt (2002) indicated that significant amounts of cement replacement in UHPC mixtures by fine quartz sand with close size and distribution improved the workability. The incorporation of ultrafine palm oil fuel ash also enhances the workability of UHPC due to the lower carbon content and specific gravity (Aldahdooh et al. 2013). The use of fillers to partially replace the cement in UHPC with steel microfibers also significantly improves the flowability (Yang et al. 2010; Yu et al. 2014).

5.2.2 Aggregates

Aggregates are mainly classified into coarse and fine aggregates. The shape, size, texture, gradation, moisture content, and density of aggregates directly affect the packing density of concrete (Wong and Kwan 2005). Generally,

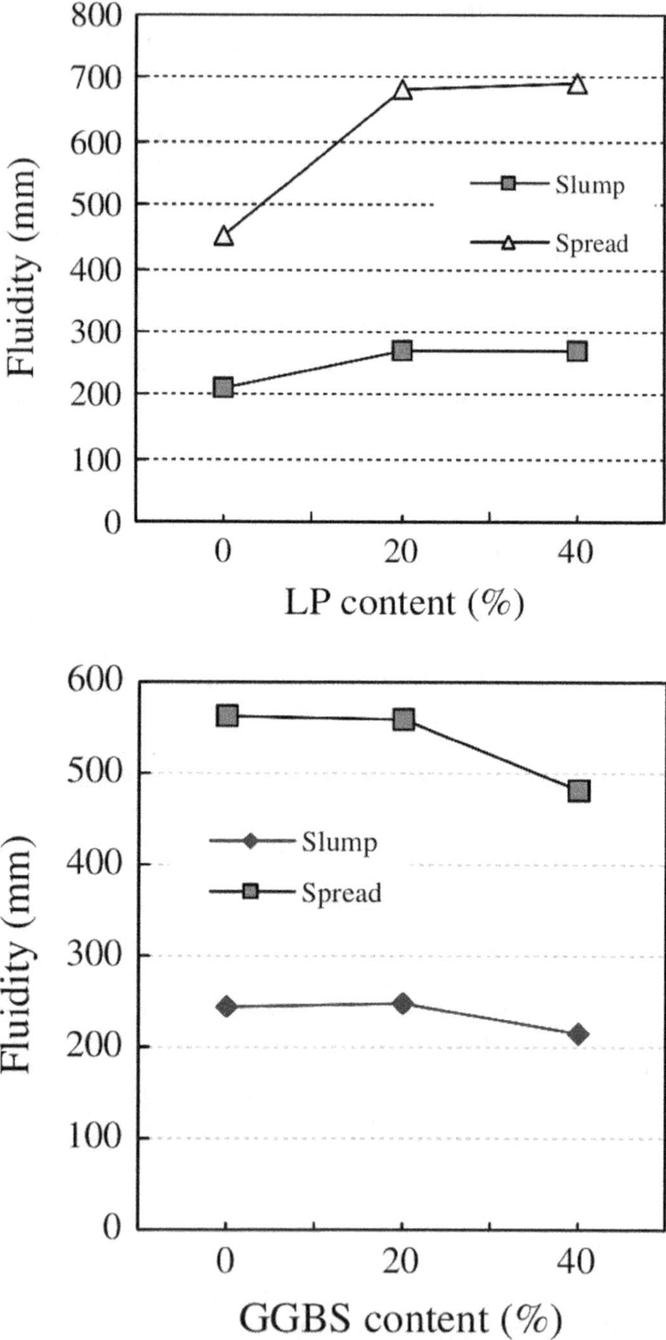

Figure 5.3 Effects of limestone power and GGBS contents on flowability of UHPC (Wang et al. 2012): (a) Limestone power; (b) GGBS.

the enhancement of packing density decreases water demand and improves the lubrication capability of binder paste, increasing the flowability of fresh UHPC (Mehta and Aitcin 1990).

Aggregates with a high surface roughness decrease the packing density due to the great interparticle frictional forces during mixing and compaction, while rounded and smooth aggregates are beneficial to the preparation of well-workable concrete (Mehta and Aitcin 1990). Yang et al. (2009) indicated that more angular geometry of sand particles reduced the flowability of fresh UHPC. Silica sand has a spherical shape surface that is similar to the particles of fine ordinary sand (FOS-I and FOS-II), while recycled glass cullet shows a sharper and more angular grain shape and a smoother surface, as illustrated in Figure 5.4. It can be seen from Table 5.2 that the UHPC mixture in the presence of recycled glass cullet exhibits lower spread flow. This can be attributed to the fact that more angular and sharp sand particles restrict the mortar to the region adjacent to the sand particles and less mortar can act as lubrication to improve flowability. Hung et al. (2020) indicated that silica sand had the largest particle size compared to other UHPC ingredients, causing an increase in the volume of free water between the aggregate grains (Arora et al. 2018), thereby improving the flowability of fresh UHPC (Kamal et al. 2019).

Figure 5.4 Optical microscopes of various types of sand with a magnification of 50 (Yang et al. 2009): (a) Silica sand; (b) Type I fine ordinary sand; (c) Type II fine ordinary sand; (d) Recycled glass cullet.

Table 5.2 Flow table test of UHPC with or without fibers

Sand type	*Fiber*	*Silica sand*	*Type I fine ordinary sand*	*Type II fine ordinary sand*	*Recycled glass cullet*
Flowability diameter (mm)	Without fibers	244	197	239	185
	With fibers	232	182	228	175

Source: Yang et al. 2009.

Lightweight aggregate is considered an effective internal curing material, as its water desorption is driven by capillary action (Lura et al. 2014; Meng et al. 2018). The water adsorption and desorption capability are closely related to the inner continuous porosity and particle size, which determines its effects on workability, mechanical properties, and durability (Jensen and Hansen 2002; Ma et al. 2019). Generally, pores above approximately 100 nm are helpful for the storage of internal curing water, while the water in pores smaller than 100 nm cannot be released into the hardened cement paste (Jensen and Lura 2006).

The workability of concrete is almost unaffected by the addition of lightweight aggregate due to the water adsorption and release process usually finished prior to setting. However, water is gradually released at a quick mixing speed, thus increasing the flow of paste (Ma et al. 2019). The slump flowability of UHPC mixtures without lightweight aggregate is increased initially, then decreases after hydration for 30 min. However, the slump flow increases from 225 to 290 mm as the age proceeds from 5 to 180 min with lightweight aggregate, as shown in Figure 5.5. This is attributed to a large amount of water being released during the first 3 h due to the water desorption of lightweight aggregate, which favors the flowability of UHPC mixtures (Shen et al. 2020). Moreover, the mini-slump flow is slightly increased from 275 to 290 mm with lightweight sand increases up to 75%, by volume of river sand (Meng and Khayat 2017).

5.2.3 Chemical admixtures

A high-range water-reducing admixture is generally incorporated into UHPC for the enhancement of workability of fresh mixtures. Compared with poly(naphthalene sulfonate), poly(melamine sulfonate), and acetone formaldehyde sulfite superplasticizer, polycarboxylate superplasticizer consisting of polycarboxylate backbone and poly(ethylene oxide) side chain works primarily by the adsorption of anchors in its main chain and steric hindrance of side chain, thus indicating the best dispersion capability (Ferrari et al. 2010; Yoshioka et al. 1997).

The dispersion efficiency of PCE in UHPC depends on the type and chemical structure. Methacrylic acid ester and allyl ether-based polycarboxylates (PCEs) are effective in dispersing cementitious particles in UHPC (Plank et al. 2009). Methacrylic acid ester–PCEs are mainly adsorbed on the surface of cement particles, while allyl ether–PCEs show a higher affinity on silica fume particles. Their combinations allow for decreasing PCE dosage by around 50%, due to the optimal adsorption and dispersion mode exhibited in the blended system (Plank et al. 2009). In general, slump diameters of paste increase with the incorporation of superplasticizers at relatively low dosages and achieve typical plateaus after the saturation dosages. When superplasticizers exceed their saturation dosage, complete surface coverage will be obtained, thus rendering stable dispersion capability.

There is an exponential relationship between the flowability of pastes and dosages (Li et al. 2017), depending on the raw materials and w/b ratio. For instance, a higher superplasticizer dosage (1.75%) is required for paste with a lower w/b ratio of 0.16 to achieve the same flow spread value of 180 mm, as illustrated in Figure 5.6 (Liu et al. 2017). The addition time of superplasticizer and its molecular structure, i.e., length of the side chains, also significantly affect the cement hydration and the formation of organomineral phases, especially for linear polyelectrolyte polymers (Flatt and Houst 2001). Superplasticizer is adsorbed on the surface of cement particles or the thin layer of hydration products formed in the prior few seconds. When it is added directly, however, its side chain is buried with hydration, causing

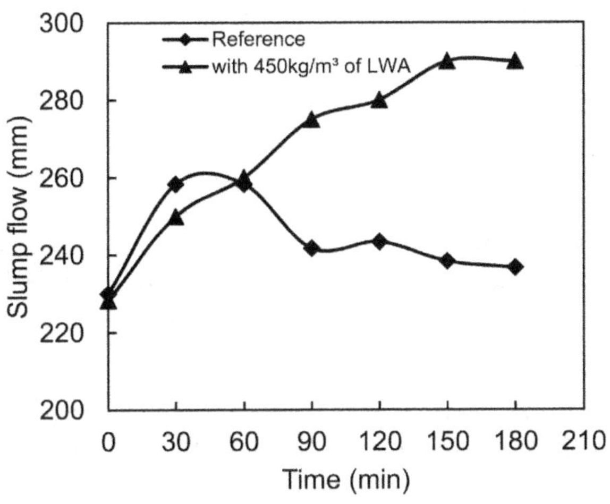

Figure 5.5 Effect of lightweight aggregate on slump flow of UHPC.

Source: Shen et al. 2020.

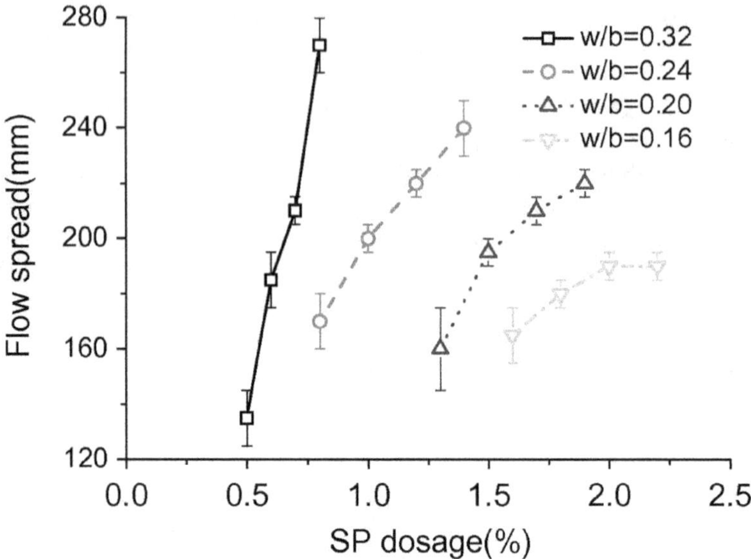

Figure 5.6 Effects of superplasticizer dosages on flow spread values of pastes with different w/b ratios.

Source: Liu et al. 2017.

a decrease in dispersion capability, as shown in Figure 5.7(a). In the case of stepwise addition, the second added superplasticizer is adsorbed on the formed hydrate products and less buried, presenting more effective dispersion (Figure 5.7(b)) (Ma et al. 2008).

5.2.4 Fibers

The incorporation of fibers enhances the ductility, impact resistance, tensile strength, and fatigue strength of hardened UHPC (Kang and Kim 2011; Yu et al. 2015) but increases the flow resistance (Rydval et al. 2016). The high-range water-reducing agent demanded for nonfibrous UHPC matrix ranges from 2% to 3% and changes from 4% to 8% for UHPC with 2% steel fibers (Wu et al. 2019).

Fiber type, shape, content, and aspect ratio affect the fresh properties of UHPC. Yu et al. (2014) found that fresh UHPC with microsteel fibers exhibited a lower slump flow value and a higher air content, but mixture with polypropylene fibers and wollastonite microfibers showed an adverse effect (Soliman and Nehdi 2012). Additionally, the polymer fibers, such as PVA, had smaller diameters and hydrophobic surfaces, magnifying their adverse

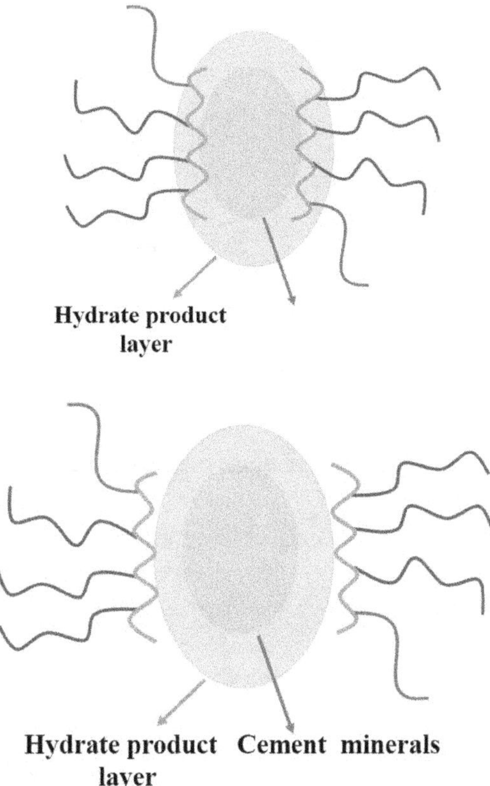

Figure 5.7 Illustration of adsorption of superplasticizers on cement particles with different addition patterns (modified based on Ma et al. 2008). (a) Direct addition. (b) Stepwise addition.

effects on the workability of fresh mixtures (Peyvandi et al. 2013). The increased fiber content decreases the flowability of UHPC. The flowability of UHPC with 1%, 2%, and 3% straight steel fibers gradually decreases by 14.9%, 25.6%, and 38.1%, respectively, as illustrated in Figure 5.8 (Wu et al. 2016). This can be attributed to the increased specific surface area and random distribution of fibers (Yu et al. 2014). Especially for fibers with a high aspect ratio, they tend to be agglomerated and decrease the flowability of concrete (Illguth et al. 2013). Li et al. (2016) stated that the incorporation of straight fibers with aspect ratios of 28 and 65 exhibited a similar influence on the spread flow of UHPC mixtures, when their volumes were 1% and 2%, respectively. However, a higher flow was obtained for UHPC with 3% fiber volume and smaller aspect ratio. Besides, the elongated,

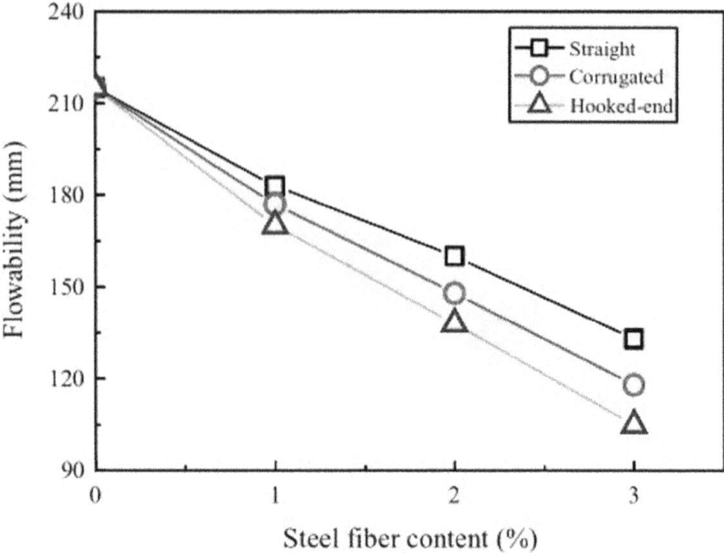

Figure 5.8 Effects of steel fiber content and shape on flowability of fresh UHPC.

Source: Wu et al. 2016.

hooked-ended, and copper-coated type steel fibers with a high volume content show higher mechanical interaction and interlock, which harmed the flowability. UHPC with hooked-end or corrugated steel fibers shows poorer flowability than that with straight fibers (Figure 5.8). The presence of deformed fibers increases the friction between fibers and aggregates and produces entangled patterns, thus hindering the flow of particles (Standard 2005; Yu et al. 2014).

5.2.5 Nanomaterials

Nanosized amorphous silica has a higher specific surface area and activity than conventional silica fume particles. Its addition causes a noticeable increase in strength development (Björnström et al. 2004) but has an adverse effect on the workability of Portland cement, depending on its size. Small-sized silica fume as fillers can displace some free water in void spaces, thus improving the workability of paste. In comparison, large agglomerates of precipitated silica with a size of 20–30 nm absorb more free water and have a negative effect on workable mixtures (Kong et al. 2013).

For UHPC, the slump flow of fresh UHPC decreases linearly with the addition of nanosilica. Compared with the reference mixture, the mixtures show a considerable decrease in the mini-slump flowability (19 mm) as the nano-SiO_2 content increases from 3 to 4 wt.%. The presence of nano-SiO_2 increases water adsorption, leading to less water available for lubrication (Muzenski et al. 2019). Ghafari et al. (2014) proposed that the 3 wt.% nano-SiO_2 was used for the achievement of the best performance in cement paste. The water requirement of normal consistency is not affected by nano-$CaCO_3$ due to the synergetic effect of liberated water and increased water demand induced by high surface area (Liu et al. 2012). Wu et al. (2016) found that the increase of both nano-SiO_2 and nano-$CaCO_3$ decreased the flowability of mixtures, which can be attributed to their different water adsorption abilities induced by particle size and specific area.

Nanomaterials with higher surface area can be adsorbed by more free water and superplasticizers, increasing the water demand. On the other hand, they can fill voids between cement particles to release some entrapped water, thus increasing the free water amount. The increased flowability of UHPC mixtures with individual or combined incorporation of nano-$CaCO_3$ and micro-$CaCO_3$ can be observed as a partial replacement for cement. The flowability values of mixtures with 2.5% micro-$CaCO_3$ and 5% nano-$CaCO_3$ are increased by 40%, while those with 2.5% nano-$CaCO_3$ and 5% micro-$CaCO_3$ show an improvement in flowability of 14.7%. This is because the lubrication and filler effects of fine $CaCO_3$ particles favor the release of entrapped water between coarser particles; thus more water available is devoted to the flowability of paste (Elkhadiri et al. 2002; Liu et al. 2012).

Unlike 0D nanoparticles, 1D fibers and 2D sheets act as reinforcing materials to bridge cracks. Thus, those nanomaterials with high aspect ratios and intrinsic strengths are preferred for use in UHPC. However, higher aspect ratios cause less lubrication effect and have adverse effects on the flowability of UHPC mixtures (Meng et al. 2018). In addition to the characteristics of materials, their content is also a key parameter affecting fresh properties, as shown in Figure 5.9. The 1D graphite nanoplatelets and carbon nanofibers play as lubrications among solid particles when their content is no more than 0.05%, which is beneficial to increasing flowability and reducing the high-range water reducer demand. However, they tend to absorb more water and high-range water reducer with more content, thus exhibiting an opposite effect on the workability of UHPC mixtures (Meng and Khayat 2016).

Carbon nanotubes can be categorized as single-walled nanotubes and multiwalled nanotubes, depending on their different structures. Carbon nanotubes are difficult to be dispersed in cementitious suspensions due to the high attractive van der Waals forces (Ko et al. 2010), thus decreasing the efficiency of carbon nanotubes in the mixtures. Collins et al. (2012) found that the mini-slump flow decreased with the increased multiwalled

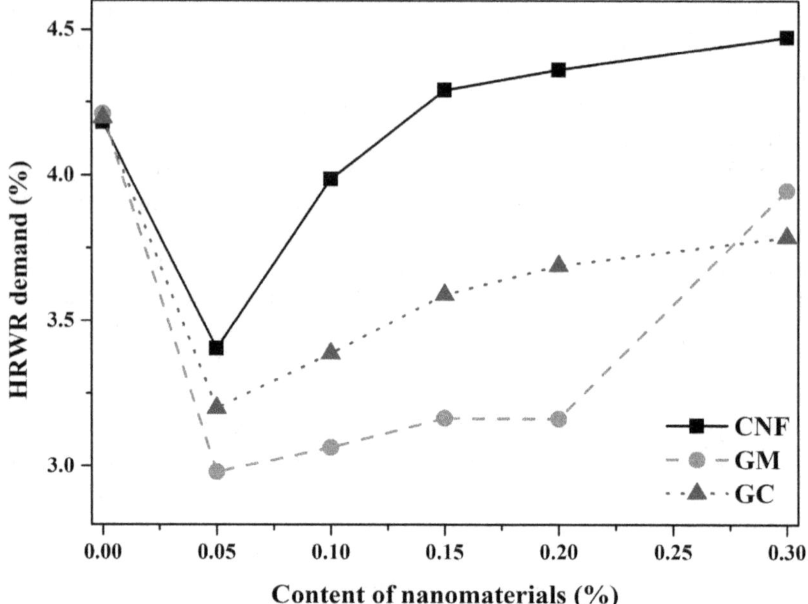

Figure 5.9 Effects of graphite nanoplatelets and carbon nanofibers on HRWR demand of mixtures with initial mini-slump of 280 ± 5 mm.

Source: Meng and Khayat 2016.

nanotube contents and exhibited slight variation with the changes in length-to-diameter ratio. For 2D graphene oxide sheet with a large surface area, it can adsorb water molecules originally contributing to the mixing process, thus increasing frictional resistance among the cement particles and the sheets. This indicates the lack of lubrication among particles and the increase of flow resistance, which is also identified as the root of poor workability (Swamy 1974). Compared with carbon nanotubes, cement paste with small amounts of graphene oxide sheets (0.02–0.05 wt.%) exhibited poorer workability (Gong et al. 2015).

5.2.6 Superabsorbent polymers

UHPC shows a high risk of autogenous shrinkage, leading to the generation of micro-cracking under restrained conditions and impairing its hardened properties (Ghafari et al. 2016; Lura et al. 2003). The superabsorbent polymer used for internal curing is an effective method to mitigate autogenous shrinkage by controlling the internal relative humidity caused by self-desiccation. However, superabsorbent polymer with high water adsorption capability decreases the slump flow (Schröfl et al. 2012) and increases the flow time of mixtures due to the decrease in the effective water content (Paiva et al. 2009).

The physical characteristics of superabsorbent polymer have a close correlation to the slump flow of paste, due to their effects on the absorption and desorption process. Superabsorbent polymer is incorporated into UHPC in a dry or a presaturated state. The mini-slump flow of UHPC increases from 275 to 305 mm as the superabsorbent polymer content increases from 0.2% to 0.6%. While the presaturated superabsorbent polymer exhibits desorption during the mixing process, UHPC without any superabsorbent polymer shows a higher flowability compared to UHPC with 0.4% dry superabsorbent polymer (Liu et al. 2020).

5.3 SLUMP RETENTION

Fresh concrete is well known to lose its workability with time. This phenomenon, called slump loss, is a critical feature of the quality of concrete, which involves chemical and physical processes (Chandra and Björnström 2002; Li et al. 2016). The retention abilities of UHPC can be greatly enhanced with a slight increase in water addition, indicating that slump retention is sensitive to the water-to-powder ratio (Li et al. 2017).

Slump loss is closely related to the superplasticizer type and dosage, due to the different adsorption behaviors on the surface of particles. When the superplasticizer achieves its saturated value, the particle surfaces are not adsorbed anymore, presenting the same flow-maintenance ability, as illustrated in Figure 5.10. Wang et al. (2012) reported that the incorporation

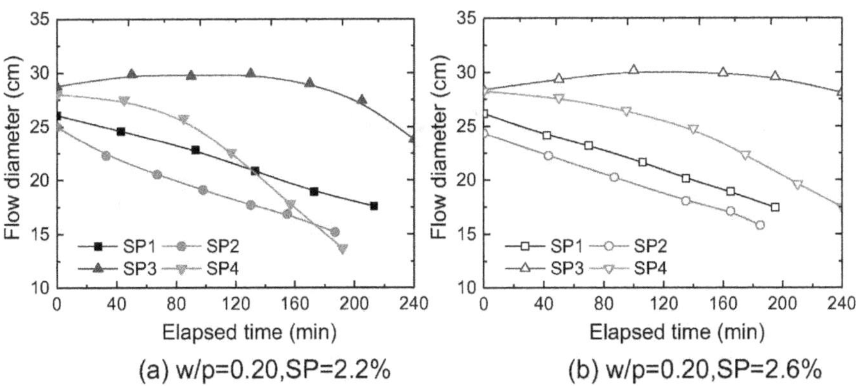

Figure 5.10 Effects of superplasticizer content and type on flow diameters of UHPC (Li et al. 2017). (a) UHPC with 2.2% superplasticizer. (b) UHPC with 2.6% superplasticizer.

of silica fume, GGBS, and limestone powder in UHPC favors less slump loss at 120 min. UHPC with 0.16 w/b ratio and 50% cementitious material, 10% silica fume, 20% GGBS, and 20% limestone powder at an appropriate superplasticizer dosage can achieve adequate workability for successful preparation.

5.4 V-FUNNEL TIME

V-funnel test is used to assess the filling ability and segregation resistance of fresh mixtures using the V-shaped apparatus, as shown in Figure 5.11. The V-funnel flow test can be carried out following ACI 238.1R (2008) and EFNARC recommendations (EFNARC 2020). The V-shape restricts the flow of mixture and increases the discharge time. It enables the evaluation of fluidity and viscosity of fresh UHPC through restricted spacing without blockage. A shorter time indicates a greater filling ability (BSEN12350-9 2010).

V-funnel time is affected by the aggregate and fiber contents. The vertical flow velocity of the fresh UHPC increases with the silica sand-to-cement ratio, as illustrated in Figure 5.12 (Hung et al. 2020). Higher silica sand-to-cement ratio increases the amount of free water but decreases the interparticle friction. The mini V-funnel flow time of UHPC with 75% lightweight sand decreases from 40 to 12 s, indicating a substantial drop in plastic viscosity (Meng and Khayat 2017).

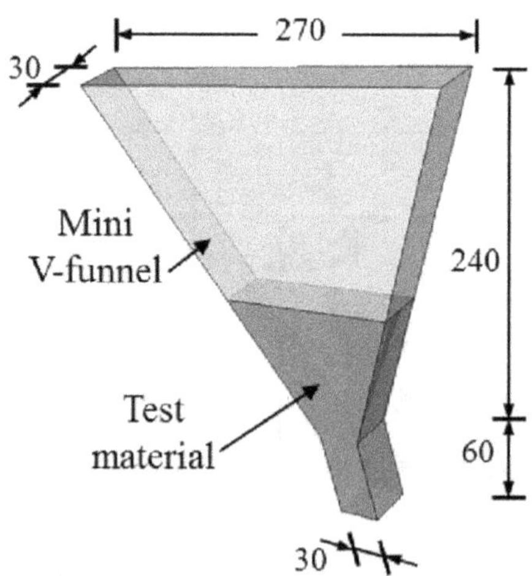

Figure 5.11 V-funnel apparatus used for workability testing of UHPC.

Source: Meng and Khayat 2017.

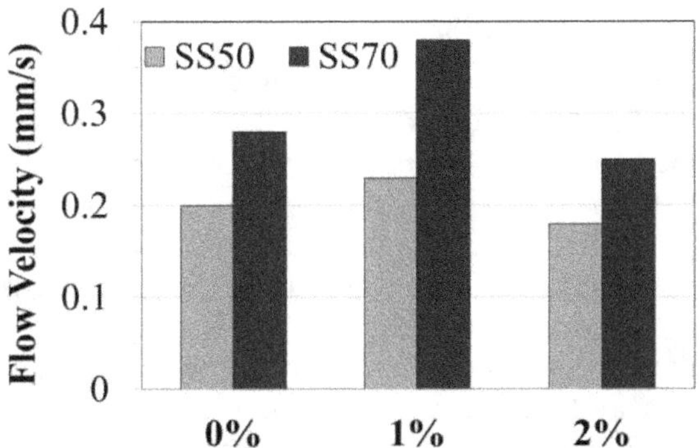

Figure 5.12 Effects of silica sand and fiber contents on vertical flow velocities of UHPC mixtures.

Source: Hung et al. 2020.

The mini V-funnel flow time gives information on the passing ability of paste and affects the plastic viscosity of paste (Meng et al. 2018). It also depends on the fiber content and cement compositions (De Schutter et al. 2008). The optimal V-funnel flow time of nonfibrous UHPC mixture is 46 ± 2 s, corresponding to the optimal plastic viscosity of 53 ± 3 Pa·s of UHPC with 2% steel fibers (Meng et al. 2018). Huang et al. (2020) found that the incorporation of 1% hooked-end steel fibers in UHPC with a silica sand-to-cement ratio of 0.7 enhanced the vertical flow velocity by 36%, while a slight increase was observed for mixture with 0.5 silica sand-to-cement ratio. The flow velocity of UHPC with 2% hooked-end steel fibers was decreased by about 10% in comparison to the nonfibrous matrix. The phenomenon is attributed to the counteracting effect between the increasing interlocking effect of steel fibers and the benefit of increased gravity load of UHPC for the vertical flow velocity. Besides, the presence of PVA fiber increases the vertical flow resistance, thus rendering a longer flow time (Meng et al. 2018).

The incorporation of dry superabsorbent polymer and extra water has a significant influence on the mini V-funnel flow time, depending on the size and state of superabsorbent polymer (Mechtcherine et al. 2015). The V-funnel time of UHPC increases from 20 to 22 s when 0.2% superabsorbent polymer with a mean particle size of 471.3 μm (S1L—superabsorbent polymer) is employed, but it decreases from 20 to 12 s for the mixture with 95.1 μm dry superabsorbent polymer (S1S—superabsorbent polymer). Nevertheless, the addition of preabsorbed superabsorbent polymer shows no serious effect on the V-funnel time compared to that with dry superabsorbent polymer (Liu et al. 2020).

5.5 TEMPERATURE AND AIR CONTENT

The temperature of fresh UHPC mixtures usually ranges from 18 to 29°C, depending on the materials used, mixing time, placing or pouring methods, and ambient conditions (Ingo et al. 2004; Kazemi and Lubell 2012). When UHPC is pumped to higher floors through long pipes, the temperature of fresh UHPC at the ground level is 23°C but increases to 40°C at the end of the nozzle (Ingo et al. 2004). This is attributed to the increased frictional heat between the UHPC and pipe wall during the pumping process.

The reported air content in UHPC mixtures varies from 0.3% to 5.4% by mixture volume, depending on mixture proportion (Wille et al. 2011). Higher w/b ratio and superplasticizer dosage increase the air content in UHPC mixtures. Furthermore, the total air content is highly dependent on the mixer type of concrete (Ingo et al. 2004). It is observed that UHPC paste prepared using laboratory mixers with a higher mixing speed shows a sticky consistency, resulting in an increased air content of 4.3%. For the same mixture composition and proportions, a relatively lower air content of 3.2% is achieved with the use of ring-type mixers installed at precast plants that apply shear forces (Ingo et al. 2004).

Ingo et al. (2004) also reported an air content below 1% using a vacuum accessory with a pressure of 50 mbar. In addition, the placing method significantly affects the air content of UHPC. For instance, the air content of mixture reduces from 2.9% to 1.3% when using a spiral pump (Ingo et al. 2004). Furthermore, the delayed addition of superplasticizer reduces the air content from 2.5% to 1% due to the decrease in viscosity of UHPC mixture (Tue et al. 2008). With the improved spread flow and enhanced properties of UHPC mixtures, the threshold air content value is less than 2% by volume (Wille et al. 2011).

5.6 SETTING TIME

Setting of concrete refers to a gradual evolution of rigidity of cementitious mixture after the addition of mixing water (ASTMC125 2018). Setting time is defined as the elapsed time from the addition of mixing water to the generation of mixture with a specified degree of rigidity. It is determined by the times required for the mortar to reach a specified value of resistance to penetration in standard (ASTM 2008). Briefly, the testing process includes the preparation of sample, sample compaction in a standard container, and then measurement of the force required by a penetrated needle into the mortar for 25 ± 2 mm depth. Then, the penetration resistances of 3.5 and 27.6 MPa are considered the initial and final setting times, respectively.

Setting of concrete is usually described as a percolation process in forming hydration products to connect the isolated or weakly bound particles (Jiang et al. 1995; Jiang et al. 1996). This process is greatly influenced by the lightweight sand content, superplasticizer dosage and molecular structure, and nanomaterial content (Bentz 2007; Bentz et al. 1999; Felekoğlu and Sarıkahya 2008). When the lightweight sand substitution, by weight of river sand, increases to 75%, both the initial and final setting times of UHPC mixture increase from 5.6 to 8.2 h and 9.0 to 16.2 h, respectively (Meng and Khayat 2017). UHPC paste with 1.2% superplasticizer (SP1) exhibits the longest initial (7.8 h) and final setting times (11.2 h), as observed in Figure 5.13. In the presence of 1.2% superplasticizer (SP2), the initial and final setting times are 4.25 and 6.5 h, respectively. A linear correlation between the final setting time and the time of maximum heat flow rate is obtained by Li et al. (2017).

A skeleton starts to develop with the processes of hydration reactions and setting (Nehdi and Soliman 2011). The addition of nano-SiO_2 particles accelerates cement hydration during both the acceleration and deceleration

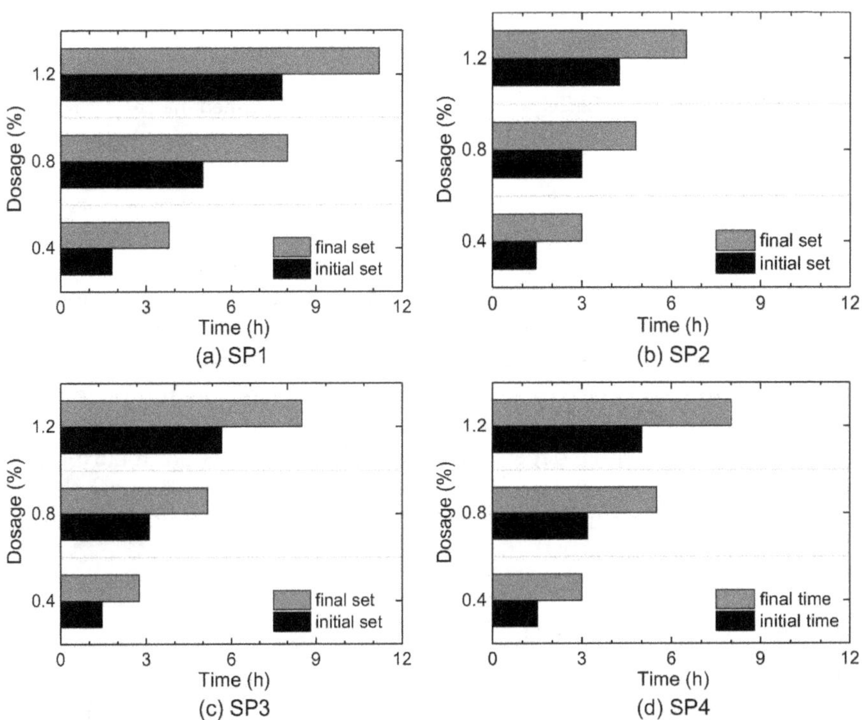

Figure 5.13 Effects of superplasticizer type and dosage on setting time of nonfibrous UHPC paste.

Source: Li et al. 2017.

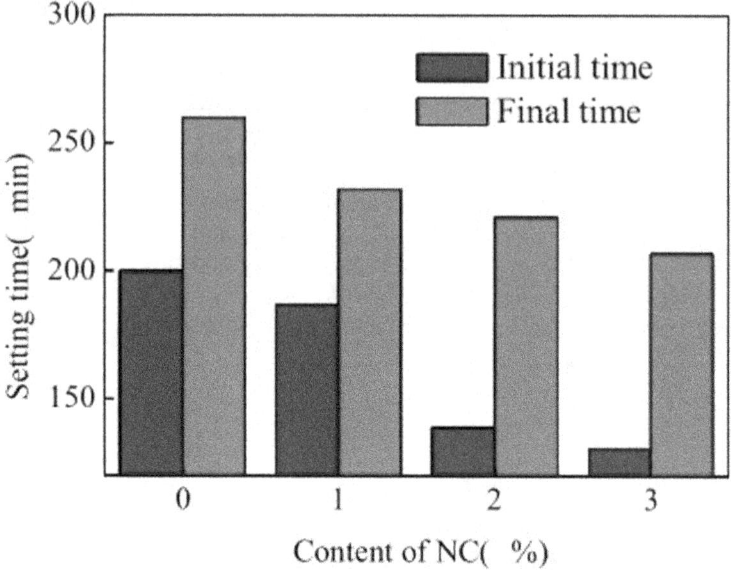

Figure 5.14 Effect of nano-$CaCO_3$ content on setting time of cement paste.

Source: Liu et al. 2012.

periods due to the consumption of $Ca(OH)_2$ and the supplement of nucleation sites (Rong et al. 2015). Thus, nano-SiO_2 helps shorten the initial and final times. In the presence of nano-$CaCO_3$, the concentrations of Ca^{2+} and OH^- produced by cement hydration decrease due to their adsorption on the surface of nano-$CaCO_3$, leading to an accelerated hydration reaction (Camiletti et al. 2013). Therefore, it shortens the setting time of UHPC, and the reduction degree depends on nano-$CaCO_3$ content and curing temperature of mixtures. The addition of 2% nano-$CaCO_3$ shortens the initial and final setting times by 61 and 39 min, respectively, as illustrated in Figure 5.14 (Liu et al. 2012).

5.7 RHEOLOGICAL PROPERTIES

Rheology refers to the science of flow and deformation of matter. It can be used to describe the workability of fresh cement-based materials quantitatively using fundamental rheological parameters, such as yield stress and plastic viscosity (Marchon et al. 2018). Yield stress is defined as the shear stress of the initial flow, which determines the stability of the suspension (Roussel 2006). Plastic viscosity, as an important parameter for robustness, describes the resistance to flow, which increases with shear rate (Wallevik and Wallevik 2011).

Fresh cement-based materials also exhibit thixotropy, which refers to a continuous decrease of apparent viscosity with time under shear force and a subsequent recovery at rest (Mewis 1979). Thixotropy describes a fresh-state structural build-up of concrete and can be quantified as an increase in yield stress over time. Rheological properties affect the mechanical properties and durability of UHPC (Deeb et al. 2012; Meng and Khayat 2017; Soliman and Tagnit-Hamou 2016). As a result, a clear understanding of rheology is essential for the preparation of UHPC to achieve excellent hardened properties.

5.7.1 Rheological behaviors of UHPC

UHPC and conventional concrete show various rheological behaviors due to their differences in raw materials and mixture proportions, as listed in Table 5.3 (Choi et al. 2016; Hu and de Larrard 1996; Mardani-Aghabaglou et al. 2013; Wallevik 2003). The Bingham linear model and nonlinear models, such as the modified Bingham and Herschel-Bulkley models, are generally applied to fit the shear stress–shear rate curves of cement-based materials (Atzeni et al. 1985; Bingham 2007; Yahia and Khayat 2003).

Four basic rheometers can be used to determine the rheological properties of cementitious materials with a low w/b ratio, including parallel disc plates, concentric cylinders, coaxial, and two-point rheometers (Banfill et al. 2020). Dils et al. (2013) found that UHPC showed shear-thinning behavior. There is a transition from shear-thinning to shear-thickening behavior when the shear rate rises to 2 s^{-1} (Artelt and Garcia 2008). This is attributed to the changed predominant force between particles with the increased shear rate (Coussot and Ancey 1999; Feys et al. 2009). Additionally, the rheological properties of UHPC are affected by fiber content and characteristics. When

Table 5.3 Typical yield stress and plastic viscosity for different concrete mixtures

Rheological parameter	*Yield stress (Pa)*	*Plastic viscosity (Pa·s)*	*References*
Conventional concrete	500–2000	50–100	Wallevik (2003)
Self-compacting concrete	5–50	100–400	Mardani-Aghabaglou et al. (2013)
High-performance concrete	50–2000	50–550	Hu and de Larrard (1996)
UHPC	10–100	20–200	Choi et al. (2016); Meng and Khayat (2017)

fiber content is over a critical concentration, the rheological parameters are significantly increased (Martinie et al. 2010). A linear variation in the relation between the shear stress and strain rate is observed for UHPC (Arora et al. 2018; Choi et al. 2016; Meng and Khayat 2017). The linear and nonlinear flow behaviors are related to the applied shear rate ranges for materials during the testing procedure and the constituents used in the studies (Assaad et al. 2003).

Some simple workability tests can also be applied to evaluate yield stress and plastic viscosity of UHPC (Choi et al. 2016). A simple equation is set to express the correlation between the variation of the spread diameter and time obtained from computational fluid dynamic analysis and rheological properties. Tregger et al. (2008) revealed the correlations between the final spread diameter and yield stress using a theoretical estimation method.

5.7.2 Cementitious materials and fillers

The cementitious materials and fillers act as significant parts of UHPC system to improve its particle packing, properties, and sustainability. The type and fineness of cement and the relative proportions of power constituents greatly affect the particle packing density (Nanthagopalan et al. 2008), reactive chemical content (Vikan et al. 2007), and interactions with chemical admixtures (Jiao et al. 2017; Mikanovic and Jolicoeur 2008).

Generally, high-fineness cement particles tend to form settlements or agglomerates, which directly affect the rheological behavior of paste (Park et al. 2005) and the hardened properties (Qing et al. 2007). Chen and Kwan (2012) found that the addition of 10%–20% superfine cement significantly increased the packing density of cementitious particles and the water film thickness of paste. Therefore, the rheological properties of paste and strength of hardened paste were improved. From Figure 5.15, it can be seen that superfine cement increased the yield stress and apparent viscosity of the mixtures with a w/b ratio greater than 0.24, while this is the opposite for the w/b ratio between 0.18 and 0.22 (Chen and Kwan 2012). The flow resistance is an exponential function of cement fineness in the case of the same cement clinker (Vikan et al. 2007). However, as cement composition varies, a poor correlation was found.

The shear-induced behavior of UHPC matrix depends on the silica fume content. The shear-thickening behavior changes to a linear Bingham behavior at higher silica fume content, as illustrated in Figure 5.16 (Wu et al. 2019). This can be attributed to the fact that the lubricating effects of fine and spherical silica fume particles decrease friction among particles (Khayat et al. 2008; Yahia 2011). This is in agreement with the results reported by Yahia (2011). For the nonfibrous UHPC matrix, its yield stress increases from 34 to 41 Pa as the silica fume content increases to 25% (Wu et al. 2019). However, these rheological parameters decrease with the

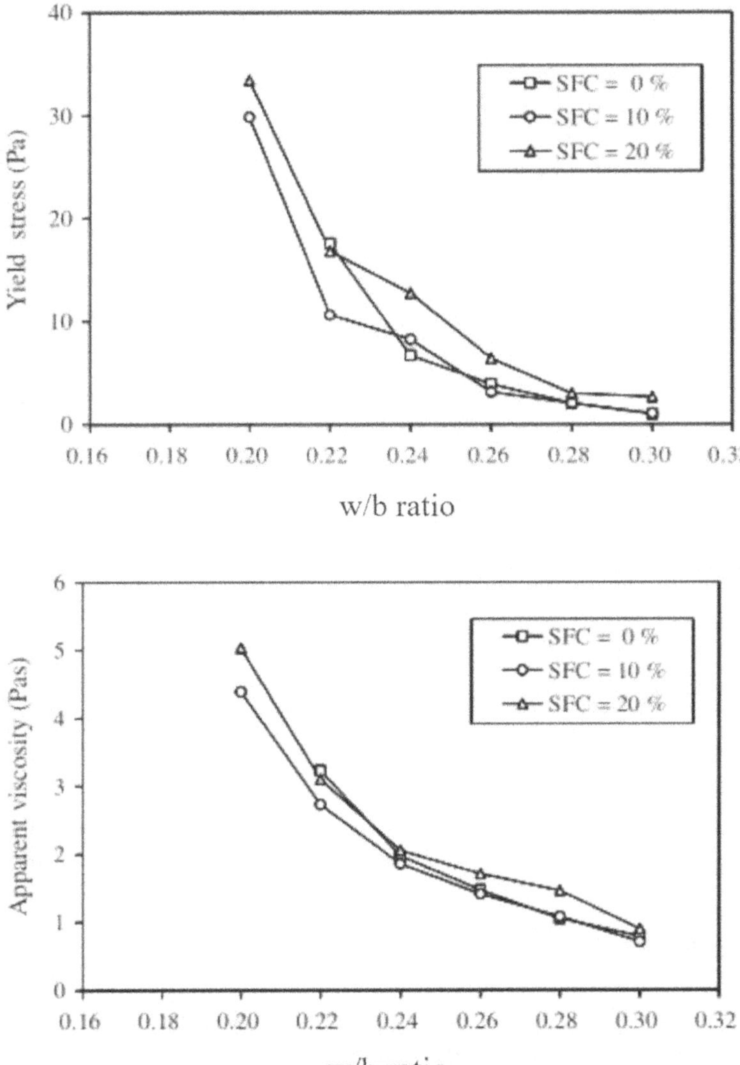

Figure 5.15 Effect of w/b ratio on rheological parameters of nonfibrous UHPC paste (Chen and Kwan 2012). (a) Yield stress. (b) Apparent viscosity.

increased replacement of fly ash due to its lubricant effects between particles (Nanthagopalan et al. 2008). With the increase of silica fume content, the plastic viscosity increases steeply and then decreases. Conversely, the yield stress shows the opposite behavior. The increased fly ash up to 30% leads

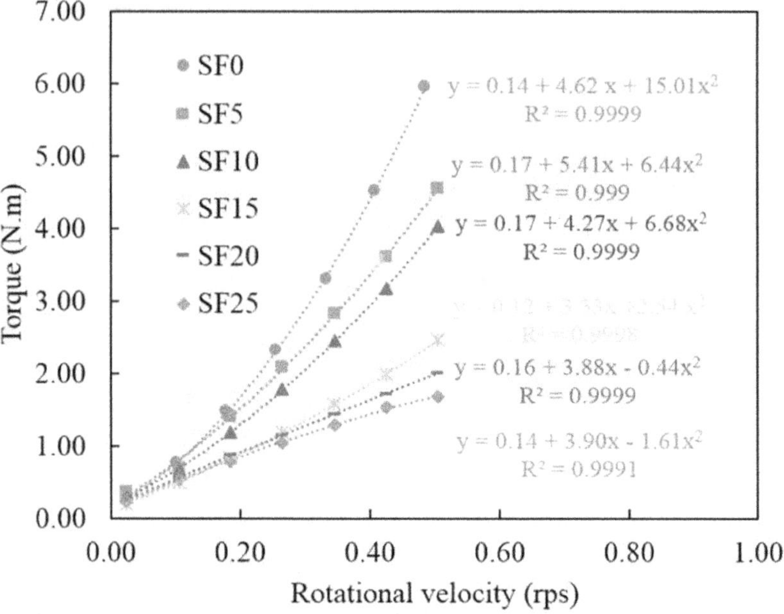

Figure 5.16 Effect of silica fume content on torque-velocity curves of nonfibrous UHPC matrix.

Source: Wu et al. 2019.

to a reduction in yield stress. However, a slight increase in the yield stress can be found with the increased level to 50% (Laskar and Talukdar. 2008).

5.7.3 Aggregates

Aggregates have a crucial effect on the rheological properties of UHPC. To obtain lower rheological properties, aggregates with larger maximum particle sizes should be employed (Westerholm et al. 2008). This is because small sand particles can increase the paste volume required to achieve desired performance due to a larger specific surface area.

Mortar made with fine aggregates from crushed bedrock exhibits high yield stress and plastic viscosity due to the higher water adsorption (Westerholm et al. 2008). Banfill (1994) reported that the yield stress of mortar increased with an increased fineness of fine aggregate. Besides, the rheological properties of paste increased with an increased level of fine aggregate, due to the increased interparticle friction (Banfill 1999). A lower cementitious materials content in UHPC mixture with the same water content decreases the internal frictions between filler materials at the microscale. This results in a lower plastic viscosity of the matrix at the macroscale (Hung et al. 2020). The slump flow of UHPC mixture increases as the lightweight aggregate content increases, and an increased V-funnel time indicates a decrease in plastic viscosity (Meng and Khayat 2017).

5.7.4 Chemical admixtures

Superplasticizers are necessary in UHPC system to enhance its flowability due to its extremely low w/b ratio and high binder content. PCE superplasticizer is considered the most effective for UHPC, due to its low dosage, high dispersion, and high water-reduction capability (Schröfl et al. 2012). In some cases, viscosity-modifying admixture is also used in UHPC to enhance fiber dispersion and orientation by adjusting the rheological properties (Khayat 1998; Lachemi et al. 2004). The combination of superplasticizer and viscosity modifying admixtures favors the improvement of workability and enhancement of the resistance to segregation of concrete. However, incompatibility between binder system and chemical admixtures should be focused on during the production and placement of UHPC (Hanehara and Yamada 1999).

5.7.4.1 Water-reducing agents

High superplasticizer dosage is normally used in UHPC to enhance its workability. However, the higher unadsorbed PCE concentration enhances the viscosity of the interstitial solution of paste (Liu et al. 2017). The higher viscosity of cement paste with a very low w/b ratio is due to its higher binder particle content, less adsorption degree of PCE, higher viscosity of the interstitial solution, and lower water film thickness (Krieger and Dougherty 1993; Struble and Sun 1995). Additionally, the shear-thickening behavior of pastes is pronounced with the increase in PCE dosage and a decrease in w/b ratio (Cyr et al. 2000; Kamibayashi et al. 2008). Yahia (2011) noted that those unabsorbed segments of superplasticizer molecules interacted with each other, thus causing the shear-thickening behavior.

Such superplasticizers used in UHPC are comb copolymers with long side chains, and their molecular structures are easily entangled under shearing forces. On the other hand, the shear forces can overcome the repulsive forces between particles under a certain shear rate and gather particles to form hydroclusters, presenting a difficulty in flowing for paste. Thus, it shows a positive correlation between the viscosity of paste and shear rate (Wagner and Brady 2009). It should be mentioned that the stepwise addition of PCEs can decrease the viscosity of paste, which is related to the cement activity and the adsorption behaviors of superplasticizers. Generally, the higher the cement reactivity is, the more effective the stepwise addition is (Ma et al. 2008).

5.7.4.2 Viscosity-modifying admixtures

In some cases, viscosity-modifying admixture is incorporated in UHPC to adjust the plastic viscosity of mixture. The increased plastic viscosity favors fiber dispersion. However, it is not beneficial for mechanical properties due to the introduction of air voids (Tosun-Felekoğlu et al. 2014). The effect of viscosity modifying admixture on the rheological properties depends on its dosage and shear rate used during the test. Meng et al. (2017) evaluated the effect of 0%–2.0% viscosity modifying admixture, by mass of binder, on the rheological properties of UHPC. It is found that the plastic viscosity and yield stress increased with the increase in viscosity-modifying admixture content. The presence of polysaccharide gum increased the viscosity and yield stress of cement paste, which is highlighted at low shear rates (Ghio et al. 1994). A shear-thinning behavior at low shear rates and shear thickening behavior at high shear rates can be obtained, due to the intermolecular cross-linking of admixture molecules at low shear and the increased repulsive forces between interparticles at a high shear rate (Bouras et al. 2012).

5.7.5 Fibers

The excellent ductility and impact resistance of UHPC can be attributed to the use of proper fibers. However, the incorporation of steel fibers increases yield stress and plastic viscosity of UHPC due to cohesive and anchoring forces between fibers and grains and interlock of fibers (Martinie et al. 2010; Wu et al. 2019). Rigid and flexible fibers are commonly used in UHPC. The presence of rigid fiber decreases the packing density and increases the yield stress of paste. The hooked fibers with higher anchorage and interlock can significantly increase the yield stress (Li et al. 2016), while flexible fibers contribute to the increase of packing density and viscosity. The rigid fibers deform easily to form entanglement. For instance, they can be entangled into an S-shaped structure to further block the movement of particles (Yamanoi and Maia 2010). Schleiting et al. (2020) reported that the workability of UHPC made with hooked-ends and high aspect ratio fibers further decreased, especially at a fiber volume greater than 1%. The plastic viscosity of fresh concrete containing circular fibers was significantly increased compared to that with straight fibers, as shown in Figure 5.17.

Compared to steel fiber, PVA fiber with hydrophilic surfaces and small diameter shows negative effects on workability of fresh paste, presenting higher yield stress and viscosity (Peyvandi et al. 2013). Meng et al. (2018) reported that the viscosity of UHPC increased with fiber volume, and the phenomenon is related to fiber type and shape. The plastic viscosity of fresh mixture with 0.5% PVA fibers was increased by 25%. UHPC with

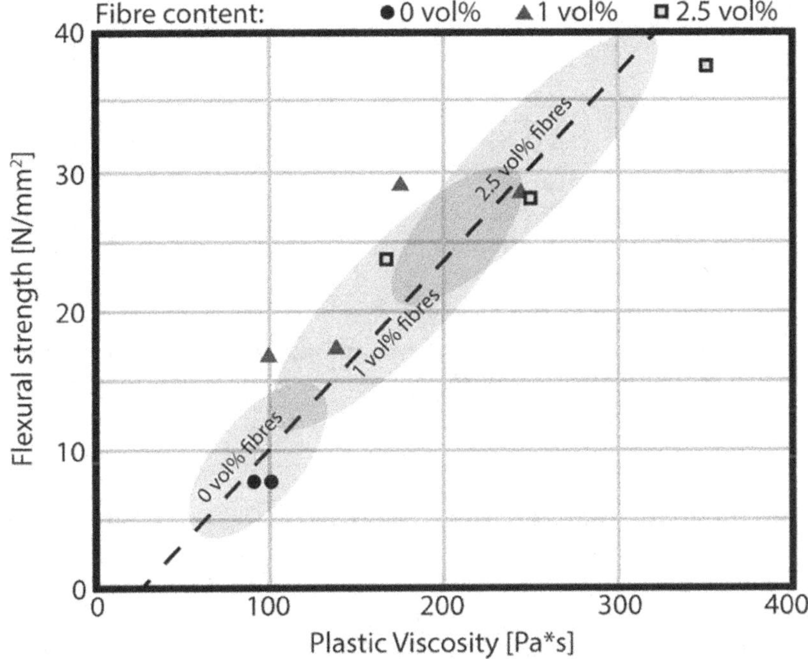

Figure 5.17 Effect of fiber content on plastic viscosity of UHPC.

Source: Schleiting et al. 2020.

2% hooked fibers showed 45% higher plastic viscosity than that with 2% straight steel fibers.

Higher fiber volume fractions decrease the packing density and increase the friction among fibers and interactions with solid materials. Therefore, cementitious materials with greater fiber volume generally exhibit higher yield stress and viscosity. Moreover, it is difficult to achieve a uniform fiber distribution (Kamal et al. 2019).

5.7.6 Nanomaterials

The effect of nanomaterials on rheological behaviors of UHPC depends on their dispersion states and contents. The addition of nanomaterials increases yield stress and viscosity due to water absorption capability and nanoparticle agglomeration. However, the voids between cement particles can be filled by nanoparticles, resulting in an increased particle packing density. This tends to decrease the yield stress and viscosity of UHPC.

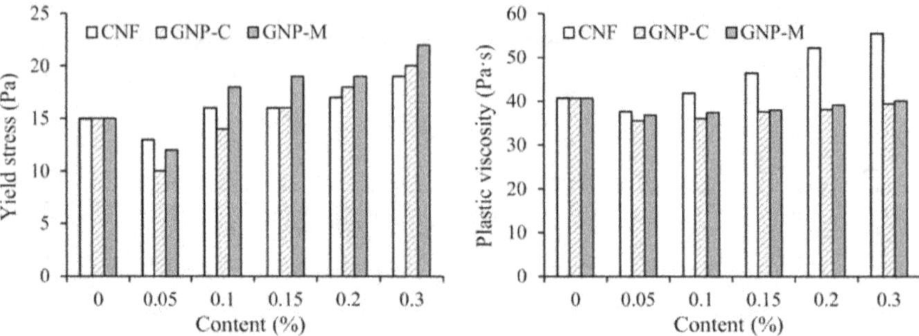

Figure 5.18 Effects of various nanomaterial types and contents on yield stress and plastic viscosity of UHPC mixtures.

Source: Meng et al. 2018.

The presence of nano-SiO_2 with a higher specific surface greatly increases the torque, yield stress, and plastic viscosity of mortars, which is obvious in its content (Berra et al. 2012; Senff et al. 2012). This is because the incorporation of nanoparticles decreases the amount of lubricating water available within the interparticle voids and increases the agglomeration and friction of cementitious particles (Senff et al. 2009). However, Collepardi et al. (2020) found that the self-compacting concrete mixture was more cohesive in the presence of nano-SiO_2, and the bleeding water and segregation were reduced. This suggests that the rheological behavior of fresh concrete with nano-SiO_2 was improved.

Typical effects of various nanomaterials and content on the yield stress and plastic viscosity of UHPC mixtures are shown in Figure 5.18. The yield stress and plastic viscosity of UHPC mixtures first decrease and then increase with the increase of nanomaterial contents (Meng et al. 2018). The lowest yield stress and plastic viscosity are found in UHPC mixtures with 0.05% nanomaterials. The observed different rheological behaviors are related to the physical characteristics of nanomaterials. The plastic viscosity of UHPC mixtures with carbon nanofibers increases from 40 to 56 Pa·s with its content increasing from 0% to 0.3%. Carbon nanofiber is adsorbed on the surface of cement particles via intermolecular forces, which partially weakens the steric hindrance and electrostatic repulsion of superplasticizer (Wille and Loh 2010). This indicates that the use of greater carbon nanofiber content accelerates the formation of the agglomeration structure of cement particles and highlights the reagglomeration action of carbon nanofiber, thus increasing the viscosity. For graphite nanoplatelets-C and graphite nanoplatelets-M, their additions increase the packing density of the cementitious materials, presenting lower plastic viscosity (Khayat 1999).

The effects of unique 2D graphene oxide sheet on the rheology of UHPC mixtures relate to its geometry. The higher viscosity of fresh pastes corresponds to the suspensions with large-scale agglomerates (Luo et al. 2015). The larger-sized graphene oxide sheet tends to intermesh, which leads to the formation of agglomerate structures and higher frictional resistance. Therefore, viscosity increases with the size of graphene oxide sheet. It should be noted that friction forces between particles are dominant for nanomaterials with different aspect ratios (Swamy 1974).

5.7.7 Superabsorbent polymers

Both adsorption and desorption processes of superabsorbent polymer affect the rheological characteristics of the material (Assaad et al. 2003; Secrieru et al. 2016), which are determined by its size and content. The initial plastic viscosity decreases from 12 to 2 Pa·s when the S1S superabsorbent polymer content increases from 0% to 0.6%. However, under 0.2% content, the addition of S1L superabsorbent polymer exhibits a higher plastic viscosity. No significant difference in the plastic viscosity is observed for UHPC with S1S and S1L superabsorbent polymer contents greater than 0.2% (Liu et al. 2020).

Paiva et al. (2009) proposed that the water absorption of superabsorbent polymer increased with the increased particle size. Typical effects of size and state of superabsorbent polymer on the rheological properties of UHPC are presented in Figure 5.19. The initial yield stresses of mixtures with dry and presaturated superabsorbent polymers (S1L- and S1S-0.4w) were similar due to their rapid water absorption and desorption kinetics before reaching a balance condition in a filtration solution, as illustrated in Figure 5.19(a).

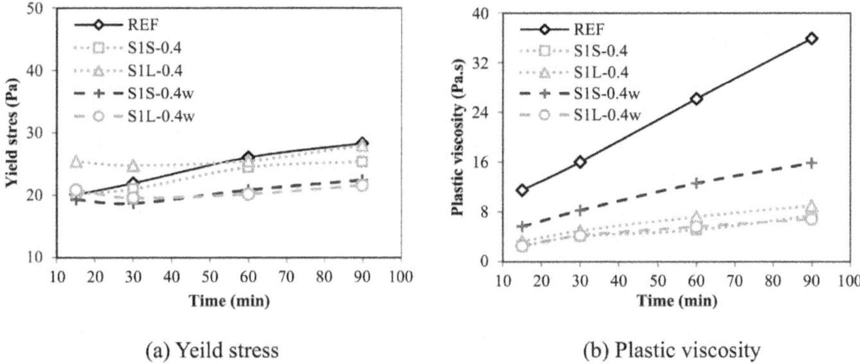

Figure 5.19 Effects of size and state of superabsorbent polymer on rheological properties of UHPC mixtures (Liu et al. 2020). (a) Yield stress. (b) Plastic viscosity.

A variation in the plastic viscosity of UHPC with dry and presaturated superabsorbent polymer can be obtained in Figure 5.19(b). The viscosity of UHPC with S1S-0.4w was higher than that with the dry S1S-0.4, while no obvious difference was observed in the presence of S1L-superabsorbent polymer (Liu et al. 2020). The rheology of UHPC is also affected by the charge density of superabsorbent polymer, due to the effect on adsorption and desorption processes (Zhong et al. 2019). Superabsorbent polymer with high anionic groups exhibits rapid adsorption and then desorption process, but no desorption is observed in the presence of superabsorbent polymer with nonionic groups or both anionic and cationic groups in cement filtrate.

5.8 SUMMARY

This chapter discusses the effect of different factors, including characteristics of cementitious materials, chemical admixture contents and types, fibers, nanomaterials, and superabsorbent polymers, on the fresh properties of UHPC mixtures. The involved fresh properties include mini-slump flow, slump retention, V-funnel time, temperature, air content, setting time, and rheological properties. Based on the above analysis, the following conclusions can be drawn:

(1) The small quantity replacement of cement by superfine cement can enhance the packing density of cement by filling voids, thus decreasing the amount of mixing water. It contributes to the enhancement of the slump flow and rheology of UHPC.
(2) Aggregates act as lubrication and their gradations directly affect the packing density of UHPC. Aggregates with high surface roughness produce large interparticle frictional forces during mixing and compaction, thus decreasing the workability. While the presence of rounded and smooth aggregate is beneficial to the preparation of well-workable concrete, the flowability, initial setting time, and final setting time of UHPC increase with lightweight aggregate content, but V-funnel time decreases, which also indicates a drop in plastic viscosity.
(3) Water-reducing agents and viscosity-modifying admixtures are generally incorporated in UHPC. The former ensures the required flowability and low yield stress, while viscosity-modifying admixtures are used to modify the viscosity of mortar and improve fiber dispersion. Superplasticizers with high dispersion capability or with stepwise addition sequence are more effective for the production of flowable mixtures. It should be mentioned that compatibility issues between the binder system and the chemical admixtures can occur during the production and placement of UHPC.
(4) Fibers cause an increase in the flow resistance of concrete mixtures and decrease the workability of UHPC, depending on their contents,

shapes, and types. The presence of deformed fibers increases the friction between fibers and aggregates and produces entangled patterns, thus hindering the flow of particles. Fibers with hooked ends and a higher aspect ratio weaken the workability, especially in the case of fiber content greater than 1%. The plastic viscosity of fresh concrete containing circular fibers increases significantly compared to that with straight fibers.

(5) An increase in slump flow and decreases in yield stress and viscosity are found in UHPC with optimal nanomaterial. However, more addition contributes to the formation of agglomerate structures and high frictional resistance, leading to higher yield stress and viscosity.

(6) The effects of superabsorbent polymers on fresh properties of UHPC mixtures relate to their adsorption and desorption capability. Superabsorbent polymers with higher water adsorption correspond to lower flowability and higher viscosity of mixtures. A decreased mini V-funnel flow time can be found with the increase of dry superabsorbent polymer content, while the addition of preabsorbed superabsorbent polymer has a limited effect on the V-funnel time.

REFERENCES

ACI Committee 238 (2008), *Report on Measurement of Workability and Rheology of Fresh Concrete, ACI 238.1R–08*, American Concrete Institute, 70.

Aldahdooh, M. A. A., Muhamad Bunnori, N., Megat Johari, M. A. (2013). Development of green ultra-high performance fiber reinforced concrete containing ultrafine palm oil fuel ash. *Constr. Build. Mater.* 48, 379–389.

Aldahdooh, M. A. A., Muhamad Bunnori, N., Megat Johari, M. A. (2013). Evaluation of ultra-high-performance-fiber reinforced concrete binder content using the response surface method. *Mater. Des.* 52, 957–965.

Arora, A., Aguayo, M., Hansen, H., Castro, C., Federspiel, E., Mobasher, B., Neithalath, N. (2018). Microstructural packing- and rheology-based binder selection and characterization for Ultra-high Performance Concrete (UHPC). *Cem. Concr. Res* 103, 179–190.

Artelt, C., Garcia, E. (2008). Impact of superplasticizer concentration and of ultra-fine particles on the rheological behaviour of dense mortar suspensions. *Cem. Concr. Res* 38(5), 633–642.

Assaad, J., Khayat, K. H., Mesbah, H. (2003). Assessment of thixotropy of flowable and self-consolidating concrete. *ACI Mater. J.* 100(2), 99–107.

ASTM C. (2008). *Standard test method for time of setting of concrete mixtures by penetration resistance.* ASTM International, West Conshohocken, PA.

ASTM C125. (2018). *Standard terminology relating to concrete and concrete aggregates.* ASTM International, West Conshohocken, PA.

ASTM C1437. (2020). *Standard test method for flow of hydraulic cement mortar.* ASTM International, West Conshohocken, PA.

ASTM C143-C143M (2012). *Standard test method for slump of hydraulic cement concrete*. ASTM International, West Conshohocken, PA.

ASTM C230/230M (2021). *Standard specification for flow table for use in tests of hydraulic cement*. ASTM International, West Conshohocken, PA.

Atzeni, C., Massidda, L., and Sanna, U. (1985). Comparison between rheological models for portland cement pastes. *Cem. Concr. Res.* 15(3), 511–519.

Banfill, P. F. G. (1994). Rheological methods for assessing the flow properties of mortar and related materials. *Constr. Build. Mater* 8(1), 43–50.

Banfill, P. F. G. (1999). *The influence of fine materials in sand on the rheology of fresh mortar*. Utilising ready mixed concrete and mortar Proceedings of an International Conference, Dundee, Scotland, UK, 411–420.

Banfill, P., Beaupré, D., Chapdelaine, F., de Larrard, F., Domone, P., Nachbaur, L., Sedran, T., Wallevik, O., Wallevik, J. (2020). *Comparison of concrete rheometers: International tests at LCPC*. National Institute of Standards and Technology, Nantes, France.

Bentz, D. P. (2007). Cement hydration: building bridges and dams at the microstructure level. *Mater. Struct.* 40(4), 397–404.

Bentz, D. P., Garboczi, E. J., Haecker, C. J., Jensen, O. M. (1999). Effects of cement particle size distribution on performance properties of Portland cement-based materials. *Cem. Concr. Res.* 29(10), 1663–1671.

Berra, M., Carassiti, F., Mangialardi, T., Paolini, A. E., Sebastiani, M. (2012). Effects of nanosilica addition on workability and compressive strength of Portland cement pastes. *Constr. Build. Mater.* 35, 666–675.

Bingham, E. C. (2007). *Fluidity and plasticity. Kessinger Publishing, LLC.*

Björnström, J., Martinelli, A., Matic, A., Börjesson, L., Panas, I. (2004). Accelerating effects of colloidal nano-silica for beneficial calcium–silicate–hydrate formation in cement. *Chem. Phys. Lett.* 392(1), 242–248.

Bornemann, R., Schmidt, M. (2002). *The role of powders in concrete. Proceedings 6th International Symposium on Utilisation of High Strength/High Performance Concrete. Leipzig*, 863–872.

Bouras, R., Kaci, A., Chaouche, M. (2012). Influence of viscosity modifying admixtures on the rheological behavior of cement and mortar pastes. *Korea-Aust. Rheol. J.* 24(1), 35–44.

BSEN12350-9. (2010). *Testing Fresh Concrete: Self-compacting Concrete V-funnel Test*, UK.

Camiletti, J., Soliman, A. M., Nehdi, M. L. (2013). Effects of nano- and micro-limestone addition on early-age properties of ultra-high-performance concrete. *Mater. Struct.* 46(6), 881–898.

Chan, Y. W., Chu, S. H. (2004). Effect of silica fume on steel fiber bond characteristics in reactive powder concrete. *Cem. Concr. Res.* 34(7), 1167–1172.

Chandra, S., Björnström, J. (2002). Influence of superplasticizer type and dosage on the slump loss of Portland cement mortars—Part II. *Cem. Concr. Res.* 32(10), 1613–1619.

Chen, J. J., Kwan, A. K. H. (2012). Superfine cement for improving packing density, rheology and strength of cement paste. *Cem. Concr. Compos.* 34(1), 1–10.

China Building Material Council. (2005). *Test methods for flowability of cement paste: GB2419—2005*. Beijing, Standards Press of China (in Chinese).

Choi, M. S., Lee, J. S., Ryu, K. S., Koh, K.T., Kwon, S. H. (2016). Estimation of rheological properties of UHPC using mini slump test. *Constr. Build. Mater.* 106, 632–639.

Collepardi, M., Jacob, J., Troli, R. (2020). *Influence of amorphos colloidal silica on the properties of self compacting concretes.* In: Proceedings of the international conference" challenges in concrete construction-innovations and developments in concrete materials and construction, Dundee, Scotland, UK, 473–483.

Collepardi, S., Luigi, C., Troli, R. (1997). *Mechanical properties of modified reactive powder concrete.* ACI Spec. Pub.

Collins, F., Lambert, J., Duan, W. H. (2012). The influences of admixtures on the dispersion, workability, and strength of carbon nanotube–OPC paste mixtures. *Cem. Concr. Compos.* 34(2), 201–207.

Coussot, P., Ancey, C. (1999). Rheophysical classification of concentrated suspensions and granular pastes. *Phys. Rev. E.* 59, 4445–4457.

Cyr, M., Legrand, C., Mouret, M. (2000). Study of the shear thickening effect of superplasticizers on the rheological behaviour of cement pastes containing or not mineral additives. *Cem. Concr. Res.* 30(9), 1477–1483.

De Schutter, G., Bartos, P., Domone, P., Gibbs, J. (2008). *Self-compacting concrete*, 1st ed. Taylor & Francis Group, USA.

Deeb, R., Ghanbari, A., Karihaloo, B. L. (2012). Development of self-compacting high and ultra high performance concretes with and without steel fibres. *Cem. Concr. Compos.* 34(2), 185–190.

Denarié, E., Houst, Y. (2011). *Cementitious matrices for high performance fibre reinforced cement composites (HPFRCC), in particular ultra-high performance fibre reinforced concretes (UHPFRC).* EP2275390AI, European Patent Office.

Dils, J., Boel, V., De Schutter, G. (2013). Influence of cement type and mixing pressure on air content, rheology and mechanical properties of UHPC. *Constr. Build. Mater.* 41, 455–463.

Dong, Z., Dasari, A., Tan, K.H. (2018). On the mechanism of prevention of explosive spalling in ultra-high performance concrete with polymer fibers. *Cem. Concr. Res.* 113, 169–177.

Duval, R., Kadri, E. H. (1998). Influence of silica fume on the workability and the compressive strength of high-performance concretes. *Cem. Concr. Res.* 28(4), 533–547.

EFNARC. (2002). *Specification and guidelines for self-compacting concrete. European federation for specialist construction chemicals and concrete systems.* Norfolk, UK, English ed.

Elkhadiri, I., Diouri, A., Boukhari, A., Aride, J., Puertas, F. (2002). Mechanical behaviour of various mortars made by combined fly ash and limestone in Moroccan Portland cement. *Cem. Concr. Res.* 32(10), 1597–1603.

Felekoğlu, B., Sarıkahya, H. (2008). Effect of chemical structure of polycarboxylate-based superplasticizers on workability retention of self-compacting concrete. *Constr. Build. Mater.* 22(9), 1972–1980.

Ferrari, L., Kaufmann, J., Winnefeld, F., Plank, J. (2010). Interaction of cement model systems with superplasticizers investigated by atomic force microscopy, zeta potential, and adsorption measurements. *J. Colloid. Interf. Sci.* 347(1), 15–24.

Feys, D., Verhoeven, R., De Schutter, G. (2009). Why is fresh self-compacting concrete shear thickening? *Cem. Concr. Res.* 39(6), 510–523.

Flatt, R. J., Houst, Y. F. (2001). A simplified view on chemical effects perturbing the action of superplasticizers. *Cem. Concr. Res.* 31(8), 1169–1176.

Ghafari, E., Costa, H., Júlio, E., Portugal, A., Durães, L. (2014). The effect of nanosilica addition on flowability, strength and transport properties of ultra high performance concrete. *Mater. Design* 59, 1–9.

Ghafari, E., Ghahari, S. A., Costa, H., Júlio, E., Portugal, A., Durães, L. (2016). Effect of supplementary cementitious materials on autogenous shrinkage of ultra-high performance concrete. *Constr. Build. Mater.* 127, 43–48.

Ghio, V. A., Monteiro, P. J. M., Demsetz, L. A. (1994). The rheology of fresh cement paste containing polysaccharide gums. *Cem. Concr. Res.* 24(2), 243–249.

Gong, K., Pan, Z., Korayem, A. H., Qiu, L., Li, D., Collins, F., Wang, C. M., Duan, W. H. (2015). Reinforcing Effects of Graphene Oxide on Portland Cement Paste. *J. Mater. Civil Eng.* 27(2), A4014010.

Hallal, A., Kadri, E. H., Ezziane, K., Kadri, A., Khelafi, H. (2010). Combined effect of mineral admixtures with superplasticizers on the fluidity of the blended cement paste. *Constr. Build. Mater.* 24(8), 1418–1423.

Hanehara, S., Yamada, K. (1999). Interaction between cement and chemical admixture from the point of cement hydration, absorption behaviour of admixture, and paste rheology. *Cem. Concr. Res.* 29(8), 1159–1165.

Hu, C., de Larrard, F. (1996). The rheology of fresh high-performance concrete. *Cem. Concr. Res.* 26(2), 283–294.

Hung, C.C., Chen, Y.T., Yen, C.H. (2020). Workability, fiber distribution, and mechanical properties of UHPC with hooked end steel macro-fibers. *Constr. Build. Mater.* 260, 119944.

Illguth, Lowke, Gehlen. (2013). *Rheology of fibre reinforced fine-grained high performance concrete for thin-walled elements—effect of type and content of steel fibres.* 1st International RILEM Conference on Rheology and Processing of Construction Materials and the 7th RILEM Conference on Self-Compacting Concrete, Paris, France.

Ingo, S., Jurgen, S. Ollver, M. (2004). *Effect of mixing and placement methods on fresh and hardened ultra-high performance concrete.* Proceeding of Ultra High Performance Concrete, Kassel, Germany, pp. 575–586.

Jensen, O. M., Hansen, P. F. (2002). Water-entrained cement-based materials: II. Experimental observations. *Cem. Concr. Res.* 32(6), 973–978.

Jensen, O., Lura, P. (2006). Techniques and materials for internal water curing of concrete. *Mater. Struc.* 39, 817–825.

Jiang, S. P., Mutin, J. C., Nonat, A. (1995). Studies on mechanism and physico-chemical parameters at the origin of the cement setting. I. The fundamental processes involved during the cement setting. *Cem. Concr. Res.* 25(4), 779–789.

Jiang, S. P., Mutin, J. C., Nonat, A. (1996). Studies on mechanism and physico-chemical parameters at the origin of the cement setting II. Physico-chemical parameters determining the coagulation process. *Cem. Concr. Res.* 26(3), 491–500.

Jiao, D., Shi, C., Yuan, Q., An, X., Liu, Y., Li, H. (2017). Effect of constituents on rheological properties of fresh concrete-A review. *Cem. Concr. Compos.* 83, 146–159.

Juenger, M. C. G., Siddique, R. (2015). Recent advances in understanding the role of supplementary cementitious materials in concrete. *Cem. Concr. Res.* 78, 71–80.

Kamal, H., Khayat, K., Meng, W., Vallurupalli, K., and Teng, L. (2019). Rheological properties of ultra-high-performance concrete—An overview. *Cem. Concr. Res.* 124, 105828.

Kamibayashi, M., Ogura, H., Otsubo, Y. (2008). Shear-thickening flow of nanoparticle suspensions flocculated by polymer bridging. *J. Colloid Interface Sci.* 321(2), 294–301.

Kang, S.H., Hong, S.G., Moon, J. (2019). The use of rice husk ash as reactive filler in ultra-high performance concrete. *Cem. Concr. Res.* 115, 389–400.

Kang, S.T., Kim, J.K. (2011). The relation between fiber orientation and tensile behavior in an Ultra High Performance Fiber Reinforced Cementitious Composites (UHPFRCC). *Cem. Concr. Res.* 41(10), 1001–1014.

Kazemi, S., Lubell, A. (2012). Influence of specimen size and fibre content on mechanical properties of ultra-high performance fibre-reinforced concrete. *ACI Mater. J.*, 109(6), 675–684.

Khayat, K. H. (1998). Viscosity-enhancing admixtures for cement-based materials—An overview. *Cem. Concr. Compos.* 20(2), 171–188.

Khayat, K. H. (1999). Workability, testing, and performance of self-consolidating concrete. *ACI Mater. J.* 96(3), 346–353.

Khayat, K. H., Yahia, A., Sayed, M. (2008). Effect of supplementary cementitious materials on rheological properties, bleeding, and strength of structural grout. *ACI Mater. J.* 105(6), 585–593.

Khayat, K., Aïtcin, P.C. (1993). *Silica fume—A unique supplementary cementitious material. In Mineral admixtures in cement and concrete.* Sarkar, S.L., and Ghosh, S.N. (eds.), 226–265. CRC Press, Taylor & Francis Group.

Ko, F. K., Gogotsi, Y., Ali, A. A., Naguib, N., Willis, P. (2010). Electrospinning of Continuous Carbon Nanotube-Filled Nanofiber Yarns. *Advanced Materials* 15(14), 1161–1165.

Kong, D., Su, Y., Du, X., Yang, Y., Wei, S., Shah, S. P. (2013). Influence of nanosilica agglomeration on fresh properties of cement pastes. *Constr. Build. Mater* 43, 557–562.

Krieger, I. M., Dougherty, Y. J. (1993). A mechanism for non-newtonian flow in suspensions or rigid spheres. *Trans. Soc. Rheol.* 37, 35–53.

Lachemi, M., Hossain, K. M. A., Lambros, V., Nkinamubanzi, P. C., Bouzoubaâ, N. (2004). Performance of new viscosity modifying admixtures in enhancing the rheological properties of cement paste. *Cem. Concr. Res* 34(2), 185–193.

Laskar, A. I., Talukdar, S. (2008). Rheological behavior of high performance concrete with mineral admixtures and their blending. *Constr. Build. Mater.* 22(12), 2345–2354.

Li, P. P., Yu, Q. L., Brouwers, H. J. H. (2017). Effect of PCE-type superplasticizer on early-age behaviour of ultra-high performance concrete (UHPC). *Constr. Build. Mater.* 153, 740–750.

Li, P., Yu, Q. L., Brouwers, H. J. H., Yu, R. (2016). Fresh behaviour of ultra-high performance concrete (UHPC), an investigation of the effect of superplasticizers and steel fibres. Proceedings of the 9th International Concrete Conference, Environment, Efficiency and Economic Challenges for Concrete Dundee, Scotland, UK, 653–644.

Liu, J., Khayat, K. H., Shi, C. (2020). Effect of superabsorbent polymer characteristics on rheology of ultra-high performance concrete. *Cem. Concr. Compos.* 112, 103636.

Liu, J., Wang, K., Zhang, Q., Han, F., Sha, J., Liu, J. (2017). Influence of superplasticizer dosage on the viscosity of cement paste with low water-binder ratio. *Constr. Build. Mater.* 149, 359–366.

Liu, X., Chen, L., Liu, A., Wang, X. (2012). Effect of Nano-$CaCO_3$ on Properties of Cement Paste. *Energy Procedia* 16, 991–996.

Luo, Z. Y., Wu, Y. Q., Hu, Q., Wang, T., Ni, M. J. (2015). Thermal and rheological properties of carbon nanotubes-oil nanofluid. *J. Chem. Eng. Chin. Univ.* 29, 35–42.

Lura, P., Jensen, O. M., van Breugel, K. (2003). Autogenous shrinkage in high-performance cement paste: An evaluation of basic mechanisms. *Cem. Concr. Res.* 33(2), 223–232.

Lura, P., Wyrzykowski, M., Tang, C., Lehmann, E. (2014). Internal curing with lightweight aggregate produced from biomass-derived waste. *Cem. Concr. Res.* 59, 24–33.

Ma, J., Orgass, M., Tue, N. (2008). *Influence of addition method of superplasticizer on the properties of fresh UHPC.* 2nd International Symposium on Ultra High Performance Concrete, Kassel, Germany.

Ma, X., Liu, J., Shi, C. (2019). A review on the use of LWA as an internal curing agent of high performance cement-based materials. *Constr. Build. Mater.* 218, 385–393.

Madandoust, R., Ranjbar, M. M., Moghadam, H. A., Mousavi, S. Y. (2011). Mechanical properties and durability assessment of rice husk ash concrete. *Biosys. Eng.* 110(2), 144–152.

Marchon, D., Kawashima, S., Bessaies-Bey, H., Mantellato, S., Ng, S. (2018). Hydration and rheology control of concrete for digital fabrication: Potential admixtures and cement chemistry. *Cem. Concr. Res.* 112, 96–110.

Mardani-Aghabaglou, A., Tuyan, M., Yılmaz, G., Arıöz, Ö., Ramyar, K. (2013). Effect of different types of superplasticizer on fresh, rheological and strength properties of self-consolidating concrete. *Constr. Build. Mater.* 47, 1020–1025.

Martinie, L., Rossi, P., Roussel, N. (2010). Rheology of fiber reinforced cementitious materials: classification and prediction. *Cem. Concr. Res* 40(2), 226–234.

Mazanec, O., Schiessl, P. (2008). *Improvement of UHPC properties through an optimized mixing procedure.* 8th International Symposium in Utilization of High-strength and High-performance Concrete, Tokyo, Japan, 307–313.

Mechtcherine, V., Secrieru, E., Schröfl, C. (2015). Effect of superabsorbent polymers (SAPs) on rheological properties of fresh cement-based mortars—Development of yield stress plastic viscosity over time. *Cem. Concr. Res.* 67, 52–65.

Mehta, P. K., Aitcin, P. C. (1990). Microstructural basis of selection of materials and mix proportions for high-strength concrete. *ACI Symposium Publication* 121, 265–286.

Meng, W., Khayat, K. H. (2016). Mechanical properties of ultra-high-performance concrete enhanced with graphite nanoplatelets and carbon nanofibers. *Compos. B Part Eng.* 107 (dec.), 113–122.

Meng, W., Khayat, K. H. (2017). Effects of saturated lightweight sand content on key characteristics of ultra-high-performance concrete. *Cem. Concr. Res* 101, 46–54.

Meng, W., Khayat, K. H. (2017). Improving flexural performance of ultra-high-performance concrete by rheology control of suspending mortar. *Compos. B Part Eng.* 117, 26–34.

Meng, W., Valipour, M., Khayat, K. H. (2016). Optimization and performance of cost-effective ultra-high performance concrete. *Mater. Struct.* 50(1), 29.

Meng, Weina, Khayat, K. H. (2018). Effect of graphite nanoplatelets and carbon nanofibers on rheology, hydration, shrinkage, mechanical properties, and microstructure of UHPC. *Cem. Concr. Res.* 105, 64–71.

Meng, Weina, Khayat, K. H. (2018). Effect of hybrid fibers on fresh properties, mechanical properties, and autogenous shrinkage of cost-effective UHPC. *J. Mater. Civil Eng.* 30, 04018030.

Mewis, J. (1979). Thixotropy—a general review. *J. Nonnewton Fluid Mech.* 6(1): 1–20.

Mikanovic, N., Jolicoeur, C. (2008). Influence of superplasticizers on the rheology and stability of limestone and cement pastes. *Cem. Concr. Res.* 38(7), 907–919.

Muzenski, S., Flores-Vivian, I., Sobolev, K. (2019). Ultra-high strength cement-based composites designed with aluminum oxide nano-fibers. *Constr. Build. Mater.* 220, 177–186.

Nadeem, A., Memon, S. A., Lo, T. Y. (2013). Mechanical performance, durability, qualitative and quantitative analysis of microstructure of fly ash and Metakaolin mortar at elevated temperatures. *Constr. Build. Mater.* 38, 338–347.

Nanthagopalan, P., Haist, M., Santhanam, M., Müller, H. S. (2008). Investigation on the influence of granular packing on the flow properties of cementitious suspensions. *Cem. Concr. Compos.* 30(9), 763–768.

Nehdi, M., Soliman, A. (2011). Early-age properties of concrete: Overview of fundamental concepts and state-of-the art research. *Constr. Build. Mater.* 164, 57–77.

NF P18-470. (2016) *Concrete—Ultra-high Performance Fibre-reinforced Concrete—Specifications, Performance, Production and Conformity*, French Standard Institute.

Paiva, H., Esteves, L. P., Cachim, P. B., Ferreira, V. M. (2009). Rheology and hardened properties of single-coat render mortars with different types of water retaining agents. *Constr. Build. Mater* 23(2), 1141–1146.

Park, C. K., Noh, M. H., Park, T. H. (2005). Rheological properties of cementitious materials containing mineral admixtures. *Cem. Concr. Res.* 35(5), 842–849.

Park, S. H., Kim, D. J., Ryu, G. S., Koh, K. T. (2012). Tensile behavior of Ultra High Performance Hybrid Fiber Reinforced Concrete. *Cem. Concr. Compos.* 34(2), 172–184.

Peyvandi, A., Sbia, L. A., Soroushian, P., Sobolev, K. (2013). Effect of the cementitious paste density on the performance efficiency of carbon nanofiber in concrete nanocomposite. *Constr. Build. Mater.* 48, 265–269.

Plank, J., Schroefl, C., Gruber, M. (2009). Use of a supplemental agent to improve flowability of ultra-high-performance concrete. *American Concrete Institute, ACI Spec. Publ.* 262, 1–16.

Plank, J., Schroefl, C., Gruber, M., Lesti, M., Sieber, R. (2009). Effectiveness of polycarboxylate superplasticizers in ultra-high strength concrete: The importance of PCE compatibility with silica fume. *J.Adv. Concr. Technol* 7, 5–12.

Qing, Y., Zenan, Z., Deyu, K., Rongshen, C. (2007). Influence of nano-SiO_2 addition on properties of hardened cement paste as compared with silica fume. *Constr. Build. Mater.* 21(3), 539–545.

Rao, G. A. (2003). Investigations on the performance of silica fume-incorporated cement pastes and mortars. *Cem. Concr. Res.* 33(11), 1765–1770.

Rong, Z., Sun, W., Xiao, H., Jiang, G. (2015). Effects of nano-SiO_2 particles on the mechanical and microstructural properties of ultra-high performance cementitious composites. *Cem. Concr. Compos.* 56, 25–31.

Rossi, P. (2001). Ultra-high-performance fiber-reinforced concretes. *Concr. Int.* 23(12), 46–52.

Roussel, N. (2006). A theoretical frame to study stability of fresh concrete. *Mater. Struct.* 39(1), 81–91.

Rydval, M., Bittner, T., Kolísko, J., Nenadálová, Š. (2016). Impact of steel fibers on workability and properties of UHPC. *Solid State Phenomena* 249, 57–61.

Sakai, E., Aizawa, K., Nakamura, A., Kato, H., Daimon, M. (2008). *Influence of superplasticizers on the fluidity of cements with different amount of aluminate phase. Second International Symposium on Ultra-high Performance Concrete.* Kassel, Germany, 85–92.

Schleiting, M., Wetzel, A., Krooß, P., Thiemicke, J., Niendorf, T., Middendorf, B., Fehling, E. (2020). Functional microfibre reinforced ultra-high performance concrete (FMF-UHPC). *Cem. Concr. Res.* 130, 105993.

Schröfl, C., Gruber, M., Plank, J. (2012). Preferential adsorption of polycarboxylate superplasticizers on cement and silica fume in ultra-high performance concrete (UHPC). *Cem. Concr. Res.* 42(11), 1401–1408.

Schröfl, C., Mechtcherine, V., Gorges, M. (2012). Relation between the molecular structure and the efficiency of superabsorbent polymers (SAP) as concrete admixture to mitigate autogenous shrinkage. *Cem. Concr. Res.* 42(6), 865–873.

Secrieru, E., Mechtcherine, V., Schröfl, C., Borin, D. (2016). Rheological characterisation and prediction of pumpability of strain-hardening cement-based-composites (SHCC) with and without addition of superabsorbent polymers (SAP) at various temperatures. *Constr. Build. Mate* 112, 581–594.

Senff, L., Hotza, D., Lucas, S., Ferreira, V. M., Labrincha, J. A. (2012). Effect of nano-SiO_2 and nano-TiO_2 addition on the rheological behavior and the hardened properties of cement mortars. *Mater. Sci. Eng. A* 532, 354–361.

Senff, L., Labrincha, J. A., Ferreira, V. M., Hotza, D., Repette, W. L. (2009). Effect of nano-silica on rheology and fresh properties of cement pastes and mortars. *Constr. Build. Mater* 23(7), 2487–2491.

Shen, P., Lu, L., Wang, F., He, Y., Hu, S., Lu, J., Zheng, H. (2020). Water desorption characteristics of saturated lightweight fine aggregate in ultra-high performance concrete. *Cem. Concr. Compos.* 106, 103456.

Shetty, M. S. (2006). *Concrete technology—theory and practice*, S. Chand and Company Ltd, New Delhi, India.

Shi, C., Wang, D., Wu, L., Wu, Z. (2015a). The hydration and microstructure of ultra high-strength concrete with cement–silica fume–slag binder. *Cem. Concr. Compos.* 61, 44–52.

Shi, C., Wu, Z., Xiao, J., Wang, D., Huang, Z., Fang, Z. (2015b). A review on ultra high performance concrete: Part I. Raw materials and mixture design. *Constr. Build. Mater.* 101, 741–751.

Shin, J. Y., Hong, J. S., Suh, J. K., Lee, Y. S. (2008). Effects of polycarboxylate-type superplasticizer on fluidity and hydration behavior of cement paste. *Korean J. Chem. Eng.* 25(6), 1553–1561.

Soliman, A., Nehdi, M. (2012). Effect of natural wollastonite microfibers on early-age behavior of UHPC. *J. Mater. Civil Eng.* 24, 816–824.

Soliman, N. A., Tagnit-Hamou, A. (2016). Development of ultra-high-performance concrete using glass powder—Towards ecofriendly concrete. *Constr. Build. Mater.* 125, 600–612.

Spiratos, N., Pagé, M., Mailvaganam, N., Malhotra, V. M., Jolicoeur, C. (2003). Superplasticizers for concrete: fundamentals, technology and practice, supplementary cementing materials for sustainable development. Inc. Editor, Ottawa, Canada.

Spiratos, N., Pagé, M., Mailvaganam, N., Malhotra, V., Jolicoeur, C. (2006). *Superplasticizers for concrete: fundamentals, technology and practice*, 2nd ed. Quebec (Canada), Marquis.

Standard, C. N. (2005). *Test methods for flowability of cement paste.* Beijing, China.

Struble, L., Sun, G.K. (1995). Viscosity of Portland cement paste as a function of concentration. *Adv. Cem. Based. Mater.* 2(2), 62–69.

Swamy, R. N. The technology of steel fibre reinforced concrete for practical applications. *Proceedings of Institution of Civil Engineers* 56, 143–159.

Tosun-Felekoğlu, K., Felekoğlu, B., Ranade, R., Lee, B. Y., Li, V. C. (2014). The role of flaw size and fiber distribution on tensile ductility of PVA-ECC. *Compos. B, Eng.* 56, 536–545.

Tregger, N., Ferrara, L., Shah, S. P. (2008). Identifying viscosity of cement paste from mini-slump-flow test. *ACI Mater. J.* 105(6), 558–566.

Van Tuan, N., Ye, G., van Breugel, K., Fraaij, A. L. A., Bui, D. D. (2011). The study of using rice husk ash to produce ultra high performance concrete. *Constr. Build. Mater.* 25(4), 2030–2035.

Vikan, H., Justnes, H., Winnefeld, F., Figi, R. (2007). Correlating cement characteristics with rheology of paste. *Cem. Concr. Res.* 37(11), 1502–1511.

Wagner, N., Brady, J. (2009). Shear thickening in colloidal dispersions. *Physics Today* 62, 27–32.

Wallevik, O. H. (2003). *Rheology—a scientific approach to develop self-compacting concrete.* Proceedings of the 3rd International RILEM Symposium on Self-compacting Concrete, Rilem, Reykjavik, 23–31.

Wallevik, O. H., Wallevik, J. E. (2011). Rheology as a tool in concrete science: The use of rheographs and workability boxes. *Cem. Concr. Res.* 41(12), 1279–1288.

Wang, C., Yang, C., Liu, F., Wan, C., Pu, X. (2012). Preparation of ultra-high performance concrete with common technology and materials. *Cem. Concr. Compos.* 34(4), 538–544.

Wang, X. H., Jacobsen, S., He, J. Y., Zhang, Z. L., Lee, S. F., Lein, H. L. (2009). Application of nanoindentation testing to study of the interfacial transition zone in steel fiber reinforced mortar. *Cem. Concr. Res.* 39(8), 701–715.

Westerholm, M., Lagerblad, B., Silfwerbrand, J., Forssberg, E. (2008). Influence of fine aggregate characteristics on the rheological properties of mortars. *Cem. Concr. Compos.* 30(4), 274–282.

Wille, K., El-Tawil, S., Naaman, A. E. (2014). Properties of strain hardening ultra high performance fiber reinforced concrete (UHP-FRC) under direct tensile loading. *Cem. Concr. Compos.* 48, 53–66.

Wille, K., Loh, K. J. (2010). Nanoengineering Ultra-High-Performance Concrete with Multiwalled Carbon Nanotubes. *Transportation Research Record* 2142(1), 119–126.

Wille, K., Naaman, A.E., Montesinos, G. (2011), Ultra-high performance concrete with compressive strength exceeding 150 MPa (22 ksi): A simpler way. *ACI Mater. J.* 108(1), 46–54.

Wong, H., Kwan, A. (2005). *Packing density: a key concept for mix design of high performance concrete.* In Proceedings of the Materials Science and Technology in Engineering Conference, HKIE Materials Division, Pittsburgh, PA, pp. 1–15.

Wu, Z., Khayat, K. H., Shi, C. (2019). Changes in rheology and mechanical properties of ultra-high performance concrete with silica fume content. *Cem. Concr. Res.* 123, 105786.

Wu, Z., Shi, C., He, W., Wu, L. (2016). Effects of steel fiber content and shape on mechanical properties of ultra high performance concrete. *Constr. Build. Mater.* 103, 8–14.

Wu, Z., Shi, C., Khayat, K. H. (2016). Influence of silica fume content on microstructure development and bond to steel fiber in ultra-high strength cement-based materials (UHSC). *Cem. Concr. Compos.* 71, 97–109.

Wu, Z., Shi, C., Khayat, K. H., Wan, S. (2016). Effects of different nanomaterials on hardening and performance of ultra-high strength concrete (UHSC). *Cem. Concr. Compos.* 70, 24–34.

Yahia, A. (2011). Shear-thickening behavior of high-performance cement grouts—Influencing mix-design parameters. *Cem. Concr. Res.* 41(3), 230–235.

Yahia, A., Khayat, K. H. (2003). Applicability of rheological models to high-performance grouts containing supplementary cementitious materials and viscosity enhancing admixture. *Mater. Struct.* 36(6), 402–412.

Yamanoi, M., Maia, J. M. (2010). Analysis of rheological properties of fibre suspensions in a Newtonian fluid by direct fibre simulation. Part 1: Rigid fibre suspensions. *J.Non-Newton. Fluid* 165(19), 1055–1063.

Yan, H., Sun, W., Chen, H. (1999). The effect of silica fume and steel fiber on the dynamic mechanical performance of high-strength concrete. *Cem. Concr. Res.* 29(3), 423–426.

Yang, I. H., Joh, C., Kim, B.S. (2010). Structural behavior of ultra high performance concrete beams subjected to bending. *Eng. Struct.* 32(11), 3478–3487.

Yang, S. L., Millard, S. G., Soutsos, M. N., Barnett, S. J., Le, T. T. (2009). Influence of aggregate and curing regime on the mechanical properties of ultra-high performance fibre reinforced concrete (UHPFRC). *Constr. Build. Mater.* 23(6), 2291–2298.

Yazıcı, H., Yardımcı, M. Y., Yiğiter, H., Aydın, S., Türkel, S. (2010). Mechanical properties of reactive powder concrete containing high volumes of ground granulated blast furnace slag. *Cem. Concr. Compos.* 32(8), 639–648.

Yazıcı, H., Yiğiter, H., Karabulut, A. Ş., Baradan, B. (2008). Utilization of fly ash and ground granulated blast furnace slag as an alternative silica source in reactive powder concrete. *Fuel* 87(12), 2401–2407.

Yoshioka, K., Sakai, E., Daimon, M., Kitahara, A. (1997). Role of Steric Hindrance in the Performance of Superplasticizers for Concrete. *J. Am. Ceram. Soc* 80(10), 2667–2671.

Yu, R., Spiesz, P., Brouwers, H. J. H. (2014). Mix design and properties assessment of ultra-high performance fibre reinforced concrete (UHPFRC). *Cem. Concr. Res.* 56, 29–39.

Yu, R., Spiesz, P., Brouwers, H. J. H. (2015). Development of Ultra-High Performance Fibre Reinforced Concrete (UHPFRC): Towards an efficient utilization of binders and fibres. *Constr. Build. Mater.* 79, 273–282.

Zhong, P., Wyrzykowski, M., Toropovs, N., Li, L., Liu, J., Lura, P. (2019). Internal curing with superabsorbent polymers of different chemical structures. *Cem. Concr. Res.* 123, 105789.

Zingg, A., Winnefeld, F., Holzer, L., Pakusch, J., Becker, S., Figi, R., Gauckler, L. (2009). Interaction of polycarboxylate-based superplasticizers with cements containing different C3A amounts. *Cem. Concr. Compos.* 31(3), 153–162.

Zingg, A., Winnefeld, F., Holzer, L., Pakusch, J., Becker, S., Gauckler, L. (2008). Adsorption of polyelectrolytes and its influence on the rheology, zeta potential, and microstructure of various cement and hydrate phases. *J. Colloid. Interf. Sci* 323(2), 301–312.

Chapter 6

Setting and hardening

6.1 INTRODUCTION

The setting and hardening behavior of UHPC are different from those of normal concrete due to the use of low water-to-binder (w/b) ratio and high content of supplementary cementitious materials (Shi et al. 2015; Wang et al. 2015). The setting in UHPC begins with the hydration of Portland cement to form calcium silicate hydrate (C-S-H) and calcium hydroxide. Calcium hydroxide can then react with pozzolanic materials, such as silica fume, to form C-S-H. The phase content in UHPC does not change after 7 d of hydration. The crystalline phase content in UHPC is considerably lower than that in normal concrete. With proper heat treatment, the pozzolanic reactions of supplementary cementitious materials in UHPC can be enhanced, resulting in an increased hydration rate, a formation of additional C-S-H, and an increased average C-S-H chain length (Shen et al. 2019). These C-S-H can fill small pores, resulting in a denser microstructure and improved mechanical properties. UHPC has a very low porosity, which often does not exceed 9%, by volume of concrete, in the pore diameter ranging from 3.75 nm to 100 μm (Schmidt and Fehling 2005). Under 150 and 200°C heat treatment, those pores in UHPC are nil (Cheyrezy et al. 1995). These characteristics are closely related to the setting and hardening processes of UHPC.

This section discusses the setting and hardening behavior of UHPC from four aspects, including heat of hydration, hydration products, microstructure development, and pore solution, to get a deep understanding of UHPC. The influences of raw materials and mixture proportions on the setting and hardening behavior are described.

6.2 HEAT OF HYDRATION

6.2.1 Hydration of portland cement in UHPC

Isothermal calorimetry is a convenient and precise method to study the early stage of hydration of cement pastes. The hydration of cement is far more complex than the sum of the hydration reactions of the individual minerals.

DOI: 10.1201/9781003203605-6

The typical depiction of a cement grain involves larger silicate particles surrounded by much smaller C_3A and C_4AF particles. The hydration of cement can be divided into several distinct periods (Figure 6.1), including initial dissolution period (I), induction period (II), acceleration period (III), deceleration period (IV), and slow reaction period (V). The more reactive aluminate and ferrite phases react first, and these reactions dramatically affect the hydration of the silicate phase.

In the first few minutes of hydration (Figure 6.1(b)), the aluminum and iron phases react with gypsum to form an amorphous gel at the surface of the cement grains, and short rods of ettringite grow. After this initial period of reactivity, cement hydration slows down and the induction period begins. After about 3 h of hydration, the induction period ends, and the acceleration period begins. From 3 to 24 h, about 30% of cement reacts to form calcium hydroxide and C-S-H. The development of C-S-H in this period occurs in 2 phases. After 10 h of hydration (Figure 6.1(c)), C_3S has produced "outer C-S-H", which grows out from the ettringite rods

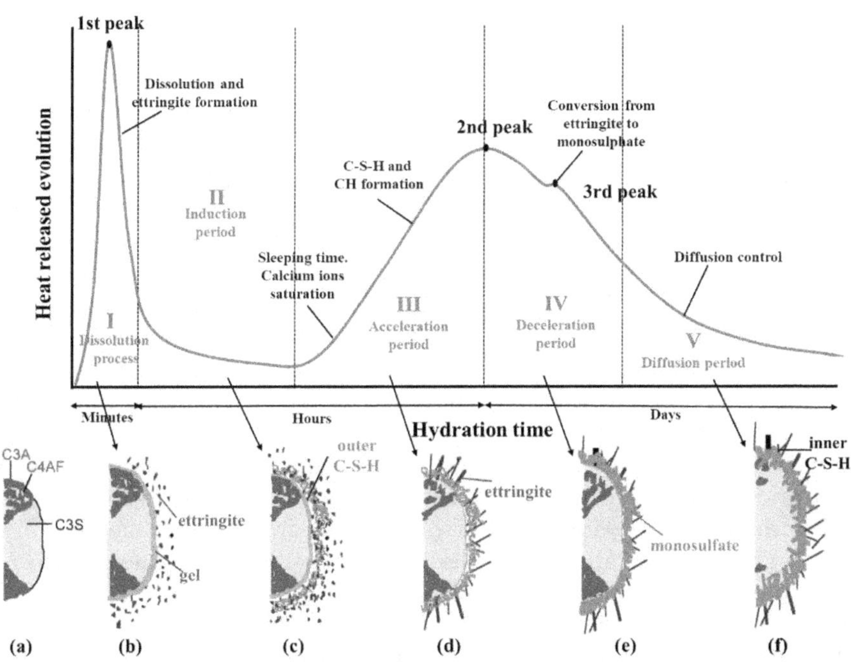

Figure 6.1 Illustration of hydration stages and schematic representation of anhydrous Portland cement particle.

Source: Adapted based on Bishop 2003.

rather than directly out from the surface of the C_3S particles. Therefore, in the initial phase of the reaction, the silicate ions must migrate through the aluminum and iron-rich phase to form C-S-H. In the latter part of the acceleration state, after 18 h of hydration, C_3A continues to react with gypsum, forming longer ettringite rods (Figure 6.1(d)). This network of ettringite and C-S-H appears to form a "hydrating shell" about 1 μm from the surface of anhydrous C_3S. A small amount of "inner C-S-H" forms inside this shell. After 1–3 d of hydration, reactions slow down and the deceleration period begins (Figure 6.1(e)). C_3A reacts with ettringite to form some monosulfate. Meanwhile, C-S-H continues to grow near the C_3S surface, narrowing the 1 μm gap between the "hydrating shell" and the anhydrous C_3S. The rate of hydration is likely to depend on the diffusion rate of water or ions to the anhydrous surface. After 2 weeks of hydration (Figure 6.1(f)), the gap between the "hydrating shell" and the grain is completely filled with C-S-H. The original "outer C-S-H" becomes more fibrous.

6.2.2 Effect of water-to-cement ratio on hydration of UHPC paste

Figure 6.2 illustrates hydration heat release rate and cumulative heat of cement paste with water-to-cement (w/c) ratio ranging from 0.2 to 0.3. During the experiment, isothermal conditions (20 ± 0.02°C) were maintained in the measuring cells. Since the very low w/c ratio paste required external mixing, the initial heat peak, occurring right after the addition of water to cement, could not be measured. A dormant period of 5–7 h, depending on w/c ratio and superplasticizer dosage, is followed by a main heat rate peak between 13 and 15 h. The main heat release peaks are delayed as w/c ratio increased, and the height of the main peaks is also increased. From Figure 6.2(b), the cumulative heat of hydration increases with the increment in w/c ratio initially but starts to decrease after about 24 h. In other words, the decrease in w/c ratio accelerates the main hydration peak and decreases the height of the main peak and total degree of hydration at later ages. The increase in the rate of heat liberation with the decrease in w/c ratio during the first hour of hydration could be explained by a higher concentration of alkalis in the pore solution (Danielson 1962). The soluble alkalis in the cement lead to a rapid pH rise that is known to accelerate cement hydration. This acceleration directly depends on the alkali concentration, which is in turn inversely dependent on the w/c ratio. This is to say, by increasing the w/c, the initial pore solution becomes diluted with respect to alkali and hydroxide ions, leading to less acceleration of cement hydration.

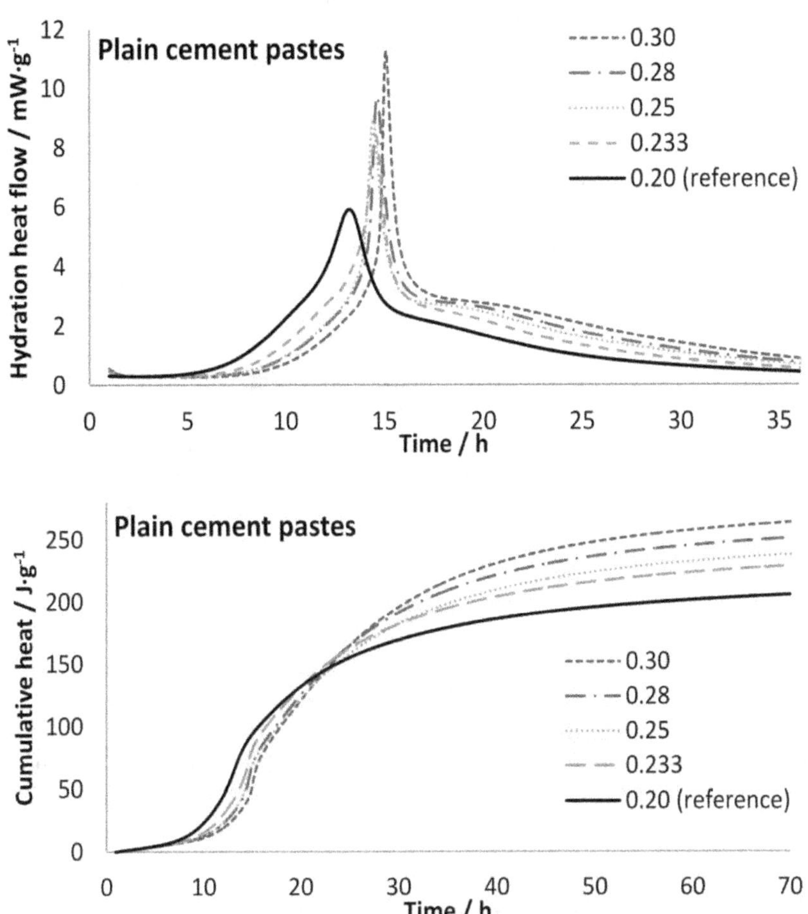

Figure 6.2 Heat of hydration of cement pastes with different w/c ratios: (a) Hydration heat release rate until 36 h; and (b) Cumulative heat until 72 h.

Source: Justs et al. 2014.

6.2.3 Effect of supplementary cementitious materials on hydration of UHPC

The addition of supplementary cementitious materials has a significant influence on the hydration of UHPC. UHPC usually contains 10%–30% silica fume, by the mass of total binder materials. It is well known that silica fume can accelerate the hydration of UHPC (Wu et al. 2016), as shown in Figure 6.3. Compared to the reference mixture without any silica fume at a

0.18 w/b ratio, the use of 10% to 25% silica fume decreases the duration of the dormant period from 12 to 9 h or less (see Figure 6.3(a)). As hydration proceeds, UHPC mixture with 25% silica fume (U25) shows accelerated hydration peak first, followed by U20, U15, and U10 with 20%, 15%, and 10% silica fume, respectively. However, the hydration peak of U0 is delayed to 28 h. The heat of hydration of UHPC mixture with silica fume evolves faster than that of the reference mixture (Wu et al. 2016). This is because the incorporation of silica fume can absorb Ca^{2+} and OH^- ions to form C-S-H, which increases the rate and amount of heat evolution (Kadri and Duval 2009). However, when the silica fume content increases to 30%, the heat evolution rate decreases, and the accelerated hydration period increases (Shi et al. 2015).

Partial replacement of cement with ground granulated blast-furnace slag (GGBS) or fly ash can delay hydration mainly during the dormant and acceleration periods (Rong et al. 2014; Shi et al. 2015), as shown in Figure 6.3(b). However, the combined use of silica fume and fly ash can promote the hydration of pure cement because silica fume is the main factor for the hydration of cement at a low w/b ratio (Langan et al. 2002). There is also a slight retardation tendency for the hydration heat of the mixtures made with silica fume and GGBS (Shi et al. 2015). Limestone powder can accelerate the hydration of cement (Wang et al. 2008; Wang et al. 2012) and improve the hydration degree of binder materials (Shi et al. 2002; Svermova et al. 2003; Tang 2004; Zhu and Gibbs 2005). The use of rice husk ash also increases the degree of hydration of cement at later ages because of its pozzolanic reaction and internal curing, which is even higher than those containing silica fume at 91 d (Kang et al. 2019; Van Tuan et al. 2011).

6.2.4 Effect of nanoparticles on hydration of UHPC

Nanomaterials can significantly enhance the performance of cement-based materials given their filling and nucleation effects and chemical reactivity. The addition of nanomaterials can shorten the dormant period and lead to an earlier acceleration period (Li et al. 2016; Wu et al. 2016). According to Sato (2006), the dormant period of the reference mixture is around 13 h, but it is shortened to about 9 h in the presence of nano-$CaCO_3$. However, the addition of 4.8% nano-$CaCO_3$, by mass of binder materials, causes an earlier and higher heat of hydration compared to the mixture with 6.4% nano-$CaCO_3$. This might be due to the dilution effect associated with the addition of nano-$CaCO_3$. Nano-SiO_2 can also accelerate the heat of hydration due to its filling effect and nucleation seeds for C-S-H precipitation (Ji 2005). Meng and Khayat (2018) found that both carbon nanofiber (CNF) and graphene nanoplatelet (GNP) increased the cumulative hydration heat. However, CNF reduced the induction period, but the use of GNP extended

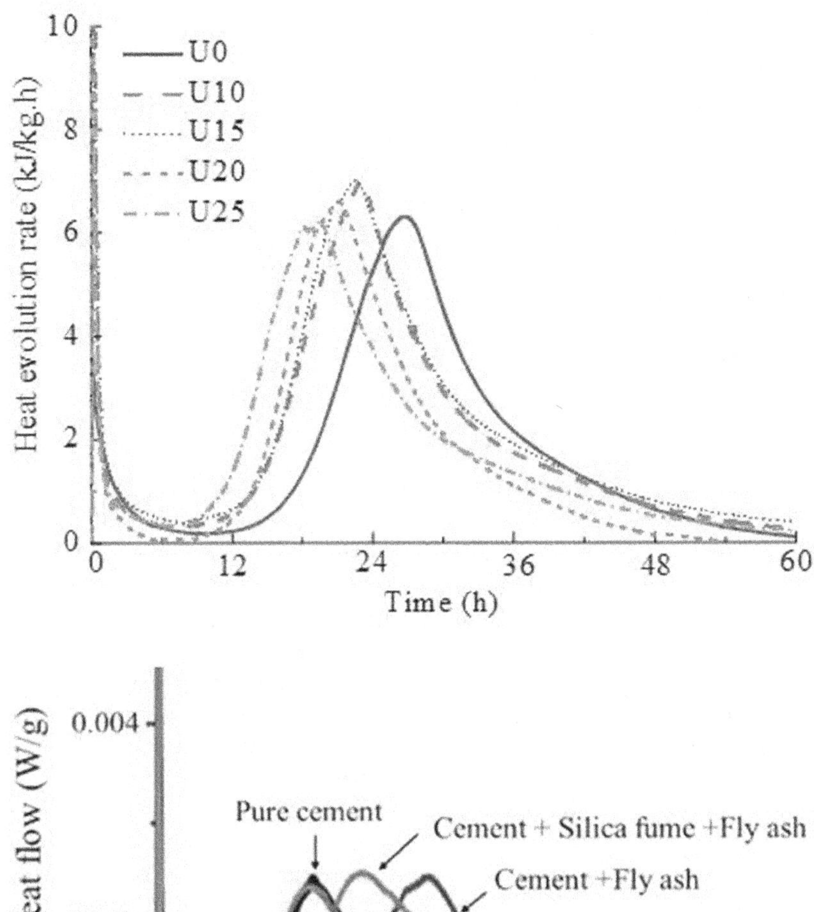

Figure 6.3 Heat evolution rate of UHPC: (a) With different silica fume contents (U0, U10, U15, U20, and U25 represent uses of 0%, 10%, 15%, 20%, and 25% silica fume, by the total mass of binder materials, respectively) (Wu et al. 2016); (b) With different contents of silica fume and fly ash (Rong et al. 2014).

the induction period. This is attributed to the less superplasticizer needed to disperse CNF. Such behavior is also found in ultra-high-strength cement-based composites with aluminum oxide nanofibers (Muzenski et al. 2019). Norhasri et al. (2019) reported that nano metaclay had a retarding effect on UHPC due to the increase in surface area.

The early hydration of binder in UHPC is usually retarded due to the high polycarboxylate superplasticizer dosage of 2%–4%, by mass of binder materials, in UHPC. The retardation is dependent on the superplasticizer type and dosage. In addition, the increase in curing temperature can accelerate the hydration of cement and promote the pozzolanic reaction between supplementary cementitious materials and calcium hydroxide (Shen et al. 2019), which has a significant influence on the hydration products as described in the following content.

6.3 HYDRATION PRODUCTS

The hydration products of binder materials in UHPC are similar to those in conventional concrete. Initially, Portland cement hydrates to form C-S-H and calcium hydroxide. Supplementary cementitious materials, such as silica fume, GGBS, and fly ash, then react with calcium hydroxide to form C-S-H (Korpa et al. 2009). The amount of calcium hydroxide in UHPC is much lower than that in conventional concrete at 28 d, as shown in Figure 6.4, although the pozzolanic reactions are still incomplete. Different types of supplementary cementitious materials have various effects on calcium hydroxide content of UHPC. When silica fume content increases from 0 to 50%, the calcium hydroxide content decreases from 7.4% to 6.25% (Shi et al. 2015). Both rice husk ash and silica fume can greatly reduce the calcium hydroxide content, and the effect of silica fume on calcium hydroxide content is greater than that of rice husk ash at later ages (Van Tuan et al. 2011). However, the effect of GGBS content has a limited effect on the calcium hydroxide content (Shi et al. 2015). Shi et al. (2015) also found that the combination of silica fume and GGBS has a negative synergistic effect on the calcium hydroxide content of UHPC.

The content of crystalline phases is considerably higher in conventional concrete, whereas less amorphous phases are observed in UHPC, as shown in Figure 6.4 (Korpa et al. 2009). No calcite is even found in UHPC after 28 d due to carbonation. The ettringite content development between the first and second days indicates some conversion of ettringite to monosulfate phase. A significant amount of aluminate may enter the amorphous C-S-H phases (Korpa et al. 2009). When UHPC contains limestone powder constituents, calcium aluminate monocarbonate is preferably formed, the hydration of C_3S is accelerated, and some carbosilicates are produced in the pastes (Kakali et al. 2000). Other literature (Nehdi et al. 1996; Rougeau and Borys 2004) showed that limestone powder in binder materials systems has a compaction filler and great dispersion effect on calcium hydroxide and

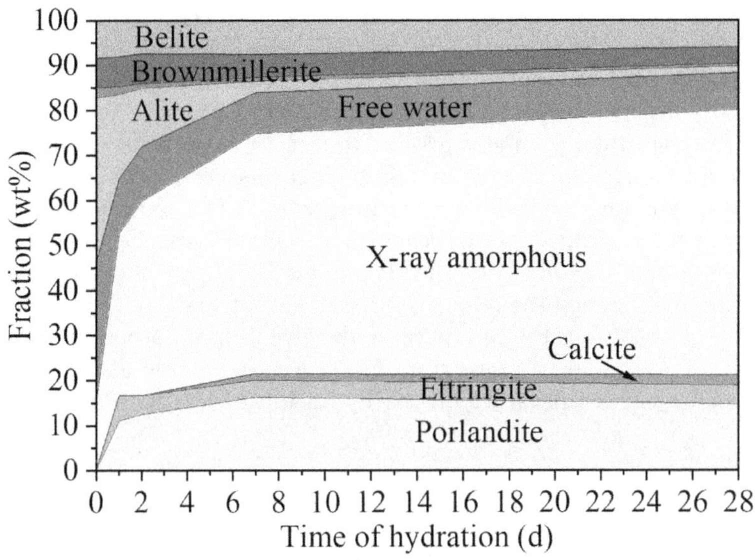

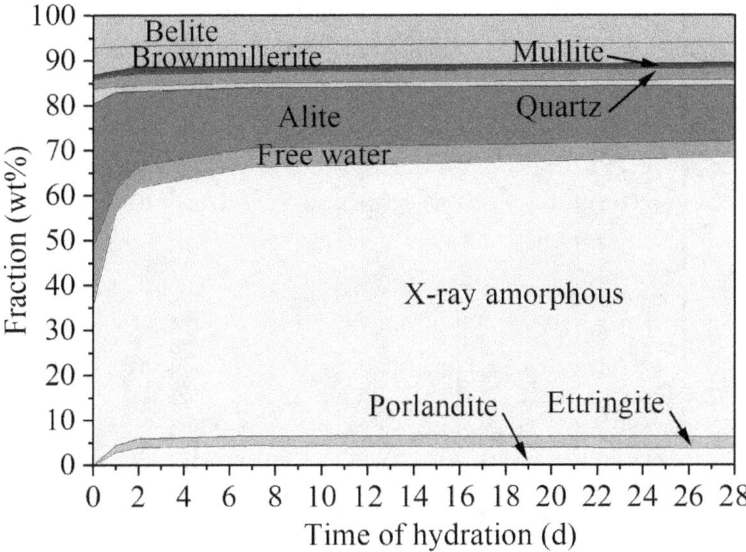

Figure 6.4 Time-dependent phase developments in conventional concrete and UHPC (adapted based on Korpa et al. 2009). (a) Conventional concrete; (b) UHPC.

C-S-H gel precipitation, and plays a core role in crystallization during the cement hydration process.

The increase in curing temperature can accelerate the hydration of cement and promote the pozzolanic reaction between supplementary cementitious materials and calcium hydroxide (Shen et al. 2019). Hydration products of cement after 90°C curing remain amorphous. The hydration of C_3S and C_2S leads to the formation of crystalline α-dicalcium silicate hydrate, while tricalcium aluminate (C_3A) and tetracalcium alumino-ferrite (C_4AF) form a hydrogarnet phase. The bond properties of these two phases are rather unfavorable. If finely ground quartz and/or other SiO_2 sources are used, a pozzolanic reaction takes place between calcium hydroxide and SiO_2, yielding crystalline tobermorite ($C_5S_6H_5$) with a diameter of 1.1 nm as the main product of reaction at temperatures between 150 and 200°C. Xonolite C-S-H (I), C-S-H (II), and a-C_2SH may also be formed at higher temperatures (Yazıcı et al. 2008). The formation of both 1.1 nm tobermorite and xonolite can increase the strength of autoclaved products (Odler 2003). It can be seen from Figure 6.5 and Table 6.1 that there is a great effect of temperature on the hydration of UHPC (Philippot et al.

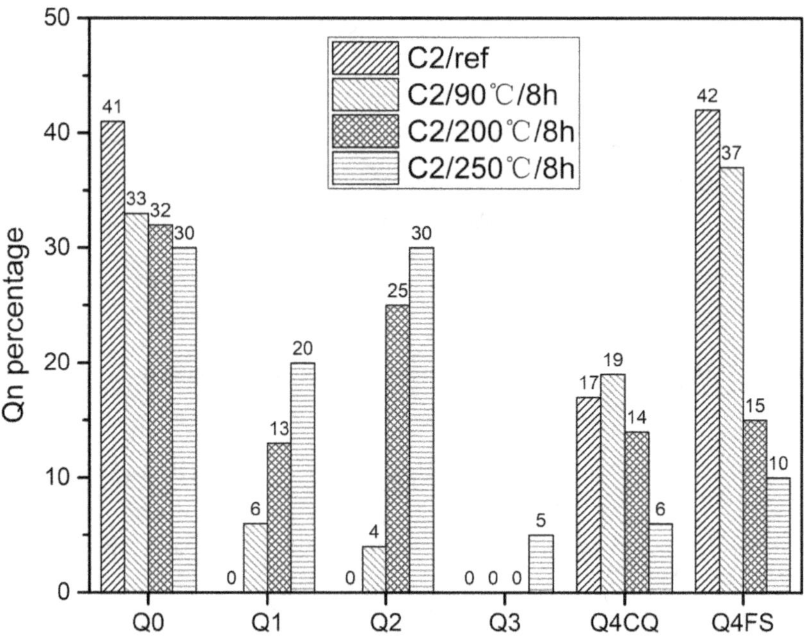

Figure 6.5 Percentages of Q^0 to Q^4 for specimens with heat treatment at temperatures of 90, 200, and 250°C for 8 h.

Notes: Q represents a SiO_4^{4-} unit and the degree of connectivity, n is related to the oxygen bonds number between the SiO_4^{4-} units (Q4FS for silica fume and Q4CQ for crushed quartz).

Source: Zanni et al. 1996.

Table 6.1 Hydrates formation ratio (H), connectivity ratio (C), average statistical chain length, pozzolanic activity of silica fume (PSF), and crushed quartz (PCQ) of samples with heat treatment at 90, 200, and 250°C

Sample	*H*	*PSF*	*PCQ*	*C*	*Statistical length*
C2/REF	0%	0%	0%	-	-
C2/90°C/8 h	10%	10%	0%	1.40	Trimer or quadrimer
C2/200°C/8 h	40%	65%	20%	1.65	Pentamer or hexamer
C2/250°C/8 h	50%	75%	65%	1.73	Hexamer or heptamer

Source: Zanni et al. 1996.

Notes: Hydrates formation ratio (H)=(Q^1 + Q^2 + Q^3); connectivity ratio (C)= (Q^1 + $2Q^2$ + $3Q^3$)/(Q^1 + Q^2 + Q^3); pozzolanic activity of silica fume (PSF)=(Q^4FS-Q^4FS_0)/Q^4FS_0; crushed quartz (PCQ) = (($QCQ4$ – $QCQ4_0$)/$QCQ4_0$).

1996; Zanni et al. 1996). The hydrate formation ratio increases from 10% to 55% when the curing temperature increases from 90 to 250°C. It is also significant to notice that Q^2 increases faster with temperature than that of Q^1 (Figure 6.5), which indicates that the C-S-H average statistical chain length increases with temperature. This is also demonstrated by the statistical average chain length that corresponds to pentamer at 200°C when it is trimer at 90 or 20°C. Between 200 and 250°C, appearance of a Q^3 peak was observed at 250°C, as shown in Figure 6.5. This peak is attributed to the presence of crystal hydrate and xonotlite, whose structure implies the presence of silicon-oxygen tetrahedra connected to three neighbor tetrahedra (Q^3 species). The H_2O to CaO (H/C) ratio of C-S-H of conventional concrete is approximately 1, while the H/C of xonotlite is 1/6, and xonotlite forms only in the inner part of concrete specimen (Cheyrezy et al. 1995). The formation of xonotlite in heat-cured UHPC is due to the local and large water vapor pressures. However, lower dynamic equilibrium vapor pressures of 3 Pa can suppress the formation of crystalline hydration products, even if no xonotlite or other crystalline hydration products form at 250°C (Feylessoufi et al. 1997).

6.4 MICROSTRUCTURE DEVELOPMENT

UHPC has a very dense and uniform microstructure due to the following fundamental effects: (1) close packing of solid particles; (2) hydration and/or pozzolanic reactions of binder materials; and (3) improvement in the interfacial transition zone between aggregates and bulk matrix (Shi et al. 2015). The internal microstructure of UHPC mainly includes unhydrated cement particles, quartz sand, and hydration products, such as C-S-H (Sorelli et al. 2008). It mainly depends on the pore structure, morphology of hydration products, and microstructure of interfacial transition zone.

6.4.1 Pore structure

UHPC has very low porosity. For UHPC at a w/b ratio of 0.20, its capillary pores become discontinuous when only 26% of cement is hydrated, instead of 54% for high-performance concrete (HPC) with 0.33 w/b ratio (Bonneau et al. 2000). Figure 6.6 shows the pore radius distribution of a high-quality class C45/55 normal-strength concrete, a class C90/105 high-strength concrete, and two UHPC mixtures (90°C heat treatment, coarse- and fine-grained mixes) measured by high-pressure mercury intrusion porosimetry (Schmidt and Fehling 2005). The pore structure of UHPC consists mainly of pores with a diameter smaller than the threshold value, and a reduction of cumulative porosity corresponds with a decrease in the threshold value (Figure 6.6). The pore size of UHPC concentrates between 2 and 3 nm, the most probable pore diameter is 2.0 nm, and its total porosity is 2.23%.

The curing method and temperature have great influences on pore structure of UHPC. Pores in the range of 3.75 to 100 nm in UHPC cured between 150 and 200°C were reported to be nil (Cheyrezy et al. 1995). When the curing temperature increases from 20 to 65°C, setting pressure does not affect the cumulative porosity in the range of 3.75 nm to 100 μm. However, setting pressure eliminates the air trapped in the sample, and

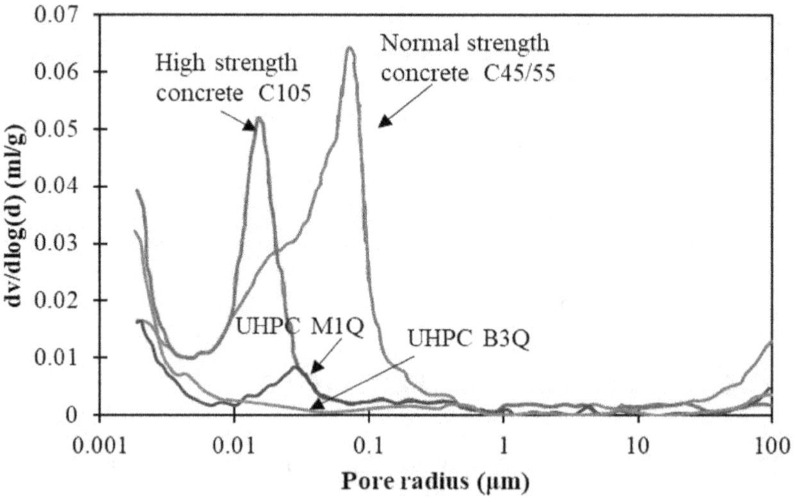

Figure 6.6 Pore radius distribution of normal-strength concrete, high-strength concrete, and two UHPC mixtures (90°C heat treatment, coarse- and fine-grained mixtures).

Source: Schmidt and Fehling 2005.

a portion of free water is also eliminated. In that temperature range, the porosity threshold is not modified. However, lower cumulative porosity, intermediate (3.75 nm to 100 μm) porosity, and threshold pore size are obtained for pressed samples at 80–200°C. The greater part (B 90°C/SC, Figure 6.7), if not the whole porosity (BQ 200°C/P, Figure 6.7), corresponds in that case to pore diameters smaller than 3.75 nm. When temperatures are achieved or above 250°C, the porosity and the threshold pore size increase (see BQ200°C/P and BQF400°C/SC in Figure 6.7).

The porosity of UHPC decreases with the increase of silica fume content due to its filling and pozzolanic effects (Shi et al. 2015). When silica fume content increases from 0% to 50%, the 3-d and 5-d porosity values decrease from 14% to 12.7% and 10.4% to 8.3%, respectively. When silica fume

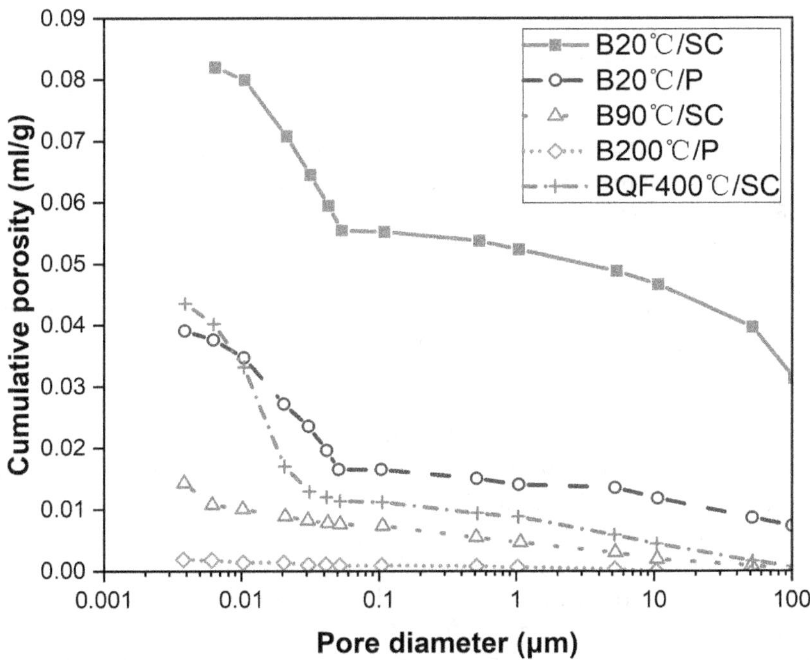

Figure 6.7 Cumulative porosity versus pore diameter of UHPC.

Notes: B20°C/SC, basic formulation heated at 20°C, soft cast (setting pressure, 1 atm); B20°C/P, basic formulation heated at 20°C, pressed (setting pressure, 30 atm); B90°C/SC, basic formulation heated at 90°C, soft cast; BQ200°C/SC, basic formulation with crushed quartz heated at 200°C, pressed (setting pressure 625 atm); BQF400°C/SC, basic formulation with crushed quartz and steel fibers heated at 400 °C, soft cast.

Source: Cheyrezy et al. 1995.

content is higher than 25% and GGBS content is less than 25%, GGBS has little effect on the porosity of UHPC at 3 d. However, GGBS at any content would increase the porosity of UHPC at 56 d. The combination of silica fume and GGBS can cause a negative synergistic effect on the porosity of UHPC (Shi et al. 2015). The total porosity of the rice husk ash-modified UHPC sample is higher than that of the silica fume-modified sample, but lower than that of the control sample because of the greater filler effect and pozzolanic reaction (Van Tuan et al. 2011). The use of 0.3% carbon nanofiber can reduce the total porosity of UHPC by approximately 35%, indicating that the presence of carbon nanofiber can refine the microstructure of UHPC (Meng and Khayat 2018).

6.4.2 Morphology of hydration products

UHPC comprises aggregates and compact matrix phases, including unhydrated particles, quartz sand, and hydration products. A reactive interface is formed between completely unhydrated core and hydration products. The core plays a role in the skeleton to matrix and greatly enhances the microstructure of the matrix phase (Long et al. 2005). Scanning electron microscope (SEM) observation indicates that the structure of hardened paste is very dense due to very low w/b, hydration of cement, and the pozzolanic action of silica fume and GGBS. The main hydration product, C-S-H gel, is homogeneous; no $Ca(OH)_2$ and ettringite could be found (Alaee 2002; Wang et al. 2012).

Figure 6.8 shows the effect of curing conditions on microstructure of UHPC. The clear unhydrated binder constituents in UHPC can be observed in Figure 6.8(b), which are adhered to several hexagonal calcium hydroxide crystals before heat curing, thus leading to a relatively loose microstructure compared to UHPC after heat curing. Visible microcracks propagating along the boundaries of the calcium hydroxide crystals are also found, as shown in Figure 6.8(b). However, the microstructures of the UHPC samples after 10 h of heat curing become much denser compared to that before heat curing, as shown in Figure 6.8(c). Meanwhile, it is less likely to observe the calcium hydroxide crystals in the sample after heat curing. The even denser microstructure of the sample after 48 h of heat curing (Figure 6.8(d)) also confirms that the chemical hydration degree is even higher. The reason is that during heat curing, the typically formed calcium with high-density nanocrystalline layers was distinctly difficult to detect the extreme value of the pore distribution curve. This indicates that heat curing can effectively accelerate the chemical hydration of UHPC and form more hydrated products to fill up the pores and voids of the internal structure, resulting in denser microstructure and smaller pores (Yu et al. 2015).

Figure 6.8 Microstructure of normal concrete and UHPC under different curing conditions: (a) Normal concrete (Bogas et al. 2020); (b) Room curing (Hu and Huang 2016); (c) 10 h of steam curing (Hu and Huang 2016); (d) 48 h of steam curing (Hu and Huang 2016); and (e) Autoclave curing (Shen et al. 2019).

6.4.3 Microstructure of interfacial transition zone

The SEM observations of microstructure of interfacial transition zone in conventional concrete and UHPC are shown in Figure 6.9. The interfacial transition zone between aggregates and paste matrix has a high porosity and calcium hydroxide content and is the weakest part of conventional concrete. However, UHPC has a very dense interfacial transition zone without noticeable pores (Wang et al. 2012). Normally, the microstructure of the interfacial transition zone is influenced by the "wall effect" in the vicinity of aggregate surfaces. This region is about 50 μm from the grain surface into the cement paste (Ollivier et al. 1995). However, sand with particle sizes ranging from 100 to 300 μm is used as the main aggregate, which can reduce the "wall effect" and thickness of the interfacial transition zone (Van Tuan et al. 2011). Moreover, such a small sand particle size and a very small thickness of the interfacial transition zone hinder the effect of supplementary cementitious materials, such as silica fume or rice husk ash, on reducing the interfacial transition zone thickness. Therefore, the thickness of the interfacial transition zone of all UHPC samples obtained is similar and very small (Van Tuan et al. 2011). However, owing to the

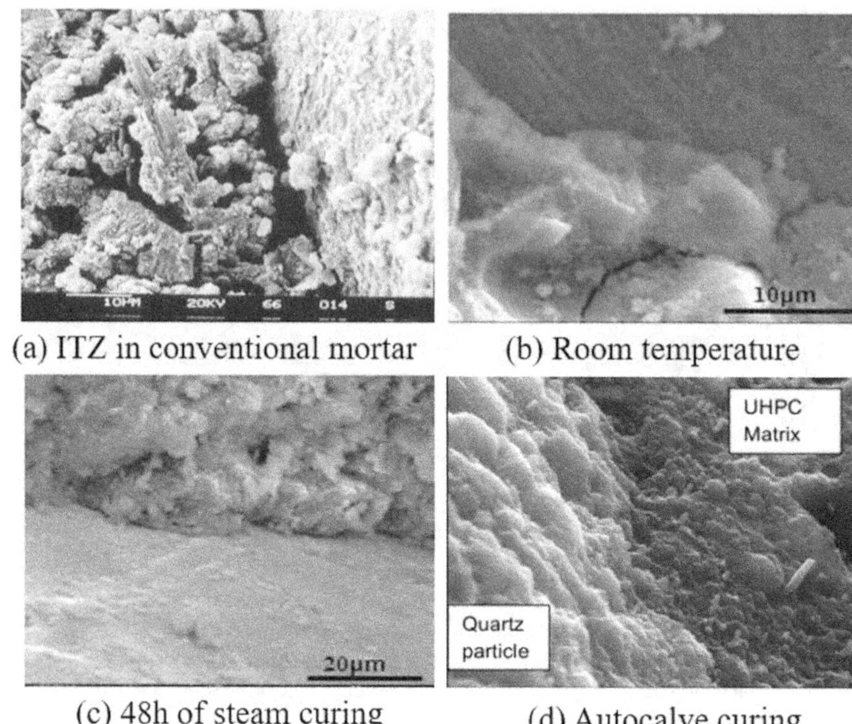

Figure 6.9 Interfacial transition zones of conventional concrete and UHPC under different curing conditions: (a) ITZ in conventional mortar (Shi et al. 2015); (b) UHPC under room curing (Hu and Huang 2016); (c) UHPC after 48 h of steam curing (Hu and Huang 2016); and (d) UHPC after autoclave curing (Shhmidt and Fehling 2007).

low w/b ratio and pozzolanic reactions between calcium hydroxide and reactive mineral admixture, which consumes most of the calcium hydroxide crystals and converts them to C-S-H (Chan and Chu 2004; Richard and Cheyrezy 1995), the interfacial transition zone in UHPC seems as dense as the matrix. The homogenous structure is important for the excellent performance of UHPC.

The denser microstructure of the interfacial transition zone in UHPC is closely related to high-temperature curing. When UHPC is cured in hot water at 90°C for 10 h, the hydration product becomes denser, calcium hydroxide cannot be observed (Figure 6.6(c)). Increased curing temperature could accelerate the hydration reaction of cement and pozzolanic activity and improve the microstructure of the interfacial transition zone (Cheyrezy et al. 1995). Autoclave curing can enhance the degree of hydration, and a homogeneous, dense microstructure consisting of close networked crystal fibers with lengths up to 1 micrometer would occur under the autoclave curing condition. Cracks and small pores are filled with crystalline C-S-H. The pore volume is lower than the heat-treated specimen, and the median pore diameter is reduced to 5 nm when autoclave curing is used. The autoclaved specimens exhibit dissolution processes around quartz grains, which produces a better cohesion between fillers and fine crystalline cement paste (Feylessoufi et al. 1996; Lehmann et al. 2009).

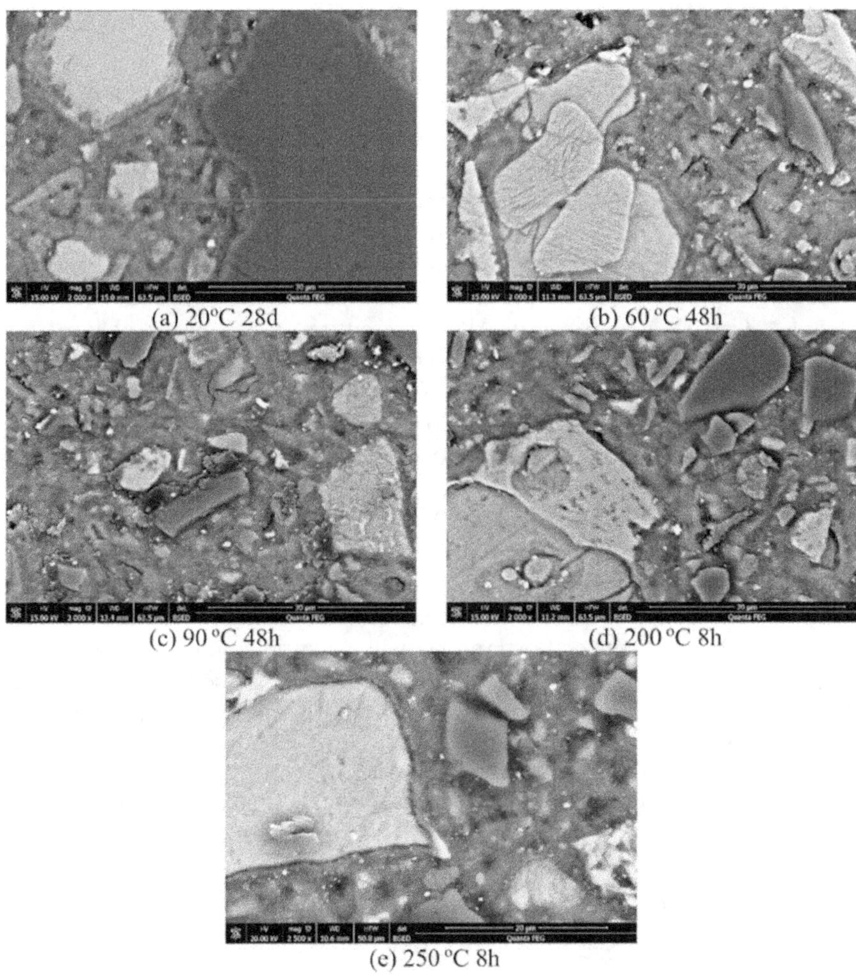

Figure 6.10 Backscattered electron (BSE) images of UHPC under different curing regimes (Shen et al. 2019). Note: (a) Standard curing at 20°C (RH 100%) for 28 d; (b) Steam curing at 60°C for 48 h; (c) Steam curing at 90°C for 48 h; (d) Autoclave curing at 200°C and 1.7 MPa pressure for 8 h; and (e) Autoclave curing at 250°C and 2.1 MPa pressure for 8 h.

Figure 6.10 shows the microtexture of polished cross-sections of UHPC samples under different curing regimes. Compared to UHPC cured at 20°C, a monolithic region surrounding the clinker can be observed in UHPC cured at 60°C. Such region is due to the increased hydration rate associated with steam curing. UHPC specimens under 90°C steam curing show a microcrack between clinker and hydrates in the form of separated hydration shells. Besides, UHPC specimens subjected to autoclave curing also show separated shells between clinker and hydration products. In addition, heat curing and autoclave curing can also improve the microstructure around the steel fiber (Shen et al. 2019).

6.5 PORE SOLUTION

The pore solution chemistry is of prime importance for understanding trace element mobility and migration process in cement. Important parameters of the cement pore solution chemistry are the ionic strength (which affects ion exchange reactions with the solid phases), pH (which determines the degree of hydrolysis for hydrolyzable ions), and ion composition (Andersson et al. 1989).

Usually, the pore solution is defined as the alkaline solution with OH^-, Na^+, K^+, Ca^{2+}, SO_4^{2-}, Si^{4+}, and AlO^{2-} presenting in the pores of hardened concrete (Vollpracht et al. 2015). The pH of this solution in ordinary Portland cement is in the range of 13 to 13.8 due to the dissolution of variable amounts of Na_2O and K_2O present in the cement. There is no relevant reference about the pore solution of UHPC because it is very difficult to obtain. Nevertheless, the pore solution of conventional concrete or HPC can provide hints for UHPC. Alkali concentrations in the pore solution depend strongly on the total alkali content of the Portland cement. This also applies to UHPC. For UHPC, the low w/b ratio can increase ion concentration. However, alkali concentrations generally decrease with the increasing replacement of Portland cement by supplementary cementitious materials. Partial replacement by supplementary cementitious materials thus reduces alkali concentrations as the alkalis incorporated in the glass fraction are slowly released compared to that from cement. In addition, the pozzolanic reaction of the supplementary cementitious materials leads to the formation of more C-S-H with lower Ca/Si ratio, which in turn results in an increased alkali binding since low Ca/Si C-S-H-phases take up more alkalis. Among the existing supplementary cementitious materials, silica fume is most effective in reducing the alkali concentration of the pore solution (Vollpracht et al. 2015).

6.6 SUMMARY

This chapter gives an overview of the setting and hardening of UHPC from four aspects: heat of hydration, hydration products, microstructure development, and pore solution. The following main conclusions can be drawn:

(1) The hydration of binders in UHPC is similar to that in conventional concrete. Silica fume and nanoparticles accelerate, whereas GGBS or fly ash retards the hydration of binder. When UHPC is cured under 90°C, the average C-S-H chain length increases. If the curing temperature is raised to 250°C, C-S-H will be dehydrated to form xonotlite.
(2) UHPC has very low porosity, especially after heat curing. Its microstructure is closely related to the addition of supplementary cementitious materials and curing conditions. The addition of silica fume or rice husk ash could decrease the porosity of UHPC due to their filling and pozzolanic effects. However, the total porosity of the rice husk ash–modified UHPC is higher than that of the silica fume–modified specimen. GGBS has little effect on the porosity of UHPC at early ages but increases the porosity of UHPC at later ages.
(3) The incorporation of nanomaterial as a partial replacement for cement appears to be effective in improving the properties of UHPC. It can produce better dispersion and interaction of the reinforcement systems and achieve significantly higher mechanical strength and durability.
(4) The curing method has a significant influence on the microstructure of UHPC. High-temperature curing can densify the microstructure due to accelerated pozzolanic reaction between calcium hydroxide and supplementary cementitious materials, such as silica fume. The autoclaved curing specimens exhibit dissolution processes around quartz grains, which produces a better cohesion between fillers and fine crystalline cement paste.

REFERENCES

Alaee, F. J. (2002). *Retrofitting of concrete structures using high performance fibre reinforced cementitious composite (HPFRCC)*. Cardiff University.

Andersson, K., Allard, B., Bengtsson, M., Magnusson, B. (1989). Chemical composition of cement pore solutions. *Cem. Concr. Res.* 19(3), 327–332.

Bishop, M. (2001). *Cement hydration inhibition and crosslinking in the guar-borate system*. Rice University.

Bogas, J. A., Carriço, A., Tenza-Abril, A. J. (2020). Microstructure of thermoactivated recycled cement pastes. *Cem. Concr. Res.* 138, 106226.

Bonneau, O., Vernet, C., Moranville, M., Aïtcin, P.C. (2000). Characterization of the granular packing and percolation threshold of reactive powder concrete. *Cem. Concr. Res.* 30(12), 1861–1867.

Chan, Y. W., Chu, S. H. (2004). Effect of silica fume on steel fiber bond characteristics in reactive powder concrete. *Cem. Concr. Res.* 34(7), 1167–1172.

Cheyrezy, M., Maret, V., Frouin, L. (1995). Microstructural analysis of RPC (Reactive Powder Concrete). *Cem. Concr. Res.* 25(7), 1491–1500.

Danielson, U. (1962). *Heat of hydration of cement as affected by water-cement ratio.* In Proceedings of the Fourth International Symposium on the Chemistry of Cement, Washington, USA, 519–526.

Feylessoufi, A., Crespin, M., Dion, P., Bergaya, F., Van Damme, H., Richard, P. (1997). Controlled rate thermal treatment of reactive powder concretes. *Advanced Cement Based Materials* 6(1), 21–27.

Feylessoufi, A., Villieras, F., Michot, L., De Donato, P., Cases, J., Richard, P. (1996). Water environment and nanostructural network in a reactive powder concrete. *Cem. Concr. Compos.* 18(1), 23–29.

Hu, G.Q., Huang, Z. H. (2016). Research on the shrinkage performance of ultra high performance concrete during heat curing (In Chinese). *Materials Review* 30(2), 113–119.

Ji, T. (2005). Preliminary study on the water permeability and microstructure of concrete incorporating nano-SiO_2. *Cem. Concr. Res.* 35(10), 1943–1947.

Justs, J., Wyrzykowski, M., Winnefeld, F., Bajare, D., Lura, P. (2014). Influence of superabsorbent polymers on hydration of cement pastes with low water-to-binder ratio. *J. Therm. Anal. Calorim*, 115(1), 425–432.

Kadri, E. H., and Duval, R. (2009). Hydration heat kinetics of concrete with silica fume. *Constr. Build. Mater.* 23(11), 3388–3392.

Kakali, G., Tsivilis, S., Aggeli, E., Bati, M. (2000). Hydration products of C_3A, C_3S and Portland cement in the presence of $CaCO_3$. *Cem. Concr. Res.* 30(7), 1073–1077.

Kang, S.-H., Hong, S.-G., Moon, J. (2019). The use of rice husk ash as reactive filler in ultra-high performance concrete. *Cem. Concr. Res.* 115, 389–400.

Korpa, A., Kowald, T., Trettin, R. (2009). Phase development in normal and ultra high performance cementitious systems by quantitative X-ray analysis and thermoanalytical methods. *Cem. Concr. Res.* 39(2), 69–76.

Langan, B. W., Weng, K., Ward, M. A. (2002). Effect of silica fume and fly ash on heat of hydration of Portland cement. *Cem. Concr. Res.* 32(7), 1045–1051.

Lehmann, C., Fontana, P., Müller, U. (2009). Evolution of phases and micro structure in hydrothermally cured ultra-high performance concrete (UHPC). In Bittnar, Z， Bartos, P. J. M.,Nemecek, J， Smilauer， V,Zeman, J., editors. *Nanotechnology in Construction 3*, Springer, Progue, Czech, 287–293.

Li, W. G., Huang, Z. Y., Zu, T. Y., Shi, C. J., Duan, W. H., Shah, S. P. (2016). Influence of nanolimestone on the hydration, mechanical strength, and autogenous shrinkage of ultrahigh-performance concrete. *J. Mater. Civ. Eng.* 28(1), 04015068.

Long, G., Xie, Y., Wang, P., Jiang, Z. (2005). Properties and micro/mecrostructure of reactive powder concrete. *J. Chin. Ceram. Soc* 33(4) 456–461.

Meng, W. N., Khayat, K. H. (2018). Effect of graphite nanoplatelets and carbon nanofibers on rheology, hydration, shrinkage, mechanical properties, and microstructure of UHPC. *Cem. Concr. Res.* 105, 64–71.

Muzenski, S., Flores-Vivian, I., Sobolev, K. (2019). Ultra-high strength cement-based composites designed with aluminum oxide nano-fibers. *Constr. Build. Mater.* 220, 177–186.

Nehdi, M., Mindess, S., Aïtcin, P.-C. (1996). Optimization of high strength limestone filler cement mortars. *Cem. Concr. Res.* 26(6), 883–893.

Norhasri, M. S. M., Hamidah, M. S., Fadzil, A. M. (2019). Inclusion of nano metaclayed as additive in ultra high performance concrete (UHPC). *Constr. Build. Mater.* 201, 590–598.

Odler, I. (2003). Hydration, setting and hardening of portland cement. *Lea's Chemistry of Cement and Concrete (Fourth Edition).* Elsevier, Amsterdam, Netherlands, 241–297.

Ollivier, J., Maso, J., Bourdette, B. (1995). Interfacial transition zone in concrete. *Adv. Cem. Based Mater.* 2(1), 30–38.

Philippot, S., Masse, S., Zanni, H., Nieto, P., Maret, V., Cheyrezy, M. (1996). ^{29}Si NMR study of hydration and pozzolanic reactions in reactive powder concrete (RPC). *Magn. Reson. Imaging* 14(7–8), 891–893.

Richard, P., Cheyrezy, M. (1995). Composition of reactive powder concretes. *Cem. Concr. Res.* 25(7), 1501–1511.

Rong, Z. D., Sun, W., Xiao, H. J., Wang, W. (2014). Effect of silica fume and fly ash on hydration and microstructure evolution of cement based composites at low water-binder ratios. *Constr. Build. Mater.* 51(2), 446–450.

Rougeau, P., Borys, B. *Ultra high performance concrete with ultrafine particles other than silica fume.* Proc., Ultra high performance concrete (UHPC). International symposium on ultra high performance concrete, 213–225.

Sato, T. (2006). *Applications of nanotechnology for the sustainable development of cement-based materials.* University of Ottawa.

Schmidt, M., Fehling E. (2005). Ultra-high-performance concrete: research, development and application in Europe. *ACI Spec. Publ* 228.1, 51–78.

Shen, P. L., Lu, L. N., He, Y. J., Wang, F. Z., Hu, S. G. (2019). The effect of curing regimes on the mechanical properties, nano-mechanical properties and microstructure of ultra-high performance concrete. *Cem. Concr. Res.* 118, 1–13.

Shhmidt, M., Fehling, E. (2007). *Grundlagen der Betontechnologie von Hochbund Ultra Hochleistungsbeton und Anwendung von UHPC im Bruckenbau.* Ultra High Performance Concrete–10 Years of Research and Development at the University of Kassel, Kassel, Germany, 70–81.

Shi, C. J., Wu, Z. M., Xiao, J. F., Wang, D. H., Huang, Z. Y., Fang, Z. (2015). A review on ultra high performance concrete: Part I. Raw materials and mixture design. *Constr. Build. Mater.* 101, 741–751.

Shi, C., Wang, D., Wu, L., Wu, Z. (2015). The hydration and microstructure of ultra high-strength concrete with cement–silica fume–slag binder. *Cem. Concr. Compos.* 61, 44–52.

Shi, Y. X., Matsui, I., Feng, N. Q. (2002). Effect of compound mineral powders on workability and rheological property of HPC. *Cem. Concr. Res.* 32(1), 71–78.

Sorelli, L., Constantinides, G., Ulm, F., Toutlemonde, F. (2008). The nano-mechanical signature of Ultra High Performance Concrete by statistical nano indentation techniques. *Cem. Concr. Res.* 38(12), 1447–1456.

Svermova, L., Sonebi, M., Bartos, P. J. M. (2003). Influence of mix proportions on rheology of cement grouts containing limestone powder. *Cem. Concr. Compos.* 25(7), 737–749.

Tang, M.-C. *High performance concrete—past, present and future. Proc., Proceedings of the International Symposium on Ultra High Performance Concrete.* Kassel, Germany, 3–9.

Van Tuan, N., Ye, G., van Breugel, K., Copuroglu, O. (2011). Hydration and microstructure of ultra high performance concrete incorporating rice husk ash. *Cem. Concr. Res.* 41(11), 1104–1111.

Vollpracht, A., Lothenbach, B., Snellings, R., Haufe, J. (2015). The pore solution of blended cements: a review. *Mater. Struct.* 49(8), 3341–3367.

Wang, C., Pu, X.C., Chen, K., Liu, F., Wu, J.H., Peng, X.Q. (2008). Measurement of hydration progress of cement paste materials with extreme-low w/b. *J. Mater. Sci. Eng.* 6, 852–857.

Wang, C., Yang, C. H., Liu, F., Wan, C. J., Pu, X. C. (2012). Preparation of Ultra-High Performance Concrete with common technology and materials. *Cem. Concr. Compos.*, 34(4) 538–544.

Wang, D. H., Shi, C. J., Wu, Z. M., Xiao, J. F., Huang, Z. Y., Fang, Z. (2015). A review on ultra high performance concrete: Part II. Hydration, microstructure and properties. *Constr. Build. Mater.* 96, 368–377.

Wu, Z. M., Shi, C. J., Khayat, K. H. (2016). Influence of silica fume content on microstructure development and bond to steel fiber in ultra-high strength cement-based materials (UHSC). *Cem. Concr. Compos.* 71, 97–109.

Wu, Z. M., Shi, C. J., Khayat, K. H., Wan, S. (2016). Effects of different nanomaterials on hardening and performance of ultra-high strength concrete (UHSC). *Cem. Concr. Compos.* 70, 24–34.

Yazıcı, H., Yiğiter, H., Karabulut, A. Ş., Baradan, B. (2008). Utilization of fly ash and ground granulated blast furnace slag as an alternative silica source in reactive powder concrete. *Fuel* 87(12), 2401–2407.

Yu, R., Spiesz, P., Brouwers, H. J. H. (2015). Development of an eco-friendly Ultra-High Performance Concrete (UHPC) with efficient cement and mineral admixtures uses. *Cem. Concr. Compos.* 55, 383–394.

Zanni, H., Cheyrezy, M., Maret, V., Philippot, S., Nieto, P. (1996). Investigation of hydration and pozzolanic reaction in Reactive Powder Concrete (RPC) using 29Si NMR. *Cem. Concr. Res.* 26(1), 93–100.

Zhu, W. Z., Gibbs, J. C. (2005). Use of different limestone and chalk powders in self-compacting concrete. *Cem. Concr. Res.* 35(8), 1457–1462.

Chapter 7

Dimensional stability

7.1 INTRODUCTION

Dimensional stability is an important engineering property of cementitious material, which significantly affects the long-term durability and serviceability of concrete structures. At early age, commonly defined as the first day, a certain amount of tensile stress will be developed in structure due to the restraints from adjunct materials or connected members when shrinkage occurs under restrained conditions. Since concrete has low strain capacity and is more sensitive to internal stresses after casting (Byfors 1980; Kasai et al. 1972), this stress may exceed tensile strength and prematurely causes microcracking or macrocracking (Bentz and Jensen 2004).

Despite the high mechanical strength and superior durability, UHPC exhibits serious autogenous shrinkage due to a very low water-to-binder (w/b) ratio and use of high quantities of binder (Shi et al. 2015; Wang et al. 2015). The rapid increase in autogenous shrinkage of UHPC at the first 24 h, which could easily exceed 1000 μm/m (Liu et al. 2019), might be responsible for the decrease in long-term durability. This shrinkage behavior of UHPC can be expected to undergo nearly half of its long-term shrinkage during the first 24 h after setting (Acker 2004). Several mitigation strategies have been proposed to reduce the autogenous shrinkage and cracking potential of cementitious materials made with low w/b, including the addition of shrinkage-reducing or expansive agents, the use of internal curing, the prolongation of moist curing durations, etc. To appropriately apply these strategies, it is important to have a full understanding of the driving forces behind autogenous shrinkage and how these shrinkage mitigation strategies work.

In this chapter, autogenous shrinkage, drying shrinkage, and creep of concrete are discussed. Testing methods and factors, such as characteristics of raw materials, w/b ratio, and curing condition, affecting autogenous shrinkage are described. Mitigation strategies, including control of hydration, addition of internal restraint, reduction of the surface tension of pore solution, formation of expansive products, and use of internal curing, for autogenous shrinkage of UHPC are finally provided.

DOI: 10.1201/9781003203605-7

7.2 AUTOGENOUS SHRINKAGE

7.2.1 Mechanisms of autogenous shrinkage

Autogenous shrinkage occurs over three different stages, i.e., liquid, skeleton formation, and hardening, within the first day after concrete mixing (Holt 2005). When concrete is still liquid, autogenous shrinkage is equivalent to chemical shrinkage. During the skeleton formation, a more rigid structure is formed as the stiffening of the matrix can partly resist chemical shrinkage stresses. At the same time, the capillary pressure starts to develop and causes shrinkage. The capillary pressure in the pore fluid is related to the water-air menisci in the partly empty pores. As water is lost from smaller pores, the water meniscus would be pulled into the capillary pores continually and generate more stress on the capillary pore walls, causing shrinkage of paste. This is similar to the phenomena described by Radocea (1992) for drying shrinkage. Once concrete reaches the hardened stage, the decreased relative humidity in the pores induced by the hydration of cement results in self-desiccation (Baroghel-Bouny 1996; Tazawa and Miyazawa 1995). Therefore, the moisture loss due to further hydration of cement after the formation of the initial structure of cement paste is called self-desiccation shrinkage. In the current publications, most researchers consider that autogenous shrinkage is similar to self-desiccation shrinkage.

There are four mechanisms commonly employed to explain the volume change in a cement-based system: capillary pressure, surface tension, disjoining pressure, and movement of interlayer water (Kovler and Zhutovsky 2006). The capillary pressure approach is advantageous among the four due to its sound mechanical and thermodynamical basis (Lura et al. 2003). It explicates the autogenous shrinkage using pore structure, relative humidity, self-stress, degree of hydration, and interface structure (Ye and Radlińska 2016). Most researchers identify that capillary pressure exists at high relative humidity of greater than 40% to 50% as menisci are established in that humidity range (Kovler and Zhutovsky 2006; Soroka 1979).

According to the capillary pressure theory, the capillary pressure in the pore fluid is related to the water-air menisci in the partly empty pores within the microstructure and the surface tension of pore solution. The formation of the liquid-vapor menisci in the pores can ultimately lead to the shrinkage of the three-dimensional microstructure. It can be seen that smaller meniscus radiuses in the low water-to-cement (w/c) ratio mixtures correspond to higher capillary pressures, resulting in a greater shrinkage. The capillary pressure can be calculated by the Young-Laplace equation (Eq. 7.1):

$$P_{cap} = \frac{2\gamma cos\theta}{r} \tag{7.1}$$

where P_{cap} is the capillary pressure; γ is the surface tension of pore fluid (N/m); θ is the liquid-solid contact angle (rad, assumed to be 0 rad); and r is the radius of curvature of the meniscus (m).

The capillary pressure can also be related to the internal equilibrium relative humidity as represented by Kelvin's equation (Eq. 7.2):

$$ln(RH) = \frac{-2\gamma V_m}{rRT} \tag{7.2}$$

where R is the universal gas constant (8.314 J/mol); T is the temperature (K); RH is the internal relative humidity; and V_m is the molar volume of pore solution ($\approx 18 \times 10^{-6}$ m^3/mol).

The linear shrinkage strain due to capillary stresses in the water-filled pores can be estimated using a modified version of Mackenzie's formula (Bentz et al. 1998):

$$\varepsilon = \frac{SP_{cap}}{3}\left(\frac{1}{K} - \frac{1}{K_S}\right) \tag{7.3}$$

where P_{cap} is the capillary tension in fluid (Pa); S is the degree of saturation of cement paste; K is the bulk modulus of paste (Pa); and K_s is the modulus of the solid skeleton inside cement paste (Pa).

7.2.1.1 Pore structure

Pore structure, including porosity, pore size distribution, and morphology of pores, plays an important role in determining the autogenous shrinkage of cement-based materials. The shrinkage stresses induced by the consumption of water in pores of different sizes are dissimilar. The connectivity of pores also directly affects the migration of moisture from saturated to unsaturated pores, thus influencing the development of shrinkage (Meddah and Tagnit-Hamou 2009). In general, a large amount of supplementary cementitious materials and a low w/c ratio lead to a refined pore structure, which accelerates the decreasing rate of the meniscus radius of capillary pores. Finer pores result in a lower relative humidity according to the Kelvin equation. Besides, the improved chemical reaction of cement paste due to the pozzolanic reactions of supplementary cementitious materials increases water consumption and the corresponding chemical shrinkage, which triggers additional surface tension. As a result, shrinkage stresses are increased.

7.2.1.2 Relative humidity

Since the equilibrium between adsorbed water and vapor pressure exist, the shrinkage mechanism can also be explained by the change in relative humidity inside the cement paste, which is connected to the degree of saturation. The relationship between the relative humidity and pore structure was discussed by some researchers (Grasley et al. 2006; Grasley and Lange 2007). They measured the relative humidity at various locations inside concrete and changes in thermal expansion and internal relative humidity of hardened cement paste. The results suggested that the number of saturated pores governed the change in relative humidity inside the concrete. From the point of thermodynamics, the reduced content of free water in capillary pores by progressive hydration of cement compounds is responsible for reducing the relative humidity in concrete. The reduction in relative humidity results in a change in the thickness of the water layer adsorbed on solid surfaces (Hagymassy et al. 1969). This is accompanied by the changes in the surface tension of solids and the disjoining pressure of adsorbed water between solid surfaces. At this early stage of hydration, the stiffness of paste is very low, and the viscous behavior is pronounced; the slightest stress acting on the system can result in large deformation.

7.2.1.3 Self-stress

Based on Hooke's law, strain and elastic modulus are physical variables determining the shrinkage stress of concrete. Li and Li (2014) established a relationship between the elastic modulus of micromatrix around capillary pores and the autogenous shrinkage using the capillary tension theory. They reported that both compressive strength and elastic modulus increased rapidly at an early age, and then the development rate slowed down, which affected the autogenous shrinkage. As a result, any factors affecting the slurry structure of the matrix at an early age can induce autogenous shrinkage. These factors can be divided into external and internal factors. Internal factors are types of aggregate and cement, w/c ratio, admixture, component size, etc., while external factors include temperature, humidity, sealing condition, and degree of restraint.

7.2.1.4 Degree of hydration

It is commonly assumed that chemical shrinkage varies directly with the degree of hydration. The degree of hydration is primarily influenced by cement type and content, supplementary cementitious materials, w/b ratio, and temperature. Many works have applied different techniques, such as microstructural analysis or thermodynamics, to measure or simulate the degree of hydration. Detailed information on the degree of hydration and the formation of C-S-H has been summarized in previous publications (Lothenbach et al. 2011; Damidot et al. 2011). Experimental results confirmed the relationship between the degree of hydration and the autogenous shrinkage of concrete by investigating the effects of raw material constitutes and curing methods on autogenous shrinkage, which will be discussed in Section 7.2.3.

7.2.1.5 Interface structure

Concrete is often considered a two-phase composite material composed of aggregates and hydrated cement paste or mortar. The interfacial transition zone serves as a bridge between them, where excess calcium hydroxide can be obtained compared to the bulk zone. Although the stiffness of each component is high, the stiffness of the composite may be reduced because of the broken bridges (i.e., voids and microcracks in the interface structure). Therefore, concrete structures depend on the combination of three components after molding, the curing age, and the degree of hydration in certain circumstances, which can affect the autogenous shrinkage. Previous studies reported an additional source of shrinkage due to the removal of restraints (calcium hydroxide crystals) caused by the pozzolanic reaction (Jensen and Hansen 1995). The ongoing pozzolanic reaction will result in a progressive decrease in relative humidity due to pore refinement and water absorption on the C-S-H gel formed. This finer pore structure should be the product of the dissolution of calcium hydroxide crystals and precipitation of C-S-H from the pozzolanic reaction.

7.2.2 Testing methods

Early-age shrinkage measurements for concrete are challenging due to the difficulties in making accurate measurements of concrete before demolding. The shrinkage must be measured immediately after casting in a mold that permits constant readings without disturbing concrete. At present, both direct and indirect testing methods can be used to measure the autogenous shrinkage of cement-based materials. Direct testing methods usually refer to volumetric or linear measurements. Indirect methods usually measure porosity or change in relative humidity of cement-based materials, which needs to establish the correlations between autogenous shrinkage and porosity or relative humidity. Different definitions for autogenous shrinkage also lead to various measurement results. Some literature define autogenous shrinkage from the initial setting. However, the times for the initial setting of different concrete mixtures are different and difficult to determine accurately. On the other hand, the shrinkage after the initial setting also includes those resulting from the hydration of cement and self-desiccation. Thus, it causes difficulty in proper determination and measurement of autogenous shrinkage. In some research, the measurement of autogenous shrinkage started 1 day after casting. Measuring autogenous shrinkage should start as soon as possible after mixing the concrete. In a word, there is no general agreement on the starting point of autogenous deformation, which is clearer and easier for measurement. This section summarizes different testing methods for autogenous shrinkage and discusses their advantages and disadvantages.

7.2.2.1 Direct testing method

Volumetric assessments are done by taking the weight of a paste or mortar sample sealed in a thin rubber membrane underwater. It measures volume changes of cement-based materials at constant temperature and the data recording starts immediately after casting. However, their limitations are apparent. That is, cement paste may be too weak to overcome the friction of tired rubber bags at an early age. Bleeding water and entrapped air are also obstructed significantly, if bleeding water is sucked back into the cement paste during the hydration of cement. Ultimate volume reduction in the immersed cement paste may be erroneously measured. Temperature control of cement paste is still yet to be successfully solved since it might cause unrepeatable results. The volumetric measurements can be classified into the membrane and capillary tube methods, which are introduced in detail as follows:

(1) Membrane method

The membrane method, or buoyancy method, was first used in the 1940s based on the change in the buoyancy of the specimen in a liquid (Yates 1941). After filling with fresh cement paste, a plastic membrane, or rubber, bag is sealed and immersed into a liquid immediately. Autogenous shrinkage is assessed by recording the weight change of the plastic membrane over time.

Water bleeding of the cement paste and changes in environmental temperature can affect the accuracy of the measurement in this method, especially when the w/c ratio is high (Ba 2003). Reabsorption of bleeding water on the surface of the specimens and change in temperature due to hydration of cement cause invalid results. A testing apparatus coupled with a temperature controller based on Yates' method was developed (Justness 1994). The film bag filled with cement paste on rotation is placed to avoid bleeding but it cannot be eliminated. The buoyancy method is easy to perform in the lab. However, there are two potential problems: (1) the frail rubber bag is only suitable for paste and mortar, not for concrete, and (2) the influences of heat release and capillary tension caused by the hydration of cement-based materials are ignored.

(2) Capillary tube method

The capillary tube method is typically used for measuring Le Chatelier contraction. A setup developed by Tazawa and Miyazawa (1995) showed that fresh cement paste is placed in tubes of different sizes. Oil or water is immediately added to the surface of the paste in the tube. Then, autogenous shrinkage is recorded with a measuring pipette inserted on top of the paste. It is thought that pure chemical shrinkage is measured when water is put on the top surface of cement paste since additional water in the sample can be supplied for the hydration of cement. However, the external water cannot reach the middle and bottom portions of the sealed paste. Self-desiccation develops from those portions. When the oil is used, recorded shrinkage includes both chemical and autogenous shrinkage. The portion of chemical shrinkage from the use of oil is smaller when compared with the use of water. Although the capillary tube method is simple and easy to perform, measured results may vary due to thermal effects, especially when the sample is large. It was found that the thickness of the specimen with a low w/c ratio shows a great effect on the measured results (Geiker 1983).

Linear measurement records the length change of cement paste, mortar, or concrete specimens cast in a rigid steel mold and recorded by sensors at one end or both ends of the specimens. The linear measurements are a more realistic representation of actual field construction and the material behavior, because thermal dilation, bleeding, setting time, and other factors are accounted for in the slab test. It is possible to convert the horizontal and vertical displacements measured on the slab test to a volumetric shrinkage. It is also possible to factor out the volume change resulting from thermal changes associated with cement hydration to measure pure autogenous shrinkage (Holt 2005). They can be classified into contact and noncontact methods according to the sensors used or horizontal and vertical methods, according to the displacement of specimens. The horizontal method is more commonly used because it can reduce the influence of gravity during the measurement. The specimens used can be either prisms or cylinders.

(a) Contact method

The contact method uses gauges or sensors embedded in or in direct contact with specimens for shrinkage measurement. The free deformation of the specimen can be ensured by embedded perpendicular thermocouples and sensors. However, the weight of the sample and the starting time of measurement can affect the measured results. Several methods for measuring the autogenous shrinkage of concrete have been reported by several researchers. Table 7.1 summarizes the common methods for measuring autogenous shrinkage at early age and their main features. According to ASTM C1698 (2014), a length-measuring gauge in contact with the specimen does not allow measuring shrinkage before the final set due to considerable errors

Table 7.1 Methods used for measuring autogenous shrinkage of cement-based materials

Measuring approach	*Main features*	*Reference*
Two dial gauges connected with bolt as data loggers	The problem of thermal-related deformations makes it reliable	Japan Concrete Institute (1999)
Inductance frequency-modulation type microdisplacement sensors	Sensors need to be chosen based on the actual measurement range to obtain more accurate measurements.	An (2001)
Fiber Bragg grating sensor	Sensors are noninvasive to the specimens due to small size, and the measurement can start at the very beginning stage after casting.	Tian et al. (2005)
Corrugated tube method	The corrugated tubes or bellows adopted can minimize the restraint on the specimens.	Jensen and Hansen (1995, 2002)

produced by the stress between the contact sensors and corrugated tube (Gao 2014). A noncontact corrugated tube system equipped with a laser sensor would be able to measure the autogenous shrinkage before setting.

(b) Noncontact methods

It is well known that contact sensors may damage cement-based materials at early stages and produce relative displacement when measuring heads are inserted. Noncontact methods, such as the laser displacement sensor and the eddy current sensor methods, are the best options to solve these problems. These sensors are suitable for bad industrial environments since they are not sensitive to oil pollution, dust, humidity, and magnetic field interference. Moreover, the results have good accuracy and stability. Figure 7.1 shows a noncontact corrugated tube system equipped with a laser sensor (Wu et al. 2019). Because the tubes are corrugated, they are much stiffer in the radial direction than they are in the longitudinal direction, thus effectively translating the volumetric deformation of paste/mortar to linear deformation.

However, this test method does not allow measuring shrinkage under drying conditions. One of the valuable methods is the laser distance sensor technique customized for a rigid mold system that can measure early-age shrinkage of UHPC under sealed or drying conditions in a contactless way (Yalçınkaya and Yazıcı 2017), as shown in Figure 7.2. In this setup, a climate-controlled cabinet is used with temperature and humidity transmitters, temperature sources (lamps), and humidity sources (ultrasonic steam diffusers). There is a humidity reducer fan to provide low relative humidity. All climatic operations and devices are controlled by a proportional-integral-derivative controller. The shrinkage

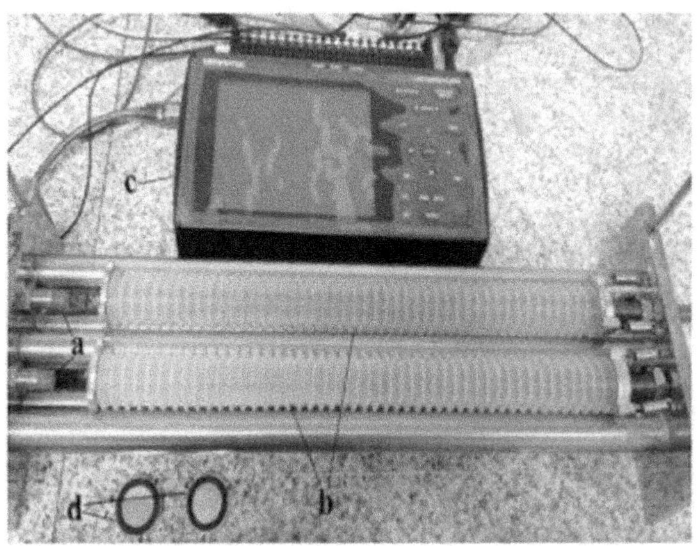

Figure 7.1 Test setup for autogenous shrinkage measurement of UHPC: (a) Noncontact sensor head; (b) Corrugated polyethylene tube; (c) Data acquisition instrument; (d) Stainless steel covers.

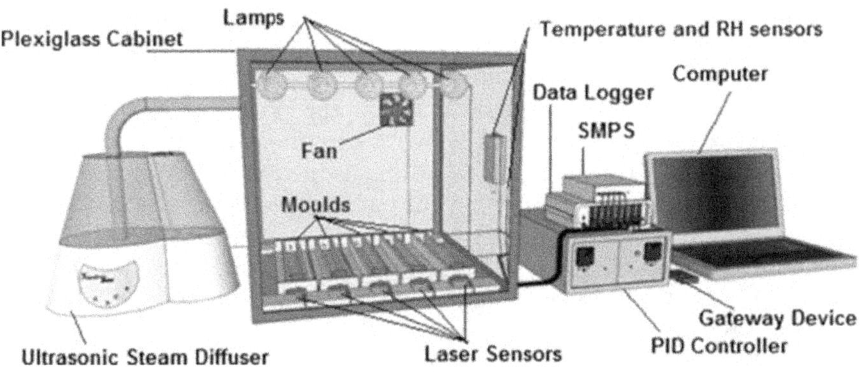

Figure 7.2 An innovative early-age shrinkage test setup.

Source: Yalçınkaya and Yazıcı 2017.

values are measured by five laser distance sensors for every five molds. The steel molds with 25 × 25 × 290 mm dimensions are covered by using plastic sheets and oiled to minimize friction. Temperature, relative humidity, shrinkage, and time data are simultaneously recorded by a computer.

In summary, the linear measurement can reflect the actual behavior of materials compared with the volumetric measurements. The measured results are more stable since studs are fixed on specimens during the measurement. Besides, the volumetric measurement can start as soon as cement paste is poured into a membrane or glass container. Among volumetric measurements, the buoyancy method is more commonly used even though the results are difficult to control. The corrugated tube method can turn volume change into length change and start the measurement immediately after casting. Thus, it is the most reasonable method for the measurement of autogenous shrinkage of cement-based materials so far.

7.2.2.2 Indirect testing method

The shrinkage and relevant mechanisms of porous cementitious systems are intimately related to their microstructure, including pore size distribution, surface area, morphology of hydrate solids, and pore solution properties (Ferraris and Wittmann 1987, Hansen 1987, Bent and Peltz 2008). Therefore, it is not surprising that the shrinkage phenomenon has been reported to vary considerably for different types of cementitious materials.

From the mechanism of internal curing, the formation of liquid-vapor menisci corresponds with the reduction in relative humidity inside cement paste. The increasing curvature of menisci leads to the development of hydrostatic tensile stresses in pore solution and compressive stresses exerted on the solid skeleton, which leads to the bulk deformation of cement paste. Thus, it is generally accepted that autogenous shrinkage is closely related to relative humidity and capillary pressure. As stated above, the capillary pressure depends on the pore size distribution of cement paste, which is affected by w/b ratio and mineral agents, and it varies with age. Therefore, the resulting behavior of shrinkage can be indirectly evaluated by examining the saturation fraction or fraction of water-filled pores and the pore structure (e.g., porosity and pore size distribution). Results from indirect methods are consistent with those from direct ones. However, chemical shrinkage at an early age cannot be measured by indirect measurements.

7.2.3 Factors affecting autogenous shrinkage of UHPC

The parameters that affect early-age shrinkage of concrete differ according to the driving forces behind each shrinkage mechanism. Although the early-age autogenous shrinkage is fully attributed to chemical shrinkage, it may behave differently concerning some factors, especially those that affect setting time and formation of a restraining structure. This section will describe the factors affecting the autogenous shrinkage from the following several aspects:

7.2.3.1 *Binder type*

(1) Cement

Cement is the core component of cement-based materials, and its composition can significantly affect the development of autogenous shrinkage. Rapid hydration of C_3A and C_3S and concentrated heat release of hydration increase the autogenous shrinkage of concrete. The impact of C_3A is the most dominant composition followed by C_3S, C_4AF, and C_2S (Van Breugel and Van Tuan 2015). Therefore, high-early-strength cement leads to greater autogenous shrinkage than that of low to moderate heat-generating cement in which higher C_2S contents are used. Besides, the higher fineness of cement promotes the development of autogenous shrinkage due to the accelerated hydration of cement. This can be attributed to the fact that an increased specific surface area of cement particles contributes to a faster reaction rate, decreasing relative humidity at a higher pace (Bazant and Chern 1985). A higher fineness of cement has a stronger ability for refinement of pores inside cement paste (Bentz 1999).

(2) Silica fume

Several types of fine supplementary cementitious materials have been used as microfillers to achieve the dense structure of UHPC and meet a need for a high degree of compactness and compressive strength. Silica fume is an industrial byproduct of ferrosilicium alloys production and has a typical average diameter of 0.2 μm, known as the main constituent of a typical UHPC mixture. The prominent effects of silica fume on the increased autogenous shrinkage are divided into three main functions: (1) filling effect, which is derived from increased nucleation sites for hydration products and increased space available for hydration product precipitation; (2) pozzolanic reaction, which leads to an increase in water consumption due to the production of additional C-S-H gel and the movement of internal restraints in the paste (represented by the calcium hydroxide crystals); and (3) refinement of pore structure, which increases the amount of fine capillary pores in the paste, thus resulting in high autogenous shrinkage of the specimens (Mazloom et al. 2004, Ghafari et al. 2016). The effects of silica fume contents on autogenous shrinkages of concrete with w/b ratios of 0.26 and 0.35 are shown in Figure 7.3. The autogenous shrinkage increases with the increase of silica fume content and is more significantly affected at a lower w/c ratio.

(3) Fly ash

The much slower hydration of fly ash than that of cement increases the effective w/b ratio by dilution effect. The initial hydration phase could be delayed so that fewer C-S-H are formed. As a consequence, the amount of fine pores is reduced, and the reduction in relative humidity is decelerated,

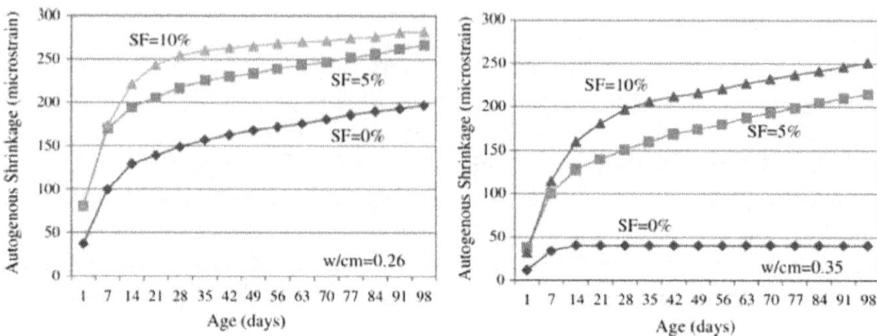

Figure 7.3 Effect of silica fume content on autogenous shrinkage of concrete.

Source: Zhang et al. 2003.

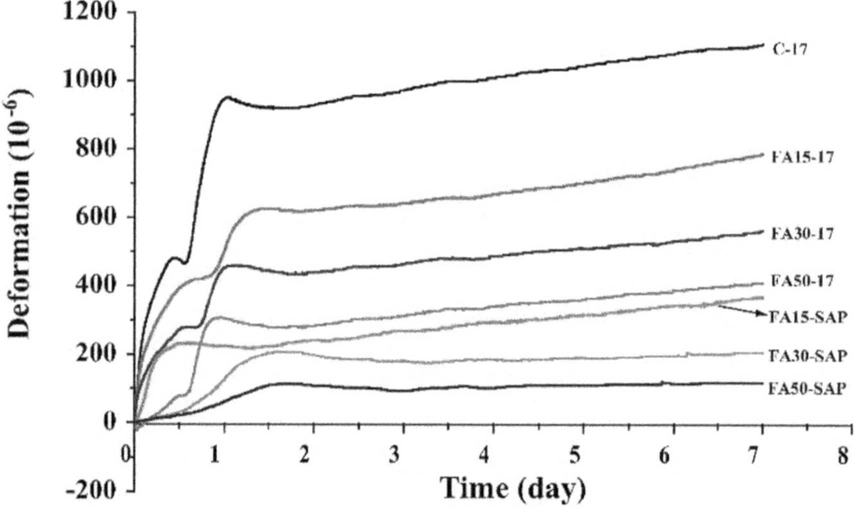

Figure 7.4 Effect of fly ash content on autogenous shrinkage of UHPC with and without internal curing.

Source: Liu et al. 2017.

which mitigates the early autogenous shrinkage. As shown in Figure 7.4 (Liu et al. 2017), the autogenous shrinkage of UHPC is reduced by about 28.9%–63.1% at 7 d as the replacement level of cement by fly ash increased from 15% to 50%. A low volume of fly ash may not inevitably reduce the autogenous shrinkage (Jiang et al. 2004). The effect of the replacement of

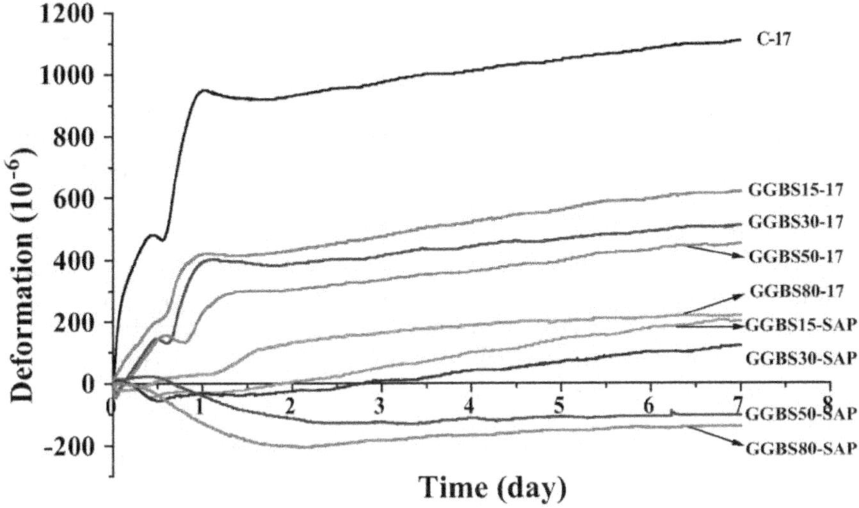

Figure 7.5 Effect of slag content on autogenous shrinkage of UHPC with and without internal curing.

Source: Liu et al. 2017.

silica fume by fly ash on the autogenous shrinkage of UHPC was investigated by Ghafari et al. (2016). Compared to the mixture without any fly ash, the 28-d autogenous shrinkage values of UHPC with 50% and 100% replacement level of fly ash were reduced by 11.1% and 16.7%, respectively.

(4) Slag

Generally speaking, there is a great controversy on the effect of slag on autogenous shrinkage. Some researchers considered that the addition of high-volume slag could accelerate the water consumption rate and rapidly decrease the critical radius of capillary pores, thus increasing the development of autogenous shrinkage (Lee et al. 2006). However, other researchers debated the increasing effect of slag on the autogenous shrinkage of concrete.

The effect of slag on the autogenous shrinkage of UHPC has been discussed in previous publications (Liu et al. 2017; Rößler et al. 2014; Yalçınkaya and Yazıcı 2017; Liu et al. 2018). Figure 7.5 shows that the 7-d autogenous shrinkage of UHPC is reduced as the replacement level of cement by slag increases to 50% (Liu et al. 2017). The trend is similar to that observed in Snoeck's work (2015). However, opposite results were also obtained by other researchers. Yalçınkaya and Yazıcı (2017) investigated the effect of the replacement of cement by 50% slag on autogenous shrinkage of UHPC. They reported that such replacement level of slag increased the 1-d autogenous shrinkage by approximately 57%. This opposite phenomenon was explained by Liu et al. (2018), who noticed the retardation of hydration at an early stage due to the low reactivity of slag. In contrast, the refined pore structure and the decrease of internal relative humidity caused by the rapid consumption of free water increased autogenous shrinkage.

The reactivity of slag relates to its fineness, and it might have a great influence on the development of autogenous shrinkage (Tazawa and Miyazawa 1997). The incorporation of slag with higher activity is probably responsible for the refinement of pore structure and greater chemical shrinkage, which induces faster and greater self-desiccation, thus resulting in greater autogenous shrinkage (Yalçınkaya and Yazıcı 2017). Besides, as the replacement ratio increases, a large amount of unhydrated slag particles acts as filler in the UHPC matrix, which would lead to fewer hydration products and lower shrinkage (Itim et al. 2011; Wu et al. 2017).

Ghafari et al. (2016) also measured the effect of slag as a substitute for silica fume on the autogenous shrinkage of UHPC. They found that when the silica fume was replaced by 50% or 100% slag, the 28-d autogenous shrinkage was decreased by about 5.6% and 8.9%, respectively. However, Pyo et al. (2017) found that using slag to replace partially silica fume showed no remarkable effect on the autogenous shrinkage developed in UHPC, although slag with an average particle size about nine times greater than silica fume was used.

(5) **Rice husk ash**

The high proportion of purity amorphous silica in rice husk ash greatly depends on the combustion temperature. Rice husk ash produced with combustion temperature at 500–700°C has high amorphous silica, but uncontrolled rice husk ash contains high carbon content (Salas et al. 2009). Van Tuan et al. (2011) found that the function of refining pore structure of UHPC by rice husk ash was greatly less than silica fume, but rice husk ash could promote the degree of hydration at later ages. Kang et al. (2019) reported that partially replacing silica fume with rice husk ash could improve pozzolanic reaction even if relatively lower amorphous silica content existed in rice husk ash compared to that in silica fume. This can be attributed to the internal curing resulting from its porous structure. Besides, they also found that rice husk ash containing high carbon content remarkably increased the pores larger than 50 nm. The relatively coarser pore structure is responsible for the lower autogenous shrinkage.

Rößler et al. (2014) showed that the total substitute of silica fume by rice husk ash greatly decreased the autogenous shrinkage of UHPC by 53.2% at 168 days. Additionally, the reduction rate of internal relative humidity in UHPC with rice husk ash was lower than that of UHPC with silica fume. Using rice husk ash to replace quartz powder appears to be more reasonable concerning the similarity in grain size and morphology. Because rice husk ash improves the pozzolanic reaction, the refinement of pore structure in UHPC can be effectively achieved because the capillary pores are filled by hydration products formed.

Ye (2012) studied the influence of the replacement ratio and fineness of rice husk ash on the autogenous shrinkage of UHPC without silica fume (Figure 7.6). Compared to a reference UHPC, the autogenous shrinkage is reduced by around 40% and 95% at 28 days when rice husk ash content is increased by 10% and 20%, respectively. It is found that a large size of rice husk ash is more effective in reducing autogenous shrinkage and a sudden expansion at 15 days occurs when using higher rice husk ash content. The former is due to the collapsed structure of rice husk ash in the process of grinding, but there is no explanation for the latter behavior.

(6) **Metakaolin**

Metakaolin obtained at a controlled temperature has comparable and even higher pozzolanic reactivity than silica fume (Salas et al. 2009), along with the acceptable cost. Metakaolin has a relatively strong water absorption capacity due to its layered microstructure (Song et al. 2018). It is found that the addition of metakaolin contributes to a better distribution of steel fiber in UHPC (Rong et al. 2018). Researchers showed conflicting results about the influence of silica fume replaced by metakaolin on autogenous shrinkage of UHPC. For example, a reduction of around 50% in autogenous shrinkage of UHPC at 6 d could be obtained when about 33% silica fume was replaced by metakaolin (Staquet and Espion 2004). However, Song et al. (2018) observed that the replacement of 40% silica fume with metakaolin significantly decreased the pore size distribution of UHPC into the range of about 5–10 nm compared to the UHPC with 0% and 100% metakaolin. This opposite phenomenon could be explained as follows: (1) the different reactivity of metakaolin used, which depends on the grain size and the content of active composition; (2) the variation in the retardation of cement hydration resulting from the different contents of superplasticizer; and (3) different proportions of raw materials. Additionally, Norhasri et al. (2019) evaluated the properties of UHPC containing nanometakaolin and compared the results to the mixtures with and without metakaolin. They found that the inclusion of nanometakaolin could delay the hydration of the cementitious pastes due to the increase in water demand. Nanometakaolin only acts as filler at an early age owing to the slow reaction. However, the addition of metakaolin accelerates the hardening process of UHPC.

(7) **Quartz powder**

The use of quartz powder as a cement replacement is expected to reduce autogenous shrinkage. Xiao (2013) reported that 20%–25% quartz powder amount demonstrated the lowest autogenous shrinkage, while UHPC

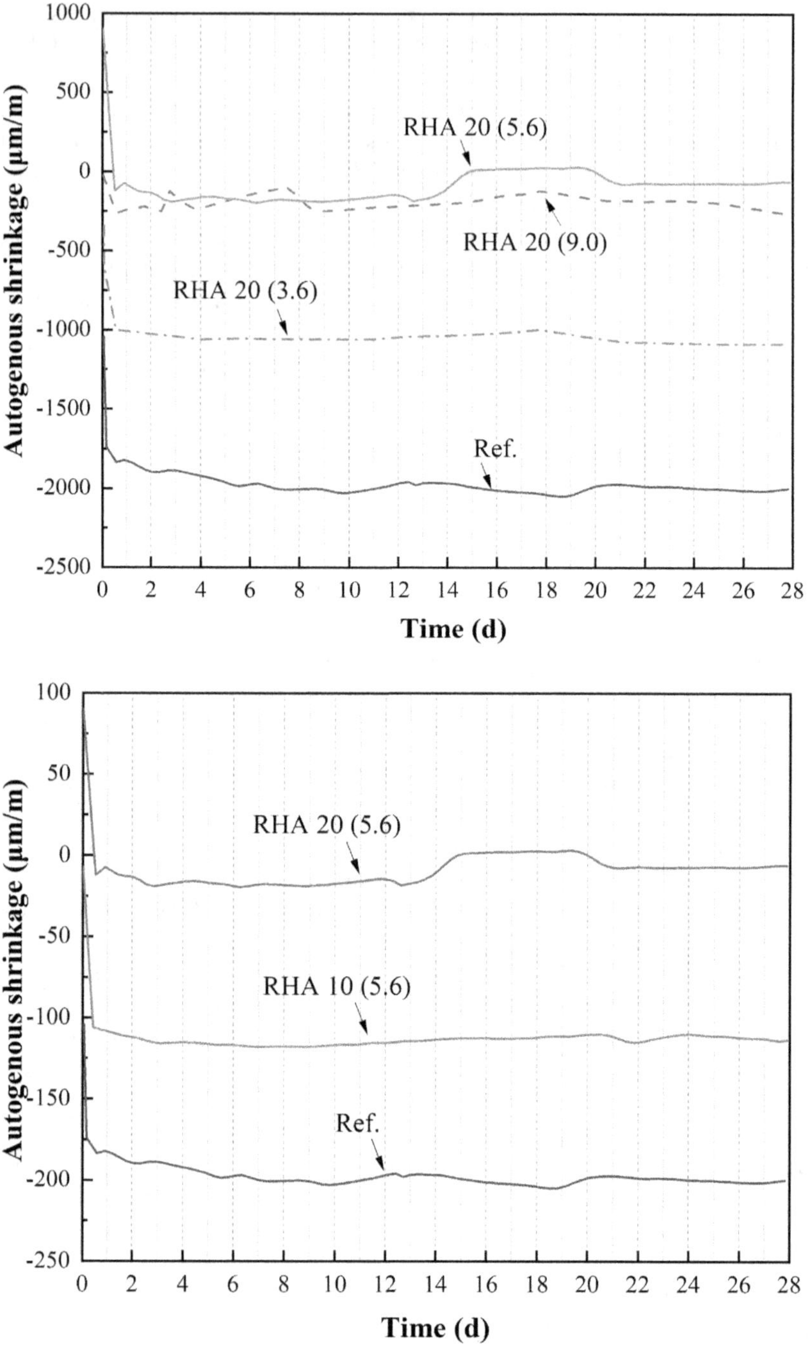

Figure 7.6 Effects of rice husk ash content and fineness on autogenous shrinkages of UHPC (modified based on Ye 2012). (a) Different amounts of RHA. (b) Different fineness values of RHA.

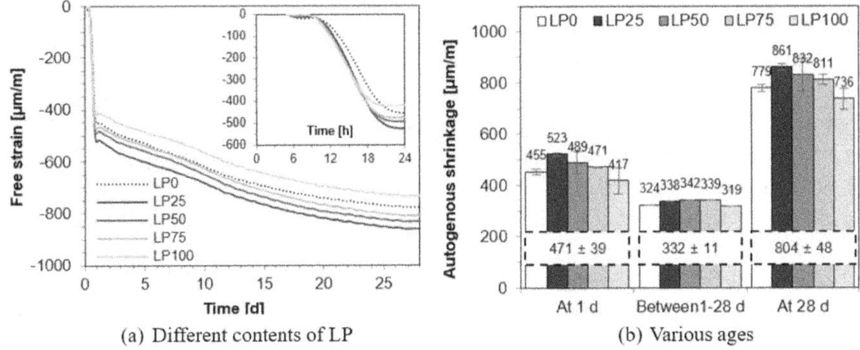

(a) Different contents of LP (b) Various ages

Figure 7.7 Effects of different limestone powder contents on autogenous shrinkages of UHPC.

Source: Kang et al. 2018.

with the lowest and highest quartz powder contents showed the greatest autogenous shrinkage. This phenomenon could be attributed to the more compacted cement pastes, which could reduce autogenous shrinkage to some extent.

The grinding of quartz would cause environmental hazards and human carcinogenic risks (Soliman and Tagnit-Hamou 2017). Using recycled materials or by-products to replace quartz powder has been carried out to develop a more sustainable UHPC. Kang et al. (2018) noticed that the development of autogenous shrinkage in UHPC was accelerated due to the filler effect within 1 day when quartz powder was replaced by limestone powder. Except for UHPC with quartz powder fully substituted by limestone powder, the autogenous shrinkage of UHPC mixtures with limestone powder was increased. However, a reduction in the increased rate of autogenous shrinkage could be observed with the increase of limestone powder, as shown in Figure 7.7. This could be explained by the reaction of C_3A with $CaCO_3$ to form calcium carboalumination like the reaction of C_3A with $CaSO_4$ to form calcium sulfoaluminate (Carlson and Berman 1960).

(8) **Wollastonite**

Soliman and Nehdi (2011) reported that the autogenous shrinkage developed in UHPC was greatly influenced by the content and aspect ratio of wollastonite. The 7-d autogenous shrinkage of UHPC incorporated 4% wollastonite with the lowest aspect ratio was reduced by 16.3%. This could be ascribed to the water displaced from cement particle clusters when using finer wollastonite.

7.2.3.2 *Water-to-binder ratio*

Water content has a major role during early-age shrinkage by controlling the microstructure and pore system, consequently affecting the capillary tension and meniscus development. The mixtures have higher cement content at the lower w/c ratio, which results in larger autogenous shrinkage due to the contribution of chemical shrinkage. On the other hand, the reduction of w/b ratio accelerates the internal relative humidity in cement paste, particularly at an early age (Liu 2019). This is because the total pore volume of UHPC reduces and pore structure is refined as the w/b ratio decreases, which induces water redistribution toward the inside of finer micropores and accelerates the reduction of the critical radius. As a result, a significant increase in capillary negative pressure is obtained.

7.2.3.3 *Aggregate*

The aggregate restraining causes stress in the matrix and results in radial and tangential microcracking around the aggregate. The rigidity of aggregate also plays a role in this response and there will be a more restraining effect on cement matrix shrinkage. Because of the restraining effect of aggregates, an increase in volume of aggregate with a larger elastic modulus can effectively restrain the autogenous shrinkage of concrete. The size of aggregates also controls the autogenous shrinkage of concrete. The finer aggregates (0–2 mm) efficiently mitigate the autogenous shrinkage of cement paste at an early age due to the relatively low modulus of elasticity, but their stiffness is not enough to resist the strains at later ages. However, the use of aggregate with a coarser size restrains the shrinkage of cement paste at later ages (Akcay 2018).

To assess the contribution of aggregate restraint to autogenous shrinkage, tests were done on neat paste, mortar, and concrete by Holt (2005). Due to the lack of aggregate providing restraint, paste exhibited greater autogenous shrinkage than a similar mortar at the same w/c ratio of 0.35. The difference in the two mixtures of this work showed the neat paste having about 1.7 times greater shrinkage compared to the mortar. Based on the mortar and paste comparisons in Figure 7.8(a), the concrete mixture has less shrinkage than that of mortar due to the higher content of aggregate restraining the shrinkable paste.

Most researchers reported that the decrease in the binder-to-sand ratio leads to a reduction in autogenous shrinkage of UHPC (Xiao 2013, 2018; Shen et al. 2018; Li et al. 2018; Meng et al. 2018). As shown in Figure 7.9, the autogenous shrinkage of the UHPC decreases with an increase in the binder-to-sand ratio due to the more significant shrinkage restraint provided by the residual binders in the UHPC prepared with a higher binder-to-sand ratio.

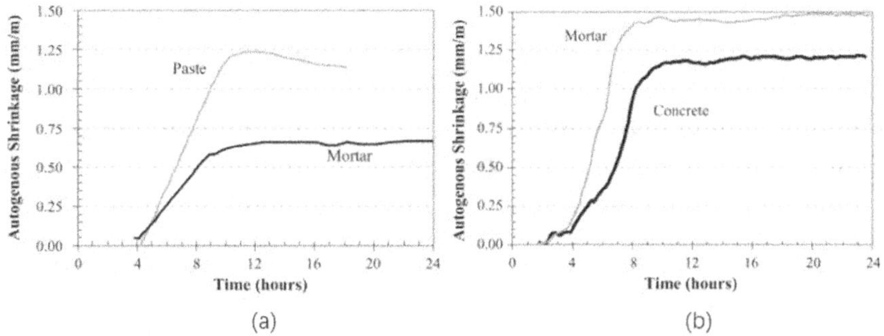

Figure 7.8 Autogenous shrinkage for: (a) Paste and mortar at w/c ratio of 0.35; (b) Mortar and concrete at w/c ratio of 0.30.

Source: Holt 2005.

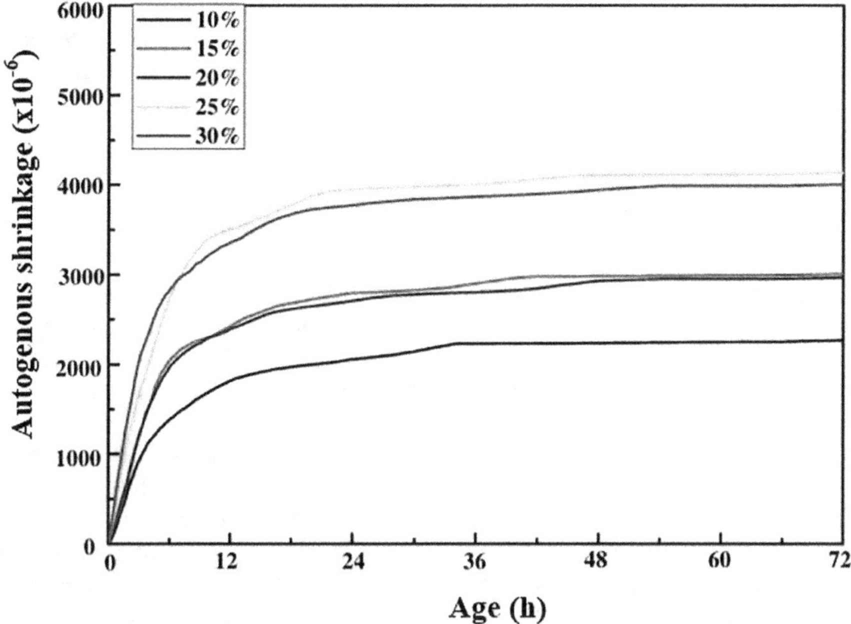

Figure 7.9 Autogenous shrinkages of UHPC with different sand-to-cement ratios.

Source: Xiao 2013.

7.2.3.4 Fibers

Fibers can resist the shrinkage deformation of the cementitious matrix with a low elastic modulus at an early age and delay the occurrence of cracking owing to the stress transfer from matrix to fibers. The efficiency of fibers in restraining shrinkage is greatly affected by their size, type and content, dispersion and orientation, elastic modulus, and aspect ratio.

(1) Steel fibers

In general, the positive effect of steel fibers on reducing autogenous shrinkage can be explained by the following mechanisms (Mangat and Azari 1984). In the mixing process of steel fiber and concrete, a few microns of water film is formed on the steel fiber surface, and the calcium hydroxide crystals grow with no constraints and arrange on the surface of steel fiber directionally, which forms a loose reticular structure in the interface of steel fiber and matrix. Besides, the high elastic modulus of steel fiber can support the skeleton in the process of free shrinkage of concrete. Due to the tensile stress of fiber, a range of bond stress distribution is created around the fiber close to the cracks, and the development of the crack tip will be inhibited due to constraints and barriers of fibers. Hence, the reduction of autogenous shrinkage is due to the smaller interspace between the microsteel fiber and the concrete matrix, denser filling, stronger bonds between fiber and concrete matrix, and a more excellent interface structure with low content of microsteel fibers. However, it should be noted that as the steel fiber content exceeds a certain amount, the cohesiveness increases, and the mobility of concrete is reduced in return. This decreases the compactness of concrete and results in the growth of pores and autogenous shrinkage in the concrete (Habel et al. 2006).

The autogenous shrinkage of UHPC is decreased with the increase of steel fiber content. Shen et al. (2018) found that when adding more than 1% of steel fibers, the decrease rate of about 5% corresponded to an increase of 1% steel fibers. However, Li et al. (2018) found no significant restraint effect of steel fibers on autogenous shrinkage of UHPC from 1 to 60 days when steel fiber content increased from 1% to 2% and 3%. This is consistent with the results reported by Wu et al. (2019). It was shown that the decreasing degree of autogenous shrinkage of UHPC was not obvious when the steel fiber content increased from 2% to 3%. The use of 2% corrugated and hooked fibers could be expected to reduce the autogenous shrinkage more efficiently in comparison with UHPC with straight fibers (Figure 7.10). This can be attributed to the enhancement of bond strength between fiber and matrix, which is also found by Fang et al. (2020). However, Meng and Khayat (2018) found that the use of 2% hooked-end fibers increased the 56-d autogenous shrinkage of UHPC due to the agglomeration of fibers. Additionally, they considered that the volume content of less than 1.5% for

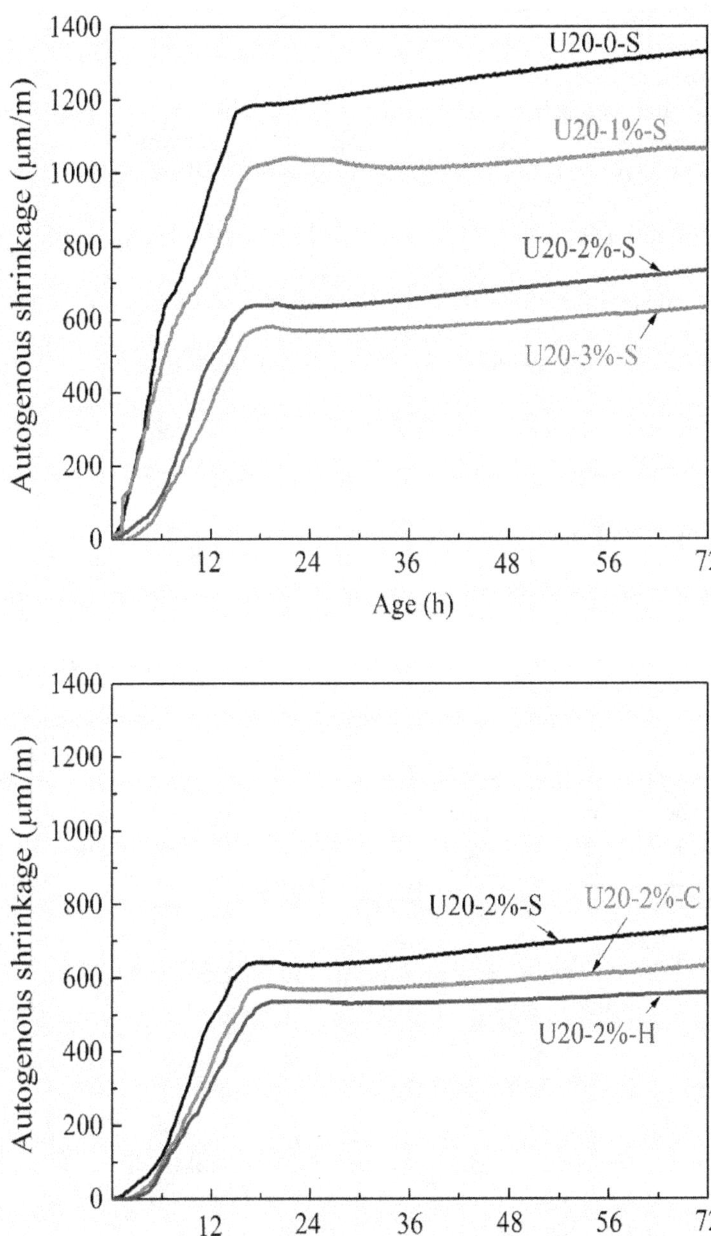

Figure 7.10 Effects of content and type of steel fiber on autogenous shrinkages of UHPC. S, C, and H represent straight, corrugated, and hooked fibers, respectively (Wu et al. 2019). (a) Various fiber volumes. (b) Various fiber shapes.

hooked-end fibers and 3% for straight fibers were suggested for use during UHPC mixing and casting.

(2) Synthetic fibers

Synthetic fibers, such as polypropylene and polyvinyl alcohol fibers, are more widely used in concrete. The hydrophilic polyvinyl alcohol fiber can absorb water in fresh concrete, thus affecting the development of autogenous shrinkage to some extent (Wu et al. 2017). Meng and Khayat (2018) found that with the addition of 1.5% steel fibers and 0.5% polyvinyl alcohol fibers, the autogenous shrinkage of UHPC was 60% of the mixture only added with steel fibers. This is due to the function of bridging microcracks of polyvinyl alcohol fibers. However, polyvinyl alcohol fibers could promote the internal connection of pores by bridging effect, which increases the permeability, as proposed by Hannawi et al. (2016). Besides, the adhesion between polyvinyl alcohol fibers and matrix is relatively weak compared to the more compact interfacial zone of steel fibers and matrix.

(3) Inorganic fiber

Meng and Khayat (2018) found that increasing nanofibers content increased the autogenous shrinkage of UHPC. This can be explained by the following aspects: (1) the high surface area of nanofibers results in an increased amount of nucleation sites for precipitating hydration products and promotes the degree of hydration of binder, and (2) the refinement of pore structure and the increase of mesopore with the size between 2 and 50 nm accelerates self-desiccation. The UHPC with graphite nanoplatelet shows less autogenous shrinkage than carbon nanofiber, due to the larger length and aspect ratio and smaller stiffness of the carbon nanofiber material compared with the graphite nanoplatelet. However, Lim et al. (2019) noticed that the incorporation of carbon nanofiber contributed to the decrease of autogenous shrinkage. Two possible explanations are as follows: (1) the carbon nanofiber dispersed homogeneously enhances the nanobridging between capillary pores and nano- or micropores, and (2) much stronger bonding properties between nanofibers and cement matrix restrains shrinkage deformation to some extent.

(4) Cellulose fibers

Cellulose fibers are other types of fibers being used in concrete industry. Kawashima and Shah (2011) reported that the addition of 1% cellulose fibers in mortar with w/b ratio of 0.28 significantly inhibited the cracks caused by autogenous shrinkage due to the internal solidification of the matrix. As a result, it was stated that the incorporation of cellulose fibers, especially at higher contents, had internal curing capabilities. However, it might harm the workability of paste, which limits its application in concrete.

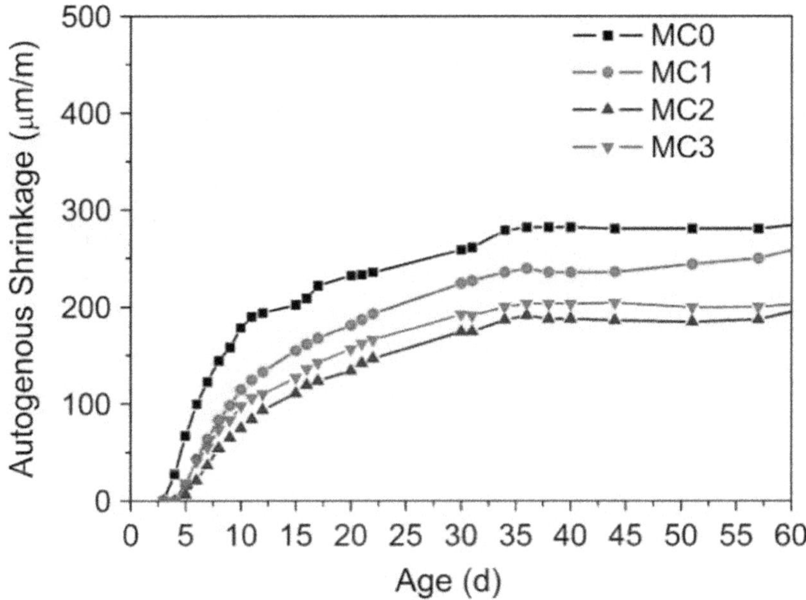

Figure 7.11 Effects of various contents of cellulose fibers on autogenous shrinkages of UHPC.

Source: Ma et al. 2019.

Ma et al. (2019) studied the effect of different contents of cellulose fibers on the autogenous shrinkage of UHPC. As exhibited in Figure 7.11, MC0, MC1, MC2, and MC3 represent the mixtures with cellulose fibers of 0, 0.7, 0.9, and 1.1 kg/m^3, respectively. It was found that the autogenous shrinkage of UHPC with 0.9 kg/m^3 cellulose fibers was decreased the most, i.e., up to about 70% and 33% lower than the reference UHPC at 7 and 60 days, respectively.

Compared to steel and synthetic fibers, cellulose fibers not only act as a reinforcement to prevent brittle failure and improve the performance of UHPC at a much lower cost but also act as an internal reservoir due to their lumen structure (Yan et al. 2014). As shown in Figure 7.12, at the center of the elementary fiber, the concentric cylinders with a small open channel in the middle called the lumen contribute to water uptake. Their highly hydrophilic feature, porous microstructure, and water-holding capacity allow them to absorb enough water regardless of the mixing process or dedicated presaturation (Ferrara et al. 2015). The obvious benefit of using cellulose fibers over the porous aggregate and superabsorbent polymer is their resistance to crack formation and propagation. Furthermore, cellulose

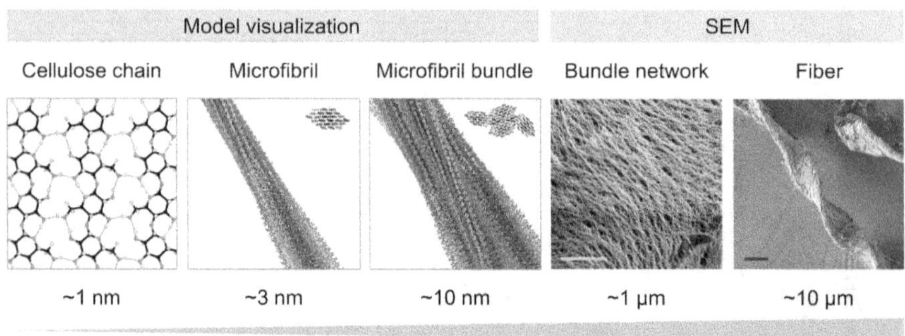

Figure 7.12 A schematic view of the multiscale structures of cellulose fibers, from left to right: molecular visualization of a layer of cellulose chains, a microfibril, and a bundle of four microfibrils, and a whole fiber. The orange and blue scale bars in the micrographs indicate 1 and 10 μm, respectively.

Source: Paajanen et al. 2019.

fibers are subjected to little volume change during water transportation, thus achieving a better internal curing effect (Ma et al. 2019). It is widely accepted that the addition of cellulose fibers also contributes to the control of cracking as induced by drying and plastic shrinkage. However, studies regarding their effect on autogenous shrinkage of UHPC are limited.

7.2.3.5 Chemical admixtures

As an essential component of concrete, different types of chemical admixtures have been widely applied to improve properties such as workability, pumpability, setting properties, mechanical performance, and durability, such as freeze-thaw resistance and shrinkage properties. This section will discuss the effects of superplasticizers, shrinkage-reducing admixtures, and expansive agents on the autogenous shrinkage of UHPC.

(1) Superplasticizer

The use of high-performance superplasticizers is critical for UHPC to achieve the desired rheology at an extremely low w/b ratio. The commonly used superplasticizer in UHPC is a comb-shaped polycarboxylate ether polymer, which can produce electrostatic repulsion and/or steric hindrance to release the entrapped water among cement particles, as illustrated in Figure 7.13 (Kong et al. 2013).

The presence of superplasticizers disturbs the process of cement hydration, thus leading to a relatively long dormant period of around 30 h for cement hydration in UHPC compared to the process of pure cement hydration with

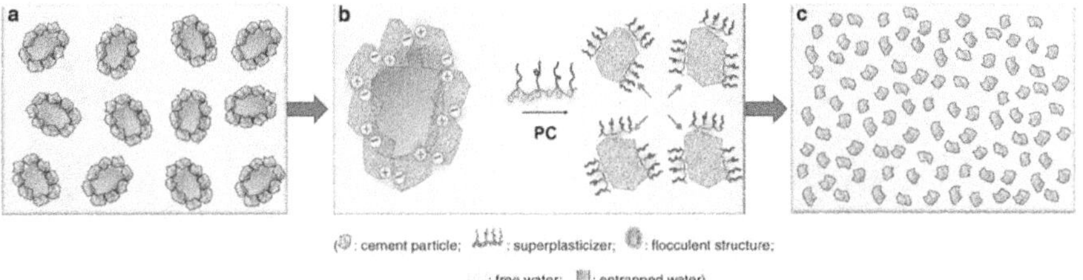

Figure 7.13 Mechanism of action of superplasticizer.

Source: Kong et al. 2013.

the dormant period lasting only about 1–2 hours (Yu et al. 2014). Rößler et al. (2014) also reported that the retardation effect induced by the high superplasticizer content resulted in the prolongation of the induction period of UHPC and prevented nucleation and growth of hydration products (Ciobanu et al. 2013).

It has been reported that the addition of 1% superplasticizer increases the autogenous shrinkage of mortar with w/c ratio of 0.3 by almost 30% (Holt 2005). The improved cement dispersion and a faster rate of hydration reactions in the matrix account for this phenomenon. On the other hand, the lower shrinkage of specimens without superplasticizer is attributed to the increased heterogeneities, such as cluster formations between aggregates and their internal restraining effect.

Adsorbed amounts of superplasticizer on cement particles resulting from charge characteristics (amount of anionic groups and charge density) and its concentrations remaining in the aqueous phase determine its dispersion efficiency. It should be noted that this action simultaneously retards cement hydration, which might mitigate the development of autogenous shrinkage to some extent. Previous publications confirmed that the variation of setting time retardation due to the addition of superplasticizer affected the development of autogenous shrinkage of UHPC (Li et al. 2017; Rößler et al. 2014; Yu et al. 2014). For the pastes with the same type of superplasticizer, the absolute autogenous shrinkage could be reduced with the increase of superplasticizer dosage, as seen in Figure 7.11(b), which means that a slower physical coagulation and chemical process within a certain period could result in a smaller autogenous shrinkage. However, there is no obvious correlation between autogenous shrinkage and setting with different types of superplasticizers from Figure 7.14(a). This is probably because the physical coagulation and chemical process have different influences on the final setting time and autogenous shrinkage.

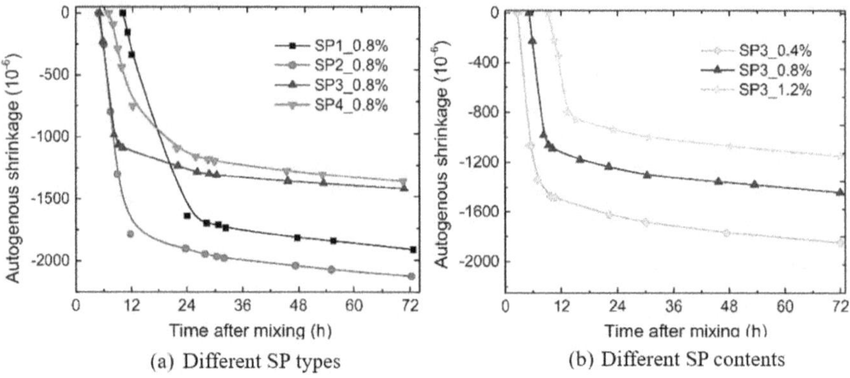

Figure 7.14 Effect of superplasticizer on autogenous shrinkage of UHPC.

Source: Li et al. 2017.

(2) Shrinkage-reducing admixtures

Shrinkage-reducing admixtures are mainly composed of amphiphilic (i.e., surfactant) molecules, known to reduce shrinkage deformation and cracks of cement-based materials. The interactions between shrinkage-reducing admixtures and pore solutions can be decisive in their effect on cement-based materials. The presence of shrinkage-reducing admixture efficiently delays the development of capillary pressure on the concave surface and lessens the shrinkage cracking of concrete at the plastic stage, especially for high-strength concrete. However, shrinkage-reducing admixtures also have potential risks of early strength drop in concrete, an extension of the aggregating time, and sometimes adverse interactions with other agents such as air-entraining agents (Tam et al. 2012).

Several papers reported that the shrinkage reduction by shrinkage-reducing admixtures in UHPC is more significant than that with the use of wollastonite (Soliman and Nehdi 2014) and expansive agents (Su et al. 2017). However, the decline rate decreases slightly with the increase of shrinkage-reducing admixtures content from 1% to 2%, as seen in Figure 7.15 (Su et al. 2017). Soliman and Nehdi (2011) investigated the efficiency of shrinkage-reducing admixtures at different temperatures and found that 40°C was more beneficial in decreasing autogenous shrinkage than those at 10°C and 20°C. A possible reason is that water is quickly consumed at high temperatures, while shrinkage-reducing admixtures remain or are consumed at a slower rate, resulting in an increased concentration of shrinkage-reducing admixtures. The temperature effect may also make some contributions. Given pure water, it is reported that the surface tension increases by a factor of approximately

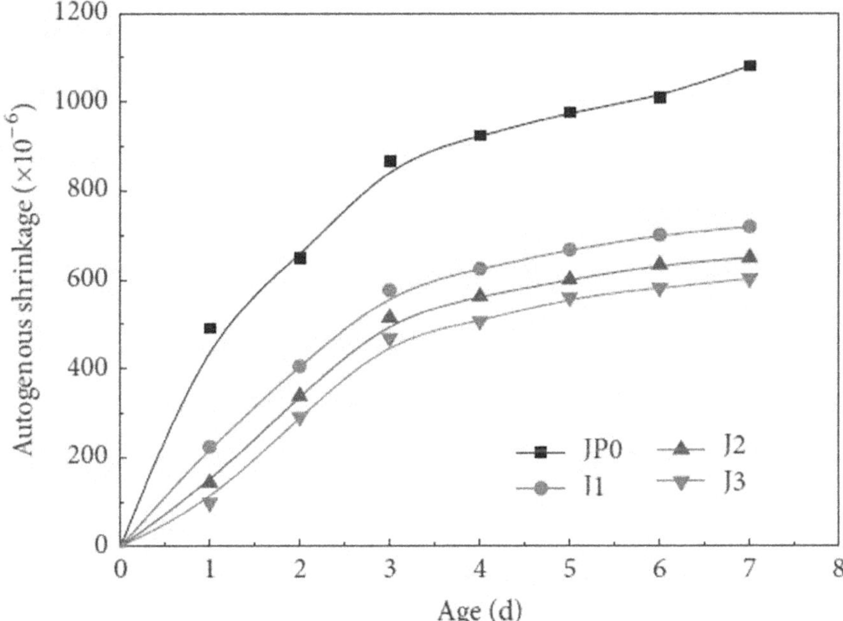

Figure 7.15 Effect of shrinkage-reducing admixtures on autogenous shrinkage of UHPC.

Source: Su et al. 2017.

1.09 when the temperature decreases from 40 to 0°C (Bentz and Jensen 2004). Given the adverse effect on the early hydration and strength development (Folliard and Berke 1997) and the loss of entrained air in the case of individual use of shrinkage-reducing admixtures (Berke et al. 2003), the hybrid use of shrinkage-reducing admixtures and superabsorbent polymers is generally recommended. Liu et al. (2019b) addressed that the use of 2% shrinkage-reducing admixtures and 0.3% superabsorbent polymers in ultra-high-strength concrete is the most effective in terms of minimum autogenous and drying shrinkage.

(3) Expansive agents

The macroscopic expansion of the system by the inclusion of expansive agents can be utilized to partially compensate for the autogenous shrinkage generated in the early stage. However, it is noticed that concrete mixed with expansive agents should be under sealed or sufficient curing conditions. Otherwise, the initial crack time could be earlier, and the ability of shrinkage cracking resistance might be reduced in dry environments

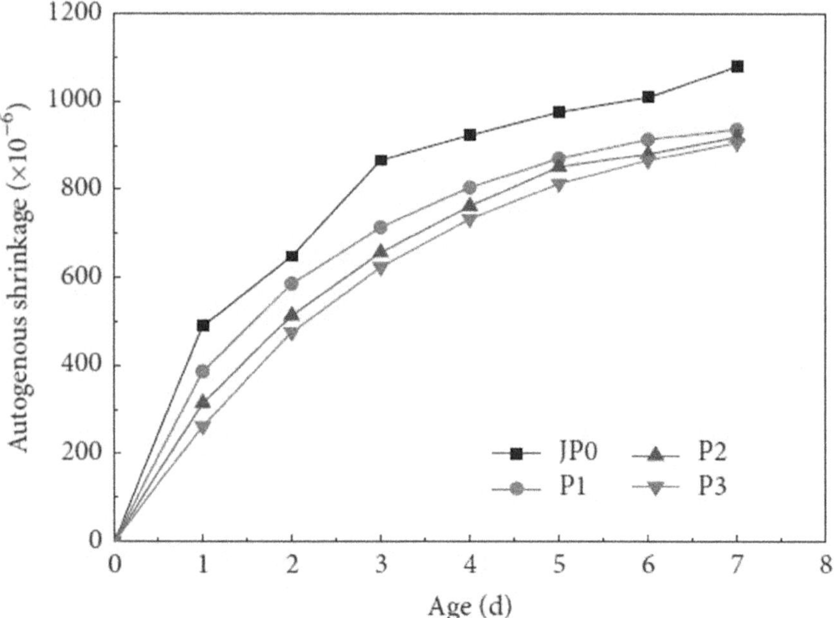

Figure 7.16 Autogenous shrinkages of UHPC with different contents of expansive agents.

Source: Su et al. 2017.

(Liu et al. 2006). In general, MgO-based expansive agents are more efficient than aluminate-based agents in cement-based materials with low w/b ratio (Mo et al. 2012). This is due to the relatively low water requirement for the hydration of MgO and its slow hydration rate.

Figure 7.16 exhibits the effect of expansive agents on autogenous shrinkage of UHPC. Compared to the use of shrinkage-reducing admixtures, as shown in Figure 7.12, it is obvious that no extremely significant improvement in autogenous shrinkage is observed when using expansive agents even if the content is increased (Su et al. 2017). This agrees with the results of Su et al. (2017). However, the combined effect of expansive agents and shrinkage-reducing admixtures could significantly reduce the autogenous shrinkage of UHPC (Park et al. 2011; Koh et al. 2011). Valipour and Khayat (2018) stated the combined use of expansive agents and saturated lightweight aggregate could significantly compensate for the autogenous shrinkage of UHPC, and the expansion phenomenon was more obvious when employing higher content.

7.2.3.6 *Curing conditions*

(1) Curing temperature

In addition to the aforementioned factors, the rate and the amplitude of autogenous shrinkage are strongly affected by the temperature history. Despite some discrepancies, most investigations have revealed that higher curing temperature causes faster development of shrinkage as compared to low curing temperature (Soliman and Nehdi 2011; Huang and Hu 2016; Lura et al. 2001). Figure 7.17 shows the effect of curing temperature on the development of autogenous shrinkage of UHPC (Soliman and Nehdi 2011). The results indicated that the autogenous shrinkage was strongly accelerated at an early age owing to the enhancement degree of cement hydration and activity of pozzolanic materials when the temperature was increased from 10 to 40°C. Later, the rate of shrinkage strain started to decrease due to the restraint effect of a solid skeleton rapidly developed at high temperatures. This is consistent with the results of Rößler et al. (2014), who found that the UHPC cured at 20°C showed lower autogenous shrinkage than the heat-treated specimens with 65 and 90°C curing for 7 d. However, no increase in autogenous shrinkage after 7 d was observed for heat-treated UHPC.

Yalçınkaya and Yazıcı (2017) reported that with the increase of temperature from 20 to 30 and 40°C, the 1-d autogenous shrinkage of UHPC with fly ash was increased by 33% and 72%, respectively. Shen et al. (2018) found that silica fume content exhibited a greater effect on autogenous shrinkage of UHPC cured at 90°C than that of cement content. However,

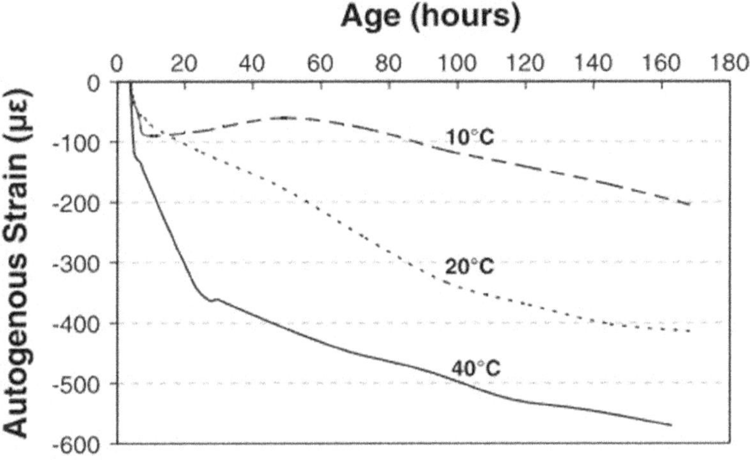

Figure 7.17 Autogenous shrinkage of UHPC cured at 10, 20, and 40°C.

Source: Soliman and Nehdi 2011.

the opposite phenomenon can be observed when UHPC was cured at room temperature.

Reducing temperature also exerts a significant influence on autogenous shrinkage. Xie et al. (2018) found that the autogenous shrinkage of UHPC was reduced as the proportion of crushed ice substituted for mixing water increased. This was due to the deceleration of the hydrate reaction at a lower temperature. Besides, the autogenous shrinkage of UHPC was greatly affected by the high temperature in different humidity conditions (Han et al. 2018). For instance, when the temperature was increased from 25 to 75°C, the 7-d autogenous shrinkage values of UHPC under soaked, sealed, and dry conditions were increased by 183.2%, 35.9%, and 71.5%, respectively

Figure 7.18(a) shows the change in shrinkage during the temperature rise period before heat curing (Huang and Hu 2016). It is found that the development of shrinkage deformation exhibited three distinctive stages: (1) the dormant stage with almost zero shrinkage lasting about 1 h due to short curing time and low curing temperature (below 50°C); (2) a gradually increased stage with shrinkage lasting around 0.5 h due to the promoted cement hydration (reaching 50°C); and (3) the acceleration stage with rapidly occurring shrinkage due to the secondary hydration of active admixtures occurred (reaching 70°C). Figure 7.15(b) exhibits that the cement hydration and pozzolanic reaction are completed after 10 h. The ultimate shrinkage value is about 450 μm/m after 60 days, which agrees well with the results concluded by Yoo et al. (2018). Heat treatment will accelerate the achievement of the ultimate shrinkage value and will effectively stabilize the UHPC against further shrinkage indefinitely.

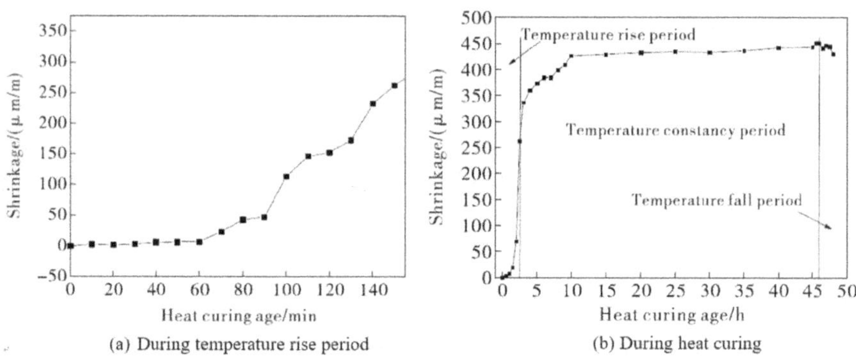

Figure 7.18 Effect of heat curing on autogenous shrinkage of UHPC.

Source: Huang and Hu 2016.

(2) Curing methods

For UHPC with a compact pore structure, external curing has little effect on the reduction of the autogenous shrinkage, while internal curing can compensate for the adverse effect induced by external curing on early shrinkage cracking of UHPC. The internal curing of concrete refers to the introduction of extra water by adding absorbent materials in concrete. Commonly used materials for internal curing include porous aggregates and polymer materials with high water absorption, e.g., superabsorbent polymer and cellulose fibers. When the internal relative humidity drops, the absorbed water can gradually be released into the matrix to compensate for the water consumption inside the cement paste. As a result, the relative humidity inside concrete can be maintained at a high level and self-desiccation can be suppressed.

Just et al. (2015) monitored that water entrained with superabsorbent polymer increased the hydration degree in a way having similarities to the increase of water-to-binder ratio, but the shrinkage reduction by the incorporation of superabsorbent polymer was much more significant than the increase of water-to-binder ratio (Figure 7.19). Nevertheless, the presence of superabsorbent polymer in UHPC with slag resulted in net expansion for 7 d (Mo et al. 2017). Kang et al. (2018) found that large and globular pores were formed in UHPC when superabsorbent polymer with stronger

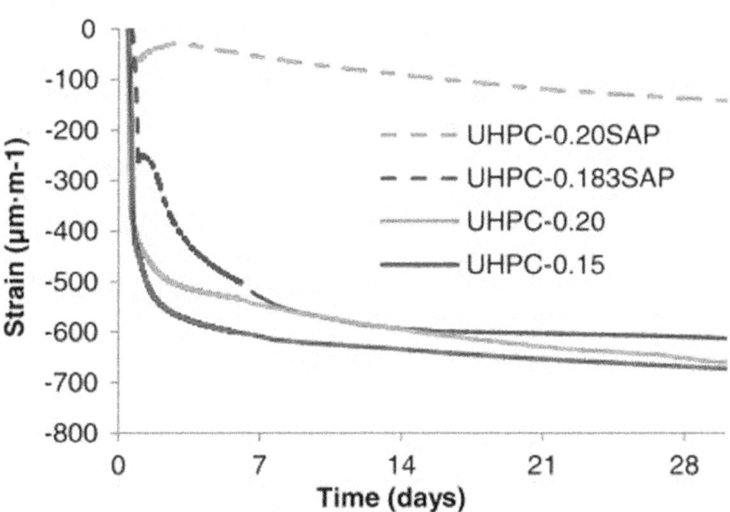

Figure 7.19 Effects of different w/b ratios and superabsorbent polymer additions on autogenous shrinkage of UHPC.

Source: Justs et al. 2015.

absorption capacity was used, which enhanced the deformability of the matrix due to the expansion. The early expansion phenomenon could be caused by the swelling of cement gel resulting from the water absorption of superabsorbent polymer or ettringite formation (Jensen and Hansen 2002; De la Varga and Graybeal 2014). Additionally, the author proposed that the internal water provided by superabsorbent polymer was remarkably effective in mitigating a considerable amount of shrinkage rapidly occurring in the heat curing period. Soliman and Nehdi (2011) pointed out that the efficiency of superabsorbent polymer was more obvious at a lower temperature (10°C). For example, the autogenous shrinkage of UHPC was decreased by approximately 71% at 10°C in comparison to the reduction in autogenous shrinkage by 24% and 28% at 20 and 40°C, respectively.

A reduction in autogenous shrinkage of UHPC by using porous aggregates has been found by some researchers (Liu et al. 2017; Wang et al. 2017). Meng and Khayat (2017; 2018) reported that the optimum substitution ratio of expanded shale for river sand was 25%. However, Liu et al. (2019c) noticed that the autogenous shrinkage of UHPC was first reduced when expanded shale content increased from 0% to 5%, 10%, and 15%, while the autogenous shrinkage of UHPC with 20% expanded shale was increased contrarily compared to that of UHPC with a 15% expanded shale. This was due to the reduction of the effective w/b ratio.

The combined use of porous aggregates and chemical shrinkage has a very significant effect on the development of autogenous shrinkage of UHPC (Valipour and Khayat 2108). Figure 7.20 shows the variations in autogenous shrinkage for UHPC with different contents of shrinkage-reducing admixtures and expansive agents. It can be seen that the incorporation of 5% MgO-based expansive agents coupled with 60% porous aggregates (EXM5LWS60) does not have a considerable effect on autogenous shrinkage with shrinkage values of 140 μm/m at 91 d compared to 35 μm/m shrinkage of mixture with only porous aggregates (LWS60). On the other hand, a greater content of 7% MgO-based expansive agents with 60% porous aggregates leads to an expansion of 250 μm/m at 28 d. Compared to the LWS60 mixture, the incorporation of shrinkage-reducing admixtures does not reduce autogenous shrinkage.

7.2.4 Mitigation strategies

Based on the factors affecting the autogenous shrinkage of UHPC, mitigation techniques can be classified into five aspects according to the restraining mechanism: (1) control of hydration reaction; (2) addition of internal restraint; (3) reduction of the surface tension of pore solution; (4) formation of expansive products; and (5) replenishment of water through internal curing (Yang et al. 2020).

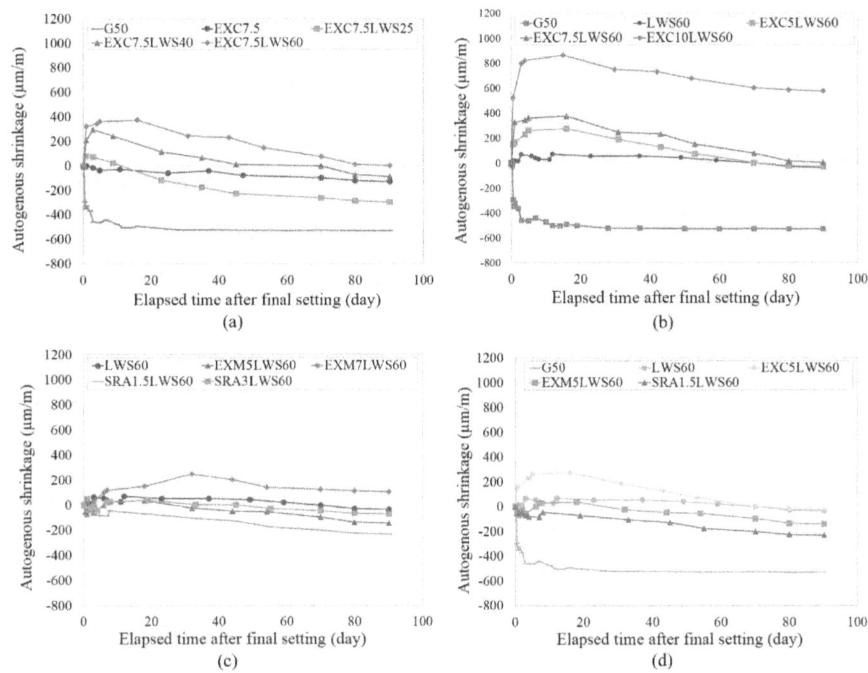

Figure 7.20 Effects of different chemical admixtures on autogenous shrinkages of UHPC.

Source: Valipour and Khayat 2108.

7.2.4.1 Control of hydration

The hydration of cement results in chemical shrinkage, which determines the pore structure of bulk paste, namely, the volume of empty porosity created, while the pore size distribution determines the size of the largest water-filled pore within the microstructure. At a given temperature, autogenous shrinkage is determined by the mixture composition of concrete, since the developed chemical shrinkage depends on the cementitious reactions. Changes in mixture composition, including the type and fineness of cement, and the addition of supplementary cementitious materials, etc., will cause variation in autogenous shrinkage due to the difference in the evolution of cement hydration. This section will discuss the reduction of autogenous shrinkage by controlling the hydration of cement from the effect of major constituents of concrete and curing methods.

(1) The replacement of cement and/or silica fume with other supplementary cementitious materials and/or inert materials

As mentioned before, silica fume is an essential constituent of UHPC. Its pozzolanic reaction accelerates cement hydration and is less sensitive to self-desiccation compared to the reaction of cement clinker (Jensen and Hansen 1996). The addition of silica fume increases pore volume in the size of 5–50 nm, which is considered to be one of the main factors influencing the autogenous shrinkage and has a significant positive correlation with the natural logarithm of autogenous shrinkage (Igarashi et al. 2005). On the other hand, the role of silica fume in refining pore structure and promoting hydration degree of UHPC has a greater effect on autogenous shrinkage than the variation in chemical shrinkage resulting from the decrease of cement content. As a result, the addition of silica fume increases the autogenous shrinkage of UHPC. The use of other supplementary cementitious materials, such as fly ash, slag, rice husk ash, and metakaolin, and inert materials to partially replace the cement and/or silica fume in UHPC is often driven by costs and improvements in long-term properties. Their effect on the development of autogenous shrinkage can be explained by the chemical properties, physical properties, replacement level, and action of refining pore structure, and they are discussed in the following section.

Chemical properties: Supplementary cementitious materials vary in chemical composition, including oxide composition, phase composition, and amorphous content. The chemistry of supplementary cementitious materials is generally characterized by lower calcium content compared to that of Portland cement. The reactivity of supplementary cementitious materials highly depends on the materials' nature and the alkalinity of the pore solution. At given pH and temperature, the dissolution of the glassy phase in supplementary cementitious materials is complex and variable, which is affected by the presence of calcium and aluminum. It is reported that calcium and aluminum can prevent the glassy phase from dissolving due to the formation of precipitated products on the surface of the glassy phase (Maraghechi et al. 2016). The addition of supplementary cementitious materials reduces the heat of hydration and thus retards the hydration process, as observed in fly ash blended systems (Hanehara et al. 2001). As a result, the chemical shrinkage and self-desiccation caused by the hydration process are reduced. However, it should be noted that with a further increase of supplementary cementitious materials, the microstructure and strength of UHPC can be negatively affected, lowering the formation of a stiff skeleton that can resist shrinkage stress (Wang et al. 2018; Soliman and Nehdi 2011).

Physical properties: The physical properties of supplementary cementitious materials also influence the development of autogenous shrinkage of UHPC. For example, the surface area and/or fineness of particles directly affect the pozzolanic reactivity and hydration degree of supplementary cementitious materials. Apart from enhanced nucleation and extra space for the growth of C-S-H during cement hydration, a reduction in particle sizes of supplementary cementitious materials increases their pozzolanic reactivity, thus increasing autogenous shrinkage. Tazawa and Miyazawa (1995) reported that the replacement of cement with up to 70% slag with Blaine fineness of 400 m^2/kg or higher gradually increased the autogenous shrinkage of paste. However, no increase in autogenous shrinkage was observed in a paste made with slag at Blaine fineness of 300 m^2/kg. The enhancement of slag activity due to higher fineness is probably responsible for the refinement of pore structure and greater chemical shrinkage, which induces faster and greater self-desiccation, thus resulting in greater autogenous shrinkage.

Besides, morphological characteristics of supplementary cementitious materials also may affect the autogenous shrinkage development of UHPC. For example, a smooth surface and a smaller surface area of fly ash easily cause the release of entrapped water, thus lowering the rapid drop of internal relative humidity, especially for the matrix with low w/b ratio. Metakaolin has a relatively strong water absorption capacity due to its layered microstructure. This may lead to delayed hydration of cementitious components in UHPC and internal curing function, causing lower autogenous shrinkage (Norhasri et al. 2019; Staquet et al. 2004). Rice husk ash provides an internal curing function due to its porous structure, which contributes to the reduction in autogenous shrinkage of UHPC (Rößler et al. 2014; Van Tuan et al. 2011; Kang et al. 2019).

Replacement level: Compared to a blend of Portland cement slag, the effect of fly ash on autogenous shrinkage of UHPC is typically more sensitive to its replacement level. When cement replacement to fly ash increases from 0% to 15%, 30%, and 50%, the autogenous shrinkage values of UHPC at 7 days are reduced by about 28.9%, 49.2%, and 63.1%, respectively. By contrast, the corresponding values are 44%, 54%, and 59% for UHPC made with slag, respectively (Liu et al. 2017). Ye (2012) found that the autogenous shrinkage of UHPC at 28 days was reduced by around 40% and 95% when rice husk ash content increased by 10% and 20%, respectively. Wang et al. (2018) noticed that there was an optimal content for the use of waste tailings, recommended to be around 20%, which minimizes the autogenous shrinkage without greatly sacrificing the mechanical properties of UHPC. This is because the further increasing replacement ratio of lead-zinc tailings (e.g., up to 30%) retards the formation of a stiff hydrated skeleton that can resist internal stresses.

Refinement of pore structure: The effect of supplementary cementitious materials on autogenous shrinkage also depends on their role in the refinement of pore structure. Each kind of supplementary cementitious material presents distinctly different effects on pore structure due to the variations in physical and chemical properties. The activity of slag is greater than that of fly ash and affected by its fineness as mentioned before. Therefore, the addition of high-volume slag might accelerate the water consumption rate and rapidly decrease the critical radius of capillary pores; thus the autogenous shrinkage is increased. This explains why some researchers report aggravated autogenous shrinkage of UHPC when slag is used (Yalçınkaya and Yazıcı2017). Similarly, Song et al. (2018) observed an increase in pore size distribution within the range of about 5–10 nm in UHPC with the combined use of 60% silica fume and 40% metakaolin. Nanomaterials like carbon nanotubes, acting as nucleation points for hydration products, are found to be capable of increasing autogenous shrinkage of cement composites as a microfiller material (Tafesse and Kim 2019).

(2) Curing temperature

As mentioned before, temperature change can affect hydration speed and autogenous shrinkage development of cement-based materials. Although most investigations have revealed that the high curing temperature increases autogenous shrinkage, compared to low curing temperature (Geiker 1983), there are some discrepancies among the current findings. Chu et al. (2012) found that high curing temperatures at early ages caused a lower magnitude of autogenous stresses at later ages compared to those subjected to relatively low curing temperatures. This phenomenon is probably because the coarse and more porous microstructure of cement paste can be formed at high temperatures (Thomas and Jennings 2002) or the restraint effect of a solid skeleton rapidly develops at high temperatures (Soliman and Nehdi 2011). However, some researchers found that the ultimate autogenous shrinkage of UHPC was not significantly affected by curing conditions (Yoo et al. 2018). It is reported that autogenous shrinkage for UHPC cured at ambient and heat conditions after 60 days was about 450 μm/m. Maruyama and Teramoto (2013) divided autogenous shrinkage of ultra-high-strength concrete into two stages based on the development rate: the earlier age followed by the later age stage. They found that low temperatures enhanced autogenous shrinkage at an early age, while high temperatures produced great autogenous shrinkage at later ages.

7.2.4.2 Addition of internal restraint

Since most aggregates are nonshrinking, ordinary concrete shrinkage depends only on the shrinkage properties of the paste and the volume fraction of the aggregates. Naturally, an increased aggregate volume fraction will result in a reduced shrinkage. However, in UHPC, the elastic modulus of the paste can approach that of the aggregates, and concrete shrinkage will thus also be influenced by the stiffness of the aggregates (Bentz and Jensen 2004) and their particle size. The autogenous shrinkage of UHPC usually decreases with the increase in aggregate size (Shen et al. 2018). However, it should be noted that the use of aggregate with a higher modulus of elasticity may enhance the local stresses, which induces potential cracking within the interfacial transition zones. Meanwhile, the increase in aggregate size negatively affects the compressive strength due to the larger size of the interfacial transition zone (Scrivener et al. 2004).

At the micrometer level, passive restraint may be provided by several components including unhydrated cement particles and calcium hydroxide crystals (Lura 2003). Soliman and Nehdi (2013) found that the addition of partially hydrated cementitious materials from recycling leftover/unused concrete could reduce the autogenous shrinkage of UHPC. The authors considered microcrystal hydration products as an internal restraint system for shrinkage reduction. However, a key issue that needs to be pointed out is the existence of calcium hydroxide crystals, which could easily lead to carbonation over time. Therefore, the reason for the reduction in autogenous shrinkage is controversial.

Because of the higher elastic modulus of steel fibers and their higher strength-to-size ratio, fibers can resist the shrinkage deformation of the cementitious matrix with a low elastic modulus at an early age and may have a significant influence on the early-age cracking of concrete. Although these fibers typically do not significantly reduce the shrinkage of concrete, they can substantially delay the occurrence of cracking owing to the stress transfer from matrix to fibers and decrease the crack width and frequency.

7.2.4.3 Reduction of the surface tension of pore solution

As shown by capillary pressure theory, a direct method for influencing autogenous shrinkage is changing the surface tension of pore solution. Shrinkage-reducing admixtures are commonly used to reduce the shrinkage of concrete and have received considerable attention in recent years. In general, the formation of shrinkage-reducing admixtures in pore solution can be classified into three parts, as shown in Figure 7.21 (Zana 2005). Part 1 represents the adsorption on the water-air interface of pore solution, playing a role in shrinkage reduction. Part 2 shows shrinkage-reducing admixtures dissolved in the pore solution. The phenomenon of aggregation of shrinkage-reducing admixtures molecules can be obtained when shrinkage-reducing admixtures concentration is over critical micelle concentration, which is not beneficial to reducing the surface tension and will affect the hydration of cement. Part 3 is a part of shrinkage-reducing admixtures that are adsorbed on the water-solid interface, which also reduces shrinkage based on the shrinkage mechanisms of surface free energy of solid gel particles (Rajabipour et al. 2008).

In addition to the reduction in capillary stress by decreasing the surface tension of pore solution, the potential role of reducing autogenous shrinkage using shrinkage-reducing agent can be due to: (1) the

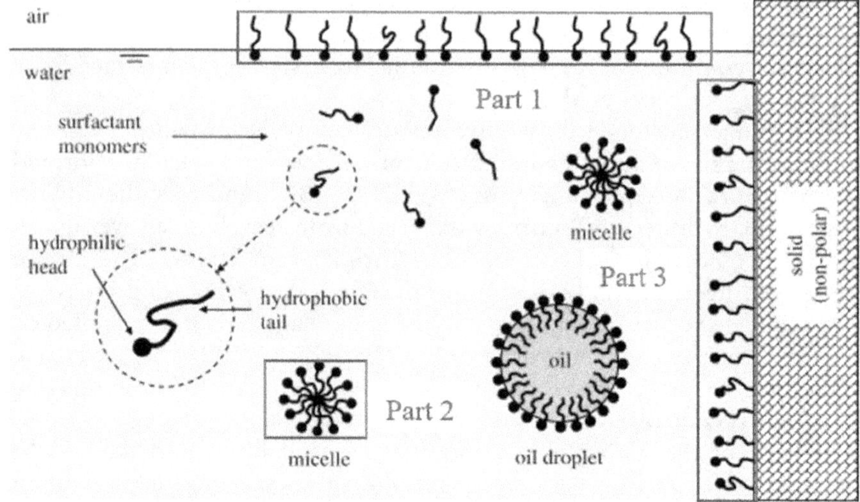

Figure 7.21 Existent form of shrinkage-reducing admixtures in cement-based materials.

Source: Zana 2005.

maintenance of internal relative humidity at a high level and decrease in self-desiccation (Bentz 2001), and (2) an increase in the concentration of Ca^{2+} and the oversaturation portlandite and ettringite (Zhan and He 2019). Consequently, early age expansion due to crystallization pressure compensates for partially autogenous stress. The retardation of hydration with the addition of shrinkage-reducing admixtures is initially believed to be due to a reduction of alkali content in the pore solution (Rajabipour et al. 2008). Later, Eberhardt (2010) noticed that the adsorption of shrinkage-reducing admixtures molecules onto portlandite limited the growth rate of crystals. The retardation of hydration of cement of course also contributes to the reduction of early autogenous shrinkage.

7.2.4.4 Formation of expansive products

Macroscopic volume expansion due to the crystallization pressure and swelling pressure can partially compensate for the autogenous shrinkage. The compensation of shrinkage by adding expansive agents is one of the effective methods widely used for crack prevention (Nagataki and Gomi 1998). Figure 7.22 (Liu et al. 2016) illustrates the principle of expansive agents to improve shrinkage behavior, which (a) describes the mechanism

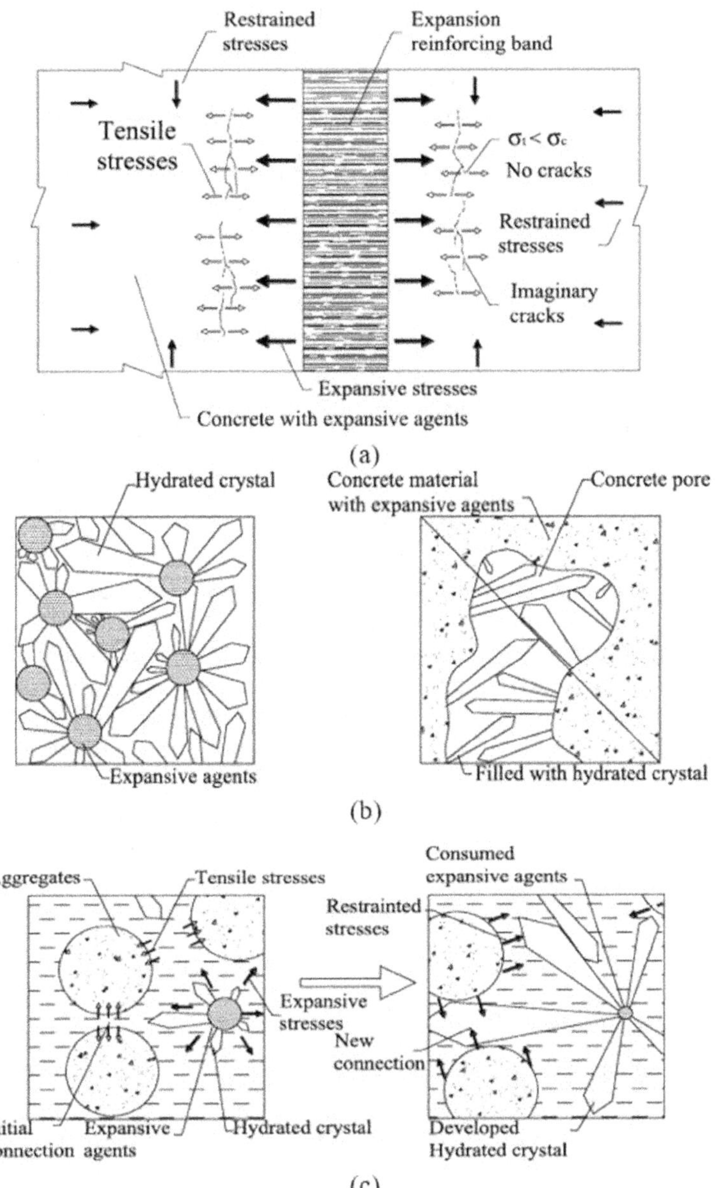

Figure 7.22 Shrinkage mitigation mechanism of expansive agents.

Source: Liu et al. 2016.

of expansive stresses and restrained stresses that compensate for tensile stresses in cracks; (b) shows the generation of expansive products and their resultant effects of filling into pores of concrete; and (c) demonstrates the process of expansion of expansive agents in concrete. It can be obviously seen that the initial connection in the hydrated matrix will be broken by expansive products, consequently resulting in a new connection.

When using expansive agents to compensate for the autogenous shrinkage, the formation of ettringite crystals might be highly localized if a uniform distribution of expansive components in the matrix is difficult to achieve. This would induce the occurrence of internal stress (Polat et al. 2015). The suitable occurrence time and degree of expansion are very important to obtain the obvious effect of shrinkage reduction. For UHPC with a very low w/b ratio, the ettringite-based expansive agents have been applied to reduce the autogenous shrinkage of UHPC (Yoo et al. 2014; Maltese et al. 2005). However, the formation of ettringite requires a large amount of free water (Mehta 1973), but the water is not enough in UHPC itself due to the very low w/b ratio (Wyrzykowski et al. 2018), which cannot well compensate for the autogenous shrinkage at later ages. This would restrain the expansion effect of expansive agents and result in less effective shrinkage reduction of UHPC. Therefore, the UHPC containing conventional ettringite-based expansive agents still shows relatively high autogenous shrinkage (Shen et al. 2017; Corinaldesi et al. 2015; Oliveira et al. 2014). Besides, the ettringite-based expansive agent might lead to the possibility of delayed ettringite formation. In this case, the shrinkage compensating effect cannot be fully performed at curing temperatures over 70°C, especially when in massive concrete (Suzuki et al. 2009). Also, the increased content of expansive agents does not significantly reduce the autogenous shrinkage of UHPC, which is probably due to the limited amount of expansive substance (Su et al. 2017). Thus, expansive agents are often used with shrinkage-reducing admixtures or internal curing substances.

7.2.4.5 Use of internal curing

External curing is the most common way to limit the shrinkage of concrete structures. In this method, the water evaporation from the concrete surface is avoided for some time immediately following placing and finishing. This can be performed by continuously wetting the exposed surfaces or preventing excessive loss of moisture from the concrete surface by covering the concrete with an impermeable membrane or applying a chemical curing agent directly on the surface. By contrast, internal water curing of concrete is the incorporation of a curing agent that serves as an internal reservoir of water and gradually releases it as the concrete dries out. The water reservoirs should be distributed uniformly and spaced close enough to provide coverage for the entire paste system.

Internal curing is complementary to the traditional external curing when complete hydration of cement is difficult to achieve, e.g., low w/c ratio concretes, concrete with supplementary cementitious materials such as fly ash, silica fume, natural pozzolans, or slag. These concrete mixtures generally present a tighter microstructure and a low superficial permeability, so water access to external curing water is impeded and insufficient mixing water remains to sustain further hydration. Therefore, internal curing has been developed as an effective technique for shrinkage mitigation of UHPC. In this method, internal curing agents such as porous aggregates, superabsorbent polymers, wood fibers, etc., are introduced into the concrete. They disperse within the concrete and absorb significant amounts of water either before or during mixing. During the hydration of cement, a system of capillary pores is formed. The chemical shrinkage, in the absence of an additional source of water, will produce a self-desiccation, as partially filled pores will be created within the microstructure (Lura et al. 2009). The capillary stresses will rapidly increase over time, as smaller and smaller pores within the hydrating cement paste empty, even as the continuing hydration further reduces the size of the remaining water-filled capillary pores (Bentz and Weiss 2011). When the relative humidity decreases due to hydration and drying, a humidity gradient arises. Since the radii of these capillary pores are smaller than the pores of an internal curing agent, these water-filled reservoirs introduced release it during cement hydration to provide a source of readily available additional water so that the capillary porosity of the hydrating cement paste remains saturated, thus minimizing the autogenous stresses and strains.

Concrete with internal curing provides the necessary additional water to prolong the time during which saturated conditions are maintained within the hydrating cement paste. To ensure the effectiveness of internal curing in cement-based materials, three key questions should be considered in the design process (Henkensiefken et al. 2009). The main factors include (1) the amount of internal curing water; (2) the characteristics of internal curing agents, e.g., the ability of water to leave the internal curing agents when needed for internal curing; and (3) the migration distance of internal curing water. These questions will be briefly addressed in this section.

(1) Amount of internal curing water

According to Powers' model, the amount of internal curing water for concrete with w/b ratio less than 0.36 is calculated using the following equation (Bentz and Snyder 1999; Jensen and Hansen 2001).

$$W_{cur} = C_f \cdot C_S \cdot \alpha_{max} \tag{7.4}$$

where W_{cur} is the amount of water needed for internal curing (kg/m^3); C_f is the binder content (kg/m^3); C_S is the chemical shrinkage of cementitious materials (kg water/kg cement); and α_{max} is the maximum degree of hydration, estimated by (w/b)/0.36.

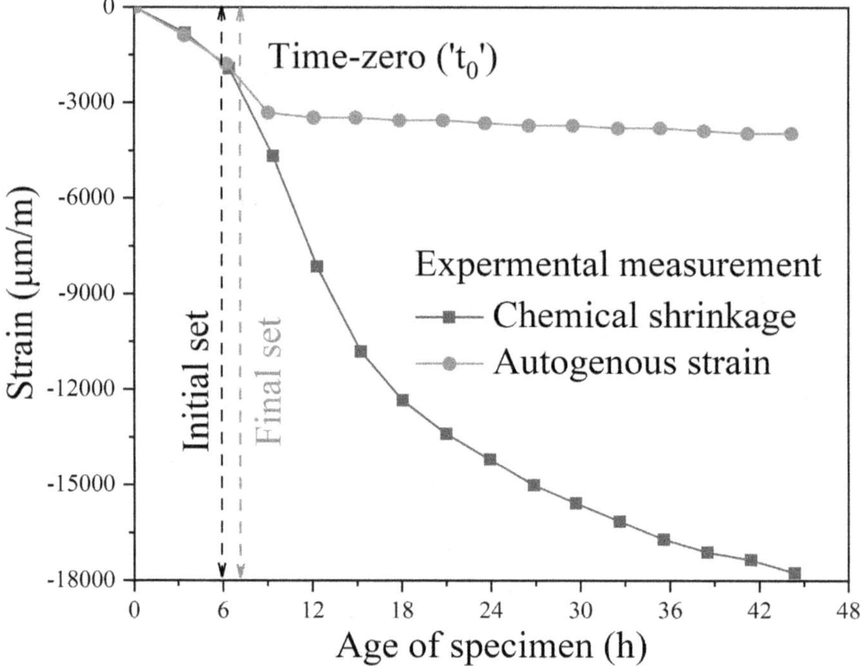

Figure 7.23 Chemical shrinkage and autogenous shrinkage volumes during hydration of a paste with a w/c ratio of 0.30.

Source: Modified based on Henkensiefken et al. 2008.

For ordinary Portland cement, C_S is usually about 0.06–0.07 mL/g. Given the higher chemical shrinkage of supplementary cementitious materials from the point of their hydration and achievable degree of hydration, C_S is complicated by the incorporation of supplementary cementitious materials. However, a value of 0.07 kg water/kg of cementitious materials is acceptable for almost all practical applications according to ASTM C1761/C1761M.

$$(w/c)_e = 0.18\ (w/c)_0 \tag{7.5}$$

where $(w/c)_e$ is the internal curing water-to-cement ratio; and $(w/c)_0$ is the initial water-to-cement ratio.

A typical chemical shrinkage result for a paste with a w/c ratio of 0.30 by a mass fraction is shown in Figure 7.23 (Henkensiefken et al. 2008). At the time of set, the volumes of chemical shrinkage and autogenous shrinkage diverge. Strictly speaking, the amount of internal curing water is the only difference between these curves; accounting for the behavior before set will generally be only a minor correction and may be unnecessary for field applications. Therefore, the volume of chemical shrinkage may be approximated as the volume of water that needs to be supplied by the internal curing agents (Henkensiefken et al. 2009).

(2) Characteristics of internal curing agents

The effectiveness of internal curing depends not only on whether there is sufficient water in the internal curing agents but also on whether it is available for the surrounding paste in time as well (Lura 2006), which is related to both the properties of internal curing agents and cement paste. The effect of characteristics of internal curing agents on the effectiveness of internal curing mainly includes the water saturation status, water absorption/desorption properties, particle size and distribution spacing, and amount of internal curing materials.

The basic principle for internal curing is that the largest pores will lose water first as the developed capillary stress will be minimized when pores are emptied in this order. The pores in internal curing agents should be generally larger than the pores of the surrounding cement paste, even at early hydration ages (less than 24 h). In this case, as the pore size of paste decreases, the driving force that pulls water out of the internal curing agents is preferred for internal curing.

Bentz and Snyder (1999) applied the protected paste volume which is defined as the volume of cement paste "protected" from frost damage by a system of air voids in the cement paste for internal curing (Natesaiyer et al. 1992). This concept assumed that if the cement paste is located within a sufficiently small distance from the surface of internal curing agents, the cement paste would be protected from self-desiccation and thus autogenous shrinkage would be mitigated. Therefore, a large number of studies attempted to reduce the particle size of internal curing agents and consequently the spacing factor to minimize the autogenous shrinkage (Lura and Van Breugel 2000; Zhutovsky et al. 2004; Lura et al. 2006; Esteves 2009). However, these experimental results suggest that the correlation between the reduction in the spacing and the minimization of autogenous shrinkage does not always exist. For example, Figure 7.24 shows a decrease in particle size of superabsorbent polymer despite the fact that the content presents a lower efficiency of reduction in autogenous shrinkage. This is due to the variations in water absorption which is strongly related to particle size. Bigger superabsorbent polymer particles have higher water absorption (Esteves 2009). Thus, the discharge of water from bigger-size superabsorbent polymer particles will take place more readily than in the smaller particles. Similar conclusions can be made about the influence of the size of porous aggregates. Figure 7.25 indicates that the shrinkage slope and autogenous shrinkage at the age of 7 days are higher for the smaller aggregate particles, although their spacing is smaller. Based on the analysis above, Zhutovsky

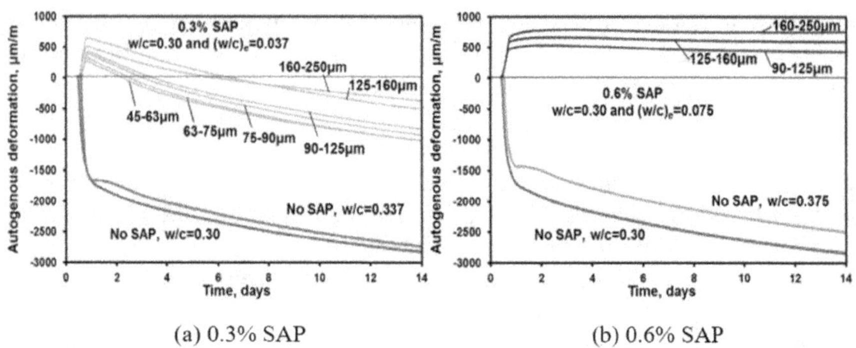

Figure 7.24 Autogenous shrinkages of cement paste with and without different contents and particle sizes of superabsorbent polymer.

Source: Lura et al. 2006.

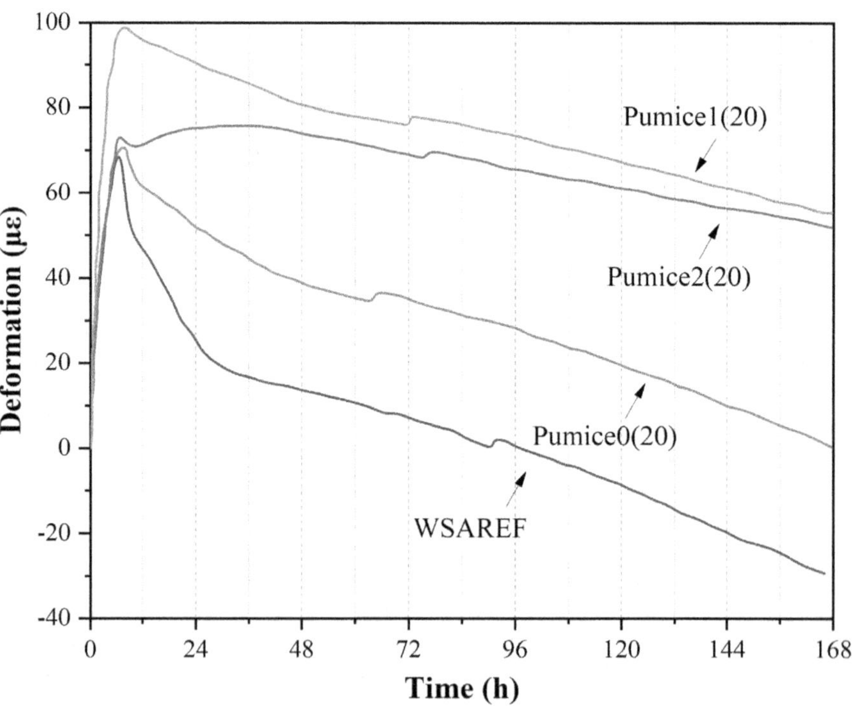

Figure 7.25 Effect of grain size on free shrinkage of mixtures with 0.33 w/c ratio.

Source: Modified based on Zhutovsky et al. 2004.

et al. (2011) pointed out that the recommendation for spacing factor in frost resistance of concrete (approximately 200 μm) cannot be extended to internal curing of high-strength concrete. To optimize the internal curing of concrete, the water introduced into the internal curing agent has to be readily available to be discharged to the surrounding cementitious matrix. This mainly depends on the structure of internal curing agents, which is of greater significance than the spacing factor.

(3) Migration distance of internal curing water

The distribution of internal curing water is responsible for the effectiveness of internal curing because it relates to the effective distance for transporting water to the surrounding matrix, namely, the proximity of cement paste to the surface of reservoirs. Alternatively, the distance between the two nearest reservoirs surfaces is called spacing. If the spacing between internal curing agents is too large and the migration distance of water is not far enough, hydrating cement paste may not be accessible to curing water within a reasonable time (Zhutovsky et al. 2004). This may restrain the diffusivity of water due to the reduction of permeability of the cement matrix caused by the denser matrix over time. If internal curing agents are well distributed throughout the matrix at a given amount, the effectiveness of internal curing may be promoted even if shorter distance is covered. This is in good coincidence with the concept of entrained air mentioned above, which suggests that small air voids will be superior to larger air voids in a finely dispersed system at equal air contents (Bentz and Snyder 1999).

Some researchers (Jensen and Lura 2006; Lura et al. 2006; Wyrzykowski et al. 2012; Mönnig 2009) have investigated the migration distance of internal curing water, and the results showed that the migration distance of internal curing water was in a range of 0.06–4 mm. The release of curing water generally occurs at an early age, i.e., during approximately the first day of hydration. Figure 7.26 describes that the relationship between the volume of protected paste and the replacement volume of lightweight aggregates varies with different travel distances of curing water. A higher travel distance can be responsible for a majority of the protected paste at a low replacement level (assuming lightweight aggregates can supply sufficient water). Nevertheless, if the water cannot travel far enough, a higher replacement level of lightweight aggregates is required to achieve a majority of protected paste. This is very important for the later ages when the microstructure of the matrix becomes so dense that the water cannot travel as far in a mature paste.

The absorption and desorption of the internal curing agents are strongly related to the materials' nature, such as microstructure and particle size, as mentioned above. This section mainly discusses the effect of the pore structure of paste on water migration. Zhutovsky et al. (2004) investigated the penetration depth of internally cured concrete with w/b ratios of 0.33 and

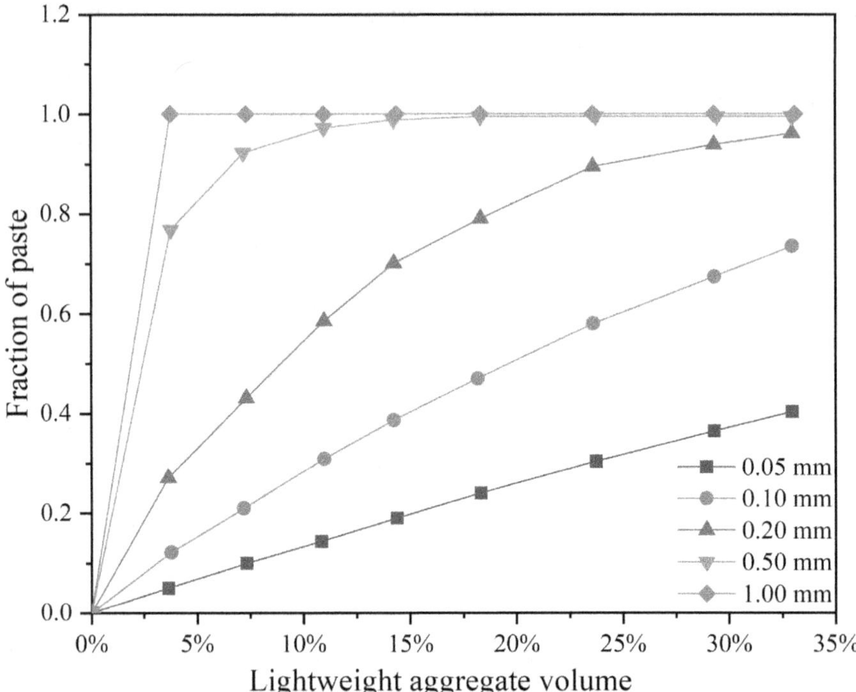

Figure 7.26 Effects of different water migration distances and volume percent of lightweight aggregates on volume of protected cement paste.

Source: Henkensiefken et al. 2009.

0.25 (with and without silica fume). The penetration depth was determined on the paste-aggregate proximity obtained by microstructure modeling. The results showed that the distance of water migration decreased with a reduction in w/b ratio. The addition of silica fume led to a more pronounced reduction in the distance of water migration. This could be explained by the tighter microstructure of the mixtures with silica fume, resulting in a smaller distance to which the internal curing water could penetrate. The presence of supplementary cementitious materials changes the chemistry of hydration reactions by providing extra space for hydration products, enhanced nucleation, and late-age pozzolanic reaction. This leads to the different pore structures of cement paste compared to the pure Portland cement systems. Therefore, the migration distance for a given internal curing agent is a function of curing time depending on the evolution of the pore structure of cement paste, which is strongly related to the w/b ratio, cement fineness, and the type of supplementary cementitious materials (Lura et al. 2006).

7.3 DRYING SHRINKAGE

Autogenous shrinkage consists of chemical shrinkage due to cement hydration and self-desiccation. Drying shrinkage is the reduction in the concrete volume due to moisture loss at constant temperature and relative humidity. It is mainly affected by total moisture loss, rate of evaporation, and bleeding. The relationship between drying shrinkage and relative humidity can be divided into three stages, as shown in Figure 7.27. It is generally agreed that stage one is attributed to the loss of capillary water, stage two results from the loss of adsorbed water on the surface of C-S-H, and stage three represents the loss of water related to the structure of C-S-H.

Since concrete with low w/b ratio typically has low bleeding, the evaporation from its surface at low relative humidity conditions will extract water from the internal mass, causing moisture loss and menisci formation (Holt 2001). Generally, more evaporation during the early stages will result in more shrinkage, especially for concrete exposure to drying conditions. It is well known that total shrinkage strains in UHPC (including both the autogenous and drying shrinkage) are expected to be higher than conventional concrete due to high quantities of binder (cement and silica fume) and low w/c ratio. Whereas numerous studies have focused on exploring the

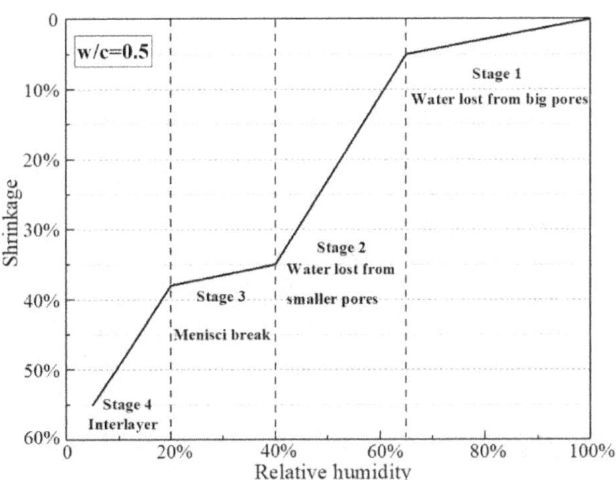

Figure 7.27 Relationship between drying shrinkage and relative humidity.

Source: Modified based on Hansen and Almudaiheen 1987.

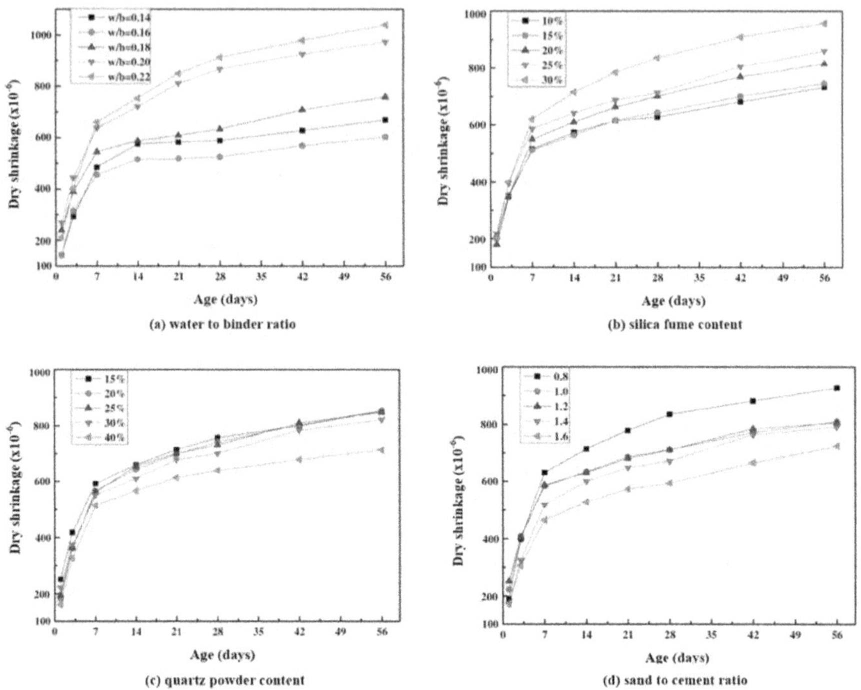

Figure 7.28 Effects of different factors on drying shrinkage of UHPC.

Source: Xiao 2013.

autogenous shrinkage of UHPC, only a limited number of studies have been conducted on drying shrinkage since autogenous shrinkage predominates the overall shrinkage. This section will show the current investigations on the drying shrinkage of UHPC from the perspective of materials and mixing proportions, the addition of chemical admixtures, and curing conditions.

Xiao (2013) investigated the effect of several factors such as sand-to-cement ratio, w/b ratio, silica fume content, and quartz powder content on the drying shrinkage of UHPC. The variations in the dry shrinkage of UHPC resulting from these factors can be summarized as follows (Figure 7.28):

(1) The dry shrinkage of UHPC increases with an increase in w/b ratio. On the one hand, this can be attributed to the relative humidity imbalance between the external environment and the concrete inside. Water available for evaporation is higher and water loss is faster in the system with a higher w/b ratio. On the other hand, a reduction in w/b ratio refines

the pore structure, which reduces the dry shrinkage due to a lower rate of water migration.

(2) The dry shrinkage of UHPC increases with an increase in silica fume content, due to its pozzolanic effect and refinement of the pore structure. As mentioned before, calcium hydroxide crystals can be regarded as a passive restraint for shrinkage. An increase in silica fume content consumes more calcium hydroxide crystals, and thus more porous C-S-H gel can be formed, which improves the shrinkage development.

(3) The restraint effect of quartz powder on dry shrinkage is limited. On the one hand, the addition of quartz powder increases the compactness of the microstructure due to the filling effect, which improves the ability of restraining shrinkage. On the other hand, water loss in capillary pores of less than 100 nm is considered to be the main factor resulting in dry shrinkage, while the smallest particle size of quartz powder used in this study is 2 μm, which cannot fill the capillary pores.

(4) The dry shrinkage of UHPC reduces with an increase in sand-to-cement ratio. The reason for this phenomenon is similar to the effect of aggregate on autogenous shrinkage. An increase in sand content means the reduction of binder content; therefore, a more restraining effect on cement matrix shrinkage can be developed.

The effect of fiber content and shape on the drying shrinkage of UHPC is shown in Figure 7.29. The optimum fiber content for restraining drying shrinkage was found to be 2%. This agrees well with the results reported by Yoo et al. (2014). The reduction in dry shrinkage can be explained by the fact that the steel fibers in the UHPC matrix act as a skeleton for the concrete matrix along with the fine aggregate to provide increased dimensional stability due to higher elastic modulus and strength-to-size ratio (Fang et al. 2020). Moreover, the incorporation of fibers disrupts the interconnection of pores, reducing the amount of water present in the capillary pores (Lei and Cao 2011). Although the straight fibers may be more evenly distributed than the deformed fibers in the matrix, a larger reduction in shrinkage is obtained with the addition of deformed fibers. This is because the incorporation of deformed fibers can produce an interlocking network among the particles due to mechanical anchorage. Greater bond strength to the cementitious matrix is more effective in restraining shrinkage. This agrees well with the findings from Ma et al. (2002), where polypropylene fibers with a Y-shaped cross-section exerted a greater effect in restraining plastic shrinkage cracking than that of the circular cross-section fibers due to their greater interfacial bond to the surrounding matrix.

In a recent study, Fang et al. (2020) reported that although the drying shrinkage of UHPC decreased with an increase in the fiber aspect ratio, the aspect ratio of fibers is not as influential as the volume fraction or the shape

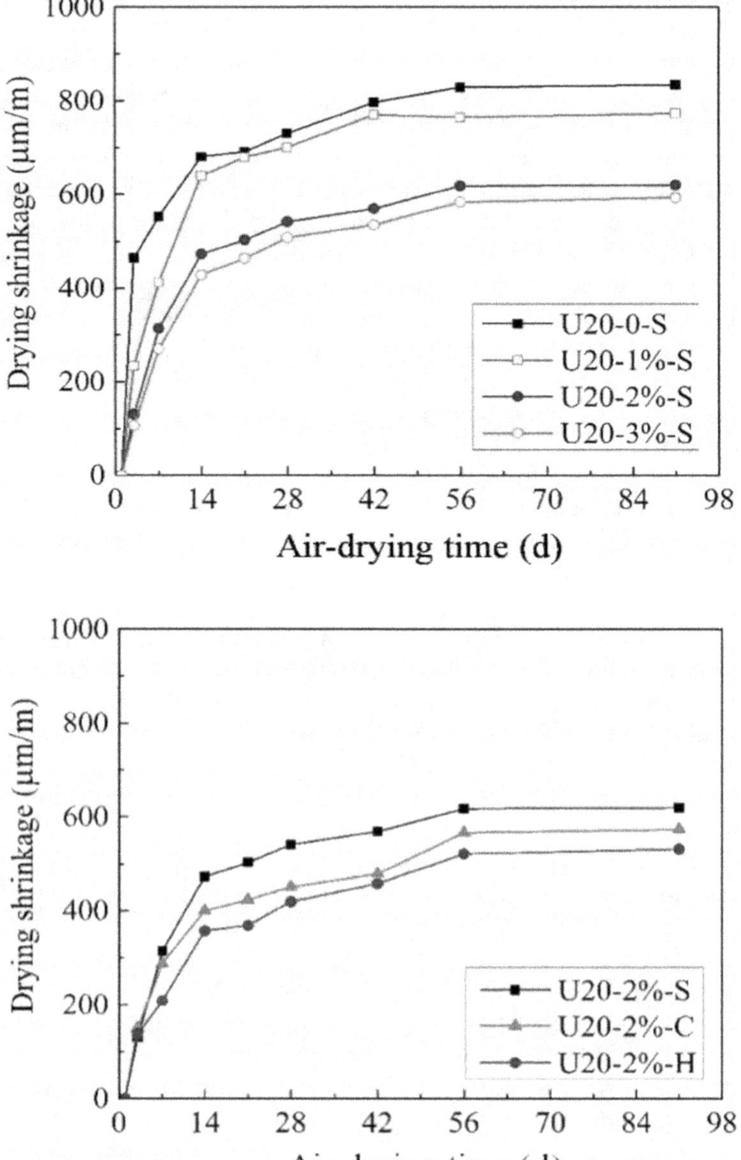

Figure 7.29 Effects of fiber content and shape on dry shrinkage of UHPC (Wu et al. 2019). (a) Fiber content. (b) Fiber shape.

of fibers on the shrinkage properties of UHPC within a certain range of the aspect ratios of fibers (i.e., 45–80).

From the perspective of drying shrinkage, the presence of sand and fiber reduces the drying shrinkage of the concrete mixture by reducing the volume of paste and restraining the shrinking of paste. The increase in the sand content and fibers content results in a decrease in the drying shrinkage of UHPC and also results in a decrease in the workability of UHPC. Therefore, the investigations aiming at understanding the underlying mechanisms governing the drying shrinkage of UHPC and its impact on performance have been carried out. For example, Li (2016) investigated the effect of different types of supplementary cementitious materials on the drying shrinkage of UHPC. It was also reported that ternary use of cement, metakaolin, and fly ash could produce UHPC paste with lower drying shrinkage than the binary use of cement and silica fume at the content of the same material. Yalçınkaya and Yazıcı (2017) investigated the effect of relative humidity and temperature on early-age drying strain of UHPC with high-volume slag and fly ash. They found that an increase in ambient temperature not only increased the drying shrinkage but also accelerated the development of shrinkage in the first few hours. High-volume replacement with supplementary cementitious materials reduced the sensitivity of UHPC to the ambient temperature in terms of drying shrinkage. However, a remarkable rise in drying shrinkage was observed as a result of the high-volume replacement of Portland cement, especially when slag was used.

Xie et al. (2018) observed that the UHPC prepared with a higher content of shrinkage-reducing admixture developed a lower drying shrinkage than their counterparts with lower shrinkage-reducing agent contents. This observation is consistent with the findings of previous studies by Yoo et al. (2013; 2014), who suggest that the shrinkage-reducing admixtures are particularly effective to mitigate the free drying shrinkage of the UHPC by reducing the surface tension in the capillary pores of the concrete during the drying process. Additionally, the introduction of superabsorbent polymers into the UHPC is possible to reduce drying shrinkage (Liu et al. 2019a; Dudziak and Mechtcherine 2008). These particles release water over time and replace the water lost due to hydration and drying, resulting in a reduction of drying shrinkage. Liu et al. (2019b) studied the properties of the hybrid system of superabsorbent polymers and shrinkage-reducing admixtures. They observed that the UHPC specimens with 0.3% superabsorbent polymers and 2% shrinkage-reducing admixtures had the minimum autogenous and drying shrinkage. This combined effect could compensate for the adverse effect of each on the properties of UHPC. The combined effect of using 60% lightweight sand and either CaO-based expansive agent, MgO-based expansive agent, or shrinkage-reducing agent can reduce drying shrinkage at 91 days by up to 700 μm/m (Valipour and Khayat 2018).

Further, Soliman and Nehdi (2011) pointed out that adequate external curing is essential to mitigate early-age deformation in UHPC, even when internal curing mechanisms are provided since it guarantees a suitable environment for shrinkage mitigation methods to work properly. In their studies, shrinkage-reducing admixtures were efficient in reducing drying shrinkage under drying conditions, while superabsorbent polymers were found to increase drying shrinkage.

7.4 CREEP

Recently, the use of UHPC for precast prestressed concrete highway bridge girders has been explored because of the potential for reduced maintenance costs relative to steel and conventional concrete girders. Time-dependent long-term stability of UHPC such as creep and shrinkage may generate long-term deformation and prestressing loss, thus reducing the enhancement provided by the prestressing (Graybeal 2005). Acker (2001; 2004) argues that UHPC completely self-desiccates between casting and the conclusion of steam treatment due to a very low w/b ratio. Thus, it exhibits no posttreatment shrinkage. The creep of concrete is closely related to shrinkage that cannot generally be uncoupled and studied separately. It is much more pronounced as the concrete is desiccating. The C-S-H phase is the only constituent in UHPC that exhibits creep. Compared with the research of long-term behavior on ordinary and high-performance concrete, existing studies on the long-term shrinkage and creep of UHPC have been relatively limited, and leading design code has not covered the creep and shrinkage design method of UHPC, which significantly limits the application of UHPC in structural engineering (Zhu et al. 2020).

The curing history of concrete and temperature of exposure has a great influence on the creep of UHPC. Graybeal (2005; 2006) conducted an unrestrained shrinkage test and compressive creep test of UHPC under various curing conditions. They found that the shrinkage of delayed steam– and standard steam–cured specimens was insignificant, while the specimens tested under tempered steam and ambient curing exhibited continued shrinkage even months after demolding. Flietstra (2011) conducted early-age creep and shrinkage tests under a compressive load and multiple thermal curing regimes to simulate current precast/prestressed plant procedures in the United States. The test results showed that delaying the thermal treatment for UHPC specimens under compressive loading has no significant effect on the final creep coefficient. Garas et al. (2009) conducted short-term tensile creep tests on UHPC and demonstrated that steel fiber reinforcement and thermal treatment can limit deformation in UHPC but concluded that further evaluation is necessary to better understand the underlying mechanisms of tensile creep in UHPC. Furthermore, Garas et al. (2010; 2012) performed long-term tensile and compressive

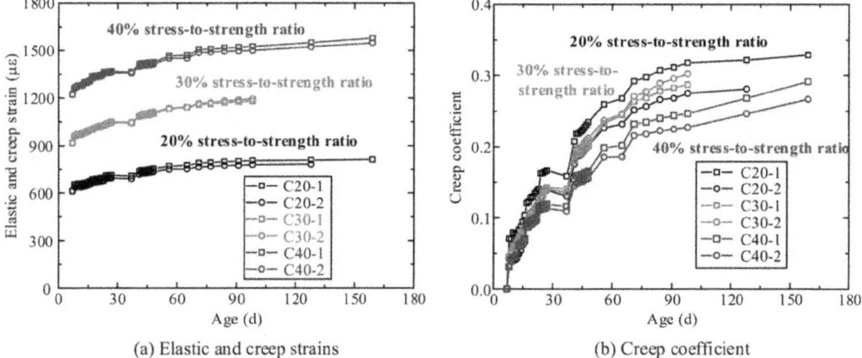

Figure 7.30 Creep behavior of UHPC.

Source: Zhu et al. 2020.

creep tests to determine the influence of thermal treatment parameters on the creep of UHPC, including temperatures lower than those recommended by the manufacturer. The test results showed that the tensile creep of UHPC was more sensitive to thermal treatment compared to tensile strength. The thermal treatment at 60 and 90°C also significantly decreased compressive creep.

The magnitude of applied stress can affect the creep of UHPC. Switek-Ray et al. (2016) tested the tensile creep and relaxation of early-age UHPC specimens under varying tensile stress ranging from 1.2 to 8.1 MPa. The test results showed a very low tensile stress level of 1.2 MPa caused a significant decrease in creep response. Nonlinear viscoelasticity was observed in several test series, and the possible underlying mechanisms were discussed. In general, existing research is focused on tensile and compressive creep of UHPC material considering concrete stress levels and curing conditions, while the current code-specified design methods have not been extended to UHPC material.

Zhu et al. (2020) studied the long-term creeps and shrinkages of UHPC with stress-to-strength ratios of 20%, 30%, and 40%, respectively (Figure 7.30). The results showed that the creep coefficients of UHPC subjected to stress-to-strength ratios ranging from 20% to 40% exhibit similar time dependence. The creep coefficient increased rapidly and continuously concerning time, especially from the 7th day (loading age) to approximately the 28th day. After about the 90th day, the change in the creep coefficient was insignificant. The final shrinkage strain was 145 μm/m, and the final creep coefficient was 0.46 based on the developed UHPC material and the test curing condition.

7.5 SUMMARY

This chapter discusses the volume deformation aspects of UHPC and can be summarized as follows:

(1) The mitigation strategies for the autogenous shrinkage of UHPC can be classified into five types based on the factors affecting autogenous shrinkage: (1) control of hydration reaction; (2) addition of internal restraint; (3) reduction of the surface tension of pore solution; (4) formation of expansive products; and (5) replenishment of water through internal curing.

(2) Different technical means of reducing autogenous shrinkage have their drawbacks. These include: (1) the compatibility between expansive agents and concrete; (2) the delay of early-age strength, prolongation of setting time, and interaction with other admixtures for the use of shrinkage-reducing admixtures; (3) fiber toughening is difficult to radically solve the problem of self-desiccation inside UHPC; (4) curing regime, including thermal curing and steam curing, can inhibit the subsequent strength development of concrete and adversely affect its durability; and (5) internal curing may have negative effects on mechanical properties due to extra pores in internal curing materials.

(3) From the perspective of the shrinkage mechanism, internal curing is a straightforward strategy to mitigate autogenous shrinkage. To ensure the effectiveness of internal curing, enough amount of internal curing water is not the only requirement. It is of great importance to ensure that the internal curing water can replenish the surrounding paste in time. The water desorption at appropriate relative humidity and the distribution of internal curing materials thus need to be considered. Besides, the distance of water migration in the cement paste also affects the results of internal curing.

(4) The creep of UHPC has not been well-characterized in the published literature. Further evaluation is necessary to better understand the underlying mechanisms of creep in UHPC.

REFERENCES

ACI 318M-14. (2014). *Building code requirements for structural concrete and commentary*. American Concrete Institute, Farmington Hills, MI.

Acker, P. (2001). *Micromechanical analysis of creep and shrinkage mechanisms*. Proceedings of ConCreep-6@MIT, Elsevier, London, 15–26.

Acker, P. (2004). Why does ultra high-performance concrete (UHPC) exhibit such a low shrinkage and such a low creep? *American Concrete Institute*. 220, 141–154.

Aitcin, P. C. (2000). *International RILEM Workshop on Shrinkage of Concrete*. 533–546.Akcay, B. (2018). Aggregate restraining effect on autogenous shrinkage of cementitious materials. *J. Civ. Eng.* 22, 3102–3111.

An, M. (2001). High performance concrete autogenous shrinkage problem. *J. Build. Mater.* 2, 163–164.

ASTM C1698-09. (2014). *Standard test method for autogenous strain of cement paste and mortar*. American Society for Testing & Materials (ASTM).

ASTM C1761/C1761M-17. (2017). *Standard specification for lightweight aggregate for internal curing of concrete*. American Society for Testing & Materials (ASTM).

Ba, H. (2003). High performance concrete early autogenous shrinkage test method research. *Industrial Construction.* 8, 66–69.

Baroghel-Bouny, V. (1996). *Texture and moisture properties of ordinary and high-performance cementitious materials*. Proceedings of Seminaire RILEM "Benton: du Materiau a la Structure", Arles, France.Bazant, Z. P., Chern, J. C. (1985). Concrete creep at variable humidity: constitutive law and mechanism. *Mater. Struct.* 18(1), 1–20.

Bentz, D. P., Garboczi, E. J., Haecker, C. J., Jensen, O.M. (1999). Effects of cement particle size distribution on performance properties of Portland cement-based materials. *Cem. Concr. Res.* 29(10), 1663–1671.

Bentz, D. P., Garboczi, E.J., Quenard, D.A. (1998). Modelling drying shrinkage in reconstructed porous materials: application to porous Vycor glass. *Model. Simul. Mater. Sci. Eng.* 6, 211–236.Bentz, D. P., Geiker, M. R., Hansen, K. K. (2001). Shrinkage-reducing admixtures and early-age desiccation in cement pastes and mortars. *Cem. Concr. Res.* 31(7), 1075–1085.

Bentz, D. P., Jensen. O. M. (2004). Mitigation strategies for autogenous shrinkage cracking. *Cem. Concr. Compos.* 26(6), 677–685.

Bentz, D. P., Peltz M. A. (2008). Reducing thermal and autogenous shrinkage contributions to early-age cracking. *ACI Mater. J.* 105, 414–420.

Bentz, D. P., Snyder, K. A. (1999). Protected paste volume in concrete: Extension to internal curing using saturated lightweight fine aggregate. *Cem. Concr. Res.* 29 (11), 1863–1867.

Bentz, D. P., Weiss, W. J. (2011). *Internal curing: a 2010 state-of-the-art review*. US Department of Commerce, National Institute of Standards and Technology, 4–7.

Byfors, J. (1980). *Plain concrete at early ages, Swed. Cem. Concr. Res.* Inst., Stockholm.

Carlson, E. T., Berman, H. A. (1960). Some observations on the calcium aluminate carbonate hydrates. *Journal of Research of the National Bureau of Standards*, 64(4), 333–341.

CEB-FIP Model Code 2010 design code Thomas Telford, (2010). Comité Euro-International du Béton-Fédération International de la Précontrainte (CEB-FIP).

Chu, I., Kwon, S. H., Amin, M. N., Kim, J. (2012). Estimation of temperature effects on autogenous shrinkage of concrete by a new prediction model. *Constr. Build. Mater.* 35, 171–182.

Ciobanu, C., Lazău, I., Păcurariu, C. (2013). Investigation regarding the effect of viscosity modifying admixtures upon the Portland cement hydration using thermal analysis. *J. Therm. Anal. Calorim.* 112(1), 331–338.

Corinaldesi, V., Nardinocchi, A., Donnini. J. (2015). The influence of expansive agent on the performance of fibre reinforced cement-based composites. *Constr. Build. Mater.* 91, 171–179.

Damidot, D., Lothenbach, B. Herfort, D., Glasser, F. (2011). Thermodynamics and cement science. *Cem. Concr. Res.* 41(7), 679–695.

Dudziak, L., Mechtcherine, V. (2008). *Mitigation of volume changes of ultra-high performance concrete (UHPC) by using super absorbent polymers. Proceedings of the 2nd International Symposium on Ultra High Performance Concrete*, Kassel University Press, Kassel, 425–432.

Eberhardt, A. B. (2010). *On the mechanisms of shrinkage reducing admixtures in self consolidating mortars and concretes.* Ph.D. diss., Bauhaus-Universitaet Weimar, Germany.

Esteves, L. P. (2009). *Internal curing in cement-based materials.* PhD diss., Universidade de Aveiro, Portugal.

Fang, C. F., Ali, M., Xie, T. Y., Visintin, P., Sheikh, A. H. (2020). The influence of steel fiber properties on the shrinkage of ultra-high-performance fiber reinforced concrete. *Constr. Build. Mater.* 242, 117993.

Ferrara, L, Ferreira, S. R., Torre, M. D. Krelani, V., Silva F. A. (2015). *Effect of cellulose nanopulp on autogenous and drying shrinkage of cement based composites.* In Nanotechnology in Construction: Proceedings of NICOM5. Sobolev, K. Shah, S. P., editors (pp. 325–330). Springer International Publishing.

Ferraris, C. F., Wittmann, F. H. (1987). Shrinkage mechanisms of hardened cement paste. *Cem. Concr. Res.* 17, 453–464.

Flietstra, J. C. (2011). *Creep and shrinkage behavior of ultra high-performance concrete under compressive loading with varying curing regimes.* Master diss., Michigan Technological University.

Folliard, K. J., Berke, N. S. (1997). Properties of high-performance concrete containing shrinkage-reducing admixture. *Cem. Concr. Res.* 27(9), 1357–1364.

Gao, P., Zhang, T., Luo, R., Wei, J., Yu, Q. (2014). Improvement of autogenous shrinkage measurement for cement paste at very early age: Corrugated tube method using non-contact sensors. *Constr. Build. Mater.* 55, 57–62.

Garas, V. Y., Kahn, L. F., Kurtis. K. E. (2009). Short-term tensile creep and shrinkage of ultra-high performance concrete. *Cem. Concr. Compos.* 31(3), 147–152.

Garas, V. Y., Kahn, L. F., Kurtis. K. E. (2010). Tensile creep test of fiber-reinforced ultra-high performance concrete. *J. Test. Eval.* 38(6), 674–682.

Garas, V. Y., Kurtis, K. E., Kahn, L. F. (2012). Creep of UHPC in tension and compression: effect of thermal treatment. *Cem. Concr. Compos.* 34(4), 493–502.

Geiker, M. (1983). *Studies of Portland cement hydration: measurement of chemical shrinkage and a systematic evaluation of hydration curves by means of the dispersion model.* PhD diss., Technical University of Denmark, Lyngby, Denmark.

Ghafari, E., Ghahari, S. A., Costa, H., Júlio, E., Portugal, A., Durães, L. (2016). Effect of supplementary cementitious materials on autogenous shrinkage of ultra-high performance concrete. *Constr. Build. Mater.* 127, 43–48.

Grasley, Z. C., Lange, D. A. (2007). Thermal dilation and internal relative humidity of hardened cement paste. *Mater. Struct.* 40(3), 311–317.

Grasley, Z. C., Lange, D. A., D'Ambrosia, M. D. (2006). Internal relative humidity and drying stress gradients in concrete. *Mater. Struct.* 39(9), 901–909.

Graybeal, B. A. (2006). *Material property characterization of ultra-high performance concrete*. US Department of Transportation, Federal Highway Administration.

Graybeal. B. A. (2005). *Characterization of the behavior of ultra-high performance concrete*. Master diss., University of Maryland.

Habel, K., Viviani, M., Denarié, E., Brühwiler, E. (2006). Development of the mechanical properties of an ultra-high performance fiber reinforced concrete (UHPFRC). *Cem. Concr. Res.* 36 (7), 1362–1370.

Hagymassy, J., Brunauer, J. R. Mikhail, R. S. (1969). Pore structure analysis by water vapor adsorption. *J. Colloid Interface Sci.* 29(3), 485–491.

Han, S., Cui, Y., Huang, H., An, M., Yu, Z. (2018). Effect of Curing Conditions on the Shrinkage of Ultra-High-Performance Fiber-Reinforced Concrete. *Adv. Civ. Eng.* 5238278.

Hanehara, S., Tomosawa, F., Kobayakawa, M., Hwang, K. (2001). Effects of water/powder ratio, mixing ratio of fly ash, and curing temperature on pozzolanic reaction of fly ash in cement paste. *Cem. Concr. Res.* 31(1), 31–39.

Hannawi, K., Bian, H., Prince-Agbodjan, W., Raghavan, B. (2016). Effect of different types of fibers on the microstructure and the mechanical behavior of ultra-high performance fiber-reinforced concretes. *Compos. Part B-Eng.* 86, 214–220.

Hansen, W. (1987). Drying shrinkage mechanisms in Portland cement paste. *J. Am. Ceram. Soc.* 70, 323–328.

Hansen, W. Almudaiheen, J. A. (1987). Ultimate drying shrinkage of concrete-influence of major parameters. *ACI Mater. J.* 84(3), 217–223.

Henkensiefken, R., Bentz, D. P., Nantung, T., Weiss, J. (2009). Volume change and cracking in internally cured mixtures with saturated lightweight aggregate under sealed and unsealed conditions. *Cem. Concr. Compos.* 31, 427–437.

Henkensiefken, R., Sant, G., Nantung, T., Weiss, W. (2008). *Comments on the shrinkage of paste in mortar containing saturated lightweight aggregate*. In: CONMOD, Delft, The Netherlands.

Holt, E. (2001). *Early age autogenous shrinkage of concrete*. Ph.D. diss., University of Washington.

Holt, E. (2005). Contribution of mixture design to chemical and autogenous shrinkage of concrete at early ages. *Cem. Concr. Res.* 35(3), 464–472.

Huang, Z., Hu, G. (2016). Study on the shrinkage performance of ultra-high performance concrete during heat curing. *Mater. Rev.* 30(4), 115–120 (in Chinese).

Igarashi, S., Watanabe, A., Kawamura, M. (2005). Evaluation of capillary pore size characteristics in high-strength concrete at early ages. *Cem. Concr. Res.* 35(3), 513–519.

Itim, A. Ezziane, K., Kadri. E. (2011). Compressive strength and shrinkage of mortar containing various amounts of mineral additions. *Constr. Build. Mater.* 25, 3603–3609.

Japan Concrete Institute (1999). *Autogenous shrinkage of concrete*. London: E&FN Spon.

Japan Concrete Institute (1999). *Autogenous shrinkage of concrete*. London, E&FN Spon.

Jensen, O. M., Hansen, P. F. (1995). A dilatometer for measuring autogenous deformation in hardening Portland cement paste. *Mater. Struct.*, 28, 406–409.

Jensen, O. M., Hansen, P. F. (1996). Autogenous deformation and change of the relative humidity in silica fume-modified cement paste. *ACI Mater. J.* 93 (6), 539–543.

Jensen, O. M., Hansen, P. F. (2001). Water-entrained cement-based materials: I. Principles and theoretical background. *Cem. Concr. Res.* 31(4), 647–654.

Jensen, O. M., Hansen, P. F. (2002). Water-entrained cement-based materials II. Experimental observations. *Cem. Concr. Res.* 32, 973–978.

Jensen, O. M., Lura, P. (2006). Techniques and materials for internal water curing of concrete. *Mater. Struct.* 39(9), 817–825.

Jiang, Z., Sun, Z., Wang, P., Wang, X. (2004). Study on self-desiccation effect of high-performance concrete. *J. Build. Mater.* 7(1), 19–24 (In Chinese).

Justness, H., Reyniers, B., Sellevold, E.J. (1994). An evaluation of methods for measuring chemical shrinkage of cementitious pastes. *Nordic Concrete Research* 2, 44–61.

Justs, J., Wyrzykowski, M., Bajare, D., Lura, P. (2015). Internal curing by superabsorbent polymers in ultra-high performance concrete. *Cem. Concr. Res.* 76, 82–90.

Kang, S. H., Hong, S. G., Moon, J. (2018). Shrinkage characteristics of heat-treated ultra-high performance concrete and its mitigation using superabsorbent polymer-based internal curing method. *Cem. Concr. Compos.* 89, 130–138.

Kang, S. H., Hong, S. G., Moon, J. (2019). The use of rice husk ash as reactive filler in ultra-high performance concrete. *Cem. Concr. Res.* 115, 389–400.

Kang, S. H., Jeong, Y., Tan, K. H. (2018). The use of limestone to replace physical filler of quartz powder in UHPFRC. *Cem. Concr. Compos.* 94, 238–247.

Kasai, Y., Yokoyama, K., Matsui, I. (1972). Tensile properties of early age concrete, Mechanical Behavior of Materials. *Society of Materials Science*, 4, 288–299.

Kawashima, S. Shah, S. P. (2011). Early-age autogenous and drying shrinkage behavior of cellulose fiber-reinforced cementitious materials. *Cem. Concr. Compos.* 33(2), 201–208.

Koh, K., Ryu, G., Kang, S., Park, J. (2011). Shrinkage Properties of Ultra-High Performance Concrete (UHPC). *Advanced Science Letters,* 4(3), 948–952.

Kong, X., Zhang, Y., Hou, S. (2013). Study on the rheological properties of Portland cement pastes with polycarboxylate superplasticizers. *Rheologica Acta*, 52(7), 707–718.

Kovler, K. Zhutovsky, S. (2006). Overview and future trends of shrinkage research. *Mater. Struct.* 39, 827–847.

Lee, K. M., Lee, H. K., Lee, S. H., Kim, G. Y. (2006). Autogenous shrinkage of concrete containing granulated blast-furnace slag. *Cem. Concr. Res.* 36(7), 1279–1285.

Lei, Y. B., Cao. X. J. (2011). Study on thermal dilation coefficient of fiber concrete at early stage. *Appl. Mech. Mater.*, 99–100, 777–781.

Li, C., Sun, Z., Yang, L., Zhang G. (2018). Study on autogenous shrinkage characteristic and mechanism of ultra-high performance cementitious composite. *Bulletin of the Chinese Ceramic Society* 324(1),012037 (In Chinese).

Li, P. P., Yu, Q. L. Brouwers, H. J. (2017). Effect of PCE-type superplasticizer on early-age behavior of ultra-high performance concrete (UHPC). *Constr. Build. Mater.* 153, 740–750.

Li, Y. Li, J. (2014). Capillary tension theory for prediction of early autogenous shrinkage of self-consolidating concrete. *Constr. Build. Mater.* 53, 511–516.

Li, Z. (2016). Drying shrinkage prediction of paste containing meta-kaolin and ultrafine fly ash for developing ultra-high performance concrete. *Mater. Today Commun.* 6, 74–80.

Lim, J. L., Raman, S. N., Safiuddin, M., Zain, M. F., Hamid, R. (2019). Autogenous Shrinkage, Microstructure, and Strength of Ultra-High Performance Concrete Incorporating Carbon Nanofibers. *Materials* 12(2), 320.

Liu, F., Shen, S. L., Hou, D. W., Arulrajah, A., S. Horpibulsuk. (2016). Enhancing behavior of large volume underground concrete structure using expansive agents. *Constr. Build. Mater.* 114, 49–55.

Liu, J. (2019). *Internal curing of SAP and FLWA in ultra-high performance concrete.* Ph.D. diss., Hunan Univ. (In Chinese).

Liu, J. M., Ou, Z. H., Mo, J. C., Wang, Y. H., Wu, H. (2017). The effect of SCM and SAP on the autogenous shrinkage and hydration process of RPC. *Constr. Build. Mater.* 155, 239–249.

Liu, J. Tian, Q., Tang, M. (2006). Influence of expansion agent and shrinkage reducing agent on shrinkage cracking of high-performance concrete. *J. Southeast Univ.* 36, 195–197 (In Chinese).

Liu, J., Farzadnia, N. Shi, C., Ma, X. (2019a). Effects of superabsorbent polymer on shrinkage properties of ultra-high strength concrete under drying condition. *Constr. Build. Mater.* 215, 799–811.

Liu, J., Farzadnia, N. Shi, C., Ma, X. (2019b). Shrinkage and strength development of UHSC incorporating a hybrid system of SAP and SRA. *Cem. Concr. Compos.* 97, 175–189.

Liu, K. Z., Yu, R., Shui, Z. H., Shui, Z. H., Li, X. H., Guo, C. H., Yu, B. L. Wu, S. H. (2019). Optimization of autogenous shrinkage and microstructure for Ultra-High Performance Concrete (UHPC) based on appropriate application of porous pumice. *Constr. Build. Mater.* 214, 369–381.

Liu, Z., El-Tawil, S., Hansen, W. (2018). Effect of slag cement on the properties of ultra-high performance concrete. *Constr. Build. Mater.* 190, 830–837.

Lothenbach, B. Scrivener, K., Hooton, R. (2011). Supplementary cementitious materials. *Cem. Concr. Res.* 41(12), 1244–1256.

Lura, P. (2003). *Autogenous Deformation and Internal Curing of Concrete.* PhD diss., Delft University. 112–125.

Lura, P., Bentz, D. P., Lange, D. A., Kovler, K., Bentur, A., van Breugel, K. (2006). Measurement of water transport from saturated pumice aggregates to hardening cement paste. *Mater. Struct.* 39(9), 861–868.

Lura, P., Bentz, D.P., Lange, D.A., Kovler, K. Bentur, A., van Breugel K. (2006). Measurement of water transport from saturated pumice aggregates to hardening cement paste. *Mater. Struct.* 39(9), 861–868.

Lura, P., Breugel, K. V., Maruyama, I. (2001). Effect of curing temperature and type of cement on early-age shrinkage of high-performance concrete. *Cem. Concr. Res.* 31(12), 1867–1872.

Lura, P., Couch, J., Jenson, O. M., Weiss, J. (2009). Early-age acoustic emission measurements in hydrating cement paste: evidence for cavitation during solidification due to self desiccation. *Cem. Concr. Res.* 39: 861–867.

Lura, P., Durand, F., Jensen, O. M. (2006). *Autogenous strain of cement pastes with superabsorbent polymers*. Technical University of Denmark, Lyngby.

Lura, P., Jensen, O. M., van Breugel, K. (2003). Autogenous shrinkage in high performance cement paste: an evaluation of basic mechanisms. *Cem. Concr. Res.* 33(2), 223–232.

Lura, P., van Breugel, K. (2000). *Moisture exchange as a basic phenomenon to understand volume changes of lightweight aggregate concrete at early age*. Baroghel-Bouny V., Aïtcin, P-C. editors. In *International RILEM Workshop on Shrinkage of Concrete*. Paris, France, 533–546.

Ma, R., Guo, L., Ye, S., Sun, W., Liu, J. (2019). Influence of hybrid fiber reinforcement on mechanical properties and autogenous shrinkage of an ecological UHPFRCC. *J. Mater. Civ. Eng.* 31(5), 04019032.

Ma, Y., Tan, M., Wu, K. (2002). Effect of different geometric polypropylene fibers on plastic shrinkage cracking of cement mortars. *Mater. Struct.* 35(3), 165–169.

Maltese, C., Pistolesi, C., Lolli, A., Bravo, A., Cerulli, T., Salvioni, D. (2005). Combined effect of expansive and shrinkage reducing admixtures to obtain stable and durable mortars. *Cem. Concr. Res.* 35(12), 2244–2251.

Mangat, P., Azari, M. M. (1984). A theory for the free shrinkage of steel fibre reinforced cement matrices. *J. Mater. Sci.* 19(7), 2183–2194.

Maraghechi, H., Rajabipour, F., Pantano, C. G., Burgos, W. D. (2016). Effect of calcium on dissolution and precipitation reactions of amorphous silica at high alkalinity. *Cem. Concr. Res.* 87, 1–13.

Maruyama, I., Teramoto, A. (2013). Temperature dependence of autogenous shrinkage of silica fume cement pastes with a very low water–binder ratio. *Cem. Concr. Res.* 50, 41–50.

Mazloom, M., Ramezanianpour, A., Brooks, J. (2004). Effect of silica fume on mechanical properties of high-strength concrete. *Cem. Concr. Compos.* 26(4), 347–357.

Meddah, M. S. Tagnit-Hamou, A. (2009). Pore structure of concrete with mineral admixtures and its effect on self-desiccation shrinkage. *ACI Mater. J.* 106(3), 241–250.

Mehta. P. K. (1973). Mechanism of expansion associated with ettringite formation. *Cem. Concr. Res.* 3(1), 1–6.

Meng, W., Khayat, K. H. (2017). Effects of saturated lightweight sand content on key characteristics of ultra-high-performance concrete. *Cem. Concr. Res.* 101, 46–54.

Meng, W., Khayat, K. H. (2018). Effect of hybrid fibers on fresh properties, mechanical properties, and autogenous shrinkage of cost-effective UHPC. *J. Mater. Civ. Eng.* 30(4), 04018030.

Meng, W., Khayat, K. H.(2018). Effect of graphite nanoplatelets and carbon nanofibers on rheology, hydration, shrinkage, mechanical properties, and microstructure of UHPC. *Cem. Concr. Res.* 105, 64–71.

Meng, W., Samaranayake, V. A., Khayat, K. H. (2018). Factorial design and optimization of ultra-high-performance concrete with lightweight sand. *ACI Mater. J.* 115(1), 129–138.

Mo, J., Ou, Z., Zhao, X., Liu, J., Wang, Y. (2017). Influence of superabsorbent polymer on shrinkage properties of reactive powder concrete blended with granulated blast furnace slag. *Constr. Build. Mater.* 146, 283–296.

Mo, L. Deng, M. Wang, A. (2012). Effects of MgO-based expansive additive on compensating the shrinkage of cement paste under non-wet curing conditions. *Cem. Concr. Compos.* 34(3), 377–383.

Mönnig, S. (2009). *Superabsorbing additions in concrete: applications, modelling and comparison of different internal water sources.* PhD diss., Stuttgart University. Germany.

Nagataki, S., Gomi, H. (1998). Expansive admixtures (mainly ettringite). *Cem. Concr. Compos.* 20(2–3),163–170.

Natesaiyer, K., Hover, K. C., Snyder, K. A. (1992). Protected-paste volume of air-entrained cement paste: part 1. *J. Mater. Civ. Eng.* 4(2), 166–184.

Norhasri, M. S., Hamidah, M. S., Fadzil, A. M. (2019). Inclusion of nano metaclayed as additive in ultra-high performance concrete (UHPC). *Constr. Build. Mater.* 201, 590–598.

Oliveira, M. J., Ribeiro, A. B., Branco. F. G. (2014). Combined effect of expansive and shrinkage reducing admixtures to control autogenous shrinkage in self-compacting concrete. *Constr. Build. Mater.* 52, 267–275.

Paajanen, A., Ceccherini, S., Maloney, T., Ketoja, J. A. (2019). Chirality and bound water in the hierarchical cellulose structure. *Cellulose.* 26, 5877–5892.

Park, J. J., Kim, S. W., Ryu, G. S., Lee, K. M. (2011). The Influence of Chemical Admixtures on the Autogenous Shrinkage Ultra-High Performance Concrete. *Key Eng. Mater.* 452–453, 725–728.

Polat, R., Demirboğa, R., Khushefati, W. H. (2015). Effects of nano and micro size of CaO and MgO, nano-clay and expanded perlite aggregate on the autogenous shrinkage of mortar. *Constr. Build. Mater.* 81, 268–275.

Pyo, S., Kim, H. K. (2017). Fresh and hardened properties of ultra-high performance concrete incorporating coal bottom ash and slag powder. *Constr. Build. Mater.* 131, 459–466.

Radocea, A. (19920. *A study on the mechanisms of plastic shrinkage of cement-based materials.* Ph.D. diss., Chalmers University of Technical, Göteborg.

Rajabipour, F., Sant, G., Weiss, J. (2008). Interactions between shrinkage reducing admixtures (SRA) and cement paste's pore solution. *Cem. Concr. Res.* 38(5), 606–615.

Rong, Z., Jiang, G., Sun, W. (2018). Effects of metakaolin on mechanical and microstructural properties of ultra-high performance cement-based composites. *Journal of Sustainable Cement-Based Materials* 7(5), 296–310.

Rößler, C., Bui, D. D., Ludwig, H. M. (2014). Rice husk ash as both pozzolanic admixture and internal curing agent in ultra-high performance concrete. *Cem. Concr. Compos.* 53, 270–278.

Salas, A., Delvasto, S., de Gutierrez, R. M. (2009). Comparison of two processes for treating rice husk ash for use in high-performance concrete. *Cem. Concr. Res.* 39(9), 773–778.

Scrivener, K. L., Crumbie, A. K., Laugesen, P. (2004). The interfacial transition zone (ITZ) between cement paste and aggregate in concrete. *Interface Sci.* 12(4), 411–421.

Shen, P., Lu, L., He, Y., Rao, M., Fu, Z., Wang, F., Hu, S. (2018). Experimental investigation on the autogenous shrinkage of steam-cured ultra-high performance concrete. *Constr. Build. Mater.* 162, 512–522.

Shen, P., Lu, L., He, Y., Wang, F., Hu. S. (2017). Hydration of quaternary phase-gypsum system. *Constr. Build. Mater.* 152, 145–153.

Shi, C. J., Wu, Z. M., Xiao, J. F., Wang, D. H., Huang, Z. Y., Fang, Z. H. (2015). A review on ultra high performance concrete: Part I. Raw materials and mixture design. *Constr. Build. Mater.* 101, 741–751.

Snoeck, D., Jensen, O. M., De Belie. N. (2015). The influence of superabsorbent polymers on the autogenous shrinkage properties of cement pastes with supplementary cementitious materials. *Cem. Concr. Compos.* 74, 59–67.

Soliman, A. M., Nehdi, M. L. (2011). Effect of drying conditions on autogenous shrinkage in ultra-high performance concrete at early-age. *Mater. Struct.* 44, 879–899.

Soliman, A. M., Nehdi, M. L. (2011). Effect of natural wollastonite microfibers on early-age behavior of UHPC. *J. Mater. Civ. Eng.* 24(7), 816–824.

Soliman, A. M., Nehdi, M. L. (2013). Effect of partially hydrated cementitious materials and superabsorbent polymer on early-age shrinkage of UHPC. *Constr. Build. Mater.*, 41, 270–275.Soliman, A. M., Nehdi, M. L. (2014). Effects of shrinkage reducing admixture and wollastonite microfiber on early-age behave or of ultra-high performance concrete. *Cem. Concr. Compos.* 46, 81–89.

Soliman, N. A., Tagnit-Hamou, A. (2017). Using glass sand as an alternative for quartz sand in UHPC. *Constr. Build. Mater.*, 145, 243–252.

Song, Q., Yu, R., Wang, X., Rao, S., Shui, Z. (2018). A novel Self-Compacting Ultra-High Performance Fibre Reinforced Concrete (SCUHPFRC) derived from compounded high-active powders. *Constr. Build. Mater.* 158, 883–893.

Soroka, I. (1979). Portland cement paste and concrete. (No Title).Staquet, S., Espion, B.(2004). *Early age autogenous shrinkage of UHPC incorporating very fine fly ash or metakaolin in replacement of silica fume[C]//International Symposium on Ultra-High-Performance Concrete*, Kessel Germany, 587–599.

Su, A., Qin, L., Zhang, S., Zhang, J., Li, Z. (2017). Effects of shrinkage reducing agent and expansive admixture on the volume deformation of ultrahigh-performance concrete. *Adv. Mater. Sci. Eng.* 6384859.

Suzuki, M., Meddah, M. S., Sato, R. (2009). Use of porous ceramic waste aggregates for internal curing of high-performance concrete. *Cem. Concr. Res.* 39(5), 373–381.

Switek-Rey, A., Denarié, E., Brühwiler, E. (2016). Early age creep and relaxation of UHPFRC under low to high tensile stresses. *Cem. Concr. Res.* 83, 57–69.

Tafesse, M., Kim, H. K. (2019). The role of carbon nanotube on hydration kinetics and shrinkage of cement composite. *Compos. Part B-Eng.* 169, 55–64.

Tam, C. M., Tam, V. W., Ng, K. M. (2012). Assessing drying shrinkage and water permeability of reactive powder concrete produced in Hong Kong. *Constr. Build. Mater.* 26(1), 79–89.

Tazawa, E., Miyazawa, S. (1995). Influence of cement and admixture on autogenous shrinkage of cement paste. *Cem. Concr. Res.* 25, 281–287.

Tazawa, E., Miyazawa, S. (1997). Influence of constituents and composition on autogenous shrinkage of cementitious materials. *Mag. Concr. Res.* 49(178), 15–22.

Thomas, J. J., Jennings, H. M. (2002). Effect of heat treatment on the pore structure and drying shrinkage behavior of hydrated cement paste. *J. Am. Ceram. Soc* 85(9), 2293–2298.

Tian, Q., Sun, W., Liao, C. W., Liu, J. P. (2005). High performance concrete autogenous shrinkage test method to explore. *J. Build. Mater.* 1, 82–89 (In Chinese).

Valipour, M., Khayat, K. H. (2018). Coupled effect of shrinkage-mitigating admixtures and saturated lightweight sand on shrinkage of UHPC for overlay applications. *Constr. Build. Mater.* 184, 320–329.

Van Breugel, K., Van Tuan, N. (2015). Autogenous shrinkage of HPC and ways to mitigate it. *Key Engineering Materials*, 629, 3–20.

Van Tuan, N., Ye, G., Van Breugel K, Copuroglu, O. (2011). Hydration and microstructure of ultra-high performance concrete incorporating rice husk ash. *Cem. Concr. Res.* 41(11), 1104–1111.

Varga, I. D. Graybeal, B. A. (2014). Dimensional stability of grout-type materials used as connections between prefabricated concrete elements. *J. Mater. Civ. Eng.* 27(9), 04014246.

Wang, D. H., Shi, C. J., Wu, Z. M., Xiao, J. F., Huang, Z. Y. Fang, Z. H. (2015). A review on ultra-high performance concrete: Part II. Hydration, microstructure, and properties. *Constr. Build. Mater.* 96, 368–377.

Wang, X., Yu, R., Shui, Z., Song, Q., Zhang, Z. (2017). Mix design and characteristics evaluation of an eco-friendly Ultra-High Performance Concrete incorporating recycled coral based materials. *J. Clea. Prod.* 165, 70–80.

Wang, X., Yu, R., Shui, Z., Zhao, Z., Song, Q., Yang, B., Fan, D. (2018). Development of a novel cleaner construction product: Ultra-high performance concrete incorporating lead-zinc tailings. *J. Clea. Prod.* 196, 172–182.

Wu, L. M., Farzadnia, N., Shi, C. J., Zhang, Z. H., Wang, H. (2017). Autogenous shrinkage of high performance concrete: a review. *Constr. Build. Mater.* 149, 62–75,

Wu, Z., Shi, C. Khayat, K. H. (2019). Investigation of mechanical properties and shrinkage of ultra-high performance concrete: Influence of steel fiber content and shape. *Compos. Part B-Eng.* 174, 107021.

Wyrzykowski, M., Lura, P., Pesavento, F., Gawin D. (2012). Modeling of water migration during internal curing with superabsorbent polymers. *J. Mater. Civ. Eng.* 24(8), 1006–1016.

Wyrzykowski, M., Terrasi, G., Lura. P. (2018). Expansive high-performance concrete for chemical-prestress applications. *Cem. Concr. Res.* 107, 275–283.

Xiao, J. (2013). *Study on preparation and properties of ultra-high performance concrete under conventional technology*. Master diss., Hunan University (In Chinese).

Xie, T., Fang, C., Ali, M. S. M., Visintin, P. (2018). Characterizations of autogenous and drying shrinkage of ultra-high performance concrete (UHPC): An experimental study. *Cem. Concr. Compos.* 91, 156–173.

Yalçınkaya, Ç., Yazıcı, H. (2017). Effects of ambient temperature and relative humidity on early-age shrinkage of UHPC with high-volume mineral admixtures. *Constr, Build. Mater.* 144, 252–259.

Yan, L., Chouw, N., Jayaraman, K. (2014). Flax fiber and its composites—A review. *Compos. Part B-Eng.* 56, 296–317.

Yang, L., Shi, C., and Wu, Z. (2020). Mitigation techniques for autogenous shrinkage of ultra-high-performance concrete—A review. *Compos. Part B-Eng.* 178, 107456.

Yates. J. C. (1941). Effect of calcium chloride on readings of a volumeter inclosing Portland cement pastes and on linear changes of concrete. *Highway Research Board Proceeding* 21, 294–304.

Ye, G. (2012). Mitigation of autogenous shrinkage of ultra-high performance concrete by rice husk ash. *J. Chin. Cera. Soci.* 40(2), 212–216 (In Chinese).

Ye, H. L. Radlińska, A. (2016). A review and comparative study of existing shrinkage prediction models for portland and non-portland cementitious materials. *Adv. Mater. Sci. Eng.* 2016, 2418219.

Yoo, D. Y., Kang, S. T., Lee, J. H., Yoon, Y. S. (2013). Effect of shrinkage reducing admixture on tensile and flexural behaviors of UHPFRC considering fiber distribution characteristics. *Cem. Concr. Res.* 54, 180–190.

Yoo, D. Y., Kim, S., Kim, M. J. (2018). Comparative shrinkage behavior of ultra-high-performance fiber-reinforced concrete under ambient and heat curing conditions. *Constr. Build. Mater.* 162, 406–419.

Yoo, D. Y., Min, K. H., Lee, J. H., Yoon. Y. S. (2014). Shrinkage and cracking of restrained ultra-high-performance fiber-reinforced concrete slabs at early age. *Constr. Build. Mater.* 73, 357–365.

Yoo, D. Y., Park, J. J., Kim, S. W., Yoon, Y. S. (2014). Combined effect of expansive and shrinkage-reducing admixtures on the properties of ultra high performance fiber-reinforced concrete. *J. Compos. Mater.* 48(16), 1981–1991.

Yoo, D. Y., Shin, H. O., Yang, J. M., Yoon, Y. S. (2014). Material and bond properties of ultra high performance fiber reinforced concrete with micro steel fibers. *Compos. B. Eng.* 58, 122–133.

Yu, R., Spiesz, P., Brouwers, H. J. (2014). Effect of nano-silica on the hydration and microstructure development of Ultra-High Performance Concrete (UHPC) with a low binder amount. *Constr. Build. Mater.* 65, 140–150.

Zana, R. (2005). *Introduction to surfactants and surfactant self-assemblies[M]// Dynamics of Surfactant Self-Assemblies.* CRC Press, 18–52.

Zhan, P., He, Z. (2019). Application of shrinkage reducing admixture in concrete: A review. *Constr. Build. Mater.* 201, 676–690.

Zhang, M. H., Tam, C. T., Leow M. P. (2003). Effect of water-to-cementitious materials ratio and silica fume on the autogenous shrinkage of concrete. *Cem. Concr. Res.* 33 (10), 1687–1694.

Zhu, L., Wang, J., Li, X. Zhao, G., Huo, X. (2020). Experimental and numerical study on creep and shrinkage effects of ultra high-performance concrete beam. *Compos. Part B-Eng.* 184, 107713.

Zhutovsky, S., Kovler, K., Bentur, A. (2004). Assessment of distance of water migration in internal curing of high-strength concrete. *ACI SP-220, Autogenous deformation of concrete*, O.M. Jensen, D.P. Bentz, P. Lura, eds., ACI, Farmington Hills, MI, 181–200.

Zhutovsky, S., Kovler, K., Bentur, A. (2004). Assessment of water migration distance in internal curing of high strength concrete. *American Concrete Institute* 220, 181–197.

Zhutovsky, S., Kovler, K., Bentur, A. (2004). Influence of cement paste matrix properties on the autogenous curing of high-performance concrete. *Cem. Concr. Compos.* 26(5), 499–507.

Zhutovsky, S., Kovler, K., Bentur, A. 2011. Revisiting the protected paste volume concept for internal curing of high-strength concretes. *Cem. Concr. Res.* 41(9), 981–986.

Chapter 8

Static mechanical properties

8.1 INTRODUCTION

Knowledge of the mechanical properties of cementitious materials at any arbitrary time is fundamental for operations, such as the removal of formwork, prestressing, or cracking control. UHPC has greater compressive strength, elastic modulus, and tensile and flexural strengths under both static and impact loading than conventional concrete and high-performance concrete. Its mechanical properties, which fundamentally impact its practical use in construction sites, are highly dependent on factors of raw materials, mixture proportion, and external conditions, such as cement type, fiber size and content, aggregate type and size, curing condition, loading rate, and specimen shape and size. Particularly, the flexural and tensile properties of UHPC are much more sensitive to fiber dispersion and orientation, which are often closely related to its workability and casting procedure. The initially developed UHPC in the early 1990s used quartz sand with a maximum particle size lower than 600 μm. The decrease in aggregate size can reduce the amount of weak interfacial transition zone between aggregate and paste, thus increasing the mechanical properties, especially elastic modulus, of UHPC. However, the use of high fines in UHPC can lead to decreased workability, high autogenous shrinkage, and increased cost. The decreased workability can adversely affect mechanical properties through intervening fiber dispersion and orientation. Recently, coarse aggregate has been added to UHPC to reduce the autogenous shrinkage and cost. The coarse aggregate type, content, and size show significant effects on the mechanical properties of UHPC. In addition, the incorporation of coarse aggregate can negatively influence the dispersion of steel fibers due to loosening and wall effects.

This chapter describes static mechanical properties, including fiber-matrix bond properties, compressive behavior, elastic modulus, flexural behavior, and tensile behavior, of UHPC. Key factors, including cement content, supplementary cement materials, curing regime, fiber characteristics and content, fiber distribution and orientation, and specimen size, affecting these properties of UHPC are considered.

DOI: 10.1201/9781003203605-8

8.2 FIBER-MATRIX BOND PROPERTIES

8.2.1 Brief introduction to fiber pullout testing methods

Fiber pullout testing is a direct method that can measure fiber-matrix shear strength under static and impact loads. Common fiber pullout testing methods can be divided into single-sided and double-sided methods in terms of the way of applying tensile force and/or fiber embedment, as illustrated in Figure 8.1. Single-sided pullout testing is relatively simple to be carried out. However, difficulty in gripping the free end of the fiber is inevitably encountered due to its very fine diameter. To secure the reliability of results, a large number of specimens are required. Furthermore, the whole fiber is fully embedded in real concrete composites, which is significantly different from the ideal test. The double-sided pullout testing was employed by some researchers (Chan and Chu 2004; Wu et al. 2016; Yoo et al. 2020). Dogbone samples with two separated halves from the sample center perpendicular to the loading direction can be used. The elimination of the adhesion of the two halves is essential to secure accurate results.

8.2.2 Typical pullout load-slip curves of straight steel fiber

The overall pullout behavior of a straight steel fiber can be described as the combination of two mechanisms: debonding of the surrounding interface and frictional slip of the fiber. The embedment fiber is fully debonded from the outer surface to the interior of the specimen, and the fiber pullout occurs under frictional resistance. The pullout behavior can be divided into four distinct regions: a well-bonded region (OA), a partially debonded region (AB),

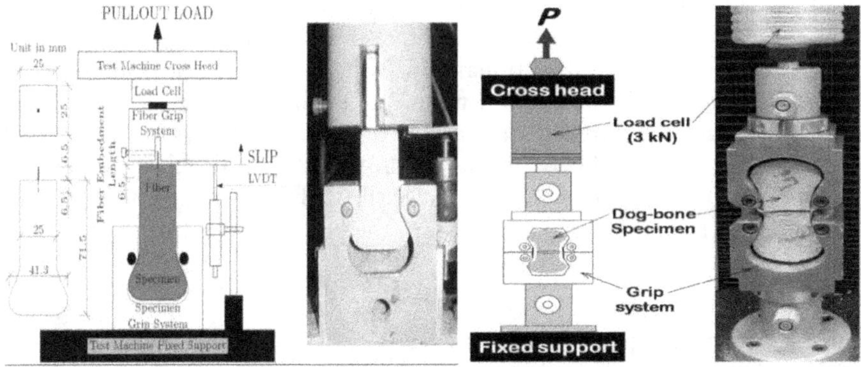

(a) Single-side pullout test (Wille and Naaman 2012a) (b) Double-side pullout test (Yoo et al. 2020)

Figure 8.1 Setups for fiber pullout tests. (a) Single-side pullout test (Wille and Naaman 2012a). (b) Double-side pullout test (Yoo et al. 2020).

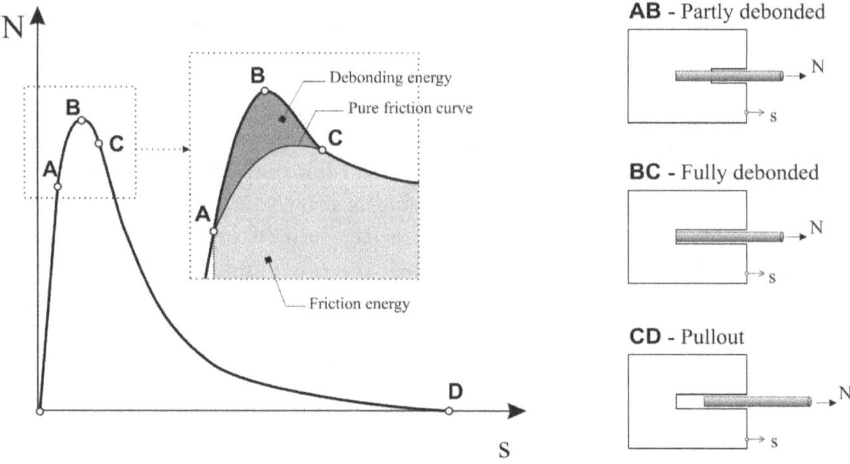

Figure 8.2 Typical pullout load-slip curve of smooth straight steel fiber.

Source: Cunha et al. (2007).

a fully debonded region (BC), and a fiber removal region (CD) (Naaman et al. 1991a, 1991b), as illustrated in Figure 8.2. At the well-bonded region, the pullout load linearly increases with the slip increase and is dominated by adhesion bonds. Before the peak load (B) is attained, a nonlinear phase in the pullout load-slip curve is observed. The prepeak branch (AB) starts with the microcracking of the interfacial transition zone, corresponding to the initiation of the debonding process. The nonlinearity observed in this part is regarded as an indication of interfacial crack propagation (Shah et al. 1995). After the maximum load is reached, the load decreases with the increase of slip, and full debonding then occurs (C). Subsequently, the fiber is pulled out under frictional slip (CD). In this region, the load further decreases with the increase of slip due to the decreased available frictional area and roughness of the failure surface. For deformed fibers, such as twisted and hooked fibers, friction and/or mechanical anchorage between the fiber and matrix play a dominant role through the slippage of fibers after attaining the full debonding stage. Thus, the bond between matrix and fiber can be effectively improved.

8.2.3 Effect of quality of matrix

The fiber-matrix bond properties are related to matrix, fiber, and fiber-matrix interface quality. The quality of the matrix is a primary factor affecting the bond properties and can be enhanced by using fine supplementary cementitious materials and high-temperature curing. Generally, higher matrix strength results in better bond. Wille and Naaman (2013) demonstrated that the average bond strength and pullout energy of straight fiber are enhanced by 100% and 66%, respectively, with the matrix strength increasing from 207 to 240 MPa. Silica fume can effectively enhance the fiber-matrix bond properties because of its filler and pozzolanic effects. The optimal silica fume content for fiber-matrix bond properties of UHPC was reported as 20%–30%, by mass of cementitious materials (Chan and Chu 2004; Wu et al. 2016b). The bond strength between straight fiber and heat-cured UHPC matrix is 4.8–5.5 MPa. Besides, the enhancement in pullout energy due to silica fume is far more significant than that of bond strength. At 30% silica fume, the pullout energy increases by approximately 100%, whereas the bond strength is enhanced by 14% (Chan and Chu 2004).

Nanomaterials, such as nano-SiO_2 and nano-$CaCO_3$, can also be used to improve the quality of fiber-matrix interface through exerting filler, nucleation, and chemical effects. The addition of 3.2% nano-$CaCO_3$ can improve the bond strength and pullout energy by 45% and 200%, respectively, compared to the reference specimen at 28 d of standard curing (Wu et al. 2018b). The use of 1% nano-SiO_2 increases the fiber bond strength and pull-out energy by about 35% and 70%, respectively (Wu et al. 2017a). However, an excessive amount of nanoparticles can increase the viscosity of fresh UHPC mixture and cause agglomeration issues, leading to more air bubbles and cracks near the fiber-matrix interface and lower bond strength.

The aggregate size and content in UHPC affect the packing density of UHPC matrix, eventually altering the bond behavior. Fine sand with a granular distribution improves the fiber-matrix bond properties due to high packing density (Wille and Naaman, 2012a). The peak pullout load can be increased by approximately 100% by appropriately adjusting the sand-to-cement ratio from 1.38 to 1.01 and increasing the small to large grain sand from 20% to 30%. Sands of SiO_2 or $ZrSiO_4$ might scratch the fiber surface during pullout, resulting in high frictional stress (Wille and Naaman 2013). The introduction of coarse aggregates in UHPC can lead to more weak interfacial transition zones between aggregate and matrix, thus causing more cracks and decreased bond strength.

The curing regime impacts the fiber-matrix interfacial bond strength by affecting the quality and quantity of hydration products. The interfacial bond strength under various curing regimes follows the order of: autoclave curing > steam curing > room curing. The bond strengths between straight fiber and UHPC matrix under water curing and autoclave curing are significantly improved by 69.1% and 156.4%, respectively, compared to that under room curing conditions. In addition, UHPC subjected to autoclave curing gains the highest interfacial bond strength of 14.2 MPa (Wang et al. 2015). This is attributed to the more hydrated C-S-H with longer average chains under autoclave curing.

8.2.4 Effects of fiber geometry and fiber surface treatment

Straight fiber can be processed to deformed fibers, such as twisted, corrugated, and hooked fibers, and used in UHPC. The deformed section of fibers can effectively enhance the bond properties between fibers and UHPC matrix. Wu et al. (2018a) evaluated the interfacial bond properties between embedded steel fibers with different shapes (i.e., straight, hooked, and corrugated fibers) and UHPC matrices proportioned with 15%–20% silica fume under water curing ages of 1, 3, 7, 28, and 91 d. The fibers have the same diameters of 0.2 mm and lengths of 13 mm. Compared to straight fibers, corrugated and hooked fibers significantly improve the bond properties by three to seven times due to mechanical anchorage and interlock (Figure 8.3). Besides, 28-d water curing can yield relatively stable bond strength and pullout energy. The bond properties remains almost stable with curing age prolonged to 91 d.

Fiber surface treatment, including brass and zinc coating and soaking in a chemical solution, can roughen the fiber surface and improve the fiber-matrix bond properties. Yoo et al. (2020) investigated the bond behavior of corroded straight steel fibers with various degrees of fiber corrosion (0%–15%) by immersing them in a 3.5% NaCl solution. The surface corrosion of straight steel fibers in UHPC effectively enhanced the pullout resistance for both aligned and inclined fibers if they were completely pulled out from the matrix. Surface corrosion on aligned steel fibers was more effective in improving the pullout resistance of UHPC than that on inclined fibers. The maximum average bond strength and pullout energy of moderately corroded

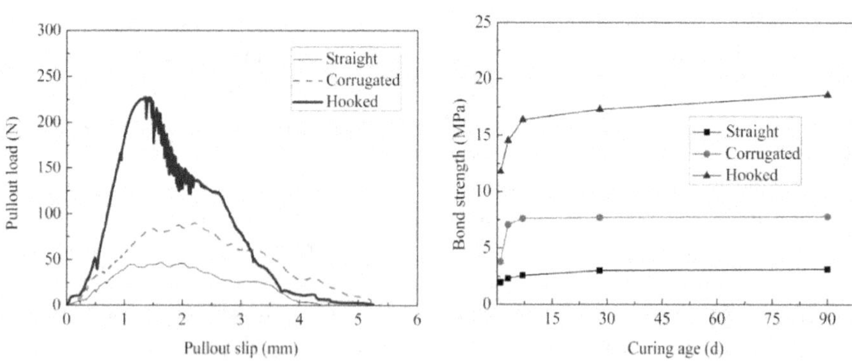

Figure 8.3 Effect of fiber geometry on fiber-matrix bond behavior of UHPC (Wu et al. 2018a). (a) Pullout load-slip curves. (b) Bond strengths at various curing ages.

fibers in UHPC are 18.5 MPa and 715.7 N·mm, respectively, which are approximately 2.7 and 1.8 times higher than those of plain fibers in the same matrix at the aligned condition.

Surface treatment using sandpapers with various grits was applied in parallel and perpendicular directions to improve the pullout resistance of straight steel fiber embedded in UHPC and achieve excellent slip-hardening characteristics (Chun et al. 2020). The surface treatment using sandpapers in both parallel and perpendicular directions effectively enhanced the pullout resistance of straight fiber in UHPC. The bond strength of plain steel fiber in UHPC was significantly improved by 31%–66% by sanding in the parallel direction. However, sanding of fibers in the perpendicular direction is recommended because of better slip-hardening characteristics.

8.2.5 Effect of fiber inclination angle

Fiber embedded in UHPC matrix with an appropriate angle can improve the pullout behavior. Yoo et al. (2019) demonstrated that the highest pullout resistances of straight, hooked, half-hooked, and twisted fibers were obtained at an inclination angle between 30° and 45°, instead of 0° or 60°. Qi et al. (2018) found that the pullout response of straight fibers showed higher sensitivity to the increased fiber-embedded angle than that of hooked-end fibers. The bond strengths between straight fiber and UHPC were increased by 19% and 53% at 30° and 45°, respectively, while the bond strengths of hooked-end fibers were increased by 10.3%–13.6% and 16.2–26.1%, respectively.

8.2.6 Effect of loading rate

Tai et al. (2016) evaluated the fiber pullout behaviors of straight, hooked, and twisted fibers from UHPC matrix under various loading rates ranging from 0.018 to 1800 mm/s using a single-fiber pullout test. With the increase in pullout rate, the peak load and total pullout work increased, as illustrated in Figure 8.4. The pullout responses of all three fibers exhibit progressively increasing rate sensitivity as the pullout speed increases and become significant during impact loading. Besides, the twisted fibers exhibited the highest average bond strength and energy dissipation capacity.

8.3 COMPRESSIVE BEHAVIOR

8.3.1 Specimen size and test methods

The compressive stress-strain relationship can reflect the stiffness degradation and ductility of concrete, except for compressive strength and initial modulus of elasticity. Cubes, prisms, and cylinders can be used to

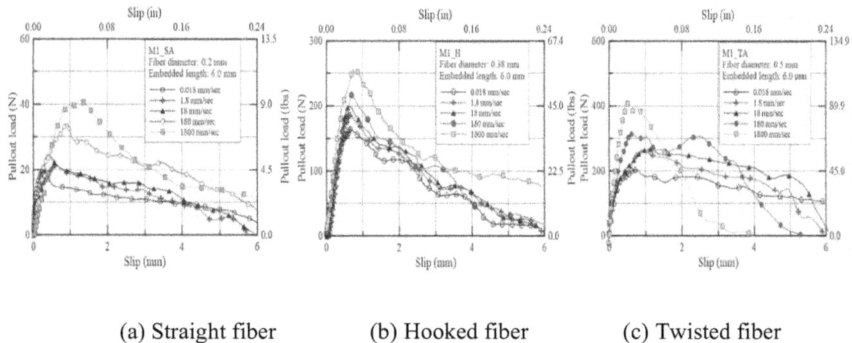

(a) Straight fiber (b) Hooked fiber (c) Twisted fiber

Figure 8.4 Pullout load-slip curves of different shapes of fibers in UHPC under various loading rates (Tai et al. 2016). (a) Straight fiber. (b) Hooked fiber. (c) Twisted fiber.

investigate the compressive stress-strain behaviors of UHPC. France, China, Switzerland, United States (USA), Japan, South Korea, etc., have formulated relevant standards and specifications on the design, testing, and applications of UHPC to facilitate its engineering construction. The minimum compressive strength for UHPC is defined in these standards. However, this value varies due to the differences in materials and specimen sizes used in each country. For example, French Standard defined UHPC with a minimum compressive strength of 130 MPa using Φ110 × 220 mm cylinders (NF P 18-470, 2017). Cubes with sizes of 100 × 100 × 100 mm are recommended for compressive behavior testing of UHPC in the standards of China (GB/T 31387, 2015) and Switzerland (prSIA 2052, 2014). The compressive strengths of UHPC should be greater than 100 and 120 MPa, respectively, according to GB/T 31387 (2015) and prSIA 2052 (2014). ASTM (United States) (2017) and KCI (South Korea) (2012) suggest the use of Φ75 × 150 mm cylinders for compressive strength measurement. The minimum compressive strengths for UHPC are determined to be 120 and 180 MPa, respectively. Besides, researchers from Japan (2006), Australia (2000), and American Concrete Institute (ACI 239R, 2018) agreed on a minimum compressive strength of 150 MPa using cylinders measuring Φ100 × 200, Φ70 × 140, and Φ102 × 204 mm, respectively.

Before compression testing, the contact surfaces of specimens should be ground and polished to minimize the influence of uneven surfaces at each end. The actual height of each specimen should be measured and recorded. Figure 8.5 illustrates the experimental setup for the uniaxial compression testing. Two linear variable differential transformers (LVDTs) are located vertically and connected to a digital transducer to capture the displacement of the specimens. The force is directly recorded by the load cell attached to the testing machine.

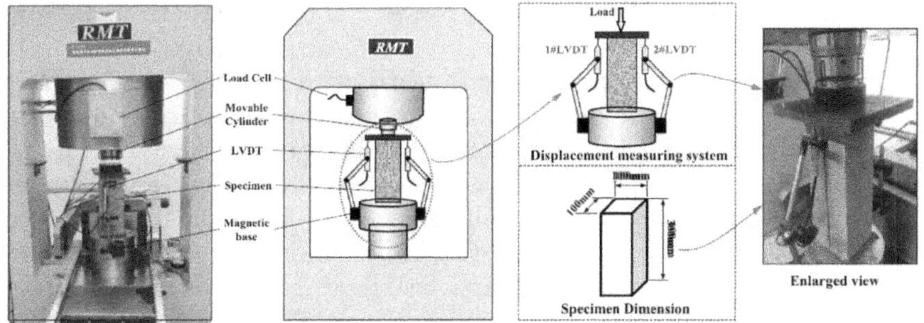

Figure 8.5 Test setup for compression testing of UHPC.

Source: Deng et al. (2020).

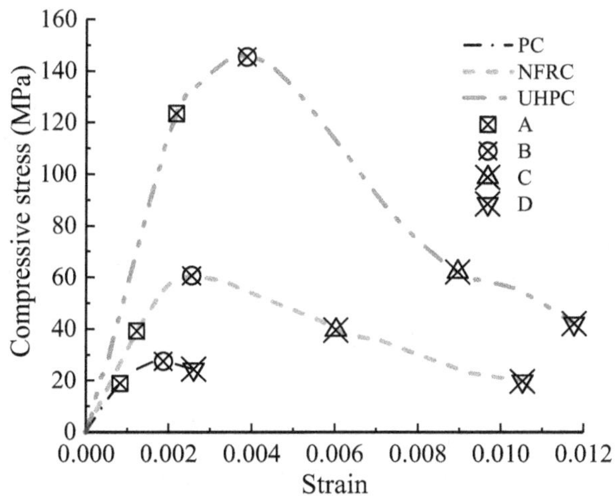

Figure 8.6 Typical compressive stress-strain curves of PC, NFRC, and UHPC.

Source: Liu et al. (2019); Wu et al. (2020).

8.3.2 Typical compressive load-deflection curves of UHPC

UHPC exhibits significantly improved compressive behavior compared to normal concrete and high-performance concrete. Figure 8.6 compares the typical compressive stress-strain curves of plain concrete (PC), normal fiber-reinforced concrete (NFRC), and UHPC. Four main stages, namely elastic deformation, initial cracking, visible cracking, and failure, can be identified, which are presented as branches OA, AB, BC, and CD, respectively.

(1) Branch OA: The stress increases linearly with the increment of strain, and an elastic deformation section is observed, where cracks initiate and propagate from intrinsic microcracks and/or pores at interfacial transition zones. The slope of the ascending part is generally regarded as the elastic modulus of concrete, which is mainly dependent on the compressive strength of concrete. PC usually exhibits a proportional limit of 30% to the ultimate compressive strength and an ultimate strain of approximately 0.002. UHPC shows a linear limit of up to 70%–80% (Graybeal 2005; Hoang and Fehling 2017) or even 90%–95% (Hassan et al. 2012) of the ultimate compressive stress.
(2) Branch AB: The stress increases nonlinearly with the increase of strain until the peak stress. A large visible crack network can be observed in the specimens. The nonlinearity is primarily attributed to the coalescence of microcracks at the paste-aggregate interfaces. PC specimens under compressive show internal and even external cracks, while fiber-reinforced concrete or UHPC specimens may not exhibit visible and external cracks due to the reinforcing effect of fibers (Li et al. 2017; Wu et al. 2020).
(3) Branch BC: After reaching the peak stress at point B, the first descending portion of the stress-strain curve of concrete can be monitored using a displacement-controlled machine. The stress reduces rapidly up to the inflection point C. The stress at C is approximately 50% of the peak stress for PC. Obvious cracks can be observed in this branch, and they continually extend with the increase of strain until forming a connected major crack in the whole section.
(4) Branch CD: After the inflection point C, the decreasing stress rate slows down as the strain increases, and the specimens then fail.

Figures 8.7 and 8.8 illustrate typical failure patterns of PC, NFRC, and UHPC under compression. As observed in Figure 8.7(a), the PC specimens under cyclic compressive loading show tensile cracks and finally collapse into several pieces (Shi et al. 2020; Xu et al. 2018). On the contrary, the failure of NFRC specimens is dominated by shear cracks with angles ranging from 45° to 60°, as shown in Figure 8.7(b–d) (Li et al. 2017; Xu et al. 2018). This is due to the restraint effect on cracking caused by the addition of fibers. Therefore, NFRC specimens demonstrate ductile failure patterns with finer cracks and relatively silent breakage (Lee and Barr 2004). For UHPC, the reference mixture without any steel fibers fails with an explosion (Figure 8.8(a)). Shear cracking and multiple cracks are observed on UHPC with 2% steel fibers, resulting in a greater ultimate strain and stress redistribution (Prabha et al. 2010). Besides, the distributions of multiple cracks are larger and wider than those of NFRC.

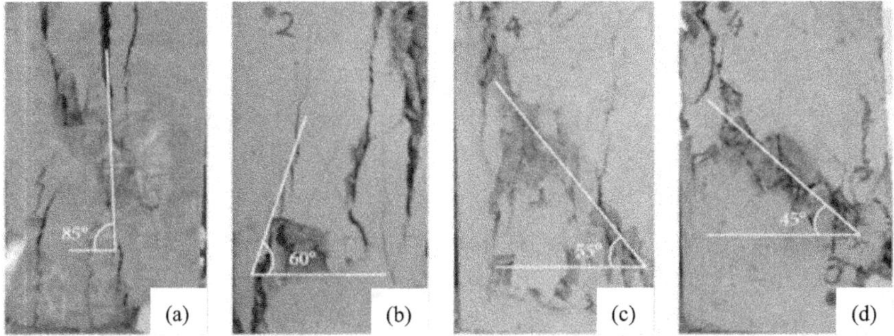

Figure 8.7 Typical failure modes of PC and NFRC under cyclic compression loading: (a) PC; (b) NFRC with 1.5% PF; (c) NFRC with 1.5% SF; and (d) NFRC with 3% hybrid PF and SF (PF: polypropylene fiber, SF: steel fiber).

Source: Xu et al. (2018).

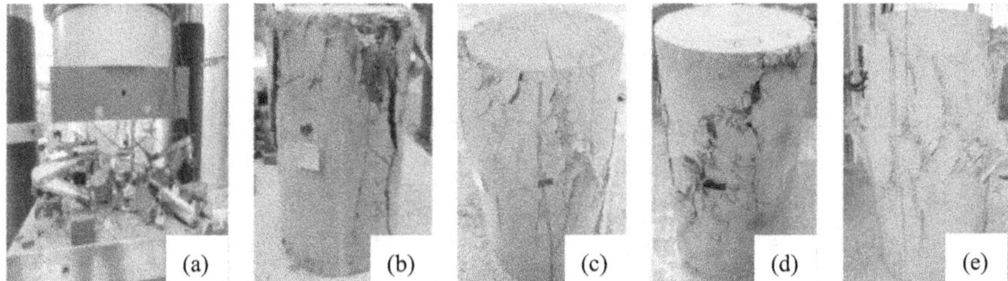

Figure 8.8 Typical failure modes of UHPC with different fiber volumes and aspect ratios under uniaxial compression: (a) UHPC matrix without any fiber; (b) UHPC with 1.5% fibers at l_f/d_f of 13/0.75; (c) UHPC with 1.5% fibers at l_f/d_f of 20/0.25; (d) UHPC with 3% fibers at l_f/d_f of 13/0.175; and (e) UHPC with 3% fibers at l_f/d_f of 20/0.25.

Source: Hoang and Fehling (2017).

8.3.3 Effect of w/b ratio

The w/b ratio is an important parameter influencing the compressive stress-strain relationships of concrete. Figure 8.9 illustrates the compressive stress-strain curves of UHPC with w/b ratios of 0.18–0.24. Generally, the decrease in w/b ratio results in an increase in peak load under compression. The use of high or moderate binder content combined with superplasticizer can ensure proper flowability of UHPC. It should be noted that excessive binder in concrete with a low w/b ratio can result in an extremely high viscosity, leading to more air bubbles and porous

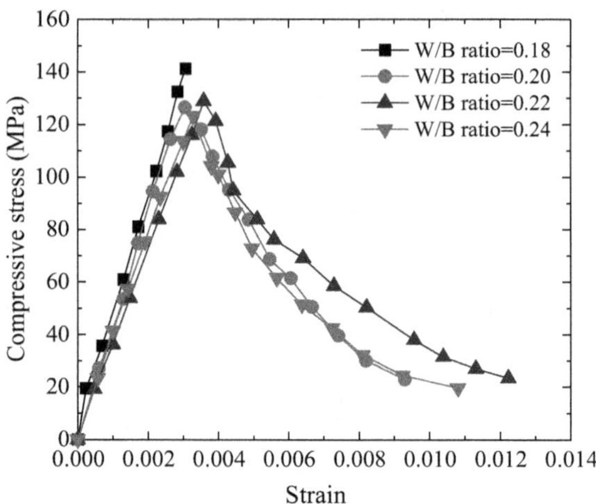

Figure 8.9 Effect of w/b ratio on compressive stress-strain curves of UHPC.

Source: Data from Huang and Tan (2007).

microstructure. Therefore, moderate binder content should be used at a fixed water content to achieve proper flowability, dense microstructure, and high mechanical properties.

Pu et al. (2002) indicated that UHPC mixtures with a 0.18 w/b ratio exhibited 28-d compressive strength of 100 MPa. Wille et al. (2011b) evaluated the effect of w/b ratio, ranging from 0.107 to 0.254, on the workability and mechanical properties of UHPC. The results indicated that the compressive strength decreased with the increase in w/b ratio from 0.107 to 0.254. Besides, air content substantially influences the compressive strength of UHPC. The reduction in w/b ratio results in a decrease in flow due to increased yield stress and viscosity of paste, causing a higher amount of entrapped air content (Wille et al. 2011b). Therefore, the strategy of increasing the strength of UHPC by reducing the w/b ratio is valid given the improved workability.

8.3.4 Effects of fiber characteristics

The addition of fibers increases the strength and elastic modulus of concrete due to the efficiency in restricting the formation and propagation of cracks. Two types of fibers, namely rigid and flexible fibers, are often used in concrete. Rigid fibers can improve the compressive behavior of concrete because of their intrinsic rigidity and fiber bridging effect. Straight steel fiber is the most commonly used one in UHPC. As the steel fiber volume increases, the increased number of fibers leads to enhanced rigidity of the whole system and eventually improved toughness of descending branch of the compressive stress-strain curve. It can result in greater postpeak ductility and energy absorption capacity of concrete. The use of 2% steel fibers enhances the elastic modulus of UHPC by 3%–7% (Alsalman et al. 2017) and increases the compressive strength by 20%–46% (Wu et al. 2017). However, excessive fiber content may impair the compressive behavior of concrete due to fiber cluster issues associated with increased plastic viscosity of the fresh mixture (Hoang and Fehling 2017). Therefore, 2% or 2.5% fiber volume is generally recommended to increase the compressive strength and ductility of UHPC.

Due to the high mechanical anchoring effect, the compressive behavior of concrete can be improved by using deformed fibers, such as hooked and corrugated fibers. Deformed fibers have higher pullout strength than straight fibers due to their better bridging effect. Wu et al. (2016a) indicated that 2% hooked-end, corrugated, and straight fibers increased the compressive strengths by 48.7%, 44.3%, and 40.3%, respectively, compared to those without any fiber. Liu et al. (2016) investigated the effects of four types of steel fiber, namely straight microfiber, spiral microfiber, hooked microfiber, and hooked macrofiber, on the compressive strengths of UHPC. The lengths and diameters of the three microfibers are 13 and 0.2 mm, whereas those of the macrofiber are 30 and 0.6 mm, respectively. Results showed that UHPC with deformed fiber exhibited greater compressive strength than that of straight fiber. However, slight differences were observed in the compressive strength of mixtures with spiral fiber, hooked microfiber, and hooked macrofiber.

Compared to rigid fibers, the addition of flexible fibers has less influence on increasing the compressive strength and toughness of fiber-reinforced concrete. Common flexible fibers used in UHPC include polymer fiber, polypropylene fiber, and polyvinyl alcohol (PVA) fiber. Generally, the addition of flexible fibers results in improved fresh properties, avoiding producing voids and honeycombs at low flowability (Dawood and Ramli 2011). Flexible fibers have a limited effect on improving the compressive strength but exert a great influence on enhancing the fracture toughness.

Hybridization of steel fibers of various sizes is beneficial for improving the compressive strength and ductility of UHPC. The highest compressive strength is achieved for UHPC with 1.5% long and 0.5% short fibers, whereas those with 2% short fibers show the worst properties (Wu et al. 2016c).

8.3.5 Effect of curing condition

Standard room temperature curing, air curing, heat curing under atmospheric pressure, and autoclave curing regimes can be used for UHPC in the laboratory and in practical applications. Standard room temperature curing is the most common, economical, and environmentally friendly method in practice. Concrete specimens are kept in a standard laboratory environment with 20°C and relative humidity greater than 95% or saturated limestone water from demolding until testing age, as specified in ASTM C31 (2021),

ASTM C511 (2019), and Chinese standard GB/T 50081-2019 (2019). Curing with a temperature of 20 ± 1°C and a relative humidity of 65% ± 8% until testing is recommended by EN 459-1: 2015 (2010).

The compressive stress-strain curves of concrete at different curing ages show various trends. Low elastic modulus, compressive strength, and very high peak strain are found at early ages. However, peak strain decreases rapidly with increasing age, while the elastic modulus and peak strength are increased. This phenomenon was noticed by Graybeal (2007), who studied the axial compressive stress-strain responses of UHPC cylinders at 18 h, 21 h, 24 h, 48 h, 71 h, 7 d, 9 d, 14 d, 28 d, and 56 d (Figure 8.10). This is because the microstructure of concrete exhibits a significant difference with increasing curing time. A more extended curing period results in further cement hydration and lower porosity in the matrix than that in the interfacial transition zone. Therefore, the paste porosity decreases at a higher rate than those of interfacial transition zone (Elsharief et al. 2003; Scrivener et al. 2004). As the compressive strength increases, the postpeak strain capacity increases, while the basic shape of the ascending branch of the compressive stress-strain curve remains unchanged. Increasing the curing duration to 71 h to 56 d increases the elastic modulus but results in a more brittle failure with little softening behavior before complete failure. A possible reason is that the accelerated heat curing increases the porosity in the cement paste matrix.

Heat curing can effectively enhance the compressive behavior of concrete because the increased temperature can speed up the chemical reactions, resulting in more hydration products (Prem et al. 2015). Graybeal and Baby (2013) evaluated the compressive properties of UHPC cylinders with

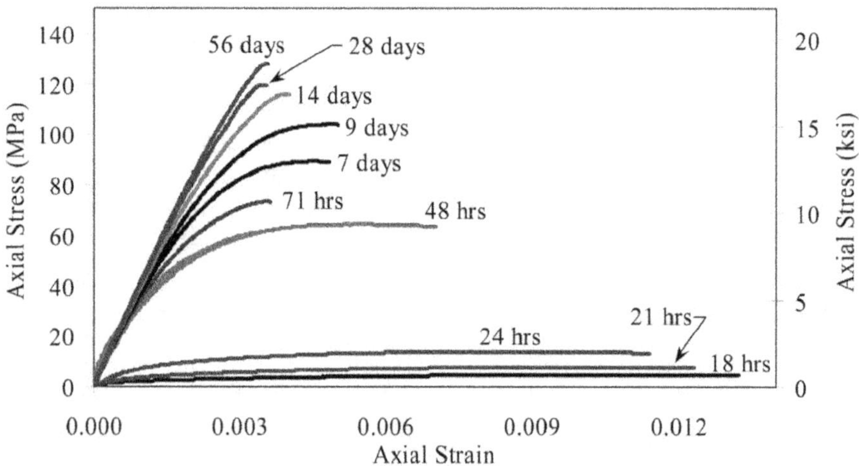

Figure 8.10 Effect of curing age on compressive stress-strain relationships of UHPC Source: Graybeal (2007).

diameters of 76 and 110 mm under different curing regimes. The compressive strengths of UHPC with 2% fibers under 48 h steam curing and 28-d room curing were 220 and 192 MPa, respectively. Abed et al. (2018) investigated the influence of curing regime on the compressive strength of reactive powder concrete (RPC). Two curing regimes were employed: (1) 24 ± 2°C water curing, and (2) 105 ± 5°C heat curing for 48 h and then 24 ± 2°C water curing until the testing age. It was found that the hot water curing had a greater effect on the early-age strength development before 7 d than that at later ages. Prem et al. (2015) evaluated the compressive properties of UHPC under water, steam, and heat curing. For steam curing, the samples were cured under 100°C at a rate of 20–30°C/h and relative humidity of 95% for 18 h. For heat curing specimens, the temperature increased to 300 ± 1°C at a rate of 2°C/min. It is shown that the compressive strengths of 14 d heat-treated specimens attained 185 MPa. UHPC after heat curing showed the highest 28-d compressive strength of 196 MPa.

Notably, the gradual increase in temperature can increase the reactivity and accelerate the hydration of cement due to more energy exerted. Noushini et al. (2016) reported that increasing the curing temperature up to 90°C improved the compressive strength of concrete. Dong et al. (2018) evaluated the effect of curing temperature on the compressive stress-strain curve of UHPC and found similar results. When the heating temperature is 200–300°C, increasing the heating temperature improves the strength during heating. Rong et al. (2020) investigated the influence of three heating temperatures of 200, 250, and 300°C on the compressive strengths of RPC. It is indicated that the compressive strength of RPC after 250°C heat curing for 8 h increased by 94%–104% compared to the reference sample. However, compressive strength is reduced when exposed to elevated temperatures greater than 300°C (Abid et al. 2019; Hou et al. 2019; Zheng et al. 2012). Increasing the crystallization degree, C-S-H size, and other hydration products can increase the porosity and decrease the strength of concrete during elevated temperatures.

8.3.6 Effect of coarse aggregate

Some research has recently demonstrated that UHPC mixed with coarse aggregate can reduce binder content and cost, improve workability, enhance projectile impact resistance, and decrease autogenous shrinkage of UHPC (Li et al. 2018; Xu et al. 2018). It should be noted that different types of aggregate exert various effects on the compressive stress-strain curves of UHPC. Figure 8.11 illustrates the effects of coarse aggregate type and content on the compressive stress-strain curves of UHPC. Specimens with different types of aggregate indicate a similar elastic modulus of the ascending segment. However, the compressive strength greatly varies with aggregate type (Wu et al. 2020). At 15% coarse aggregate content, UHPC made with granite shows the highest compressive strength, followed by basalt, limestone, and

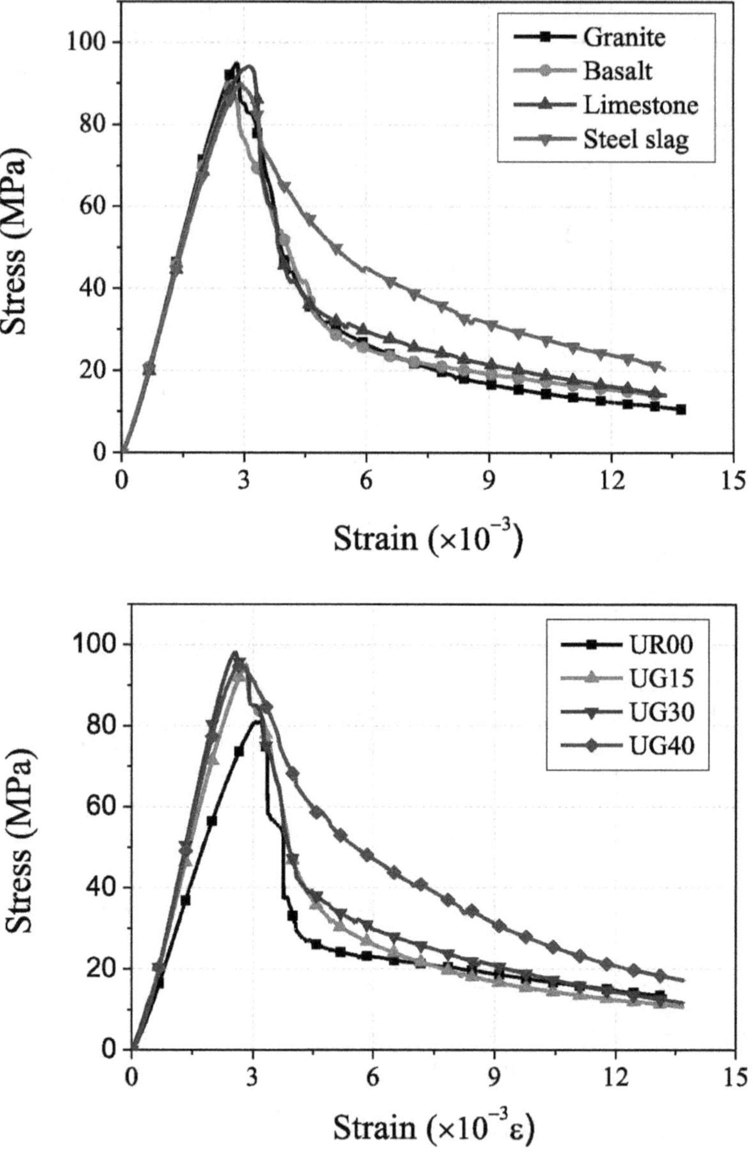

Figure 8.11 Effects of coarse aggregate type and content on compressive stress-strain curves of UHPC: (a) 15% content; (b) Various granite contents.

Source: Wu et al. (2020).

then steel slag. The microstructure of specimens with granite aggregate is more homogeneous, and a distinct boundary between aggregate and matrix is observed compared with the other types of aggregate.

Conversely, steel slag has worse homogeneity and spherical morphology, resulting in stress concentration, as observed in Figure 8.12. Moreover, the elastic modulus and peak strength increase with the increasing granite content. When reaching a critical content, the strength of the mixture is reduced. The reduction of strength may be attributed to the reduced concrete flowability associated with a substantial increase in content, resulting in more defects within concrete (Xu et al. 2019). The moderate addition of aggregate reduces the slope of the descending branch and improves the ductility (Figure 8.12(b)).

8.4 ELASTIC MODULUS

Elastic modulus, a critical parameter for the design and analysis of concrete structures, can affect the deformation of concrete at the elastic stage and early-age cracking. UHPC usually has a high elastic modulus in the range

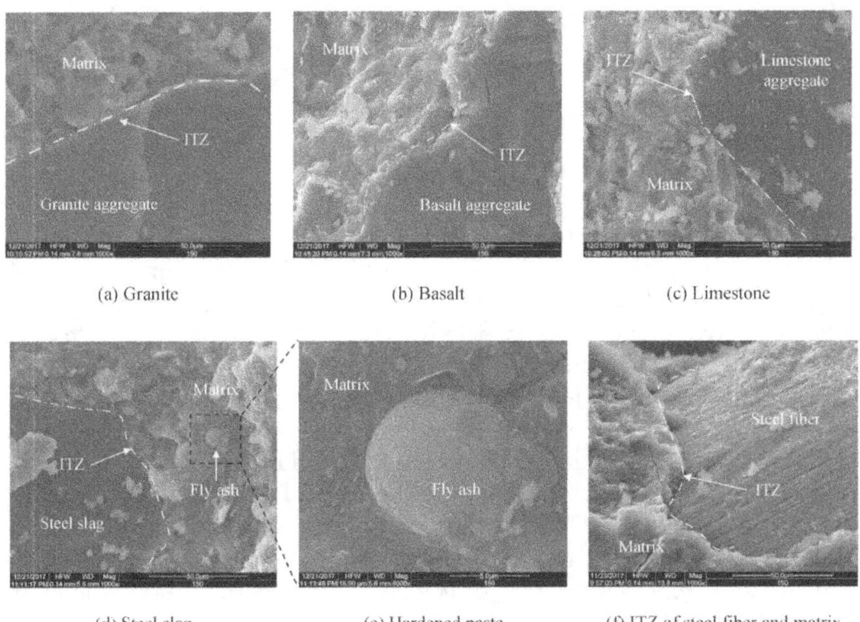

Figure 8.12 SEM images of microstructures of UHPC with different coarse aggregate types.

Source: Wu et al. (2020).

from 40 to 60 GPa, which is 20%–70% higher than those of normal concrete and high-performance concrete (Ahmad et al. 2015; Hannawi et al. 2016; Wang et al. 2015). Empirical equations can be used to determine the elastic modulus of concrete based on measured compressive strength (Alsalman et al. 2017; Graybeal 2007; Ma et al. 2004). The empirical equations can give a quantitative estimation of the elastic modulus.

8.4.1 Effect of binder

Silica fume with small particle size and high activity is one of the indispensable raw materials for the preparation of UHPC. The elastic modulus of UHPC increases with the increase of silica fume content up to 20% and then decreases when silica fume content is greater than 20% (Gesoglu et al. 2016). This is mainly because of the high pozzolanic activity of silica fume. It can entirely or partially consume the calcium hydroxide produced by cement hydration in the matrix or interfacial transition zone to produce hydrated calcium silicate. In addition, its smaller particle size produces a filling effect, resulting in a denser UHPC microstructure and improved mechanical properties (Wang et al. 2015). The calcium hydroxide content and porosity of UHPC specimens mixed with 15%–25% silica fume at 28 d standard curing are 3% and 5%–8%, respectively (Wu et al. 2016b; Wu et al. 2019a). At 30% silica fume content, the fluidity is significantly reduced, and the plastic viscosity of UHPC is increased, thus decreasing the compressive strength and elastic modulus of UHPC.

Since w/b ratio is low in UHPC, high content of unreacted cement clinker and supplementary cementitious materials remains in the UHPC matrix (Zhao and Sun 2014). The elastic modulus values of cement clinker and fly ash are 145 GPa (Velez et al. 2001) and 83.7 GPa (Zhao and Sun 2014), respectively, which are higher than those of hydration products. Thus, fly ash and unhydrated cement particles can play a "micro aggregate effect" to enhance the elastic modulus of UHPC (Zhao and Sun 2014). However, the elastic modulus of slag is 5.7–25.4 GPa (Zhao et al. 2005), and too much residue slag will reduce the elastic modulus. Figure 8.13 shows the effects of fly ash and slag contents on the elastic modulus of UHPC with 25% replacement rate of silica fume to cement. The addition of fly ash increases the elastic modulus of UHPC, while the slag shows the opposite behavior (Yazici et al. 2009). When the fly ash and slag contents are higher than 20% and the total replacement rate of cement is greater than 45%, the elastic modulus decreases greatly.

Nanomodification is used to modify the performance of cement-based materials. When using nanomodification, many low-density hydrates are formed in the early stage, which effectively improves the microstructure of the matrix and the bond strength of matrix and aggregate and eventually increases the compactness and early mechanical properties of UHPC

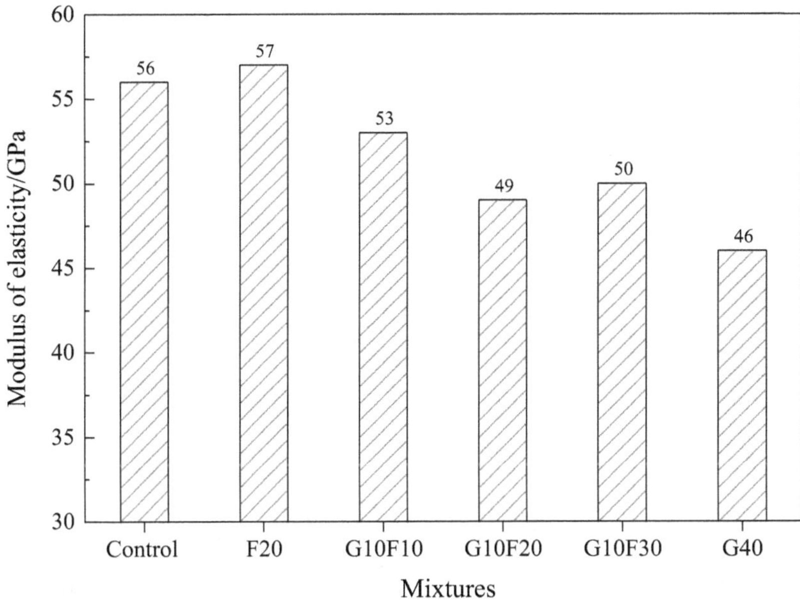

Figure 8.13 Effects of slag and/or fly ash contents on modulus of elasticity of mixtures (autoclave curing). (Note: G10F20 means cement was replaced with 10% granulated blast furnace slag and 20% fly ash.)

Source: Data from Yazici et al. (2009).

(Ghafari et al. 2015). Nano-SiO_2 with small particle size and high activity can accelerate the early hydration reaction and generate many hydration products, thus improving the fiber-matrix bond performance and microstructure (Wu et al. 2017a). When mixing 1% nano-SiO_2 and 10% silica fume, the 90-d compressive strength and elastic modulus of UHPC are increased by 8% and 7%, respectively (Gesoglu et al. 2016). The mechanical properties and the compactness of the interfacial transition zone near the aggregate are also significantly improved with the addition of nano-SiO_2, which increases the elastic modulus of the interfacial transition zone from 50% to 80% compared to that of the matrix (Xu and Wang 2018).

Figure 8.14 depicts elastic moduli of UHPC mixtures made with different w/b ratios and basalt aggregate contents. With the decrease in w/b ratio, the elastic modulus of UHPC increases. UHPC mixture at 0.14 w/b ratio shows the highest elastic modulus of 54.9 GPa. This is because a lot of unhydrated cement particles act as filler materials in the cement paste, eventually resulting in improved elastic modulus (Alsalman et al. 2017).

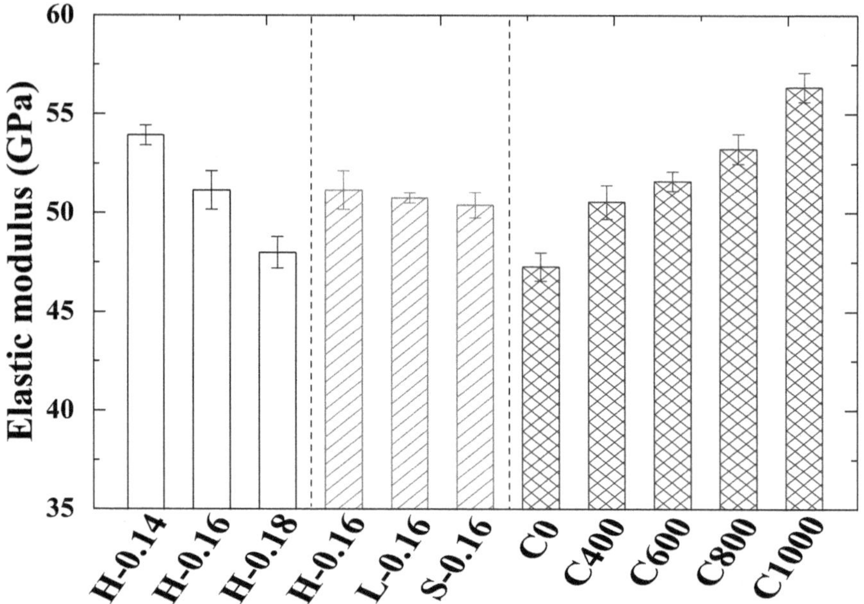

Figure 8.14 Measured elastic modulus values of different UHPC mixtures.

Source: Ouyang et al. (2020).

8.4.2 Effect of coarse aggregate

The commonly used coarse aggregates in UHPC can be divided into two categories according to their chemical composition, including carbonate and silicate. Carbonate includes limestone, marble, and dolomite, while silicate contains basalt, sandstone, granite, and diabase. Limestone is mainly used to prepare ordinary concrete. In contrast, basalt, granite, and diabase with high strength and high elastic modulus are used in high-strength concrete. Their physical and mechanical properties are shown in Table 8.1.

The addition of coarse aggregate into UHPC can reduce the amount of cementitious material. When the maximum particle size of coarse aggregate is 8 and 16 mm, the cementitious material content can be reduced to 800 and 700 kg/m^3, respectively (Li et al. 2018). The use of coarse aggregate can provide an excellent skeleton for UHPC, thus enhancing the elastic modulus of UHPC (Chen et al. 2017).

The influence of coarse aggregate on the elastic modulus of UHPC is mainly related to its elastic modulus, content, particle size, and the characteristics of the interfacial transition zone around coarse aggregate. Vishalakshi et al. (2018) found that conventional concrete made with

Table 8.1 Physical and mechanical properties of aggregate

Type	*Density (kg/m³)*	*Compressive strength (MPa)*	*Modulus of elasticity (GPa)*	*Porosity (%)*	*Water absorption (%)*
Limestone	1800–2600	70–128	21–84	0.53–27.0	0.1–4.45
Marble	2700	70–140	10–34	0.10–6.00	0.1–0.8
Dolomites	2100–2700	40–120	13–34	0.30–25.0	-
Basalt	2600–3100	160–250	43–106	0.30–21.8	0.30
Sandstone	2200–2600	47–180	27.8–54	1.60–28.3	0.2–7.0
Granite	2630–2800	75–120	54.3–69	0.04–2.80	0.1–0.7
Diabase	2530–2970	160–180	69–79	0.29–1.13	0.8–5.0

different types of aggregate showed significant variation in modulus of elasticity compared to high-strength concrete due to its higher volumetric proportion of aggregate and obvious weak interfacial transition zones. Basalt with high strength and siliceous content is mixed into UHPC as coarse aggregate. Its content and particle size influence the elastic modulus of UHPC. Ouyang et al. (2020) indicated that UHPC mixtures with coarse basalt aggregate exhibit higher elastic modulus than that of the reference mixture without any aggregate (Figure 8.14). With the basalt content increasing from 0 to 1000 kg/m^3, the elastic modulus increases from 47 to 57 GPa. When UHPC is mixed with basalt with a maximum particle size of less than 8 mm, the compressive strength of UHPC first increases and then decreases as coarse aggregate content increases, while the elastic modulus increases linearly. The mixture with 800 kg/m^3 basalt content reaches the maximum elastic modulus of 58.35 GPa, increasing by 25.6% compared to UHPC without any basalt (Huang and Li 2018). However, for basalt with a particle size of 5–10 mm, the increased basalt content has little effect on the compressive strength of UHPC (Chen et al. 2017). When the basalt content is 480 kg/m^3, the elastic modulus of UHPC after 90-d standard curing increases by 7.8%, but the flexural strength decreases. Figure 8.15 shows effects of three particle sizes of 5.0, 9.5, and 50 mm and different coarse aggregate volumes on the elastic modulus of UHPC. It can be observed that the elastic modulus of UHPC increases first with the coarse aggregate volume increasing to 25% and then decreases (Haile et al. 2019).

8.4.3 Effect of fiber type and content

The addition of steel fiber slightly affects the compressive strength and elastic modulus of UHPC. The axial compressive stress-strain curves of UHPC and nonfibrous UHPC matrix are shown in Figure 8.16. The elastic section of

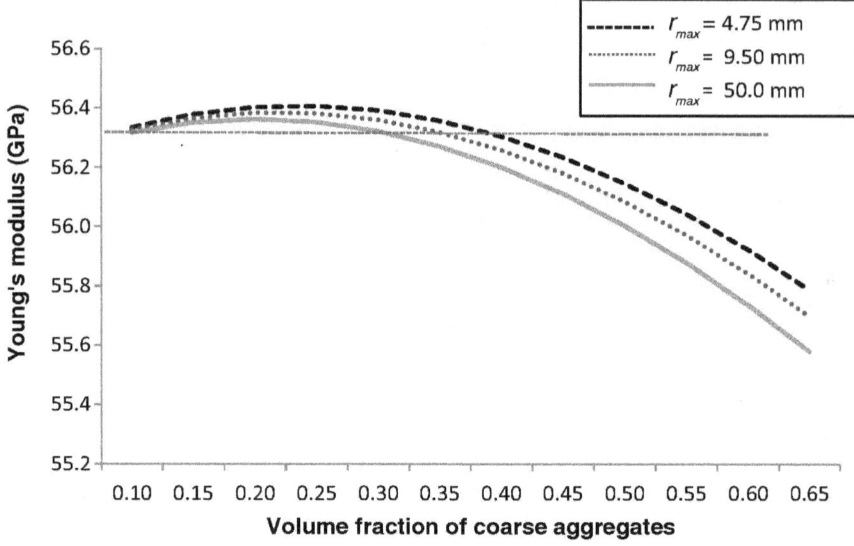

Figure 8.15 Change in elastic modulus of UHPC with coarse aggregate content at maximum sizes of 5.0, 9.5, and 50.0 mm.

Source: Haile et al. (2019).

the ascending branch becomes longer and steeper due to the action of fiber, and the elastic modulus of UHPC increases slightly (Hassan et al. 2012). The steel fiber volume has little effect on the elastic modulus of UHPC. When the fiber volume is 2%, the elastic modulus increases by 3%–7% (Alsalman et al. 2017). However, some research showed that the elastic modulus of UHPC is not significantly improved when steel fiber volume increases from 0 to 2.5% or even decreases when straight steel fiber volume is more than 1% (Ren et al. 2018). This is because of the fiber cluster issue, which easily occurs during mixing at high fiber content, thus reducing the compressive strength and elastic modulus of UHPC (Yoo et al. 2013). In addition, the shape of steel fiber, including thin, straight shape, hook-end shape, wavy shape, and spiral shape, does not significantly improve the elastic modulus of UHPC (Ahmad et al. 2019; Yoo et al. 2013).

The mixture proportion of UHPC should be suitably optimized to achieve a higher elastic modulus. Figure 8.17 shows the effects of main factors, including w/b ratio, coarse aggregate content, sand content, cement content, silica fume content, slag content, and fiber volume, on the elastic moduli of UHPC (Ouyang et al. 2020). The development rates of elastic moduli vary with the factors. It is indicated that w/b ratio has the greatest effect on the elastic modulus of UHPC, followed by coarse aggregate content, sand

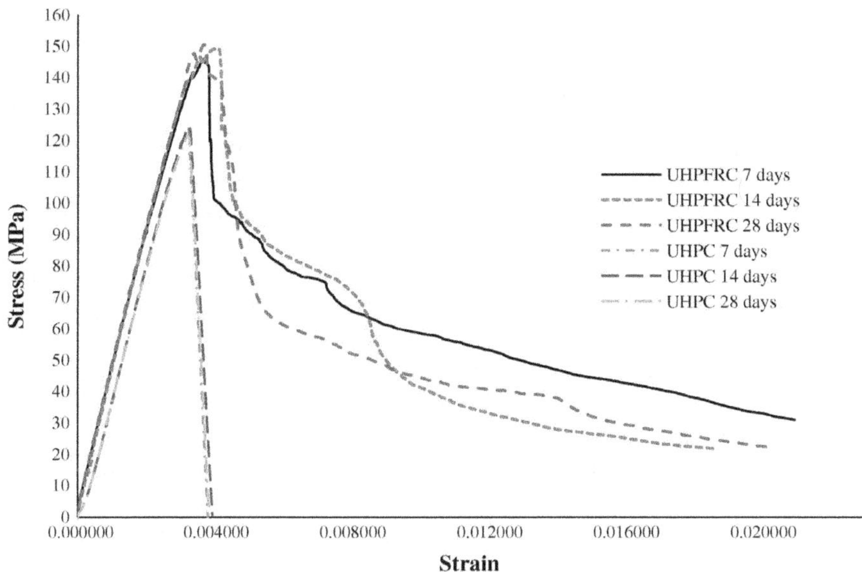

Figure 8.16 Compressive stress-strain curve of UHPC with and without steel fiber at different curing ages.

Source: Hassan et al. (2012).

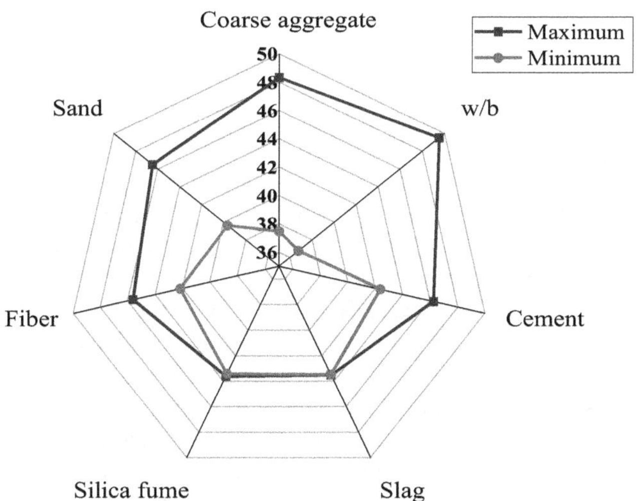

Figure 8.17 Maximum and minimum predicted elastic moduli of UHPC (unit in GPa).

Source: Ouyang et al. (2020).

content, and then cement content, while slag content shows the slightest effect.

8.4.4 Effect of curing condition

Common curing methods used in the UHPC are standard curing, steam curing, hot water curing, and autoclave curing (Zhang et al. 2019). Compared with standard curing, steam curing increases the pozzolanic activity of mineral admixture and accelerates the hydration reaction rate, preventing microcracks development (Wang et al. 2018). The main hydration product in the UHPC under steam curing is ultra-high-density (UHD) C-S-H, while high-density (HD) C-S-H accounts for a large amount in the hydration products under standard curing. As shown in Figure 8.18, when the curing temperature increases from 20 to 250°C, the UHD C-S-H volume increases from 11.5% to 100%. The low-density (LD) C-S-H disappears when the curing temperature exceeds 90°C. Under autoclaved curing at 250°C, the composition and structure of C-S-H are changed, and a large amount of dehydration forms toxonotlite (Yazici et al. 2009). The increase of UHD C-S-H volume improves the microelastic properties and the macroelastic modulus.

In addition, high-temperature curing renders denser matrix and lower porosity due to the presence of high contents of unhydrated cement clinker and supplementary cementitious materials in low w/b ratio UHPC. The pore content with a pore diameter between 3.75 nm and 100 μm in UHPC is less than 9%. Moreover, high-temperature curing can also enhance the quality of interfaces caused by the addition of coarse aggregate, resulting in an improvement in the elastic modulus of UHPC.

8.4.5 Empirical equations for predicting elastic modulus of UHPC

The elastic modulus of concrete can be linked to its compressive strength using empirical equations. Experience gained during the past years showed that this is almost true for ordinary concrete. Usually, the very porous hydrated cement paste constitutes the weakest link of concrete, thus influencing the compressive strength and the elastic modulus. Therefore, the hydrated cement paste affects concrete's compressive strength and elastic modulus. This is why all national codes present direct relationships linking the elastic modulus to the compressive strength of ordinary concrete using equation $E_c = \alpha(f_c)^n$. Unfortunately, due to the difference in materials and setting parameters, engineers have not yet been able to end up with a universal relationship. Recently, some empirical equations have been established for UHPC, as summarized in Table 8.2.

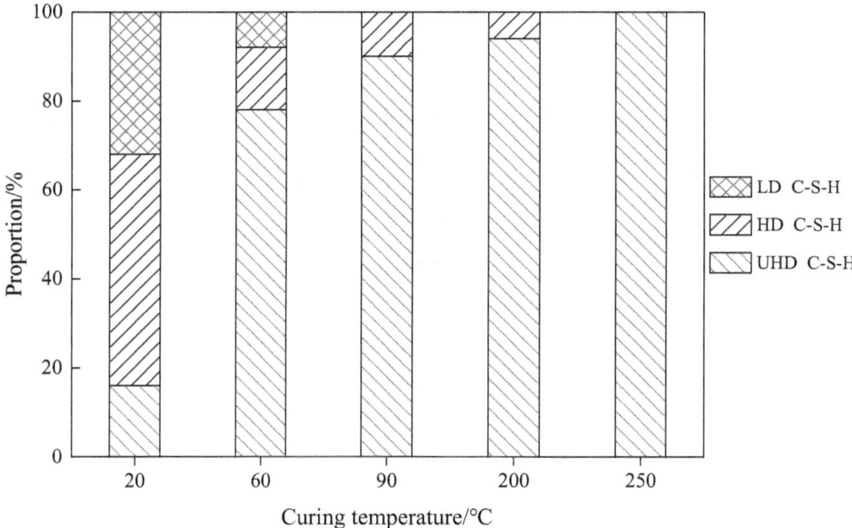

Figure 8.18 Effect of curing regime on volume fractions of hydration phases.

Source: Shen et al. (2019).

8.4.6 Analytical methods for predicting elastic modulus of UHPC

Several analytical methods, including ensemble-volume average method (EVA), generalized self-consistent scheme (GSCS), and Mori-Tanaka scheme, have been used to predict the elastic modulus of cementitious materials. The simplest models are two-phase models involving aggregate and hydrated cement paste. In these models, the constituents are supposed to carry either the same strain (in the Voigt model) or develop the same stress (in the Reuss model), as shown in Figure 8.19.

The Voigt model assumes that the elastic modulus of composites concrete can be predicted using Eq. (8.1):

$$E_c = E_0 g_0 + E_1 g_1 \quad \text{with } g_0 + g_1 = 1 \tag{8.1}$$

where E_0 is the elastic modulus of the mortar; E_1 is the elastic modulus of coarse aggregate; and g_0 and g_1 are the relative volumes of mortar and coarse aggregate, respectively.

The calculated elastic modulus value predicted using the Reuss model is as follows:

Table 8.2 Expressions of elastic modulus with respect to compressive strength

No.	*Equations*	*Notes*	*Ref.*
1	$E_c = 9.5\sqrt[3]{f_{cu}}$	Heat-cured UHPC, $f_{cu} \geq$ 140 MPa	Interim Recommendations (2002)
2	$E_c = 19\sqrt[3]{\frac{f_{cu}}{10}}$ $E_c = 21.902\sqrt[3]{\frac{f_{cu}}{10}}$	UHPC without coarse aggregates UHPC with basalt coarse aggregates	Ma et al. (2004)
3	$E_c = \frac{1}{0.0172 + 0.8364/f_{cu}}$	60 MPa $\leq f_{cu} \leq$ 220 MPa	Guo et al. (2017)
4	$E_c = 8.01 f_{cu}^{0.36}$	31 MPa $\leq f_{cu} \leq$ 235 MPa 25 GPa $\leq E_c \leq$ 68.3 GPa	Alsalman et al. (2017)
5	$E_c = 4.2\sqrt{f_{cu}}$	UHPC with crushed aggregate, f_{cu} = 177 MPa	Sritharan et al. (2003)
6	$E_c = 4.36\sqrt{f_{cu}}$	$f_{cu} \approx$ 130 MPa	Ahmad et al. (2015)
7	$E_c = 3.84\sqrt{f_{cu}}$	$126 \leq f_{cu} \leq$ 193 MPa	Graybeal (2007)

Note: f_{cu} is the compressive strength of UHPC (MPa); E_c is the static elastic modulus of UHPC (GPa).

$$\frac{1}{E_c} = \frac{g_0}{E_0} + \frac{g_1}{E_1} \quad \text{with } g_0 + g_1 = 1 \tag{8.2}$$

More complicated and complex models can be found in the literature, such as the one proposed by Hansen (1965):

$$\frac{E_c}{1-2v_c} = \frac{E_1}{1-2v_1} \frac{\left[g_0 \frac{E_0}{1-2v_0} + \left(\frac{1+v_0}{2(1-2v_0)} + g_1\right)\frac{E_1}{1-2v_1}\right]}{\left(1+\frac{1+v_0}{2(1-2v_0)} g_1\right)\frac{E_0}{1-2v_0} + \left(\frac{1+v_0}{2(1-2v_0)} g_0\right)\frac{E_1}{1-2v_1}} \tag{8.3}$$

where v_c, v_0, and v_1 are the Poisson's ratios of concrete, mortar, and coarse aggregates, respectively.

GSCS was employed to determine the interphase elastic modulus of concrete (Hashin and Monteiro 2002). Dhar et al. (2018) used GSCS to obtain the effective modulus of recycled aggregate concrete considering the effect

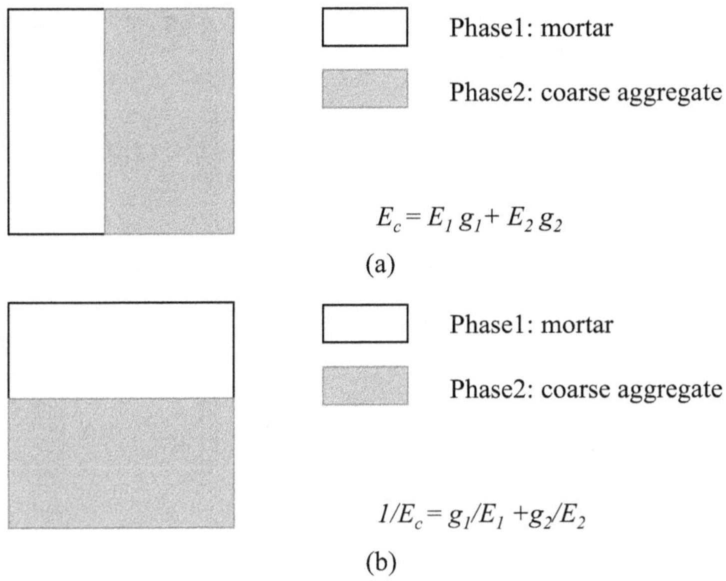

Figure 8.19 (a) The Voigt model and (b) Reuss model.

of the new interfacial transition zone. The effective bulk elastic modulus of composites can be expressed as follows:

$$K = K_{itz} + \frac{f_{ag}\left(K_{ag} - K_{itz}\right)\left(3K_{itz} + 4G_{itz}\right)}{3K_{itz} + 4G_{itz} + 3(1 - f_{ag})\left(K_{ag} - K_{itz}\right)} \tag{8.4}$$

Additionally, the shear elastic modulus is obtained using the solution of a quadratic equation, given as follows:

$$A\left(\frac{G}{G_{itz}}\right)^2 + 2B\left(\frac{G}{G_{itz}}\right) + C = 0 \tag{8.5}$$

Therefore, the solution is expressed as:

$$\frac{G}{G_0} = \frac{-B + \sqrt{B^2 - AC}}{A} \tag{8.6}$$

where K and G are the bulk and shear elastic moduli of concrete, respectively; K_{itz} and G_{itz} are the bulk and shear elastic moduli of interfacial transition zone; K_{ag} and G_{ag} are the bulk and shear elastic moduli of aggregate; f_{ag} is the volume fraction of aggregates; and A, B, C, η_1, η_2, η_3 are parameters and can be determined as follows:

$$
\begin{aligned}
A = {} & 8\left(\frac{G_{ag}}{G_{itz}} - 1\right)(4 - 5v_{itz})\eta_1 f_{ag}^{10/3} - 2\left[63\left(\frac{G_{ag}}{G_{itz}} - 1\right)\eta_2 + 2\eta_1\eta_3\right] f_{ag}^{7/3} \\
& + 252\left(\frac{G_{ag}}{G_{itz}} - 1\right)\eta_2 f_{ag}^{5/3} \\
& - 50\left(\frac{G_{ag}}{G_{itz}} - 1\right)(7 - 12v_{itz} + 8v_{itz}^2)\eta_2 f_{ag} + 4(7 - 10v_{itz})\eta_2\eta_3
\end{aligned}
\tag{8.7}
$$

$$
\begin{aligned}
B = {} & -4\left(\frac{G_{ag}}{G_{itz}} - 1\right)(4 - 5v_{itz})\eta_1 f_{ag}^{10/3} + 2\left[63\left(\frac{G_{ag}}{G_{itz}} - 1\right)\eta_2 + 2\eta_1\eta_3\right] f_{ag}^{7/3} \\
& - 252\left(\frac{G_{ag}}{G_{itz}} - 1\right)\eta_2 f_{ag}^{5/3} \\
& + 75\left(\frac{G_{ag}}{G_{itz}} - 1\right)(3 - v_{itz})v_{itz}\eta_2 f_{ag} + 3(15v_{itz} - 7)\eta_2\eta_3
\end{aligned}
\tag{8.8}
$$

$$
\begin{aligned}
C = {} & 4\left(\frac{G_{ag}}{G_{itz}} - 1\right)(5 - 7v_{itz})\eta_1 f_{ag}^{10/3} - 2\left[63\left(\frac{G_{ag}}{G_{itz}} - 1\right)\eta_2 + 2\eta_1\eta_3\right] f_{ag}^{7/3} \\
& + 252\left(\frac{G_{ag}}{G_{itz}} - 1\right)\eta_2 f_{ag}^{5/3} \\
& + 25\left(\frac{G_{ag}}{G_{itz}} - 1\right)(v_{itz}^2 - 7)\eta_2 f_{ag} - (7 + 5v_{itz})\eta_2\eta_3
\end{aligned}
\tag{8.9}
$$

$$
\eta_1 = \left(\frac{G_{ag}}{G_{itz}} - 1\right)(49 - 50v_{ag}v_{itz}) + 35\left(\frac{G_{ag}}{G_{itz}}\right)(v_{ag} - 2v_{itz}) + 35(2v_{ag} - v_{itz})
\tag{8.10}
$$

$$
\eta_2 = 5v_{ag}\left(\frac{G_{ag}}{G_{itz}} - 8\right) + 7\left(\frac{G_{ag}}{G_{itz}} + 4\right)
\tag{8.11}
$$

$$
\eta_3 = \left(\frac{G_{ag}}{G_{itz}}\right)(8 - 10v_{itz}) + (7 - 5v_{itz})
\tag{8.12}
$$

where v_{ag} and v_{itz} are the volume fractions of aggregate and interfacial transition zone, respectively.

Since cementitious material is a porous multiphase composite, the GSCS approach may overestimate the contribution of pores to mechanical properties. Thus, the Mori-Tanaka scheme (Mori and Tanaka 1973) is proposed to predict elastic modulus. This model can effectively reflect the effects of inclusions and interface by selecting the prevailing phase as the matrix (Sorelli et al. 2008).

The Mori-Tanaka method can be used to numerically determine the elastic modulus of C-S-H, cement paste, mortar, and concrete. However, the elastic modulus and volume fraction of each phase should be known to serve as inputs for the Mori-Tanaka scheme. Representative volume element (RVE) is thus used as a homogenous method in the multiscale scheme. The macroscopic strain tensor ($\overline{E}$) is related to the microscopic strain (ε_i) by the average volume.

$$\overline{E} = \langle \varepsilon_i \rangle_V = \frac{1}{V}\int_V \varepsilon_i dV \tag{8.13}$$

where $\overline{E}$ is the macroscopic strain tensor; and ε_i is the microscopic strain.

Meanwhile, it is assumed in linear continuum micromechanics that the macroscopic strain $\overline{E}$ can be linked to the microscopic strain ε_i by means of a linear strain localization condition:

$$\varepsilon_i = A_i : \overline{E} \tag{8.14}$$

where A_i is a four-order localization tensor; and the symbol ":" stands for double contraction. The simplest form of strain tensor and localization tensor is a constant, that is, $\varepsilon_i = \overline{E} \Leftrightarrow A_i = I$. I is the four-order unit tensor.

Eshelby inclusion theory encounters the morphology at different levels of composites and an estimate A_i^{est} of the localization tensor is given below (Eshelby 1957):

$$A_i^{est} = \left[I + S_i C_0^{-1}(C_i - C_0)\right]^{-1} \tag{8.15}$$

where C_0 and C_i are the stiffness tensors of the matrix and the inclusion, respectively; S_i is Eshelby's tensor for the inclusion (Eshelby 1957), which depends on C_0, inclusion shape, and orientation of phase. Matrix is generally treated as an isotropic material, and the inclusions are simplified as spherical particles, which ensures the isotropic of the local and global variables.

$$C_i = 3K_i J + 2G_i \xi; C_0 = 3K_0 J + 2G_0 \xi \tag{8.16}$$

where K_i, G_i, $K_{0,}$ and G_0 are the bulk and shear moduli of inclusion and matrix, respectively;$J_{ijkl} = (1/3)\,\delta_{ij}\delta_{kl}$ is the volumetric of the four-order unit tensor I; and $\delta_{ij(kl)}$is the Kronecker delta; ξ is the deviator. As for spherical inclusions, the Eshelby tensor can be applied:

$$S_i = \alpha J + \beta\xi \tag{8.17}$$

$$\alpha = \frac{3K_0}{3K_0 + 4G_0},\ \beta = \frac{6(K_0 + 2G_0)}{5(3K_0 + 4G_0)}. \tag{8.18}$$

The macroscopic properties are characterized by the homogenization of different local properties, in combination with the volume-averaging relation linking the microscopic stress $\bar{\Sigma}$ and the macroscopic stress σ_i:

$$\bar{\Sigma} = \langle \sigma_i \rangle_V. \tag{8.19}$$

The linear elastic constitutive law for each phase at microscale, i.e., $\sigma_i = C_i\varepsilon_i$ together with Eq. (8.15), i.e., $\sigma_i = (C_i : A_i) : \bar{E}$, results in the linear homogenization formula for the macroscopic elasticity tensor (C_{hom}), given as follows.

$$\bar{\Sigma} = C_{\text{hom}} : E \tag{8.20}$$

$$C_{\text{hom}} = \langle C_i : A_i \rangle_V = \sum_i f_i C_i : A_i \tag{8.21}$$

Eq. (8.20) is an exact theoretical definition of C_{hom}, and the practical determination of C_{hom} is generally based on the estimation of the localization tensor for each phase A_i^{est}. The assumption of $A_i^{est} = I$ implies that all phases have the same strain so that the Voigt-Reuss mixture rule can be applied.

$$C_{\text{hom}}^{est} = \langle C_i : I \rangle_V = \sum_i f_i C_i \tag{8.22}$$

Then, the estimation of macroscopic homogeneous elastic tensor C_{hom}^{est}can be obtained as follows.

$$C_{\text{hom}}^{est} = \left\langle C_i : \left[I + S_i C_0^{-1}(C_i - C_0)\right]\right\rangle_V \tag{8.23}$$

The homogeneous bulk and shear moduli can be obtained:

$$C_{\text{hom}}^{est} = 3K_{\text{hom}}^{est}J + 2G_{\text{hom}}^{est}\xi \tag{8.24}$$

$$K_{\text{hom}}^{est} = \sum_i f_i K_i \left(1+\alpha\left(\frac{K_i}{K_0}-1\right)\right)^{-1} \times \left[\sum_i f_i \left(1+\alpha\left(\frac{K_i}{K_0}-1\right)\right)^{-1}\right]^{-1} \tag{8.25}$$

$$G_{\text{hom}}^{est} = \sum_i f_i G_i \left(1+\beta\left(\frac{G_i}{G_0}-1\right)\right)^{-1} \times \left[\sum_i f_i \left(1+\beta\left(\frac{G_i}{G_0}-1\right)\right)^{-1}\right]^{-1} \tag{8.26}$$

where K_{hom}^{est} and G_{hom}^{est} are the homogeneous bulk and shear modulus values, respectively, and f_i is the volume fraction of the multiphase.

The Mori-Tanaka method appropriately shows a relationship between the morphology and mechanical properties of two-phase composite, as shown in Eqs. (8.27) and (8.28).

$$\alpha^{est} = \alpha_0 = \frac{3K_0}{3K_0 + 4G_0}; \beta^{est} = \beta_0 = \frac{6(K_0 + 2G_0)}{5(3K_0 + 4G_0)} \tag{8.27}$$

The homogenized elastic modulus and Poisson's ratio can be written below.

$$E_{\text{hom}}^{est} = \frac{9K_{\text{hom}}^{est}G_{\text{hom}}^{est}}{3K_{\text{hom}}^{est} + G_{\text{hom}}^{est}}; \nu_{\text{hom}}^{est} = \frac{3K_{\text{hom}}^{est} - 2G_{\text{hom}}^{est}}{6K_{\text{hom}}^{est} + 2G_{\text{hom}}^{est}}. \tag{8.28}$$

where $\alpha_{\text{hom}}^{est}$ and β_{hom}^{est} are the homogeneous parameters; and E_{hom}^{est} and ν_{hom}^{est} are the homogeneous elastic modulus and Poisson ratio, respectively.

8.5 FLEXURAL BEHAVIOR

8.5.1 Brief introduction to specimen size and bending testing methods

Flexural performance is one of the most important properties of concrete. Three- or four-point bending tests are often used to determine the flexural properties of UHPC. These tests are more straightforward to execute than direct tensile testing. Due to the use of fibers, UHPC can exhibit greater flexural strength compared to conventional concrete. Hence, standards for conventional concrete are less applicable for measuring the flexural strength of UHPC. ASTM C1609 (2012) is the most frequently applied standard for testing the effects of fibers on the flexural strength of UHPC. This standard is initially formulated for fiber-reinforced concrete with normal strength using prisms (length > 350 mm) with a size of 100 × 100 × 400 mm. It is required that the depth and width of specimens should be at least three times greater than the fiber length. Other researchers investigated the flexural properties of UHPC according to standards, such as GB/T 17671-1999 (1999) and EN 196-1(2016). In both standards, small prisms of 40 × 40 × 160 mm are used. The use of such small test specimens excludes the use of macrofibers since there is no room for the free orientation of the fibers.

Many material parameters influence the flexural performance of UHPC. The parameters include strength, stiffness, shape, aspect ratio, Poisson's ratio of fibers, matrix performance, and the physico-chemical and frictional bond properties at the fiber-matrix interface. Under the assumption that an identical UHPC matrix is used, the fiber type is the most governing factor among the parameters mentioned above. Thus, the material, geometry, and volume contents of fiber should be very carefully determined to improve the flexural performance of UHPC.

8.5.2 Flexural load-deflection response curves

As the fundamental characteristic, the flexural property has been extensively studied. The flexural behavior can be categorized into deflection hardening and softening behaviors, according to the change of load-carrying capacity after the first cracking, as shown in Figure 8.20. The first-cracking point of UHPC is defined as the limit of proportionality (LOP). Flexural strength, also known as modulus of rupture (MOR), or bend strength, or transverse rupture strength, is defined as the stress before it yields.

8.5.3 Effects of fiber characteristics

The addition of steel fibers remarkably improves the flexural strength and toughness of UHPC. The flexural properties of UHPC vary significantly with the fiber type, content, shape, dispersion, and orientation (Deng et al. 2018; He et al. 2020).

8.5.3.1 Fiber content

Fibers enable the concrete to sustain structural integrity toward tensile load after the first-cracking point by bridging cracks and transferring the load across the cracks. Steel fiber content has a limited effect on the first-cracking strength and first-crack deflection of flexural load-deflection curve of UHPC but has a considerable effect on the peak load and toughness (Park et al. 2017; Wu et al. 2016a; Yoo et al. 2017b). The first-cracking strength is strongly dependent on the tensile cracking strength of the cementitious

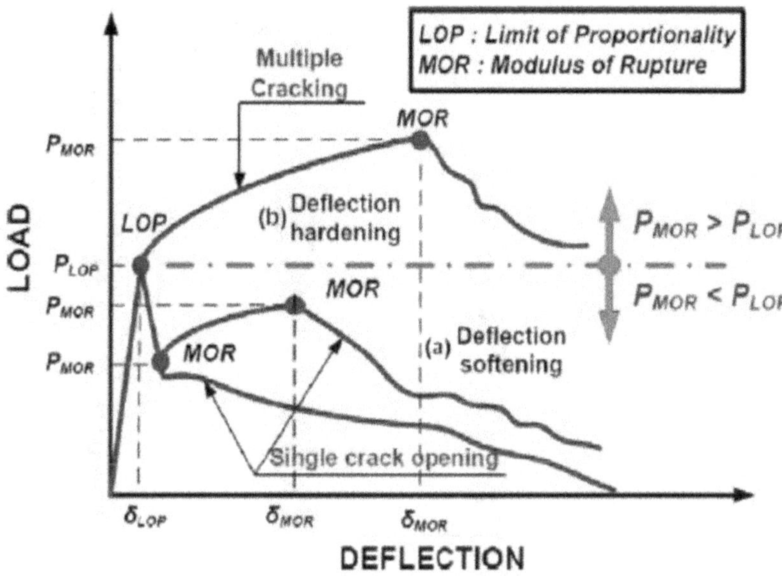

Figure 8.20 Typical flexural load-deflection curves of UHPC.

Source: Kim et al. (2011).

matrix. Some researchers reported improvement in first-cracking strength with fiber content (Jin et al. 2018), even up to 6% (Abbas et al. 2015). This effect can be explained by the formation of multiple microcracks that delay the growth of macrocracks, leading to higher first-cracking strength (Abbas et al. 2015).

The roles of fibers are mainly exerted after reaching the first-cracking strength (Prem et al. 2015). Figure 8.21 presents the 28-d flexural strength of UHPC made with 0%–3% straight, corrugated, and hooked steel fibers. Fibers have a diameter of 0.2 mm and a length of 13 mm. With the increase of steel fiber content, the flexural strength gradually increases. The flexural strength of UHPC matrix without any fiber is 24.8 MPa and it increases to 30.3, 39.7, and 42.6 MPa, respectively, when 1%, 2%, and 3% straight fibers are used. Such values are approximately 22%, 60%, and 72% greater than that of the matrix (Wu et al. 2019b). A possible reason is that at higher fiber content, fibers are more uniformly dispersed to help sustain loads (Yoo et al. 2017c).

However, a large number of fibers is prone to cause fiber cluster issues associated with reduced workability, leading to nonuniform fiber dispersion and orientation and eventually impairing flexural properties of UHPC. Thus, there exists a critical fiber content, usually ranging from 2% to 4%.

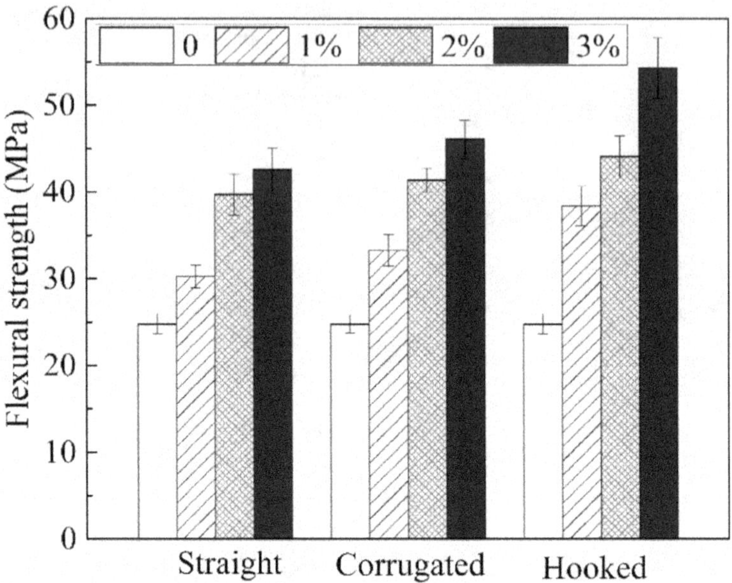

Figure 8.21 Changes in 28-d flexural strengths of UHPC with fiber content and shape.

Source: Wu et al. (2019b).

Meng and Khayat (2018) observed fiber cluster issues for UHPC with fiber content greater than 3%. Yoo et al. (2013) reported that the flexural behavior of UHPC was enhanced with an increase in fiber content up to 4%.

8.5.3.2 Fiber type and geometry

In addition to fiber content, fiber type and geometry can affect the performance of UHPC. Yoo et al. (2017c) investigated the effect of fiber volume (0.5%, 1.0%, 1.5%, and 2.0%), fiber shape (straight, hooked, and twisted fibers), and fiber aspect ratio (65, 97.5, and 100) on flexural behavior of UHPC. Fibers with a diameter of 0.3 mm and a length of 30 mm were used. The authors stated that at low fiber content (≤1%), UHPC mixtures made with twisted fibers developed the highest flexural strength. When the fiber volume exceeds 1.5%, comparable flexural strength and lower toughness are observed compared to the mixture made with straight fibers. Given the fiber shape, the flexural performance of UHPC is improved with the increase in the aspect ratio of fiber. Wille et al. (2012c) examined the effect of fiber geometry, including straight fiber ($d_f = 0.2$, $l_f = 13$ mm), hooked fiber ($d_f = 0.38$, $l_f = 30$ mm), and twisted fiber ($d_f = 0.3$, $l_f = 30$ mm), on the tensile properties of UHPC. It is reported that the use of 1.5% deformed fibers leads to tensile strength and tensile strain of 13 MPa and 0.6%, respectively, which are 60% and 300% higher than those with straight fibers. Given the fiber aspect ratio, the addition of 1%–3% corrugated and hooked fibers increases the flexural strengths by 4%–10% and 10%–27%, respectively, compared to the straight fiber (Wu et al. 2019b). Deformed fibers can enhance flexural strength because of the improvement in fiber-matrix bond strength associated with mechanical anchorage from the deformed section.

8.5.3.3 Fiber hybridization

Fiber hybridization is a promising method to improve the flexural performance of UHPC, which makes macro- and microfibers or fibers with different shapes and types play a role at different levels. Fiber hybridization in ordinary-strength matrices produces a favorable hybridization effect on the ductility in comparison to that with monofiber. Kim et al. (2011) found that the enhancements in modulus of rupture, deflection capacity, and energy absorption capacity were different according to the types of macrofiber with the increasing amount of microfiber blended. The flexural performance of hybrid UHPC according to the types of macrofiber follows the order of hooked fiber > twisted fiber > long smooth fiber. Hai et al. (2016) stated that steel and polyvinyl alcohol fibers efficiently improved the bending properties of UHPC. The flexural-to-compression and tension-and-compression ratios increase by 40%.

8.5.3.4 Fiber dispersion and orientation

Placing methods and workability affect the fiber dispersion and orientation in UHPC, hence affecting the flexural performance. Techniques used to study the distribution characteristics of fiber in concrete mainly include image analysis, transmission X-ray photography, AC-IS, etc. Image analysis is often used since it is a direct, reliable, and destructive method. Parameters include fiber distribution coefficient, fiber quantity per unit area, fiber image packing density, and fiber orientation.

Fiber alignment corresponding to the direction of principal tensile stress can significantly increase the effectiveness of fibers in bridging cracks, thus enhancing the flexural properties of UHPC. Kang et al. (2011) evaluated the effect of fiber distribution characteristics on the flexural strength of UHPC under two different placing directions. It is found that the first-cracking and ultimate flexural strengths of UHPC placed parallel to the longitudinal direction of mold are 5.5% and 61%, respectively, greater than that placed transversely. Barnett (2010) studied the effect of casting methods on flexural strength by pouring three-round panel specimens in three different ways. Fibers tend to align perpendicular to the flow of concrete. The fiber orientation is found to have a very significant effect on the flexural strength of panel specimens. Panels poured from the center are found to have the highest strength. The flow of concrete outward from the center of the panel leads to fibers tending to line up perpendicular to the radius of the panel, hence increasing the total number of fibers to bridge the radial cracks formed during testing.

Adequate plastic viscosity is necessary to secure appropriate fiber dispersion and orientation since fibers can segregate and agglomerate in the suspension of low plastic viscosity. Teng et al. (2020) stated that the optimal plastic viscosities to enable the greatest fiber orientation of UHPC mortar made with 1%, 2%, and 3% steel fibers are 36, 52, and 66 Pa s, respectively. Mixtures having lower yield stress are more flowable, hence resulting in more favorable fiber alignment. The enhancement in yield stress can enhance the risk of fiber clusters and hinder fiber rotation during the mixture flow, thus leading to decreased fiber orientation (Boulekbache et al. 2010).

Steel fibers are randomly dispersed in UHPC when prepared by a direct cast method. To improve the flexural properties of UHPC, an L-shaped device with a narrow horizontal channel is developed to control the flow of fresh mixture and then the orientation of steel fibers (Huang et al. 2018). Figure 8.22 illustrates the flexural load-deflection curves of UHPC prepared with different w/b ratios at various fiber volumes, in which the fiber was oriented or not. The flexural properties are significantly enhanced by the orientation. Compared to specimens prepared using direct casting, this flow control method improves fiber orientation number and fiber orientation coefficient from 0.6 to 0.7, and 0.4–0.6 to 0.7–0.9, respectively. The greatest flexural strength and toughness are obtained at 1% fiber volume. The flexural strengths of UHPC made with 0.20, 0.22, and 0.24 w/b ratios are enhanced by 50.9%, 64.3%, and 55.0%, respectively, when compared to those prepared by direct casting. The toughness values are improved by 56.7%, 52.9%, and 65.1%, respectively.

8.5.4 Effect of supplementary cementitious materials and curing regime

The use of supplementary cementitious materials and curing regime can significantly improve the flexural strength and toughness of UHPC. Under standard curing, the toughness of UHPC specimens mixed with 20%–40% slag or fly ash is 18%–46% higher than that of specimens using 100% Portland cement (Yazici et al. 2009). Yang et al. (2009) studied the influence of curing temperatures of 90 and 20°C on the compressive strengths of UHPC. It is shown that the flexural strength and fracture energy of UHPC cured at 20°C for 7 d are 10% and 15% lower than those at 90°C.

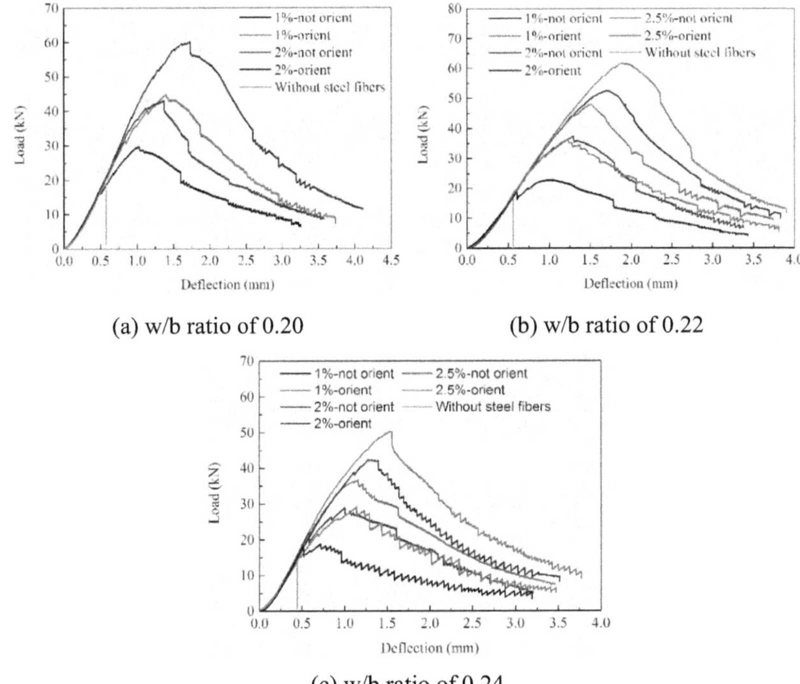

Figure 8.22 Flexural load-deflection curves of UHPC with different fiber volumes and orientations (Huang et al. 2018). (a) w/b ratio of 0.20; (b) w/b ratio of 0.22; (c) w/b ratio of 0.24.

8.5.5 Effect of coarse aggregate

Coarse aggregate type and content have a slight effect on flexural load-deflection curves of UHPC, along with a significant effect on the cracking and postcracking stages, as shown in Figure 8.23. At a given coarse aggregate content, coarse aggregate type exerts a great effect on the load-deflection curves of UHPC. For instance, mixtures with 15% limestone demonstrate the highest peak load. Specimens with 30% granite aggregate show the highest peak load and the most voluminous descending stage. The coarse aggregate content is found to have a limited effect on the elastic modulus (Figure 8.23(d)). However, great differences are observed at the cracking and postcracking stages. UHPC with higher granite aggregate content has lower flexural strength and inferior ductility. The weakened postcracking flexural behavior is largely attributed to the reduction in utilization efficiency of steel fiber after incorporating coarse aggregate (Xu et al. 2019).

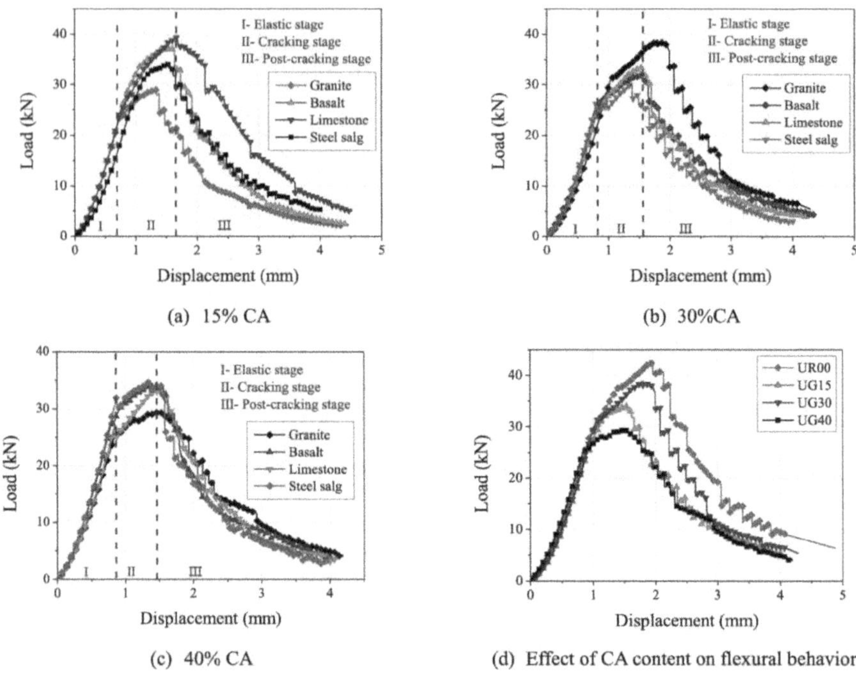

Figure 8.23 Effects of coarse aggregate content and type on flexural load-displacement curves of UHPC.

Source: Xu et al. (2019).

Coarse aggregate type has a greater effect on flexural strength than that of the first-crack strength. Generally, when coarse aggregate content is lower than 30%, UHPC mixtures with basalt and granite aggregates show higher first-crack and ultimate flexural strengths than those containing limestone and steel slag. Mixtures with 15% and 30% basalt aggregate have the highest first-crack strengths. The smallest first-crack and flexural strengths are observed in UHPC mixtures with steel slag. However, opposite results are observed for UHPC mixtures with 40% coarse aggregate. The first-crack and flexural strengths of specimens containing steel slag and limestone are slightly higher than those with granite and basalt aggregates.

8.5.6 Specimen size

Specimen size also affects the flexural behavior of UHPC. Reineck and Frettlöhr (2010) studied the size effects of flexural behaviors of UHPC and found that the aspect ratio (b/h) of specimens is a critical parameter.

As the b/h increases from 1 to 5, the ultimate tensile stress is significantly decreased. Yoo et al. (2016) evaluated the size effect of flexural behavior of UHPC beams with sizes of 50 × 50 × 250, 100 × 100 × 400, and 150 × 150 × 550 mm by conducting four-point flexural testing. UHPC mixture is prepared with 2% short (d_f = 0.2 mm, l_f = 13 mm), long smooth (d_f = 0.3 mm, l_f = 30 mm), and long twisted (d_f = 0.3 mm, l_f = 30 mm) steel fibers. It is indicated that the higher the aspect ratio of the used fibers, the better the flexural properties and the lower the sensitivity of the flexural properties of UHPC related to beam size. Nguyen et al. (2013) investigated the size effect of four-point flexural behavior of UHPC with hybrid fibers. Two UHPC mixtures and three types of beams with sizes of 50 × 50 × 150, 100 × 100 × 300, and 150 × 150 × 450 mm were used. One mixture contained 1.0% twisted (d_f = 0.3 mm, l_f = 30 mm) steel macrofibers and 0.5% short (d_f = 0.2 mm, l_f = 13 mm) smooth steel microfibers (UHPC1), while the other incorporated 1.0% twisted steel macrofibers and 1.0% short smooth steel microfibers (UHPC2). The flexural strength, normalized deflection, and normalized energy absorption capacity were analyzed. It is found that the size effect of UHPC1 is more significant than that of UHPC2. UHPC with greater toughness indicates a less size effect.

8.6 TENSILE BEHAVIOR

UHPC tensile properties are distinct from those of conventional concrete due to the increased tensile cracking capacity of the cementitious composite matrix and the crack-bridging behavior of the fibers. In contrast to fiber-reinforced conventional concretes, UHPC can exhibit significant, sustained postcracking tensile capacity before crack localization, fiber pullout, and loss of tensile capacity. Figure 8.24 shows a schematic of the three distinct tensile behaviors that UHPC can exhibit: (1) linear-elastic behavior before cracking; (2) postcracking strain hardening behavior and dispersed discrete cracking; and (3) softening behavior during strain localization across specific cracks (Habel et al. 2006). The strain-hardening tensile behavior of UHPC is of interest in the design of structural components, while softening behavior is of little practical interest as crack localization and fiber pullout coincide with the loss of tensile capacity and component failure unless secondary load paths are available.

8.6.1 Brief introduction to specimen size and tensile testing methods

The direct tensile test is generally considered to be the most accurate and reliable method to evaluate the tensile properties of concrete (Sarfarazi and Schubert 2017). Compared to bending or wedge splitting tests, the uniaxial

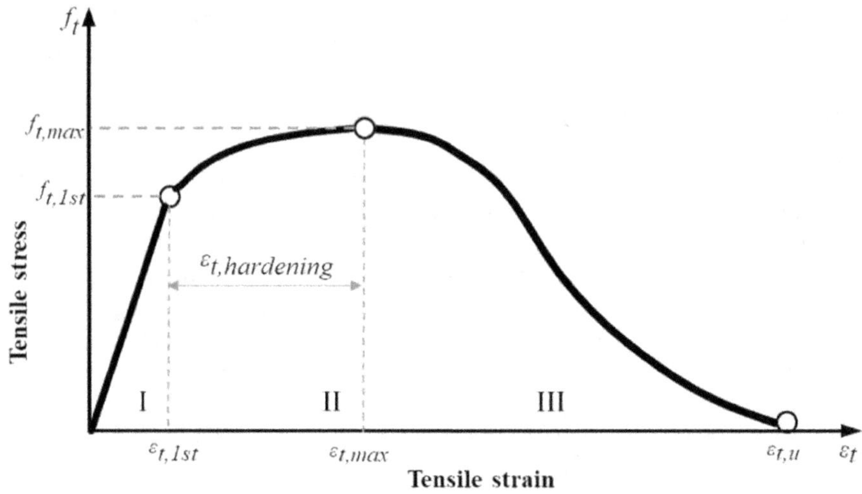

Figure 8.24 Tensile behavior of UHPC.

tensile tests can directly provide material tensile behavior over three stages, namely elastic, strain hardening, and softening, without the need for high computational effort for backward calculation. Due to the increased tensile cracking capacity of the matrix and the crack-bridging effect of fibers, the tensile strengths of UHPC are distinct from those of plain concrete (Russell and Graybeal 2013). AFGC (2002) and JSCE (2006) both provide recommendations on performing uniaxial tensile tests of UHPC materials. However, detailed information regarding testing conditions, specimen geometry, and analytical procedures are not available. Besides, direct tensile tests are challenging to perform. Difficulties in obtaining evenly distributed stresses throughout the cross-section and controlling a stable load versus displacement-crack opening response will be encountered for cementitious-based materials. The predetermination of crack development by notching the specimen in direct tension is commonly utilized to simplify the test setup, deformation control, measurement system, and data analysis. However, this is not appropriate for capturing the strain-hardening behavior of UHPC. Therefore, unnotched specimens are often used to capture the strain-hardening behavior under direct tensile test, while notched specimens are usually preferred for investigating strain-softening behavior (Wille et al., 2014). Pinned-end conditions facilitate specimen alignment and even stress distribution before cracking but do not support an even crack opening throughout the cross-section compared to a fixed boundary condition.

Tables 8.3 to 8.5 summarize commonly used specimens, including unnotched (dogbone-shaped and prism/cylinder specimens) and notched specimens for evaluating the tensile behavior of UHPC. The specimens are glued, anchored, or clamped to the machine, and the attachment is either fixed or pinned. Unnotched dogbone-shaped specimens geometries and test setups used for direct tensile tests of UHPC or other types of concrete with compressive strength over 120 MPa are listed in Table 8.3. The central rectangular shanks of dogbone-shaped specimens have a cross-section ranging from 13 × 30 to 120 × 120 mm and a length varying from 80 to 385 mm. Liem et al. (2014) pointed out that the increases in gauge length, section area, and specimen volume have little impact on the postcracking strength of UHPC. In contrast, it induces a significant decrease in strain capacity and energy absorption capacity and a great increase in crack spacing. Additionally, fiber orientation in specimens can be considered to be a randomly 3D structure, providing that the gauge dimensions in two perpendicular planes of the specimens are at least three times higher than the fiber length. Otherwise, it can be regarded as a randomly 2D structure (Li and Wang 2003). Wille et al. (2014) concluded that the three general characteristics of dogbone-shaped specimens should have the following characteristics: (1) a middle part with a constant or nearly constant area to promote multiple cracking behaviors; (2) larger cross-sectional areas at the supports to avoid support failure; and (3) smooth transition from the supported area to the middle part to reduce the effects of stress concentrations.

To avoid eccentric load, at least three specimens in each group should be secured to ensure the accuracy of the results. The loading rate usually remains constant throughout the testing, ranging from 0.01 to 3 mm/min. Besides, the piecewise loading rate, i.e., 0.02 mm/min in the hardening domain and then increasing to 0.2 mm/min in the softening domain (Denarié 2006), is also used. Furthermore, strain rates of 1×10^{-5} s^{-1} (Hafiz and Denarié 2020) or 1×10^{-4} s^{-1} (Wille et al. 2012b) are also applied to obtain quasistatic conditions.

Unnotched prism/cylinder specimen geometries and test setups used in direct tensile tests are listed in Table 8.4. The loading rate for tensile testing ranges from 0.005 to 3 mm/min. The cross-section and gauge length of unnotched prism/cylinder specimens range from 12 × 40 to 50 × 50 mm and 76.2 to 360 mm, respectively. The specimens are glued, clamped, or pinned to the machine. Before testing, sheets, such as aluminum sheets, are usually glued onto the ends of the test specimens to avoid failure outside of the gauge length. At least three specimens should be employed to ensure the accuracy of the results.

Notched prism/cylinder specimen geometries and test setups used in direct tensile tests are shown in Table 8.5. Notches are installed on both sides of or around the specimen to induce a critical crack at the center of specimens.

Table 8.3 Direct tension test setups – unnotched dogbone-shaped specimens

Specimen size (mm)	*Materials*	*Grip/ attachment*	*Specimen number*	*Loading rate*	*Ref.*
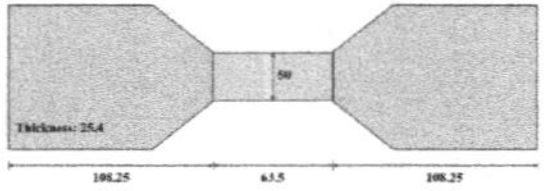	UHPC			0.127 mm/min was used until the postpeak load drop to 90% of peak load. The displacement rate was then increased to 1.27 mm/min until failure.	Abokifa and Moustafa (2021)
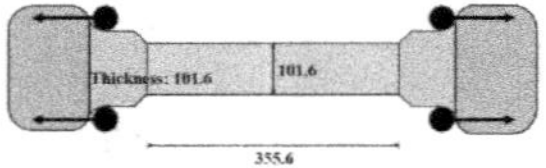	UHPC	Anchored/ pinned	≥4	-	Aghdasi et al. (2016)
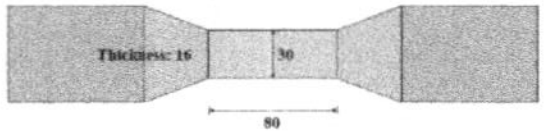	UHPC	-	-	0.2	Aldahdooh et al. (2013)
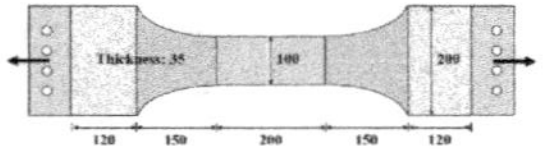	UHPC	Side glued/ pinned	-	-	Benson and Karihaloo, (2005)
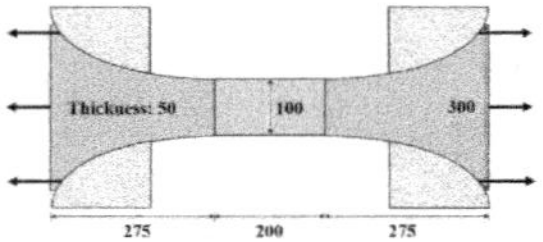	UHPC	Top glued/ anchored	-	0.1	Boulay et al. (2004)
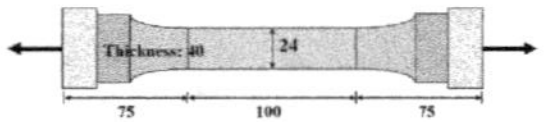	HS-SHCC	Side glued	4	3	Curosu et al. (2017)
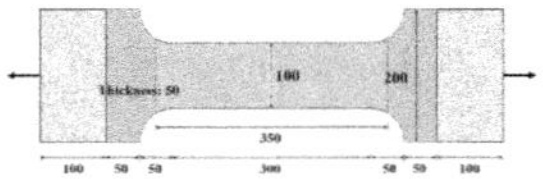	UHPC	Side glued	≥5	0.02 mm/min in the hardening domain and 0.2 mm/min in the softening domain.	Denarié (2006)

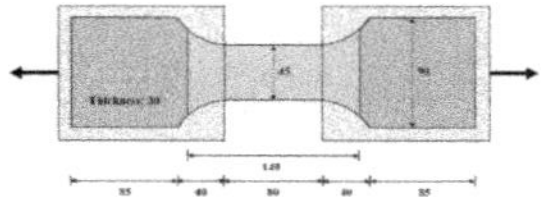	UHPC	Anchored	-	0.5	Donnini et al. (2021)
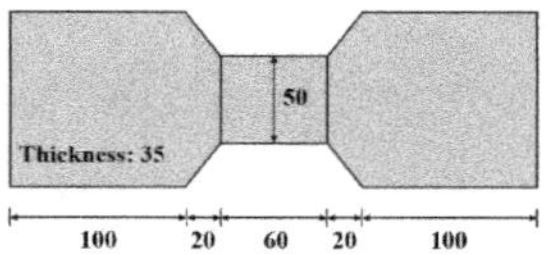	UHPC	-	3	-	Gangwar et al. (2019)
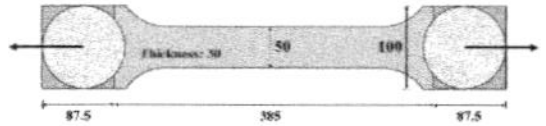	UHPC	Side glued	-	1×10^{-5} s^{-1}	Hafiz and Denarié (2020)
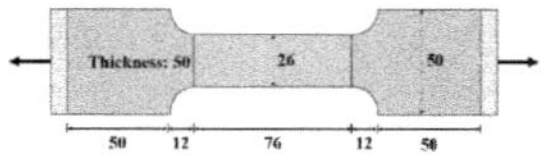	UHPC	Top glued	-	0.4	Hassan et al. (2012)
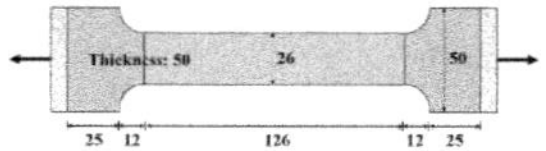	UHPC	Top glued	-	0.4	Hassan et al. (2012)
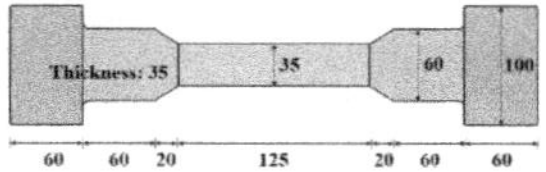	UHPC	-	3	0.5	Hung et al. (2019)
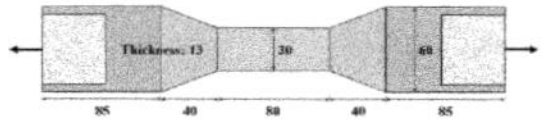	UHP-SHCC	Fixed	-	-	Kamal et al. (2008); Kunieda et al. (2010)
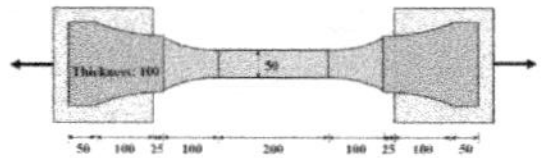	UHPC	Anchored	-	0.05	(Kravanja and Sovják 2018)

(Continued)

Table 8.3 (Continued)

Specimen size (mm)	*Materials*	*Grip/ attachment*	*Specimen number*	*Loading rate*	*Ref.*
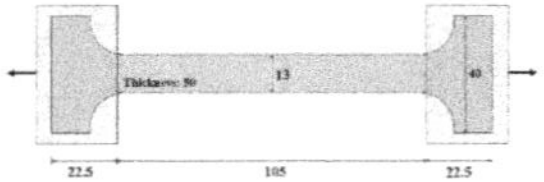	UHPC	Anchored	6	0.042	(Lampropoulos et al. 2016)
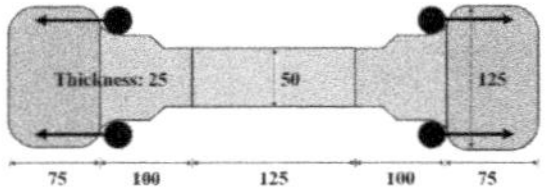	UHPC	Anchored/ pinned	-	1	(Liem et al. 2014)
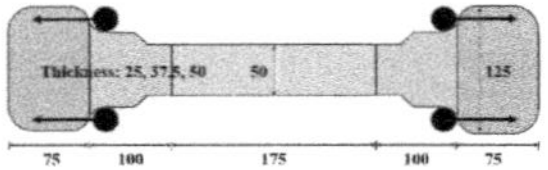	UHPC	Anchored/ pinned	-	1	(Liem et al. 2014)
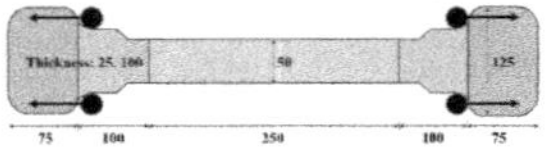	UHPC	Anchored/ pinned	-	1	(Liem et al. 2014)
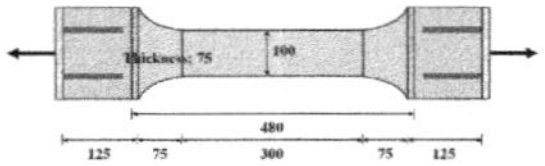	UHPC	Top pinned	-	The initial rate of 0.12 mm/min up to the inception of the first crack. After cracking, the displacement rate was gradually increased to 0.2, 0.5, 0.75, 1, 1.5, 2, 3, and 5 mm/min until the crack opened up to 1.5, 2, 2.5, 3, 4, 6, and 9 mm, respectively.	(Miletić 2016)
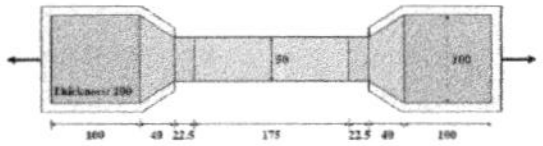	UHPC	Anchored	-	0.4	Park et al. (2012)

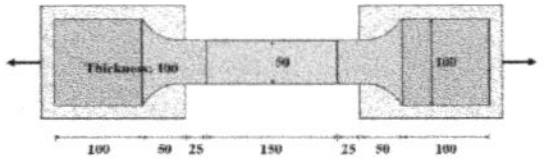	UHPC	Anchored	-	0.15	Qiu et al. (2020)
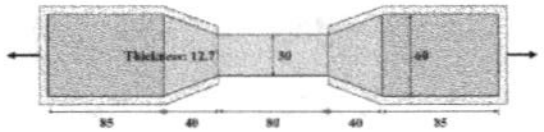	HSHDC	Anchored	4	0.3	Ranade et al. (2011)
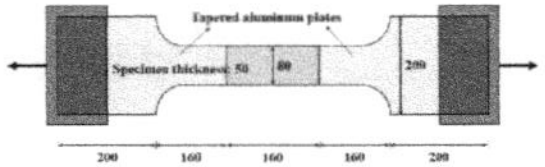	UHPC	Side glued	5	0.05 mm/min in the prepeak domain; 0.1 mm/min in the postpeak domain	Shen and Brühwiler (2020)
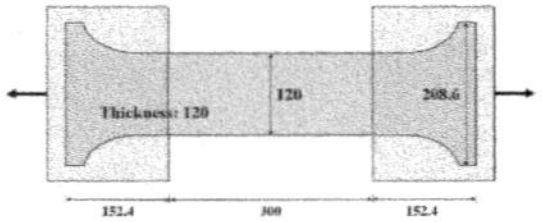	UHPC	Anchored	-	0.01	Singh et al. (2017)
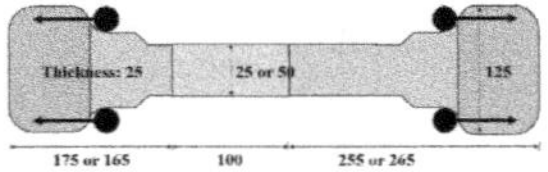	UHPC	Anchored/ pinned	-	1	Tran et al. (2015)
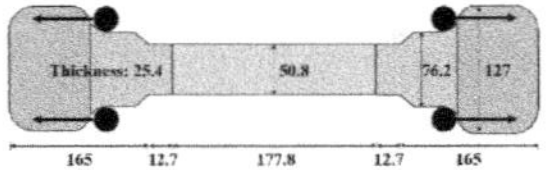	UHPC	Anchored/ pinned	3	-	Wille et al. (2011a)
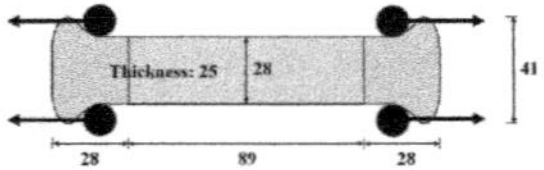	UHPC	Anchored/ pinned	3–6	$1 \times 10^{-4}\ s^{-1}$	Wille et al. (2012b)
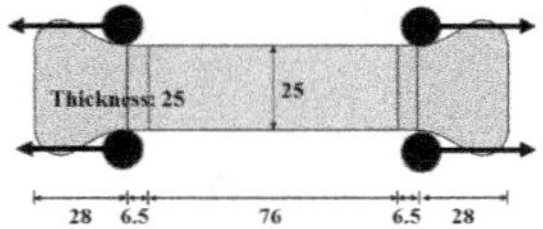	UHPC	Anchored/ pinned	3	$1 \times 10^{-4}\ s^{-1}$	Xu and Wille (2015)

(Continued)

Table 8.3 (Continued)

Specimen size (mm)	*Materials*	*Grip/ attachment*	*Specimen number*	*Loading rate*	*Ref.*
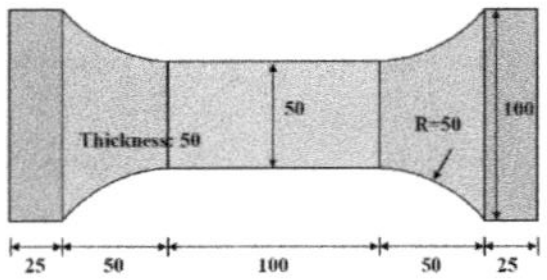	UHPC	-	5	-	Yang et al. (2017)
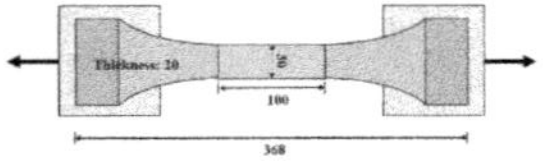	ECC	Anchored	6	0.5	Yu et al. (2020)
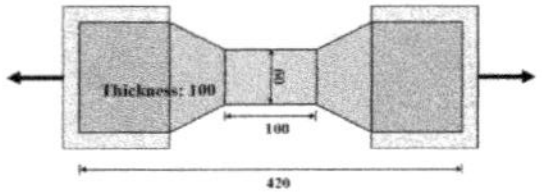	ECC	Side glued	6	0.5	Yu et al. (2020)
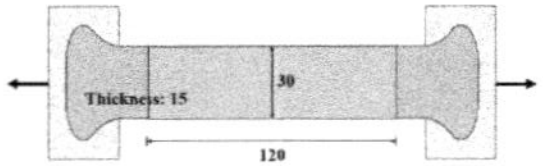	HS-ECC	Anchored	-	0.5	Zhang et al. (2021)
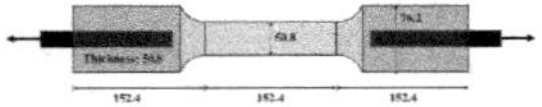	UHPC	Top pinned	3	0.254	Zhou and Qiao (2018)

Note: All dimensions in mm; the gauge region is marked with purple color.

HS-SHCC = high-strength strain-hardening cementitious composites; HS-ECC = high-strength engineered cementitious composites; UHP-SHCC = ultra-high-performance strain-hardening cementitious composites; HSHDC = high-strength, high-ductility concrete.

Table 8.4 Direct tension test setups—unnotched prism/cylinder specimens

Specimen size (mm)	*Materials*	*Grip/ attachment*	*Specimen number*	*Loading rate (mm/min)*	*Ref.*
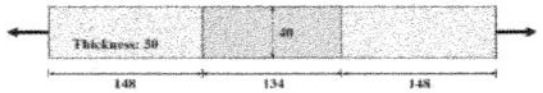	UHPC	Side glued	6	0.12	Abrishambaf et al. (2017)
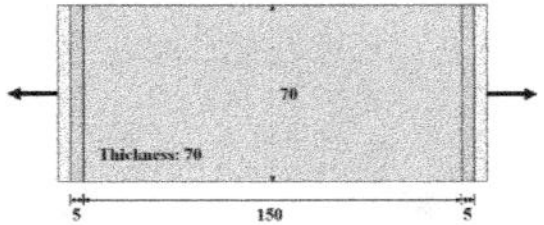	UHPC	Top glued	-	0.006	AFGC (2002); Chanvillard and Rigaud (2003)
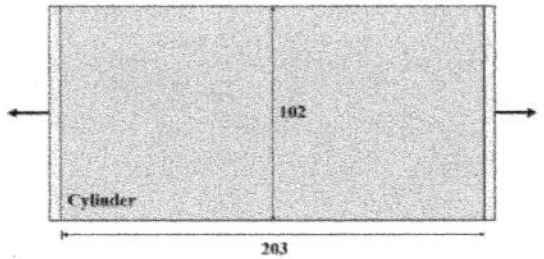	UHPC	Top glued	-	0.005	Benjamin A. (2006)
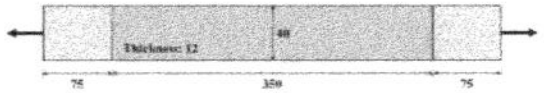	UHPC	Side glued	-	0.01	Brameshuber et al. (2016); Ganesh and Murthy (2019)
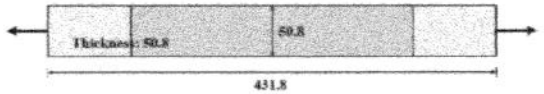	UHPC	Side glued	-	0.326	Bridi Valentim et al. (2019)
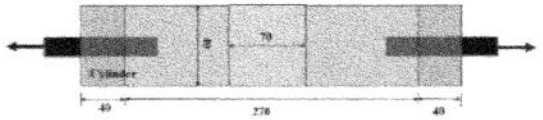	HS-SHCC	Top pinned	-	3	Curosu et al. (2016)
	UHP-FRC	Side glued + clamped/ pinned	-	0.152	Graybeal and Baby (2013)

(*Continued*)

Table 8.4 (Continued)

Specimen size (mm)	*Materials*	*Grip/ attachment*	*Specimen number*	*Loading rate (mm/min)*	*Ref.*
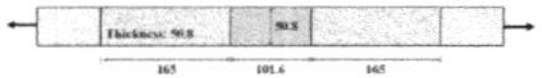 Thickness: 50.8; 50.8; 165, 101.6, 165	UHP-FRC	Side glued + clamped/ pinned	5	0.152	Graybeal et al. (2012); Graybeal and Baby (2013)
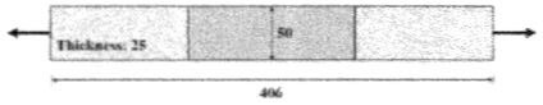 Thickness: 25; 50; 406	UHP-FRC	Side glued	≥3	0.5	Roy et al. (2017)
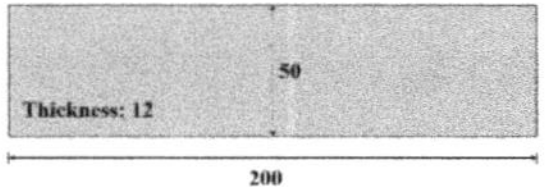 Thickness: 12; 50; 200	UHPC	-	5	0.1	Toledo Filho et al. (2012)
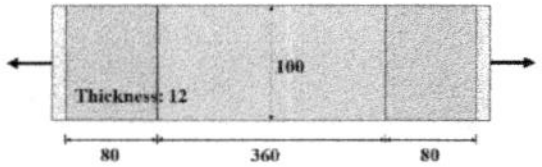 Thickness: 12; 100; 80, 360, 80	UHPC	Top glued	-	1×10^{-6} s^{-1}	Zhang et al. (2015)
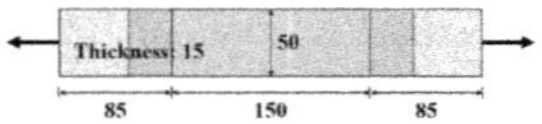 Thickness: 15; 50; 85, 150, 85	UHPC	Side glued	6	0.1	Zhao et al. (2020)

Note: All dimensions in mm; the gauge region is marked with purple color.

UHP-FRC = Ultra high performance-fiber reinforced concrete.

Table 8.5 Direct tension test setups—notched prism/cylinder specimens

Specimen size (mm)	*Materials*	*Grip/ attachment*	*Number of specimens*	*Loading rate (mm/min)*	*Ref.*
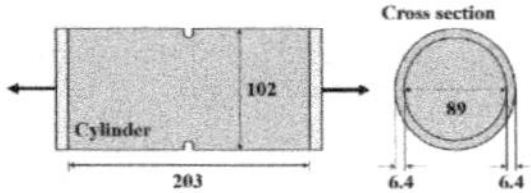	UHPC	Top-glued	-	0.005	Benjamin (2006)
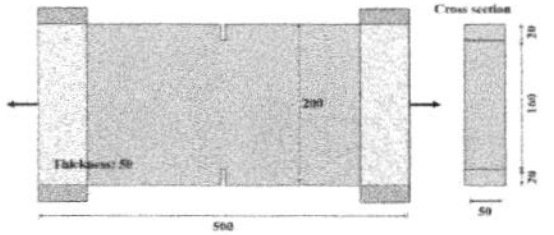	UHPC	Side glued + anchored (greased)	-	0.02	Charron et al. (2008); Denarié et al. (2003)
	UHPC	Side glued	6	0.6 mm/min up to a displacement of 2 mm; 3 mm/min until the completion of the test	Le Hoang and Fehling (2017)
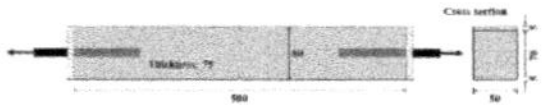	UHPC	Anchored/ Top pinned		0.18 kN/min	Islam (2021)
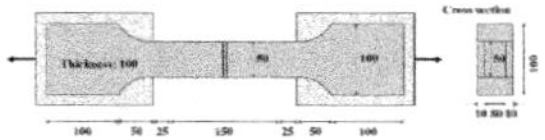	UHPC	Anchored	-	0.2	Kang and Kim (2011)
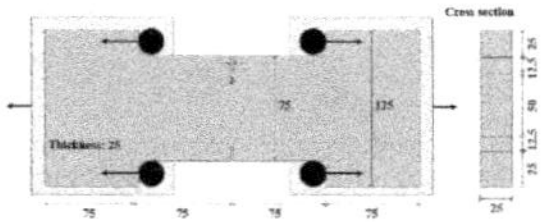	UHPC	Anchored/ pinned	-	-	Lim and Hong (2016)

(*Continued*)

Table 8.5 (Continued)

Specimen size (mm)	*Materials*	*Grip/ attachment*	*Number of specimens*	*Loading rate (mm/min)*	*Ref.*
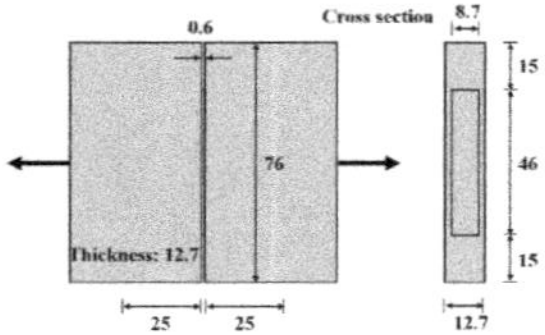	HSHDC	Side glued	6	0.5	Ranade et al. (2013)
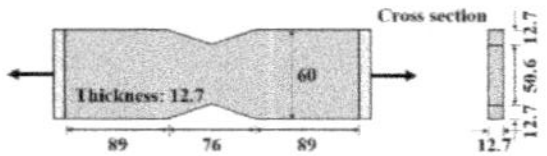	UHS-GFRC	Top glued	5	0.13	Roth et al. (2010)
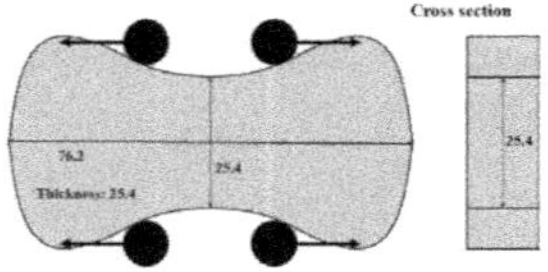	UHPC	Anchored/ pinned	-	0.025	Engineers (2001); Shafieifar et al. (2017)
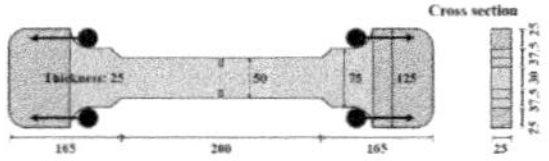	UHPC	Anchored/ pinned	-	1	Tran et al. (2016)
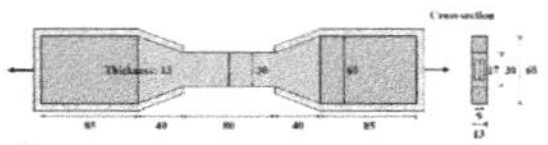	UHDCC	Anchored	-	-	Yu et al. (2017)

Note: UHS-GFRC = ultra-high-strength, glass fiber-reinforced concrete; UHDCC = ultra-high ductile cementitious composites.

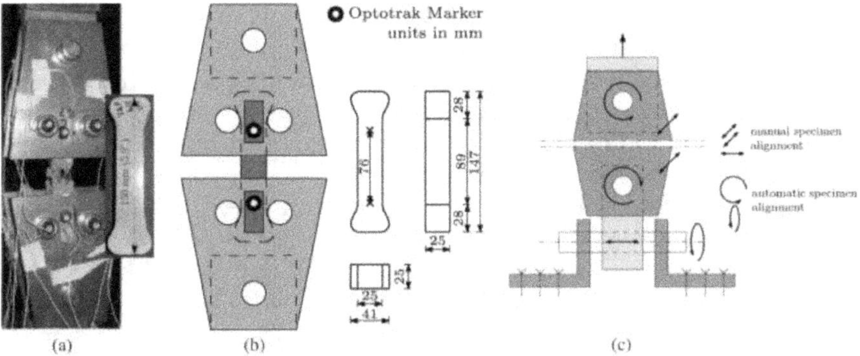

Figure 8.25 Direct tensile test setup: (a) Test setup; (b) Specimen geometry and fixture; and (c) Grip system.

Source: Wille et al. (2014).

Due to various notched specimen geometries, the notched depth varies from 2 to 20 mm. The notched area of notched prism/cylinder specimens ranges from 153 to 8000 mm^2. Furthermore, the notched area to remain ratio ranges from 0.39 to 0.94. The grip/attachment of notched prism/cylinder specimens is similar to unnotched dogbone-shaped specimens or prism/cylinder specimens. A displacement control from 0.005 to 1 mm/min or force control of 0.18 kN/min is used in tensile testing. In some literature, the piecewise loading rate with a loading rate of 0.6 mm/min up to crack opening at 2 mm and then increased to 3 mm/s to speed up the test was also applied (Le Hoang and Fehling, 2017). At least five specimens are employed to attain the accuracy of results.

Figure 8.25 shows the tensile test setup specified in AASHTO T 132-87 (2009). According to the recommendations of the French UHPC design regulations (2002), it is suggested that the maximum particle size of constituents should be limited to one-fifth the smallest size of the specimens, which is 5 mm. This permits testing paste and mortar-like materials, such as UHPC, with a maximum particle size of 0.8 mm. Compared to the fiber length (13–30 mm), specimen molds' height and width (25 and 25 mm) are small. This leads to fiber alignment in the direction of the applied tensile load, which is further enhanced by pouring the specimens into layers.

8.6.2 Effects of fiber content and shape

The direct tensile strength of UHPC is improved with the increase in fiber content (Kang et al. 2010). A large number of studies showed that the direct tensile strength of UHPC is about 7–10 MPa when 3% or less short straight steel fibers are added. The first-cracking strength only slightly increases with increased fiber volume fraction. The mixture with smooth fiber leads to about 15% higher first-cracking tensile strength than that with deformed fibers. This is because the smaller diameter of the used smooth fibers increases the bond's average contribution at the onset of matrix cracking, thus leading to a higher first-cracking strength.

A considerable improvement in the direct tensile strength of UHPC made with deformed fibers is commonly observed when compared to straight fibers (Liu et al. 2016; Meng and Khayat 2018; Wille et al. 2014). However, a decrease in tensile strength of UHPC made with two types of macro hooked fibers (>40%) and macro twisted fibers (15%) is observed compared to the straight microfibers (l = 13 mm) (Yoo et al. 2019). This could be explained by the fiber cluster issues and matrix damage resulting from high bond strength (Yoo et al. 2019). Besides, major differences in tensile strength are

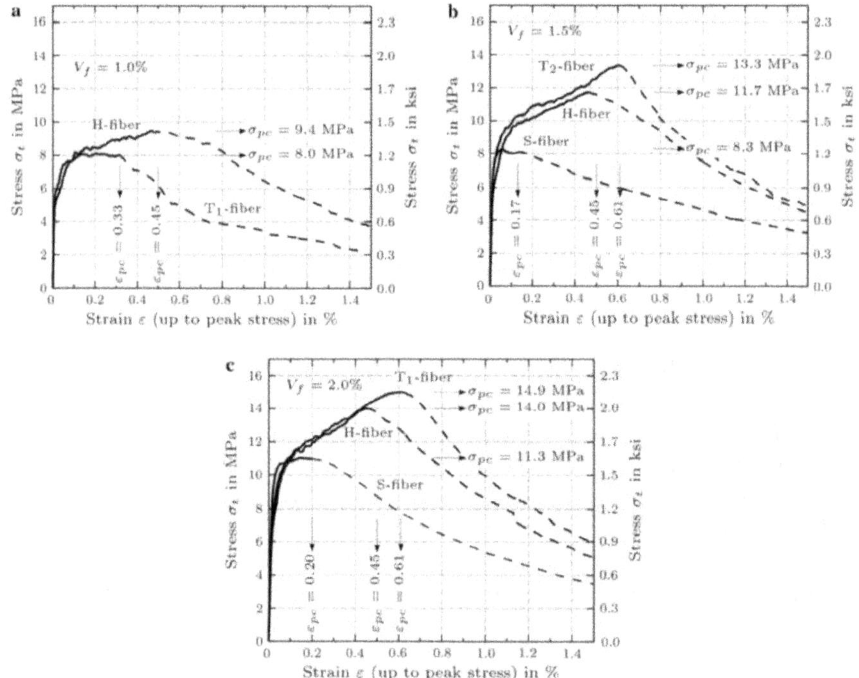

Figure 8.26 Direct tensile response of UHC made with different types of high-strength steel fibers: (a) V_f = 1.0%; (b) V_f = 1.5%; and (c) V_f = 2.0%.

Source: Wille et al. (2011a).

also observed for UHPC with 1% hooked fibers at various fiber lengths of 30 and 62 mm (Park et al. 2012). The decreased capacity is explained by the reduced frictional bond and number of long fibers at the same fiber volume.

Figure 8.26 compares the tensile behavior of UHPC made with smooth fiber, hooked fiber, and twisted fiber at a fixed volume of 1.0%, 1.5%, and 2.0%, respectively. The tensile strength and the maximum postcracking strain are significantly improved by using deformed steel fibers instead of smooth fibers. Moreover, the path of the stress-strain curves of UHPC with hooked and twisted fiber is similar to the peak load of UHPC with hooked fiber. UHPC with hooked fiber starts to show strain-softening behavior at peak strain equal to 0.46%, while twisted fiber keeps increasing the tensile stress up to peak strain equal to 0.6%.

8.6.3 Effect of fiber hybridization

Although the tensile strength can be improved by further increasing fiber content, it will cause reduced flowability and poor construction performance of UHPC, as well as increased cost of materials. To improve the tensile strength, researchers blended fibers with different sizes and shapes (Abdallah et al. 2018; Chun and Yoo 2019; Park et al. 2012; Yoo et al. 2019). Park et al. (2012) investigated the effect of a combination of macrofiber and microfiber on the tensile behavior of UHPC. The microfiber volume varied from 0 to 1.5%, while the macrofiber volume was fixed at 1.0%. The type of macrofiber plays a significant role in determining the overall shape of tensile stress-strain curves. UHPC made with twisted macrofibers shows the best tensile behavior, while those with long, smooth macrofibers exhibit the worst behavior. Yoo et al. (2019) developed UHPC with a direct tensile strength of up to 17 MPa when using shaped fibers 30 mm in length and straight fiber 13 mm in length.

8.6.4 Effect of nanoparticles

Nanomaterials are also used to improve the tensile behaviors of UHPC (He et al. 2017; Meng and Khayat 2016). He et al. (2017) coated polyethylene fiber surfaces using carbon nanotubes to improve the fiber-matrix adhesion strength and found that the direct tensile strength and strain capacity of UHPC are increased by 15% and 20%, respectively.

8.6.5 Effect of fiber alignment and orientation

The tensile properties of UHPC are significantly varied with mixture proportion and locations due to the different fiber orientation distributions caused by casting direction, flow distance, and structural forms of construction. The specimens used in the test are generally poured one by one from a

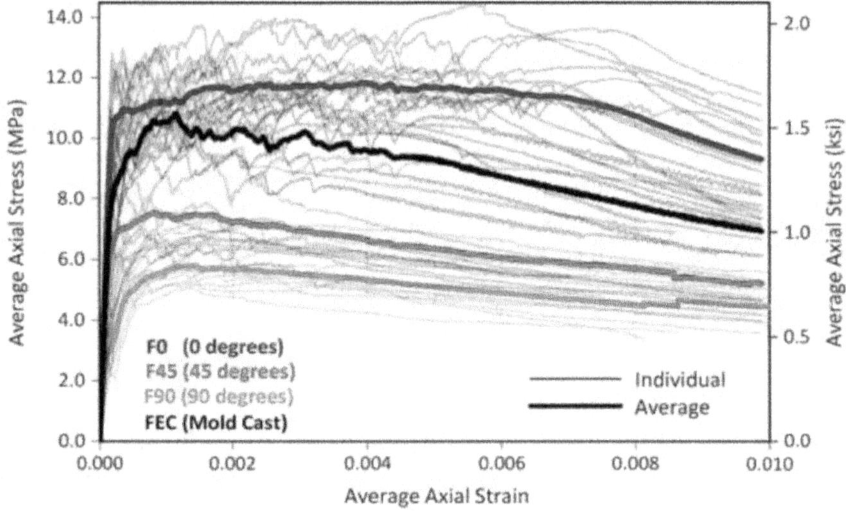

Figure 8.27 Tensile responses of UHPC prisms with four orientations.

Source: Duque and Graybeal (2017).

slender mold. The fibers tend to orient in the direction of the principal tensile stress. Fiber alignment corresponding to the direction of principal tensile stress can significantly increase the effectiveness of fibers in bridging cracks, thus enhancing the tensile properties of UHPC.

Figure 8.27 shows the tensile stress-strain curves of UHPC prepared with various fiber alignments. The F0 specimens extracted parallel to the material flow direction overestimate the average cracking stress, average multicracking stress, strain at localization, and elastic modulus. Significant reductions in the average first-cracking and multicracking stress are observed for F45 and F90 specimens. This suggests a strong influence of the fiber orientation distribution on the postcracking response and the first cracking stress. The average strains at crack localization for F45 and F90 are found to be around 50% of the strain for F0. For those materials and from a structural design standpoint, a maximum value of 0.0025 is usually adopted for the ultimate tensile strain in the constitutive tensile law, which is of the order of the maximum value observed for F45 and F90. The results confirm the anisotropy in the tensile mechanical properties, primarily due to fiber orientation effects induced by the casting conditions and the material flow.

The effect of fiber orientation is quite well evidenced by the stress-strain curves in Figure 8.28. Three main stages, including elastic, hardening, and softening, can be identified. The notable exceptions are the "non-oriented"

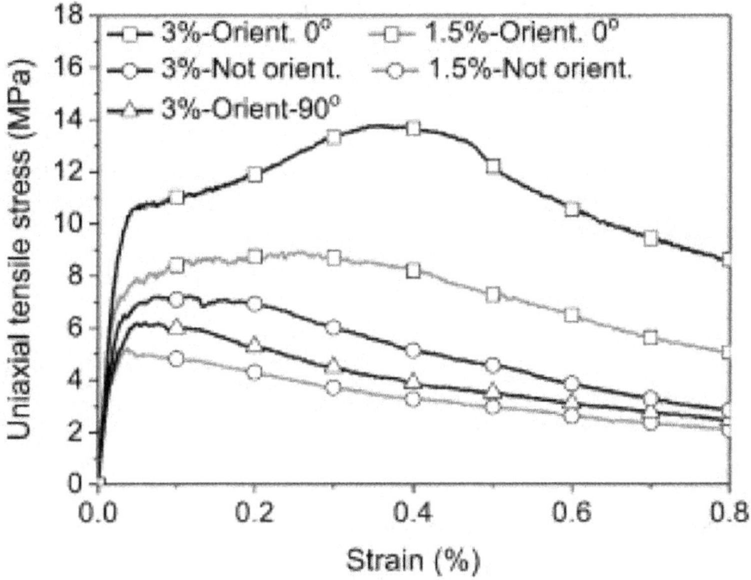

Figure 8.28 Uniaxial tensile stress-strain curves of UHPC with different fiber volumes and orientations.

Source: Abrishambaf et al. (2017).

specimens with 1.5% fibers, in which the hardening branch is not observed. The different fiber orientation curves are practically overlapped in this stage since the fiber distribution and orientation only slightly influence the elastic deformation. After the matrix cracking, a plateau or an increase in stress is observed, accompanied by the formation of multiple microcracks. The behavior during this stage depends highly on how the fibers are distributed and oriented throughout the cementitious material, which justifies the differences between the curves and the scattering observed among different specimens of the same series. The softening stage is governed by fiber pullout behavior.

8.7 SUMMARY

This chapter discusses the fiber-matrix bond, compressive, flexural, and tensile properties of UHPC under static loading. Specimen sizes and test methods for these properties are first mentioned, and factors affecting those properties are discussed. Based on the discussion from this chapter, the following conclusions can be summarized.

(1) Fiber-matrix bond properties play a dominant role in enhancing tensile/flexural strength and toughness of UHPC. The main factors affecting this behavior involve matrix strength, use of supplementary cementitious materials, curing regime, fiber geometry, fiber dispersion and orientation, and loading rate. The densification of the matrix is the most fundamental and practical approach to enhancing the fiber-matrix bond properties. However, its efficiency to enhance the fiber-matrix bond and mechanical properties of UHPC is much lower than those improved with the use of deformed fibers.
(2) The fiber type and volume significantly affect the static mechanical properties of UHPC. With the increase in fiber volume, the compressive, flexural, and tensile strengths are gradually enhanced. Blending fibers of various sizes and types is an effective method to enhance the tensile/flexural strengths and strain capacity of UHPC due to the combined functions exerted by each type of fiber. Deformed fiber can significantly increase flexural and tensile strengths. However, excessive fiber content can lead to high cost and fiber cluster issues associated with reduced workability, thus reversely impairing the mechanical properties of UHPC. Therefore, the design UHPC incorporated with appropriate fiber content and adequate rheology is essential to ensure proper fiber dispersion and orientation and eventually enhance mechanical properties and sustainability.
(3) UHPC with appropriate content of coarse aggregate shows better workability, higher compressive strength, and elastic modulus. However, the incorporation of coarse aggregate can disturb the dispersion and orientation of fibers, bringing a disadvantage to bond strength and fiber strengthening and toughening effects.

REFERENCES

AASHTO T 132. (2009). Standard method of test for tensile strength of hydraulic cement mortars. American Association of State and Highway Transportation Officials.

Abbas, S., Soliman, A. M., Nehdi, M. L. (2015). Exploring mechanical and durability properties of ultra-high performance concrete incorporating various steel fiber lengths and dosages. *Constr. Build. Mater.* 75, 429–441.

Abdallah, S., Rees, D. W. A., Ghaffar, S. H., Fan, M. Z. (2018). Understanding the effects of hooked-end steel fibre geometry on the uniaxial tensile behaviour of self-compacting concrete. *Constr. Build. Mater.* 178, 484–494.

Abed, M., Nasr, M., Hasan, Z. (2018). Effect of silica fume/binder ratio on compressive strength development of reactive powder concrete under two curing systems. *MATEC Web Conf.* 162, 02022.

Abid, M., Hou, X., Zheng, W., Hussain, R. R. (2019). Effect of fibers on high-temperature mechanical behavior and microstructure of reactive powder concrete. *Materials (Basel, Switzerland)* 12(2), 1–30.

Abokifa, M., Moustafa, M.A., (2021). Mechanical characterization and material variability effects of emerging non-proprietary UHPC mixes for accelerated bridge construction field joints. *Constr. Build. Mater.* 308, 125064.

Abrishambaf, A., Pimentel, M., Nunes, S. (2017). Influence of fibre orientation on the tensile behaviour of ultra-high performance fibre reinforced cementitious composites. *Cem. Concr. Res.* 97, 28–40.

ACI 318-14. (2014). *Building Code Requirements for Structural Concrete (ACI 318-14): Commentary on Building Code Requirements for Structural Concrete*, American Concrete Institute, Farmington Hills, MI.

AFGC, 2002. Interim Recommendations. (2002). Ultra high performance fiber-reinforced concretes. AFGC Publication, France.

Aghdasi, P., Heid, A.E., Chao, S.H., (2016). Developing ultra-high-performance fiber-reinforced concrete for large-scale structural applications. *ACI Mater. J.* 113, 559–569.

Afroughsabet, V., Biolzi, L., Ozbakkaloglu, T. (2016). High-performance fiber-reinforced concrete: A review. *J. Mater. Sci.* 51(14), 6517–6551.

Ahmad, S., Rasul, M., Adekunle, S. K., Al-Dulaijan, S. U., Maslehuddin, M., Ali, S. I. (2019). Mechanical properties of steel fiber-reinforced UHPC mixtures exposed to elevated temperature: Effects of exposure duration and fiber content. *Compos. Pt. B-Eng.* 168, 291–301.

Ahmad, S., Zubair, A., Maslehuddin, M. (2015). Effect of key mixture parameters on flow and mechanical properties of reactive powder concrete. *Constr. Build. Mater.* 99, 73–81.

Aldahdooh, M.A.A., Muhamad Bunnori, N., Megat Johari, M.A., (2013). Evaluation of ultra-high-performance-fiber reinforced concrete binder content using the response surface method. *Mater. Des.* 52, 957–965.

Alsalman, A., Dang, C. N., Prinz, G. S., and Hale, W. M. (2017). Evaluation of modulus of elasticity of ultra-high performance concrete. *Constr. Build. Mater.* 153, 918–928.

ASTM C1609/C1609M-12. (2012). Standard test method for flexural performance of fiber-reinforced concrete (using beam with third-point loading). ASTM International. West Conshohocken, PA, USA.

ASTM C511. (2019). Standard specification for mixing rooms, moist cabinets, moist rooms, and water storage tanks used in the testing of hydraulic cements and concretes. ASTM International. West Conshohocken, PA, USA.

ASTM C31/C31M-21A. (2021). Standard practice for making and curing concrete test specimens in the field. ASTM International, West Conshohocken, PA, USA.

ASTM C1856/C1856M-17. (2017). Standard practice for fabricating and testing specimens of ultra-high performance concrete. ASTM International, West Conshohocken, PA, USA.

Aydin, S., Yazici, H., Yardimci, M. Y., Yiğiter, H. (2010). Effect of aggregate type on mechanical properties of reactive powder concrete. *ACI Mater. J.* 10(5), 441–449.

Barnett, S.J., Lataste. J., Parry, T., Millard, S. G., Soutsos, M. N. (2010). Assessment of fibre orientation in ultra high performance fibre reinforced concrete and its effect on flexural strength. *Mater. Struct.* 43(7), 1009–1023.

Benjamin A., G., (2006). Material property characterization of ultra-high performance concrete. Federal Highway Administration.

Benson, S.D.P., Karihaloo, B.L., (2005). CARDIFRC®—Development and mechanical properties. Part III: Uniaxial tensile response and other mechanical properties. *Mag. Concr. Res.* 57, 433–443.

Bentz, D. P., Ardani, A., Barrett, T., Jones, S. Z., Lootens, D., Peltz, M. A., Sato, T., Stutzman, P. E., Tanesi, J., Weiss, W. J. (2015). Multi-scale investigation of the performance of limestone in concrete. *Constr. Build. Mater.* 75, 1–10.

Boulay, C., Rossi, P., Tailhan, J.L., (2004). Uniaxial tensile test on a new cement composite having hardening behaviour. 6th RILEM Symp. Fibre-Reinforced Concr. 61–68.

Boulekbache, B., Hamrat, M., Chemrouk, M., Amziane, S.(2010). Flowability of fibre-reinforced concrete and its effect on the mechanical properties of the material, *Constr. Build. Mater.* 24 (9), 1664–1671.

Bridi Valentim, D., Aaleti, S., Amirkhanian, A., Kreger, M.E., (2019). Experimental evaluation of test methods to characterize tensile behavior of UHPC, in: Second International Interactive Symposium on UHPC. pp. 1–10.

BS EN 459-1:2015. 2010. Building Lime Part: Definitions, Specifications and Conformity Criteria, British Standard Institution, London.

CEB-FIP Model code. 1993. The European Concrete Committee (CEB)

CECS 13: 2009. 2009. *Standard test methods for fiber reinforced concrete*, P.R. China.

Chan, Y. W., Chu, S. H. (2004). Effect of silica fume on steel fiber bond characteristics in reactive powder concrete. *Cem. Concr. Res.* 34(7), 1167–1172.

Chanvillard, G., Rigaud, S., (2003). Complete characterization of tensile properties of DUCTAL UHP-FRC according to the French recommendations, in: Naaman, A.E., Reinhardt, H.W. (Eds.), Fourth International Workshop on High Performance Fiber Reinforced Cement Composites. pp. 21–34.

Charron, J.P., Denarié, E., Brühwiler, E., (2008). Transport properties of water and glycol in an ultra high performance fiber reinforced concrete (UHPFRC) under high tensile deformation. *Cem. Concr. Res.* 38, 689–698.

Chen, J., Liu, J., Liu, J., Zhang, Q., Zhang, L., Lin, W., Han, F. (2017). An experimental study and a mechanism analysis on mechanical properties of ultra-high performance concrete with coarse aggregate. *Mater. Repor.* 31, 115–119.

China National Standards, Method of Testing Cements—Determination of Strength, GB/T 17671-1999, China, Beijing, 1999 (in Chinese).

Chinese Standard-GB/T 31387-2015. (2015). Reactive Powder Concrete. China Architecture & Building Press, Beijing, China, 2015.

Chun, B., Yoo, D. Y. (2019). Hybrid effect of macro and micro steel fibers on the pullout and tensile behaviors of ultra-high-performance concrete. *Compos. Part-B Eng.* 162(APR.1), 344–360.

Chun, B., Yoo, D. Y., Banthia, N. (2020). Achieving slip-hardening behavior of sanded straight steel fibers in ultra-high-performance concrete. *Cem. Concr. Compos.* 113, 103669.

Cunha, V. M., Barros, J. A., Sena-Cruz, J. (2007).Pullout behaviour of hooked-end steel fibres in self-compacting concrete. Universidade do Minho. Departamento de Engenharia Civil (DEC).

Curosu, I., Liebscher, M., Mechtcherine, V., Bellmann, C., Michel, S., (2017). Tensile behavior of high-strength strain-hardening cement-based composites (HS-SHCC) made with high-performance polyethylene, aramid and PBO fibers. *Cem. Concr. Res.* 98, 71–81.

Curosu, I., Mechtcherine, V., Millon, O., (2016). Effect of fiber properties and matrix composition on the tensile behavior of strain-hardening cement-based composites (SHCCs) subject to impact loading. *Cem. Concr. Res.* 82, 23–35.

Dawood, E. T., Ramli, M. (2011). High strength characteristics of cement mortar reinforced with hybrid fibres. *Constr. Build. Mater.* 25(5), 2240–2247.

Denarié, E., (2006). SAMARIS D25b-guidance for the use of UHPFRC for rehabilitation of concrete highway structures. MCS/EU.

Denarié, E., Habel, K., Brühwiler, E., (2003). Structural behavior of hybrid elements with Advanced Cementitious Materials (HPFRCC). In: Proceedings of the Fourth International Workshop on High Performance Fiber Reinforced Cement Composites (HPFRCC 4), RILEM, Ann Arbor, USA, 1–12.

Deng, F., Ding, X., Chi, Y., Xu, L., Wang, L. (2018). The pull-out behavior of straight and hooked-end steel fiber from hybrid fiber reinforced cementitious composite: Experimental study and analytical modeling. *Compos. Struct.* 206, 693–712.

Deng, F., Xu, L., Chi, Y., Wu, F., Chen, Q. (2020). Effect of steel-polypropylene hybrid fiber and coarse aggregate inclusion on the stress-strain behavior of ultra-high performance concrete under uniaxial compression. *Compos. Struct.* 252.

Dhar, A., Rajasankar, J., Anandavalli, N. (2018). A mathematical formulation to find effective bulk and shear moduli of recycled aggregate concrete. *Constr. Build. Mater.* 168(20), 747–757.

Dong, S., Han, B., Yu, X., Ou, J. (2018). Constitutive model and reinforcing mechanisms of uniaxial compressive property for reactive powder concrete with super-fine stainless wire. *Compos. Part-B Eng.* 166, 298–309.

Donnini, J., Lancioni, G., Chiappini, G., Corinaldesi, V., (2021). Uniaxial tensile behavior of ultra-high performance fiber-reinforced concrete (UHPFRC): Experiments and modeling. *Compos. Struct.* 258, 113433.

Duque, L. F. M., Graybeal, B. (2017). Fiber orientation distribution and tensile mechanical response in UHPFRC. *Mater. Struct.* 50(1), 1–17.

Elsharief, A., Cohen, M. D., Olek, J. (2003). Influence of aggregate size, water cement ratio and age on the microstructure of the interfacial transition zone. *Cem. Concr. Res.* 33, 1837–1849.

EN 1992-1-1. (2004). Eurocode 2: Design of concrete structures: Part 1: General rules and rules for buildings. Brussels, Belgium.

EN 196-1:2016 (2016). Methods of testing cement – Part 1: Determination of strength. European Standard.

Engineers, U.A.C. (2001). CRD-C 260-01 Standard Test Method for Tensile Strength of Hydraulic Cement Mortars. COE Stand. 1–8.

Eshelby, J. D. (1957). The determination of the elastic field of an ellipsoidal inclusion, and related problems. *Proceedings of the Royal Society A: Mathematical, Physical and Engineering Sciences* 241(1226), 376–396.

NF P 18-470, (2017). Ultra-high performance fibre-reinforced concrete (UPHFRC)—specification, performance, production and conformity, French Standard Institute, France.

Ganesh, P., Murthy, A.R., 2019. Tensile behaviour and durability aspects of sustainable ultra-high performance concrete incorporated with GGBS as cementitious material. *Constr. Build. Mater.* 197, 667–680.

Gangwar, S., Mishra, S., Sharma, H.K., 2019. An experimental study on mechanical properties of ultra-high-performance fiber-reinforced concrete (UHPFRC), in: Sustainable Construction and Building Materials. Singapore, 873–886.

Gesoglu, M., Guneyisi, E., Asaad, D. S., Muhyaddin, G. F. (2016). Properties of low binder ultra-high performance cementitious composites: Comparison of nanosilica and microsilica. *Constr. Build. Mater.* 102, 706–713.

Gesoglu, M., Guneyisi, E., Muhyaddin, G. F., Asaad, D. S. (2016). Strain hardening ultra-high performance fiber reinforced cementitious composites: Effect of fiber type and concentration. *Compos. Part-B Eng.* 103, 74–83.

Ghafari, E., Costa, H., Julio, E. (2015). Critical review on eco-efficient ultra high performance concrete enhanced with nano-materials. *Constr. Build. Mater.* 101, 201–208.

Gowripalan N, Gilbert R I. Design guidelines for ductal prestressed concrete beams. Reference Article, The University of NSW, 2000.

Graybeal, B. A. (2005). Characterization of the behavior of ultra-high performance concrete.Ph.D. Dissertation, University of Maryland, USA.

Graybeal, B. A. (2007). Compressive behavior of ultra-high-performance fiber-reinforced concrete. *ACI Mater. J.* 104(2), 146–152.

Graybeal, B.A., Baby, F., Marchand, P., Toutlemonde, F., 2012. Direct and flexural tension test methods for determination of the tensile stress-strain response of UHPFRC, in: Schmidt, M., Fehling, E. (Eds.), Proceedings of Hipermat 2012, 3rd International Symposium on UHPC and Nanotechnology for High Performance Construction Materials. Kassel, Germany, March 7–9, pp. 395–402.

Graybeal, B. A., Baby, F. (2013). Development of direct tension test method for ultra-high-performance fiber-reinforced concrete. *ACI Mater. J.* 110(2), 177–186.

Guo, X., Kang, J., Zhu, J. (2017). Constitutive relationship of ultrahigh performance concrete under uni-axial compression. *J. Southeast Univ.* 47(2), 369–376.

Habel, K., Denarié, E., Brühwiler, E. (2006). Structural response of elements combining ultrahigh-performance fiber-reinforced concretes and reinforced concrete. J. Struct. Eng. 132(11), 1793–1800.

Hafiz, M.A., Denarié, E., 2020. Tensile response of UHPFRC under very low strain rates and low temperatures. *Cem. Concr. Res.* 133, 106067.

Hai, R., Liu, J., Zhang, M., Zhang, L. (2016). Performance of hybrid steel-polyvinyl alcohol fiber reinforced ultra high performance concrete. Concrete (China), 319, 95–97.

Haile, B. F., Jin, D. W., Yang, B., Park, S., Lee, H. K. (2019). Multi-level homogenization for the prediction of the mechanical properties of ultra-high-performance concrete. *Constr. Build. Mater.* 229, 116797.

Hannawi, K., Bian, H., Prince-Agbodjan, W., Raghavan, B. (2016). Effect of different types of fibers on the microstructure and the mechanical behavior of Ultra-High Performance Fiber-Reinforced Concretes. *Compos. Part-B Eng.* 86, 214–220.

Hansen, T. C. (1965). Influence of Aggregate and Voids on Modulus of Elasticity of Concrete, Cement Mortar, and Cement Paste. *Am. Concrete Inst. J.* 62(2), 193–216.

Hashin, Z., Monteiro, P. J. M. (2002). An inverse method to determine the elastic properties of the interphase between the aggregate and the cement paste. *Cem. Concr. Res.* 32, 1291–1300.

Hassan, A. M. T., Jones, S. W., Mahmud, G. H. (2012). Experimental test methods to determine the uniaxial tensile and compressive behaviour of ultra high performance fibre reinforced concrete (UHPFRC). *Constr. Build. Mater.* 37, 874–882.

He, B., Zhu, X., Ren, Q., Jiang, Z. (2020). Effects of fibers on flexural strength of ultra-high-performance concrete subjected to cryogenic attack. *Constr. Build. Mater.* 265(30), 120323.

He, S., Qiu, J. S., Li, J. X., Yang, E. H. (2017). Strain hardening ultra-high performance concrete (SHUHPC) incorporating CNF-coated polyethylene fibers. *Cem. Concr. Res.* 98, 50–60.

Hoang, A. L., Fehling, E. (2017). Influence of steel fiber content and aspect ratio on the uniaxial tensile and compressive behavior of ultra high performance concrete. *Constr. Build. Mater.* 153, 790–806.

Hou, X., Ren, P., Rong, Q., Zheng, W., Zhan, Y. (2019). Effect of fire insulation on fire resistance of hybrid-fiber reinforced reactive powder concrete beams. *Compos. Struct.* 209, 219–232.

Huang, H. H., Gao, X. J., Li, L. S., Wang, H. (2018). Improvement effect of steel fiber orientation control on mechanical performance of UHPC. *Constr. Build. Mater.* 188, 709–721.

Huang, Z., Cao, F. (2012). Effects of nano-materials on the performance of UHPC. *Mater. Repor.* 26(18), 136–141.

Huang, Z., Li, S. (2018). Study on mechanical properties of ultra high performance concrete with coarse aggregate. *J. Hunan Univ. (Natural Sciences).* 45(3), 47–54.

Huang, Z.Y, Tan, B. (2007). Research on stress-strain curves of reactive powder concrete with steel-fiber under uniaxial compression. *J. China Three Gorges Univ.* (*Natural Sciences*). 27(9): 415–420 (In Chinese).

Hung, C.C., Lee, H.S., Chan, S.N. (2019). Tension-stiffening effect in steel-reinforced UHPC composites: Constitutive model and effects of steel fibers, loading patterns, and rebar sizes. *Compos. Part-B Eng.* 158, 269–278.

Islam, M.M.U., 2021. Investigation of tensile creep for Ultra-High-Performance Fiber Reinforced Concrete (UHPFRC) for the long-term. *Constr. Build. Mater.* 305, 124752.

Jin, L., Zhang, R., Tian, Y., Dou, G., Du, X. (2018). Experimental investigation on static and dynamic mechanical properties of steel fiber reinforced ultra-high-strength concretes. *Constr. Build. Mater.* 178, 102–111.

Japan Society of Civil Engineers (JSCE). (2006). Recommendations for Design and Construction of Ultra High Strength Fiber Reinforced Concrete Structures (Draft). JSCE Guidelines for Concrete No 9, 2006.

Kamal, A., Kunieda, M., Ueda, N., Nakamura, H., 2008. Evaluation of crack opening performance of a repair material with strain hardening behavior. *Cem. Concr. Compos.* 30, 863–871.

Kang, S.T., Lee, Y., Park, Y.D., Kim, J.K. (2010). Tensile fracture properties of an ultra high performance fiber reinforced concrete (UHPFRC) with steel fiber. *Compos. Struct.* 92(1), 61–71.

Kang, S. T., Lee, B. Y., Kim, J. K., Kim, Y. Y. (2011). The effect of fibre distribution characteristics on the flexural strength of steel fiber-reinforced ultra high strength concrete. *Constr. Build. Mater.* 25(5), 2450–2457.

Kang, S.T., Kim, J.K.(2011). The relation between fiber orientation and tensile behavior in an ultra high performance fiber reinforced cementitious composites (UHPFRCC). *Cem. Concr. Res.* 41, 1001–1014.

KCI-M-12-003.(2012). Design guidelines for ultra high performance concrete K-UHPC structure. Korea Concrete Institute.

Kim, D. J., Park, S. H., Ryu, G. S., Koh, K. T. (2011). Comparative flexural behavior of Hybrid Ultra High Performance Fiber Reinforced Concrete with different macro fibers. *Constr. Build. Mater.* 25(11), 4144–4155.

Kong, L., Du, Y. (2015). Interfacial interaction of aggregate-cement paste in concrete. *Journal of Wuhan University of Technology-Materials Science Edition* 30(1), 117–121.

Kravanja, S., Sovják, R., (2018). Ultra-high-performance fibre-reinforced concrete under high-velocity projectile impact. Part I. experiments. Acta Polytech. 58, 232–239.

Kunieda, M., Kozawa, K., Ueda, N., Nakamura, H., (2010). Fracture analysis of strain hardening cementitious composites by means of discrete modeling of short fibers. In: Byung Hwan Oh, Choi, O.C., Chung, L., Korea Concrete Institute (Eds.), Jeju Oh BH, editor. In: Proceedings of FraMCos-7. Fracture mechanics of concrete and concrete structures. Seoul, Korea. p: 501–508..

Lampropoulos, A.P., Paschalis, S.A., Tsioulou, O.T., Dritsos, S.E. (2016). Strengthening of reinforced concrete beams using ultra high performance fibre reinforced concrete (UHPFRC). *Eng. Struct.* 106, 370–384.

Le, A.H., (2022). An experimental evaluation of direct tensile strength for ultra-high performance concrete, in: Serna, P., Llano-Torre, A., Martí-Vargas, J.R., Navarro-Gregori, J. (Eds.), Fibre Reinforced Concrete: Improvements and Innovations II. Springer, pp. 958–964.

Le Hoang, A., Fehling, E. (2017). Influence of steel fiber content and aspect ratio on the uniaxial tensile and compressive behavior of ultra high performance concrete. *Constr. Build. Mater.* 153, 790–806.

Lee, K. M., Park, J. H. (2008). A numerical model for elastic modulus of concrete considering interfacial transition zone. *Cem. Concr. Res.* 38(3), 396–402.

Lee, M. K., Barr, B. I. G. (2004). An overview of the fatigue behaviour of plain and fibre reinforced concrete. *Cem. Concr. Compos.* 26(4), 299–305.

Lee, J.H., Hong, S.G., Joh, C., Kwahk, I., Lee, J.W., (2017). Biaxial tension–compression strength behaviour of UHPFRC in-plane elements. *Mater. Struct. Constr.* 50, 1–17.

Li, V.C., Wang, S., (2003). On high performance fiber reinforced cementitious composites, in: Proceeding of the JCI Symposium on Ductile Fiber Reinforced Cementitious Composites, DFRCC. Tokyo, Japan, pp. 13–23.

Li, B., Xu, L. H., Chi, Y., Huang, B., Li, C. N. (2017). Experimental investigation on the stress-strain behavior of steel fiber reinforced concrete subjected to uniaxial cyclic compression. *Constr. Build. Mater.* 140, 109–118.

Li, P. P., Yu, Q. L., Brouwers, H. J. H. (2018). Effect of coarse basalt aggregates on the properties of Ultra-high Performance Concrete (UHPC). *Constr. Build. Mater.* 170, 649–659.

Liem, D., Sung, G., Taek, K., Joo, D., 2014. Size and geometry dependent tensile behavior of ultra-high-performance fiber-reinforced concrete. *Compos. Part-B Eng.* 58, 279–292.

Lim, W.Y., Hong, S.G., 2016. Shear tests for ultra-high performance fiber reinforced concrete (UHPFRC) beams with shear reinforcement. *Int. J. Concr. Struct. Mater.* 10, 177–188.

Liu, J., Han, F., Cui, G., Zhang, Q., Lv, J., Zhang, L., Yang, Z. (2016). Combined effect of coarse aggregate and fiber on tensile behavior of ultra-high performance concrete. *Constr. Build. Mater.* 121, 310–318.

Liu, X., Wu, T., Liu, Y. (2019). Stress-strain relationship for plain and fibre-reinforced lightweight aggregate concrete. *Constr. Build. Mater.* 225, 256–272.

Ma, J., Orgass, M., Dehn, F., Schmidt, D., Tue, N. V. (2004). Comparative investigations on ultra-high performance concrete with and without coarse aggregates. Proceedings of International Symposium on Ultra High Performance Concrete(UHPC), Kassel, Germany, 205–212.

Meng, W. N., Khayat, K. H. (2016). Mechanical properties of ultra-high-performance concrete enhanced with graphite nanoplatelets and carbon nanofibers. *Compos. Part-B Eng.* 107, 113–122.

Meng, W., Khayat, K. H. (2018). Effect of hybrid fibers on fresh properties, mechanical properties, and autogenous shrinkage of cost-effective UHPC. *J. Mater. Civ. Eng.* 30(4), 04018030.

Miletić, M., (2016). Modeling of localized deformation in high and ultra-high performance fiber reinforced cementitious composites. Kansas State University.

Mori, T., Tanaka, K. (1973). Average stress in matrix and average elastic energy of materials with misfitting inclusions. *Acta Meta.* 21(5), 571–574.

Naaman, A.E., Namur, G.G., Alwan, J.M., Najm, H.S. (1991a). Fiber pullout and bond slip I: analytical study. *J. Struct. Eng.* 117 (9) 2769–2790.

Naaman, A.E., Namur, G.G., Alwan, J.M., Najm, H.S. (1991b). Fiber pullout and bond slip II: experimental validation. *J. Struct. Eng.* 117 (9) 2791–2800.

Nguyen, D. L., Kim, D. J., Ryu, G. S., Koh, K. T. (2013). Size effect on flexural behavior of ultra-high-performance hybrid fiber-reinforced concrete. *Compos. Part-B Eng.* 45(1), 1104–1116.

Noushini, A., Aslani, F., Castel, A., Gilbert, R. I., Uy, B., Foster, S. J. (2016). Compressive stress-strain model for low-calcium fly ash-based geopolymer and heat-cured Portland cement concrete. *Cem. Concr. Compos.* 73, 136–146.

Ouyang, X., Shi, C., Wu, Z., Li, K., Shan, B., Shi, J. (2020). Experimental investigation and prediction of elastic modulus of ultra-high performance concrete (UHPC) based on its composition. *Cem. Concr. Res.* 138, 106241.

prSIA 2052.(2014). Bétons fibrés Ultra-Performant: Matériaux, dimensionnement et exécution (UHPC: Material, dimensioning and construction). Swiss Society of Engineers and Architects.

Park, S. H., Kim, D. J., Ryu, G. S., Koh, K. T. (2012). Tensile behavior of ultra high performance hybrid fiber reinforced concrete. *Cem. Concr. Compos.* 34, 172–184.

Park, J.J., Yoo, D.Y., Park, G.J., Kim, S.W. (2017). Feasibility of reducing the fiber content in ultra-high-performance fiber-reinforced concrete under flexure. *Mater.* 10(2), 118.

Prabha, S. L., Dattatreya, J. K., Neelamegam, M., Rao, M. V. S. (2010). Study on stress-strain properties of reactive powder concrete under uniaxial compression. *Inter. J Eng. Sci. Tech.* 2(11), 6408–6416.

Prem, P. R., Murthy, A. R., Bharatkumar, B. H. (2015). Influence of curing regime and steel fibres on the mechanical properties of UHPC. *Mag. Concr. Res.* 67(18), 988–1002.

Pu, X., Wang, Z., Wang, C., Yan, W., Wang, Y. (2002). Mechanical properties of super high-strength and high performance concrete. *J. Build. Struct.* 23(6), 49–55.

Qi, J., Wu, Z., Ma, Z. J., Wang, J. (2018). Pullout behavior of straight and hooked-end steel fibers in UHPC matrix with various embedded angles. *Constr. Build. Mater.* 191, 764–774.

Qiu, M., Zhang, Y., Qu, S., Zhu, Y., Shao, X., (2020). Effect of reinforcement ratio, fiber orientation, and fiber chemical treatment on the direct tension behavior of rebar-reinforced UHPC. *Constr. Build. Mater.* 256, 119311.

Ranade, R., Li, V.C., Stults, M.D., Rushing, T.S., Roth, J., Heard, W.F., (2013). Micromechanics of High-Strength, High-Ductility Concrete. *ACI Mater. J.* 110, 375–384.

Ranade, R., Stults, M.D., Li, V.C., Rushing, T.S., Roth, J., Heard, W.F., 2011. Development of high strength high ductility concrete. In: 2nd International RILEM Conference on Strain Hardening Cementitious Composites, Rio de Janeiro, Brazil, 1–8.

Recommendation of RILEM TC 232-TDT. (2016). Test methods and design of textile reinforced concrete: Uniaxial tensile test: test method to determine the load bearing behavior of tensile specimens made of textile reinforced concrete. Mater. Struct. 49, 4923–4927.

Ren, G. M., Wu, H., Fang, Q., Liu, J. Z. (2018). Effects of steel fiber content and type on static mechanical properties of UHPCC. *Constr. Build. Mater.* 163, 826–839.

Reineck, K. H., Frettlöhr, B. (2010). Tests on scale effect of UHPFRC under bending and axial forces. In: Proceedings of the Third International fib Congress and Exhibition Incorporating the PCI Annual Convention and National Bridge Conference. Washington, DC, USA. Compact Disc, Paper 54.

Rong, Q., Hou, X., Ge, C. (2020). Quantifying curing and composition effects on compressive and tensile strength of 160–250 MPa RPC. *Constr. Build. Mater.* 241, 117987.

Roth, M.J., Eamon, C.D., Slawson, T.R., Tonyan, T.D., Dubey, A., (2010). Ultra-high-strength, glass fiber-reinforced concrete: Mechanical behavior and numerical modeling. *ACI Mater. J.* 107, 185–194.

Roy, M., Hollmann, C., Wille, K., (2017). Influence of volume fraction and orientation of fibers on the pullout behavior of reinforcement bar embedded in ultra high performance concrete. *Constr. Build. Mater.* 146, 582–593.

Russell, H. G., Graybeal, B. A. (2013). *Ultra-high performance concrete: A state-of-the-art report for the bridge community*. Federal Highway Administration. Office of Infrastructure Research and Development, USA.

Sarfarazi, V., Schubert, W. (2017). Numerical simulation of tensile failure of concrete in direct, flexural, double punch tensile and ring tests. *Period. Polytech-Civ*. 61(2), 176–183.

Scrivener, K. L., Crumbie, A. K., Laugesen, P. (2004). The interfacial transition zone (ITZ) between cement paste and aggregate in concrete. *Interface Sci*. 12, 411–421.

Shafieifar, M., Farzad, M., Azizinamini, A., (2017). Experimental and numerical study on mechanical properties of ultra high performance concrete (UHPC). *Constr. Build. Mater*. 156, 402–411.

Shah, S. P., Swartz, S. E., Ouyang, C. (1995). Fracture mechanics of concrete: applications of fracture mechanics to concrete, rock and other quasi-brittle materials. John Wiley & Sons.

Shen, P., Lu, L., He, Y., Wang, F., Hu, S. (2019). The effect of curing regimes on the mechanical properties, nano-mechanical properties and microstructure of ultra-high performance concrete. *Cem. Concr. Res*. 118, 1–13.

Shen, X., Brühwiler, E., (2020). Influence of local fiber distribution on tensile behavior of strain hardening UHPFRC using NDT and DIC. *Cem. Concr. Res*. 132, 106042.

Shi, C., Wu, Z., Xiao, J., Wang, D., Huang, Z., Fang, Z. (2015). A review on ultra high performance concrete: Part I. Raw materials and mixture design. *Constr. Build. Mater*. 101, 741–751.

Shi, X., Park, P., Rew, Y., Huang, K., Sim, C. (2020). Constitutive behaviors of steel fiber reinforced concrete under uniaxial compression and tension. *Constr. Build. Mater*. 233, 117316.

Singh, M., Sheikh, A.H., Mohamed Ali, M.S., Visintin, P., Griffith, M.C., (2017). Experimental and numerical study of the flexural behaviour of ultra-high performance fibre reinforced concrete beams. *Constr. Build. Mater*. 138, 12–25.

Sorelli, L., Constantinides, G., Ulm, F.-J., Toutlemonde, F. (2008). The nano-mechanical signature of Ultra High Performance Concrete by statistical nanoindentation techniques. *Cem. Concr. Res*. 38, 1447–1456.

Sritharan, S., Bristow, B., Perry, V. Characterizing an ultra-high performance material for bridge applications under extreme loads. Proc., 3rd International Symposium on High Performance Concrete, PCI, Orlando, Florida.

Tai, Y. S., El-Tawil, S., Chung, T. H. (2016). Performance of deformed steel fibers embedded in ultra-high performance concrete subjected to various pullout rates. *Cem. Concr. Res*.89, 1–13.

Teng, L., Meng, W., Khayat, K.H. (2020). Rheology control of ultra-high-performance concrete made with different fiber contents. *Cem. Concr. Res*. 138, 106222.

Toledo Filho, R.D., Koenders, E.A.B., Formagini, S., Fairbairn, E.M.R. (2012). Performance assessment of ultra high performance fiber reinforced cementitious composites in view of sustainability. *Mater. Des*. 36, 880–888.

Tran, N.T., Tran, T.K., Jeon, J.K., Park, J.K., Kim, D.J. (2016). Fracture energy of ultra-high-performance fiber-reinforced concrete at high strain rates. *Cem. Concr. Res.* 79, 169–184.

Tran, N.T., Tran, T.K., Kim, D.J. (2015). High rate response of ultra-high-performance fiber-reinforced concretes under direct tension. *Cem. Concr. Res.* 69, 72–87.

Velez, K., Maximilien, S., Damidot, D., Fantozzi, G., Sorrentino, F. (2001). Determination by nanoindentation of elastic modulus and hardness of pure constituents of Portland cement clinker. *Cem. Concr. Res.* 31(4), 555–561.

Vishalakshi, K. P., Revathi, V., Reddy, S. S. (2018). Effect of type of coarse aggregate on the strength properties and fracture energy of normal and high strength concrete. *Eng. Fract. Mech.* 194, 52–60.

Wang, D., Shi, C., Wu, Z., Xiao, J., Huang, Z., Fang, Z. (2015). A review on ultra high performance concrete: Part II. Hydration, microstructure and properties. *Constr. Build. Mater.* 96, 368–377.

Wang, L., Yang, H. Q., Zhou, S. H., Chen, E., Tang, S. W. (2018). Hydration, mechanical property and C-S-H structure of early-strength low-heat cement-based materials. *Mater. Lett.* 217, 151–154.

Wille, K., Kim, D. J., Naaman, A. E. (2011a). Strain-hardening UHP-FRC with low fiber contents. *Mater. Struct.* 44(3), 583–598.

Wille, K., Naman, A. E., Parra-Montesinos, G. J. (2011b). Ultra-high performance concrete with compressive strength exceeding 150 MPa (22ksi): A Simpler Way. *ACI Mater. J.* 108, 46–53.

Wille, K., Naaman, A. E. (2012a). Pullout behavior of high-strength steel fibers embedded in ultra-high-performance concrete. *ACI Mater. J.* 109(4), 479–487.

Wille, K., El-Tawil, S., Naaman, A.E., (2012b). Strain rate dependent tensile behavior of ultra-high performance fiber reinforced concrete. RILEM Bookseries 2, 381–387.

Wille, K., Naaman, A. E., El-Tawil, S., Parra-Montesinos, G. J. (2012c). Ultra-high performance concrete and fiber reinforced concrete: achieving strength and ductility without heat curing. *Mater. Struct.* 45(3), 309–324.

Wille, K., Naaman, A. E. (2013). Effect of ultra-high-performance concrete on pullout behavior of high-strength brass-coated straight steel fibers. *ACI Mat. J.* 110(4), 451.

Wille, K., El-Tawil, S., Naaman, A. E. (2014). Properties of strain hardening ultra high performance fiber reinforced concrete (UHP-FRC) under direct tensile loading. *Cem. Concr. Compos.* 48, 53–66.

Wu, F., Xu, L., Chi, Y., Zeng, Y., Deng, F., Chen, Q. (2020). Compressive and flexural properties of ultra-high performance fiber-reinforced cementitious composite: The effect of coarse aggregate. *Compos. Struct.* 236, 111810.

Wu, Z., Shi, C., He, W., Wu, L. (2016a). Effects of steel fiber content and shape on mechanical properties of ultra high performance concrete. *Constr. Build. Mater.* 103, 8–14.

Wu, Z., Shi, C., Khayat, K., H. (2016b). Influence of silica fume content on microstructure development and bond to steel fiber in ultra-high strength cement-based materials (UHSC). *Cem. Concr. Compos.* 71, 97–109.

Wu, Z., Shi, C., He, W., Wang, D. (2016c). Uniaxial compression behavior of ultra-high performance concrete with hybrid steel fiber. *J. Mat. Civil Eng.* 28(12), 06016017.

Wu, Z., Khayat, K. H., Shi, C. (2017a). Effect of nano-SiO_2 particles and curing time on development of fiber-matrix bond properties and microstructure of ultra-high strength concrete. *Cem. Concr. Res.* 95, 247–256.

Wu, Z., Shi, C., He, W. (2017b). Comparative study on flexural properties of ultra-high performance concrete with supplementary cementitious materials under different curing regimes. *Constr. Build. Mater.* 136, 307–313.

Wu, Z., Shi, C., He, W., Wang, D. (2017c). Static and dynamic compressive properties of ultra-high performance concrete (UHPC) with hybrid steel fiber reinforcements. *Cem. Concr. Compos.* 79, 148–157.

Wu, Z., Khayat, K. H., Shi, C. (2018a). How do fiber shape and matrix composition affect fiber pullout behavior and flexural properties of UHPC? *Cem. Concr. Compos.* 90, 193–201.

Wu, Z., Shi, C., Khayat, K. H. (2018b). Multi-scale investigation of microstructure, fiber pullout behavior, and mechanical properties of ultra-high performance concrete with nano-$CaCO_3$ particles. *Cem. Concr. Compos.* 86, 255–265.

Wu, Z., Shi, C., Khayat, K. H., Xie, L. (2018c). Effect of SCM and nano-particles on static and dynamic mechanical properties of UHPC. *Constr. Build. Mater.* 182, 118–125.

Wu, Z., Khayat, K. H., Shi, C. (2019a). Changes in rheology and mechanical properties of ultra-high performance concrete with silica fume content. *Cem. Concr. Res.* 123, 105786.

Wu, Z., Shi, C., Khayat, K. H. (2019b). Investigation of mechanical properties and shrinkage of ultra-high performance concrete: Influence of steel fiber content and shape. *Compos. Part-B Eng.* 174, 107021.

Xu, M., Wille, K., (2015). Fracture energy of UHP-FRC under direct tensile loading applied at low strain rates. *Compos. Part-B Eng.* 80, 116–125.

Xu, J., Wang, X. (2018). Effect of nano-silica modification on interfacial transition zone in concrete and its multiscale modelling. *J Chin. Ceram. Soci.* 46(8), 1053–1058.

Xu, L., Li, B., Chi, Y., Li, C., Huang, B., Shi, Y. (2018). Stress-strain relation of steel-polypropylene-blended fiber-reinforced concrete under uniaxial cyclic compression. *Adv. Mater. Sci. Eng.* 2018, 1–19.

Xu, L., Wu, F., Chi, Y., Cheng, P., Zeng, Y., Chen, Q. (2019). Effects of coarse aggregate and steel fibre contents on mechanical properties of high performance concrete. *Constr. Build. Mater.* 26(10), 97–110.

Xu, Y., Liu, J., Liu, J., Zhang, P., Zhang, Q., Jiang, L. (2018). Experimental studies and modeling of creep of UHPC. *Constr. Build. Mater.* 175, 643–652.

Yang, S.L., Millard, S.G., Soutsos, M.N., Barnett, S.J., Le, T.T. (2009). Influence of aggregate and curing regime on the mechanical properties of ultra-high performance fibre reinforced concrete (UHPFRC). *Constr. Build. Mater.* 23(6): 2291–2298.

Yang, J., Su, J., Chen, B., Luo, X., Shen, X., (2017). The optimized design of dogbones for tensile test of ultra-high performance concrete, in: AFGC-ACI-Fib-RILEM

Int. Symposium on Ultra-High Performance Fibre-Reinforced Concrete. Montpellier, France, 175–182.

Yazici, H., Yardimci, M. Y., Aydin, S., Karabulut, A. S. (2009). Mechanical properties of reactive powder concrete containing mineral admixtures under different curing regimes. *Constr. Build. Mater.* 23(3), 1223–1231.

Yi, S. T., Kim, J. K., Oh, T. K. (2003). Effect of strength and age on the stress-strain curves of concrete specimens. *Cem. Concr. Res.* 33(8), 1235–1244.

Yoo, D.-Y., Lee, J.-H., Yoon, Y.-S. (2013). Effect of fiber content on mechanical and fracture properties of ultra high performance fiber reinforced cementitious composites. *Compos. Struct.* 106, 742–753.

Yoo, D. Y, Banthia, N, Kang, S. T., Yoon, Y. S.(2016). Size effect in ultra-high-performance concrete beams. *Eng. Fract. Mech.*, 2016, 157 : 86–106.

Yoo, D.Y., Kim, M. J., Kim, S.W., Park, J.J. (2017a). Development of cost effective ultra-high-performance fiber-reinforced concrete using single and hybrid steel fibers. *Constr. Build. Mater.* 150, 383–394.

Yoo, D.Y., Kim, S.W., Park, J.J. (2017b). Comparative flexural behavior of ultra-high-performance concrete reinforced with hybrid straight steel fibers. *Constr. Build. Mater.*132, 219–229.

Yoo, D.Y., Kim, S., Park, G.J., Park, J.J., Kim, S.-W. (2017c). Effects of fiber shape, aspect ratio, and volume fraction on flexural behavior of ultra-high-performance fiber-reinforced cement composites. *Compos. Struct.* 174, 375–388.

Yoo, D. Y., Kim, S., Kim, J. J., Chun, B. (2019). An experimental study on pullout and tensile behavior of ultra-high-performance concrete reinforced with various steel fibers. *Constr. Build. Mater.* 206, 46–61.

Yoo, D. Y., Gim, J. Y., Chun, B. (2020). Effects of rust layer and corrosion degree on the pullout behavior of steel fibers from ultra-high-performance concrete. *J Mater. Res. Tech.* 9(3), 3632–3648.

Yu, K., Wang, Y., Yu, J., Xu, S., (2017). A strain-hardening cementitious composites with the tensile capacity up to 8%. *Constr. Build. Mater.* 137, 410–419.

Yu, K., Ding, Y., Zhang, Y.X., (2020). Size effects on tensile properties and compressive strength of engineered cementitious composites. *Cem. Concr. Compos.* 113, 103691.

Zhang, H., Ji, T., Lin, X. (2019). Pullout behavior of steel fibers with different shapes from ultra-high performance concrete (UHPC) prepared with granite powder under different curing conditions. *Constr. Build. Mater.* 211, 688–702.

Zhang, L., Liu, J., Liu, J., Zhang, Q., Han, F. (2018). Effect of steel fiber on flexural toughness and fracture mechanics behavior of ultrahigh-performance concrete with coarse aggregate. *J. Mater. Civil Eng.* 30(12), 04018323.

Zhang, Z., Xu-dong, S., Li, W., Zhu, P., Hong, C., (2015). Axial tensile behavior test of ultra high performance concrete. *China J. Highw. Transp.* 28, 50–58(in Chinese).

Zhang, Z., Liu, S., Yang, F., Weng, Y., Qian, S. (2021). Sustainable high strength, high ductility engineered cementitious composites (ECC) with substitution of cement by rice husk ash. *J. Clean. Prod.* 317, 128379.

Zhao, Q., Sun, W., Zheng, K., Jiang, G. (2005). Comparison for elastic modulus of cement, ground granulated blast-furnace slag and fly ash particles. *J Chine. Ceram. Soci.* 7, 837–841.

Zhao, S., Sun, W. (2014). Nano-mechanical behavior of a green ultra-high performance concrete. *Constr. Build. Mater.* 63, 150–160.

Zhao, X., Li, Q., Xu, S., (2020). Contribution of steel fiber on the dynamic tensile properties of hybrid fiber ultra high toughness cementitious composites using Brazilian test. *Constr. Build. Mater.* 246, 118416.

Zheng, W., Li, H., Wang, Y. (2012). Compressive stress-strain relationship of steel fiber-reinforced reactive powder concrete after exposure to elevated temperatures. *Constr. Build. Mater.* 35, 931–940.

Zhou, G. (2005). Preparation of the diabase gravel and performance of concrete. *J Wuhan Univ. Technol. Mater.* 27(11), 85–87.

Zhou, Z., Qiao, P., 2018. Direct tension test for characterization of tensile behavior of ultra-high performance concrete. *J. Test. Eval.* 48, 2730–2749.

Chapter 9

Dynamic mechanical properties

9.1 INTRODUCTION

The dynamic mechanical properties of concrete materials refer to the mechanical strength and deformation properties under dynamic loads, such as vibration, impact, and explosion. The research on the dynamic performance of concrete can be traced back to the 1920s. In 1917, Abrams (1917) carried out a compression test on ordinary concrete with strain rates of 8×10^{-6} and 2×10^{-4} s^{-1}, respectively. It was found that the compressive strength of concrete has strain rate sensitivity. In general, the dynamic mechanical properties of UHPC can be characterized by intrinsic frequency, dynamic elastic (shear) modulus, damping coefficient, etc. Practice shows that UHPC is a kind of strain rate–sensitive material. Its mechanical properties and failure modes under dynamic loads are different from those under static loads. Therefore, studying the dynamic response of UHPC under various dynamic loading conditions has important theoretical and engineering significance for the evaluation of the dynamic strength and stability of civil and military structures.

UHPC is a material with high mechanical properties and has broad application prospects in earthquake engineering, ocean engineering, military engineering, protective engineering, etc. UHPC structures may be subjected to loads with different strain rates in their service period. Figure 9.1 shows typical strain rates for various types of loading. Generally, the creep strain rate is lower than 10^{-6} s^{-1}, the strain rate under static loading is about 10^{-6} to 10^{-5} s^{-1}, and the strain rate of structural response under seismic loading is about 10^{-3} to 10^{-2} s^{-1}. Besides, the strain rate under impact loading is about 10^{0} to 10^{1} s^{-1}, and the strain rate under explosion loading is more than 10^{2} s^{-1}.

The dynamic mechanical properties of UHPC, including compressive, tensile, bending, and ballistic properties, can be evaluated using experiments and numerical simulation. The test methods mainly include hydraulic impact, drop-hammer, Split Hopkinson Press Bar (SHPB), bullet

DOI: 10.1201/9781003203605-9

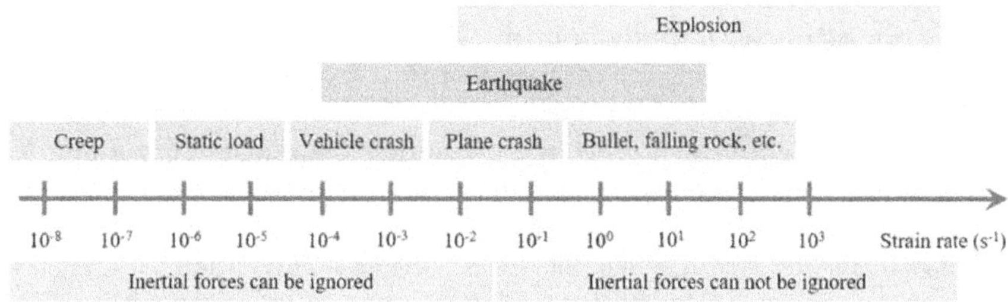

Figure 9.1 Typical strain rates for various types of loading.

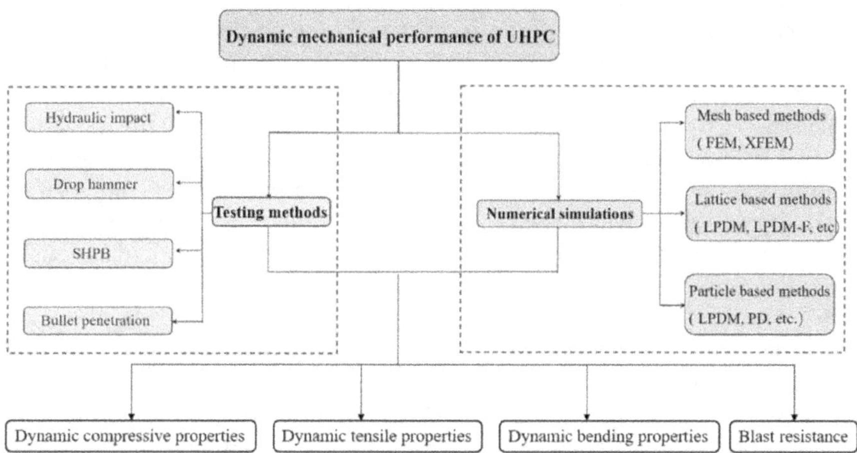

Figure 9.2 Research methods for dynamic mechanical properties of UHPC.

penetration, ballistic, etc. According to different algorithm principles, the numerical simulation method can be divided into mesh-based, lattice-based, and particle-based methods, as shown in Figure 9.2.

This chapter describes the recent work on experiments and numerical simulations of dynamic mechanical properties of UHPC. The testing methods and their measuring principles are summarized. The effects of material composition, strain rate, impact time, and extreme conditions on dynamic mechanical properties of UHPC are discussed. The reasons for the strain rate sensitivity behavior of UHPC are described. Finally, some suggestions for future works on dynamic mechanical properties of UHPC are put forward.

9.2 TEST METHODS

As mentioned in the introduction, the dynamic mechanical properties of UHPC can be tested using a hydraulic method, drop weight method, SHPB method, projectile penetration method, etc. The ranges of the strain rates that apply to the abovementioned test methods are summarized in Table 9.1. It is generally considered that the strain rate sensitivity is almost independent of the strain when the strain rate is less than 10^{-5} s^{-1}. It decreases with the increase in plastic strain when the strain rate is greater than 10^{-2} s^{-1}. The strain rate ranging between 10^{-5} and 10^{-3} s^{-1} is called quasistatic or low strain rate (Zukas et al. 1982). To reduce the experimental error, the contact surfaces of tested specimens should be polished before testing (Wilhelm et al. 2016).

Hydraulic impact machines are widely used in the field of crushing, rock drilling, pile driving, etc. Figure 9.3(a) is a schematic diagram of typical, most commonly used hydraulic impact machine. It consists of oil supply pressure, hydrogen chamber, piston, hydraulic cylinder, etc. Hydraulic cylinder power comes from pressurized hydraulic oil. Its operational principle is using hydraulic pressure as the source of power to push the movement of the piston under the action of hydrogen and oil pressure until the end of the impact and into the next cycle. The whole impact process relies on the reversing valve to switch the oil circuit and regulate the movement of the piston. In general, the force on the piston rod can be calculated using the following equation:

$$F = P \times A \tag{9.1}$$

where F is the force on the piston rod (N); P is the pressure in the cylinder(Pa); and A is the piston area (m^2). However, a significant limitation of the test is that only a single loading mode is possible if the gas-liquid hydraulic impact machine is used.

A hydraulic loading system was proposed to obtain static-dynamic composite loading and includes an impact oil cylinder, an accumulator system, an impact cylinder control system, an oil source system of static load, an oil source system of dynamic load, a super-large flow quick-opening valve, and an oil source system of middle beam, as demonstrated in Figure 9.3(b). During the test, the impact cylinder is driven by high pressure and a large-flow liquid from the hydraulic system. The fluid movement is transferred to the piston rod, which exerts an impact force on the sample.

In the basic experimental research on dynamic properties of UHPC, drop weight test and SHPB test are commonly used. The schematic diagram of the drop weight impact device is shown in Figure 9.4. An impact test is completed by dropping the hammer with a weight of m from the height h to the specimen, and then recording its energy consumption and strength when

Table 9.1 Different testing methods for dynamic mechanical properties of UHPC and their applicable strain rates

Testing device	*Hydraulic*	*Drop hammer*	*SHPB*	*Bullet*
Applicable strain rate (s^{-1})	10^{-5}–1	1–10	1–10^4	10^3–10^5
Device characteristics	Electrohydraulic servo system Automatic control	Hammer falls freely after lifting to a certain height	Pressure bar with transient pulse stress	Projectile ejection
Loading method	Multiaxial tension and compression loading	Uniaxial or multiaxial impact compression	Split tension Uniaxial impact compression Impact compression with side pressure	Impact compression
Advantages	Strain rate can be controlled Multiaxis loading	Low cost and simple test principle	Easy to collect data, load waveform adjustment	Large strain rate range
Disadvantages	High experiment cost Technically difficult	Load value is not easy to measure, and energy calculation is often used	Dispersion effect High requirements for sample uniformity	High requirements for data testing

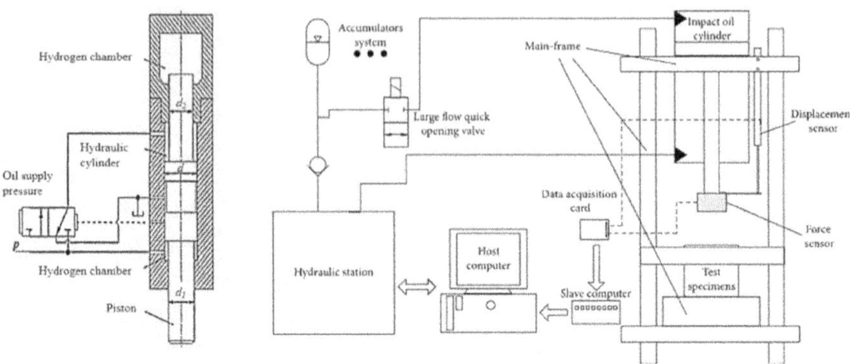

Figure 9.3 Schematic of hydraulic impact test machine. (a) Gas-liquid hydraulic impact machine (Chen et al. 2011). (b) Hydraulic loading system (Wang et al. 2019).

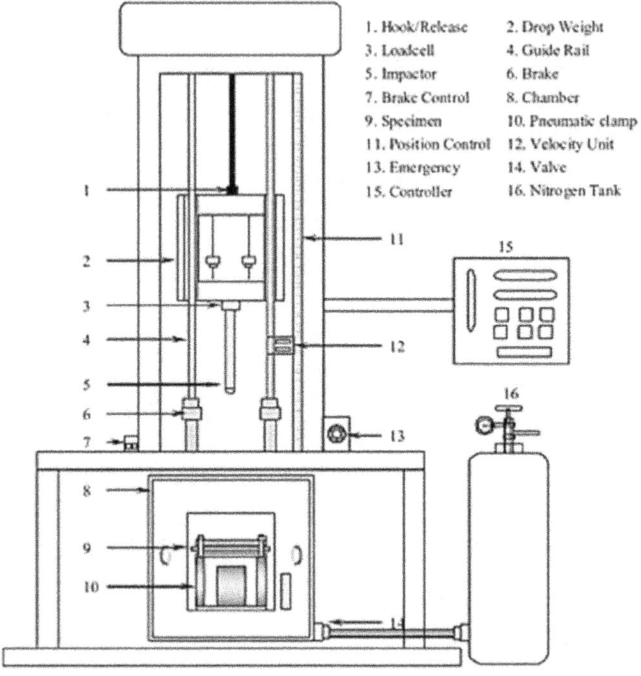

Figure 9.4 Schematic diagram of drop weight impact machine.

Source: Lee et al. (2013).

concrete is damaged. The results of the drop weight test depend on many factors, including mass and velocity of hammer, specimen size, stiffness of bearing, and definition of damage of concrete. The data obtained are generally used for qualitative comparison. In 1949, the SHPB machine came out and was improved by Davis et al. (1963) and Kolsky (1949) in the middle of the 20th century, separating the original Hopkinson pressure rod into an incident bar and a transmission bar. In 1963, Lindholm pasted strain gauges on the incident bar and the projection bar, successfully replacing them, which greatly improved the applicability of SHPB to the dynamic mechanical properties testing of concrete. SHPB test device is mainly composed of a striker, an incident, a transmitter, an absorbing bar, a test and data acquisition system, a data processing system, etc., as illustrated in Figure 9.5. It can record the curves of stress-strain, stress-time, and strain rate–time conveniently. The accuracy of the dynamic stress-strain curves of UHPC measured by SHPB is mainly affected by Poisson's effect, radial inertia effects, end effects, and nonlinear friction.

Projectile penetration is generally used for bullet impact tests of protective structures with special requirements. The schematic diagram of projectile penetration is indicated in Figure 9.6. The impact resistance performance of UHPC is obtained using different types of bullets, changing the striking distances and the velocities of bullets. After the penetration test, the projectile velocity, crater diameter, penetration depth, crack number, length, and width are important indexes to evaluate the antipenetration performance of UHPC.

Antiknock performance is an important index in the application of UHPC in military engineering. The current antiexplosion test mainly puts TNT explosives at a certain distance from UHPC, as shown in Figure 9.7. There are two methods to test the antiexplosive performance of UHPC. One is to visually evaluate the damage of UHPC board after the test and then measure its weight loss. The second method is to arrange measuring points on the UHPC board to measure the propagation time of ultrasonic waves before and after the explosion. In addition, the laser displacement sensor is often used to detect the deformation of UHPC board, as indicated in Figure 9.7.

To understand whether different loading methods affect the experimental results, Cotsovos (2008) summarized the dynamic mechanical properties of concrete measured by different instruments, as shown in Figure 9.8. Results found that the use of different loading techniques is an important factor leading to the dispersion of experimental data. However, the discreteness of the experimental data does not draw a clear conclusion on the effect of the type of loading technique on the behavior of specimens. Different from the experimental data shown in Figures 9.8(a and b), the results presented in Figure 9.8(c) show that the increase of strength commenced suddenly at a strain rate of around 10 s^{-1} and beyond. For strain rates greater than 10 s^{-1}, the specimen strength increases rapidly with the increase in loading rate.

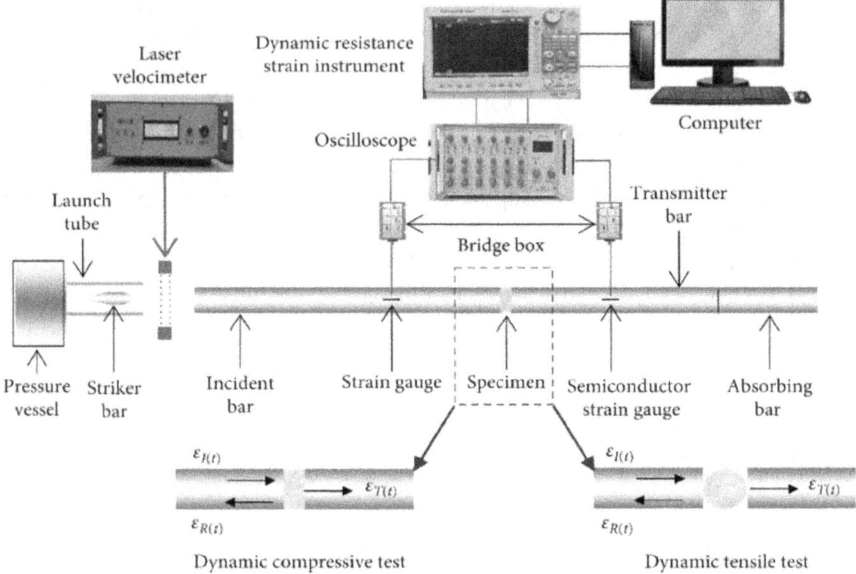

Figure 9.5 Schematic diagram of SHPB.

Source: Ren et al. (2011).

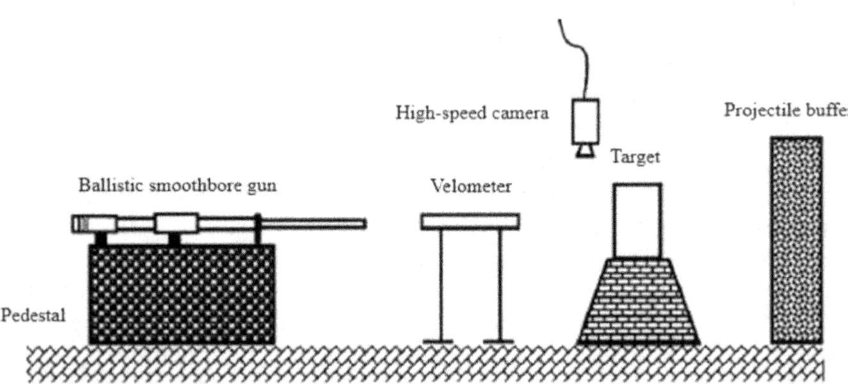

Figure 9.6 Schematic diagram of projectile penetration.

Source: Zhang et al. (2019).

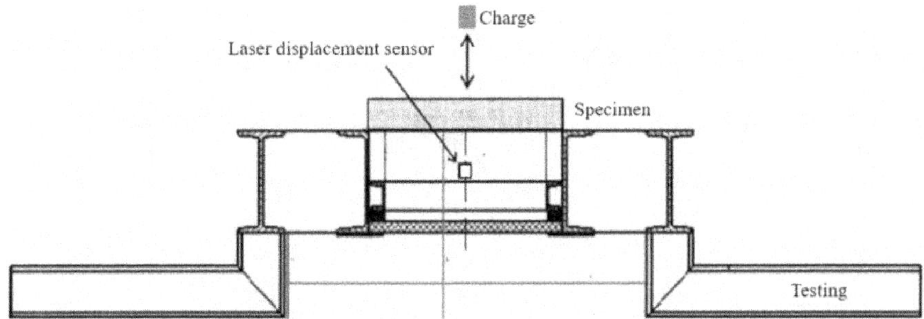

Figure 9.7 Schematic diagram of antiexplosion test.

Source: Bibora et al. (2017).

In addition, the discreteness of SHPB experimental data is significantly less than that of the other test methods. The abovementioned test devices are summarized in Table 9.1.

9.3 NUMERICAL SIMULATIONS METHODS

Numerical simulation is an effective way to investigate the dynamic mechanical properties of UHPC further. It can be divided into mesh-based, lattice-based, and particle-based methods. Mesh-based methods include finite element method (FEM) (Zienkiewiczet al. 1977) and extended finite element method (XFEM) (Moës & Belytschko, 2002). Typical lattice-based methods are lattice discrete particle model (LDPM) (Cusatis et al. 2011a, 2011b), lattice discrete particle model-fiber (LDPM-F), etc. Particle-based methods include the smoothing particle method (SPH) (Liu et al. 2003), the element-free Galerkin method (EFG) (Dolbow and Belytschko 1998), and the materials point method (MPM) (Bardenhagen 2000). The mesh-based method is the most widely used method to study dynamic mechanical properties, while lattice models, in which spring networks are needed without meshes remeshing, lead to a high computational efficiency when investigating UHPC subjected to low-velocity impact force. The strength, damage, strain rate, and state equation should be considered in the constitutive model of UHPC. The main features of those methods are summarized in Table 9.2.

The implicit finite element method is generally used to simulate the static mechanical behaviors of UHPC, while the explicit finite element method usually is adopted to simulate the UHPC under dynamic loads, as shown in Figure 9.9. For quasistatic problems, both implicit and explicit finite element methods can be used to obtain effective results.

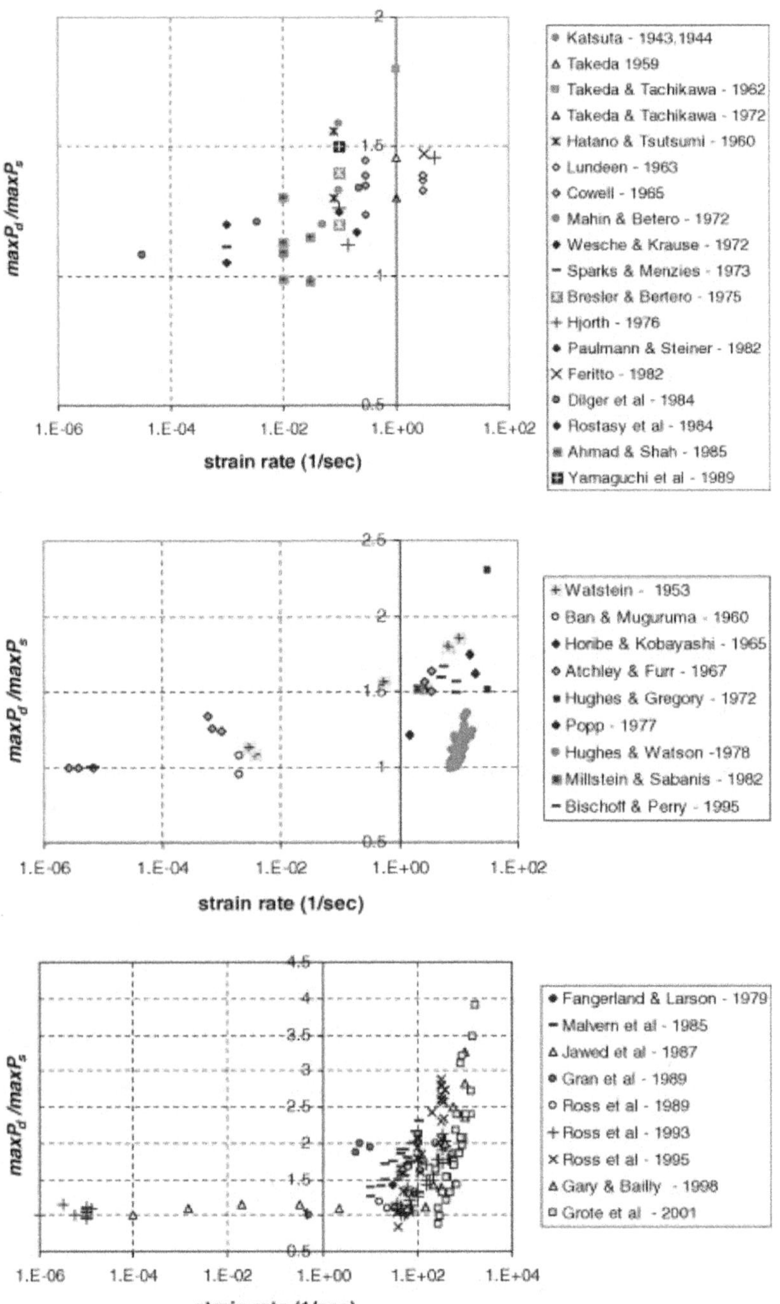

Figure 9.8 Experimental data obtained from different test devices (Cotsovos et al. 2018). (a) Experimental data obtained from hydraulic loading. (b) Experimental data obtained from drop-hammer experiments. (c) Experimental data obtained from SHPB experiments.

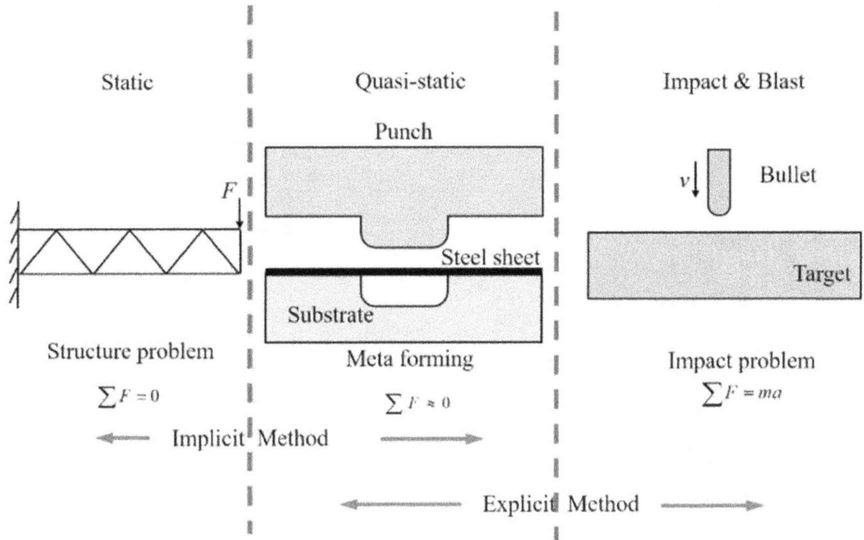

Figure 9.9 Scope of application of implicit and explicit finite element methods.

Commonly used software for studying the dynamic mechanical properties of UHPC are Abaqus/Explicit, ANSA, Autodyn, LS-DYNA, etc. LS-DYNA is the most widely used software among them due to its rich constitutive model of concrete. Now Tylor-Chen-Kuszmaul (TCK) model (Wang et al. 2007), RHT concrete model (Borrvall and Riedel 2011), Holmquist Johnson Cook (HJC) model (Ren et al. 2017), Karagozian and Case (K&C) model (Markovich et al. 2011), and Continuous Surface Cap Model (CSCM) (Schwer and Murray 2002) have been developed. K&C and CSCM models can be applied to analyze the dynamic mechanical properties of UHPC. A previous study found that the CSCM model is more suitable for analyzing UHPC behaviors at low-velocity impact simulation than the KCC model (Bo et al. 2011). This is attributed to the difficulty to exhibit the tensile strain hardening of UHPC by the KCC model.

9.4 DYNAMIC PROPERTIES OF UHPC

9.4.1 Dynamic compressive properties of UHPC

The evaluation index of dynamic compressive properties of UHPC mainly includes impact times, fracture energy, shape and size of damage surface of UHPC, whether there is perforation or not, and the number of fragments when the specified degree of damage of UHPC is reached (Li and Wu 2018). By comparing with quasistatic mechanical properties of UHPC, the degree of

Table 9.2 Main features of different numerical methods for dynamic properties of UHPC

Method	*Advantages*	*Disadvantages*	*Models/ algorithms*	*Main principle(s)*	*Applications*	*Ref.*
FEM	1. Widely used methods; there are many models for researchers to choose from and many examples for new researchers to learn 2. UHPC models can be imported from other software, such as CAD. 3. Implemented to model multiple field problems of materials 4. Easily implemented with high-order element	1. Complex meshing techniques are required for mesoscale model and multiscale model 2. Improper meshes result in a poor computation accuracy 3. The computational cost is high when considering fibers and interfacial transition zone in UHPC models	CSCM	Revising the strain rate parameters in tension and compression by data fitting	Capture the tensile strain hardening behavior of UHPC	Guo et al. (2018)
			Modified Kent-Park model	Considering a four-parameter function (including the peak stress and the strain, resident strength to peak stress, and the beginning strain of resident strength)	Evaluate the compressive strength of high-strength steel-confined UHPC	Malvar et al. (2000)

KCC model HJC model RHT model	KCC model: considering three independent strength surfaces – yield strength surface, the maximum strength surface, and the residual strength surface HJC model: mainly includes the yield surface equation, the damage evolution equation, and the state equation RHT model: considering three ultimate surfaces – failure surface, elastic limit surface, and residual strength surface	Widely used to model the dynamic behavior of UHPC.	Malvar et al. (2000); Du et al. (2020); Sovjak et al. (2015)
Cohesive zone model for dynamic	Zero-thickness cohesive elements and a strain-rate sensitive traction-separation constitutive model	Applied to model the dynamic behavior of UHPC and the debonding behavior of fibers under impact load	Ellis et al. (2014)

(Continued)

Table 9.2 (Continued)

Method	*Advantages*	*Disadvantages*	*Models/ algorithms*	*Main principle(s)*	*Applications*	*Ref.*
XFEM	1. Based on LSM and PUFEM and independent of physical meshes 2. Developed for discontinued problems, such as interface, crack, or inclusion 3. For UHPC, it captures the interface between fiber and cement matrix at mesoscale	1. Difficult to build enrichment function for multifield (such as chemo-diffusion-mechanic coupled discontinues problem) 2. Extra degree of freedom needs to be added at discontinues, resulting in an increased computational time	LSM and PUFEM	LSM: capture the discontinues; PUM: construction of a global approximation space and the evaluation of discretization	Implemented to model the arbitrary crack of concrete	Eftek-hari et al. (2014)
			LSM and Debonding enrichment function	Capture the discontinuity in the displacement field between fiber and cement matrix with LSM	Implemented to study the failure behaviors of UHPC	Roth et al. (2015)
RSBN	Low computational cost	The behavior of fiber limits to elastic deformation	Implicit time integration scheme	Voronoi polygonal mesh	Mechanical behaviors of UHPC	Bolander et al. (1997)
LDPM	1. Take advantage of CSL and DPM (including DEM) and better simulate the mechanical response of concrete under cyclic load 2. Volume and compaction effect of UHPC can be better considered 3. Improved stress resolution at mesoscale	High computational cost	DPM and CSL model	DPM: define the particle-particle interactions in periodic boundary domains CSL: determine the lattice connections with Delaunay triangulation	Modeling the pervasive failure and fragmentation of UHPC	Cusatis et al. (2011a)

	4. Enhanced representation of mesoscale crack and damage distribution					
LDPM-F	Consider the interactions between fibers and particles through embedded model without extra degree of freedom				Modeling the failure of UHPC	Smith et al. (2014)
DEM	1. Based on Newton's second law, it is not necessary to satisfy the displacement continuity and deformation compatibility equations 2. Mesh-free, can be used to solve large displacement and nonlinear problems 3. Intuitively reflect the damage evolution of concrete 4. Superiority in computational complexity, computational speed, and storage space	Lack of theoretical basis, there are assumptions in particle motion, displacement, and deformation, which greatly affect the calculation accuracy	Projected algorithm/spring-beam particle model	Projected fiber length on the disordered two-dimensional system Fibers were modeled using the elastic spring-beam particle model	Modeling the damage evolution of concrete	Wang et al. (2007)

(Continued)

Table 9.2 (Continued)

Method	*Advantages*	*Disadvantages*	*Models/ algorithms*	*Main principle(s)*	*Applications*	*Ref.*
PD	1. Based on integrate function, not PDE 2. Take advantage of MD and mesh-free method 3. High accuracy and efficiency in the analysis of discontinuities at macro-/microscale 4. No singular problems for discontinues such as failure	1. The calculation efficiency of PD is lower than that of FEM when simulating undamaged elastic problems 2. Convergence and error issue	BBPD	Reformulation of the terms of integrodifferential equations; the pairwise force is linearly dependent on the bond stretch	Applied to simulate the damage and failure behavior of concrete without considering singular field	Mi et al. (2017)
MD	1. Can represent the nanostructure of material 2. Mesh-free, can be used to deal with nonequilibrium problems	High computational cost	Inter-atomic force laws. Based on Newtonian mechanics and statistics and thermodynamics	Coupled with other numerical methods (FEM and XFEM) to provide a way to study the properties of materials on multiscale	Applied to acquire the properties of materials at nanoscale	Abraham et al. (1997); Celis (1982)

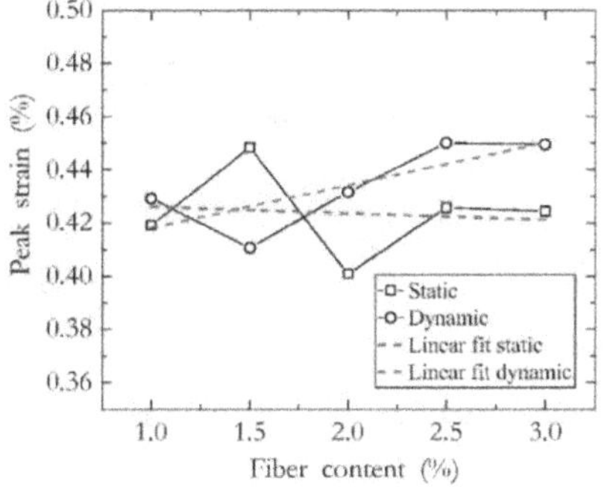

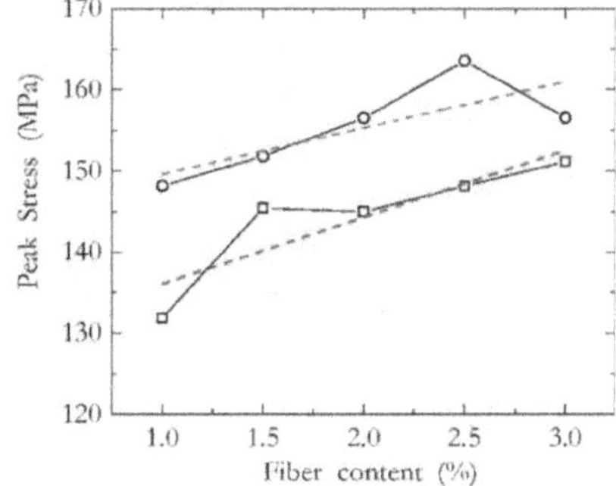

Figure 9.10 Peak strains and peak stresses of UHPC with different fiber contents.

Source: Krahl et al. (2018).

improvement of the impact resistance of UHPC can be obtained. It is generally believed that the impact resistance of UHPC is much higher than that of normal concrete (Máca et al. 2014). This is because of the fact that steel fiber in UHPC is mainly to inhibit the occurrence of internal microcracks and prevent the development of macrocracks. On the one hand, the crack resistance of steel fibers improves the impact resistance of UHPC. On the other hand, tensile failure may occur in UHPC from the point of view of the stress wave. The content and type of fiber have significant effects on the dynamic compressive properties of UHPC. Previous research found that with the increase in the content of steel fiber, the peak stress, peak strain, and dynamic compressive strength of UHPC are increased, as shown in Figure 9.10 (Krahl et al. 2018). In addition, the dynamic compressive properties of UHPC can be successfully analyzed using lattice discrete particle model-fiber (LDPM-F) (Smith et al. 2014). Results show that the dynamic compression inertia effect and crack propagation of UHPC are changed under different strain rates. The dynamic compressive strength of UHPC increases by 1.24 times when the strain rate increases from 0.001 to 10 s^{-1}.

UHPC is a strain rate-sensitive material (Tran et al. 2016). When the strain rate is less than the sensitive threshold, the dynamic compressive strength is lower than the static compressive strength of UHPC. When the strain rate is increased, the dynamic compressive strength increases faster, and the strain rate hardening effect of the material is obvious (Máca et al. 2014). When the strain rate reaches the sensitive threshold, the dynamic compressive strength reaches or slightly exceeds the static compressive strength of UHPC. The impact damage of UHPC with and without fiber is shown in Figure 9.11. It can be found that when the strain rate is less than 100 s^{-1}, obvious blocky fragments can be observed. However, when

Figure 9.11 Impact damge of UHPC with and without steel fiber (a) UHPC matrix without fiber; (b) UHPC with 4% steel fibers.

Source: Máca et al. (2014).

the strain rate is greater than 100 s^{-1}, the specimen will be destroyed into powder. Nevertheless, the powder fragments are still connecting with each other after impact for the mixture with 4% steel fibers, reflecting the strong bridging effect of steel fiber.

The addition of hybrid fibers can improve the mechanical performance of UHPC. Wu et al. (2016) studied the dynamic compressive properties of UHPC with a combination of hybrid long and short steel fibers by SHPB. The variation of dynamic increase factor (DIF) with strain rate was obtained by combining it with the quasistatic compressive strength of UHPC. Results showed that the dynamic compressive strengths of UHPC with 1.5% long and 0.5% short fibers were the highest when the impact velocities were 8.9, 11.7, and 13.9 m s^{-1}, which were 59.1%, 43.5%, and 39.5% higher than those of UHPC without any fiber, respectively, as illustrated in Figure 9.12. With the increase in the volume fraction of short fibers, the dynamic compressive properties of UHPC decrease, and the number of fracture fragments increases (Wu et al. 2016). This is because the long steel fibers are hard to pull out and can better transmit and withstand impact loads than short steel fibers. In addition, the distribution angle of fiber has a great influence on the mechanical properties of UHPC. The deformation ability of UHPC decreases first and then increases when the distribution angle increases from 0° to 90°. Results indicated that the dynamic increase factor (DIF) increases, but the enhancing effect of dynamic compressive strength of UHPC decreases with the increase of strain rate. This is because more microcracks will be generated in UHPC under higher strain rates. When the strain rate exceeds a certain range, the bridging ability of the microcracks is weakened due to the limited number of steel fibers. Different types of fibers play different roles in improving the performance of UHPC. For example, the compressive strength of UHPC with ultra-high-molecular-weight polyethylene (UHMWPE) fibers is increased by 23% and the ultimate strain is

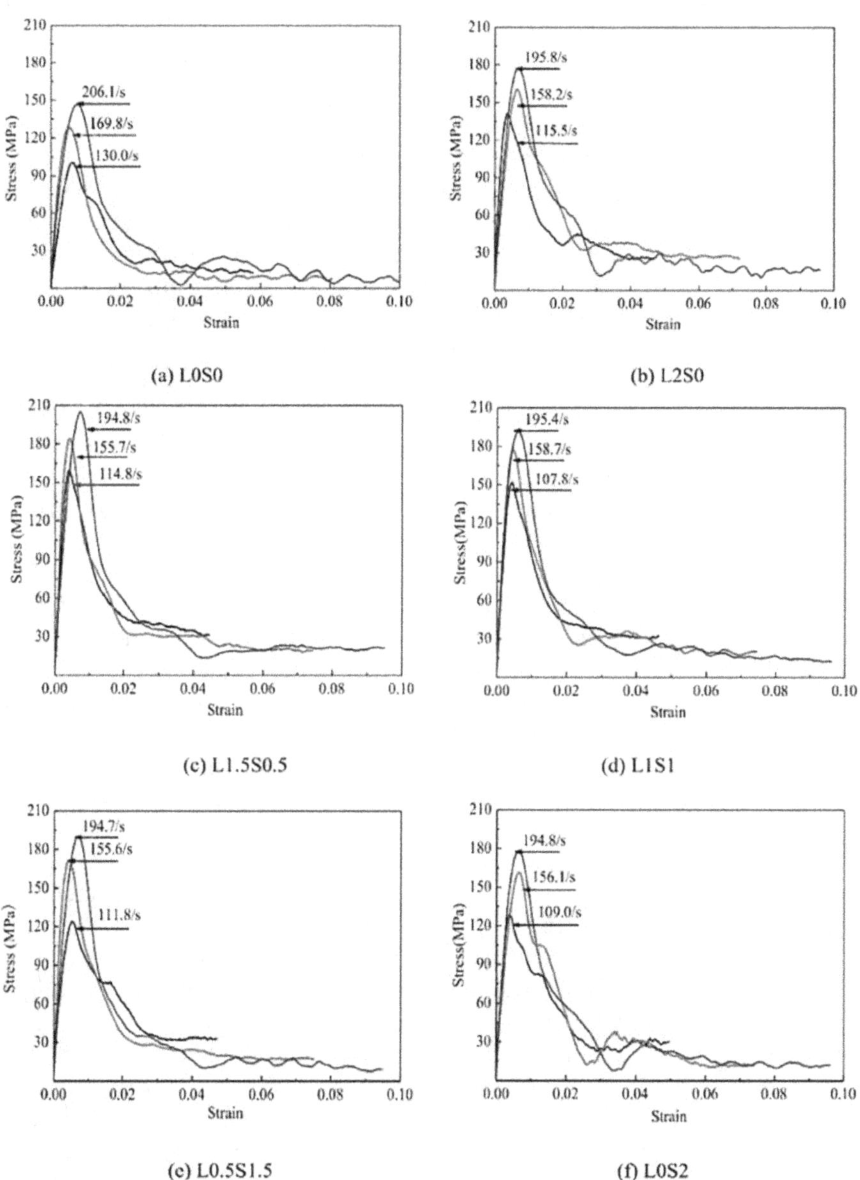

Figure 9.12 Dynamic compressive stress-strain curves of UHPC with hybrid fibers.

Source: Wu et al. (2016).

increased by at least 17.5% compared to that with polyvinyl alcohol (PVA) fibers.

9.4.2 Dynamic tensile properties of UHPC

The properties evaluation indexes of dynamic tensile properties mainly include tensile strength, elastic modulus, and energy absorption. With the increase in strain rate, the failure stress, failure strain, and elasticity modulus of UHPC increase, which can be featured by DIF. The dynamic tension-compression ratio is significantly higher than the static tension-compression ratio. Zhang et al. (2019) studied the dynamic tensile strength of UHPC at different impact velocities in the range of 4–14 m s^{-1} and compared it with the quasistatic tensile strength of UHPC. Results show that with the increase of impact velocity, the dynamic tensile strength of UHPC increases significantly, performing a high strain sensitivity. The minimum dynamic tensile strength is less than the quasistatic tensile strength of UHPC with high fiber content, and the dynamic tensile strength of UHPC is greater than that of UHPC matrix.

The content and type of fibers play an important role in the dynamic tensile properties of UHPC. Wu et al. (2018) studied the dynamic tensile behaviors of UHPC with 0%, 1.0%, and 2.0% hybrid short and long steel fibers. Results showed that the dynamic tensile strength of UHPC increases with the increase of strain rates. The highest DIF of 14.89 is obtained for the UHPC with 2.0% hooked steel fibers at the strain rate of 110.7 s^{-1}, as shown in Figure 9.13. The effects of single addition of steel cotton fiber, copper-plated steel fiber, hooked steel fiber, and polypropylene fiber on the spalling strengths of UHPC using variable cross-section SHPB were also investigated (Zhang 2008). It is shown that copper-plated steel fiber and hooked steel fiber both have obvious effects, while polypropylene fibers and steel cotton fibers have little effect on the spalling strengths of UHPC.

Wille et al. (2016) studied the dynamic uniaxial tensile properties of UHPC at different strain rates (0.0001–0.1 s^{-1}). Different types (straight, hooked, and twisted fibers) and different contents (1.5%, 2%, and 3%) of fibers were considered. The DIF of UHPC with hooked fiber is slightly higher than that of UHPC with straight or twisted fibers. When the strain rate is 0.1 s^{-1}, the DIFs of UHPC with hooked, straight, and twisted fibers are 1.25, 1.20, and 13, respectively. This is mainly because hooked fibers can better improve the fiber-matrix bond behaviors, compared to straight and twisted fibers. Previous studies showed that the bond strength and toughness of UHPC with hooked fibers are about 7 and 4 ti mes greater than those of UHPC with straight fibers, respectively (Wu et al. 2018). A new, improved strain energy impact test machine (I-SEFIM) was used to study the effect of the strain rate on the tensile behaviors of UHPC at a wide range of

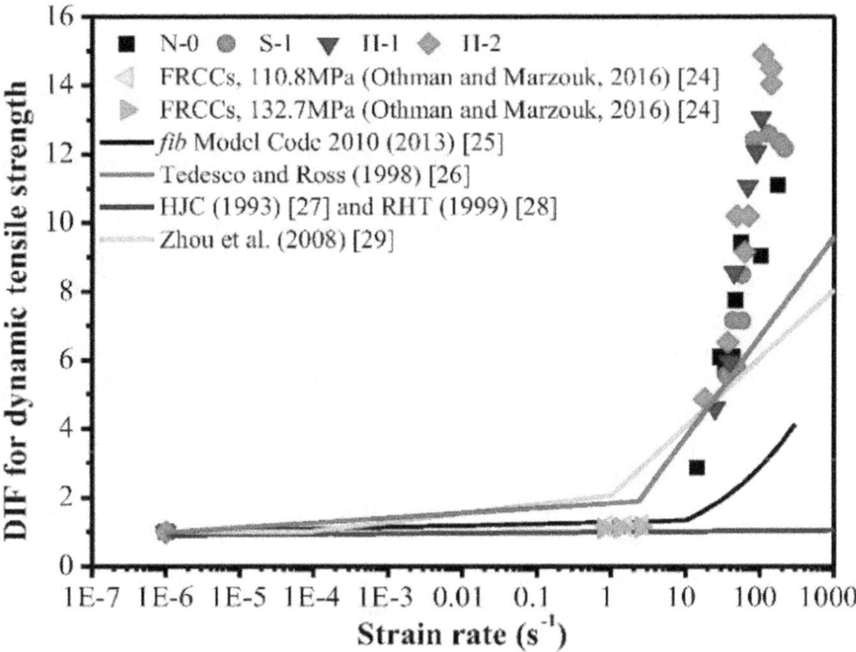

Figure 9.13 DIFs for dynamic tensile strengths of UHPC.

Source: Wu et al., (2018).

high strain rates (10–150 s^{-1}) (Park et al. 2016). The results showed that the postcracking strength of UHPC with 2% straight steel fibers is three times greater than that of quasistatic one.

SHPB and HJC models were adopted to study the dynamic tensile properties of UHPC (Rong and Sun 2012). Only microcracks can be found in UHPC at a high strain rate (30 s^{-1}) when the volume fraction of steel fiber is 4%, while UHPC without fiber reinforcement is completely broken. Moreover, an increase in the steel fiber volume can greatly improve the dynamic tensile properties of UHPC. Brazilian test and XFEM can also be used to study the fracture behaviors and tensile strength of UHPC (Wang and Zhang 2019). The dynamic strength of UHPC is 19 MPa at the strain rate of 30 s^{-1}. The crack propagates along the axial direction according to the results of experiment and numerical simulation. Su et al. (2017) simulated the dynamic tensile behaviors of UHPC by using a 3D mesoscale numerical model. UHPC matrix and steel fibers are considered. Results show that the proposed model has better accuracy than the homogenous model.

9.4.3 Dynamic bending properties of UHPC

The bending performance of UHPC is generally tested by a three-point bending drop weight test. Data recording system is used to record the velocity of drop hammer, impact load, specimen deformation, and energy consumption. Habel et al. (2008) studied the dynamic bending properties of UHPC using the dynamic drop weight test and particle-spring model. The flexural strength of UHPC increases with the increase of the strain rate. When the strain rate reaches the maximum strain rate of 2 s^{-1}, the flexural strength of UHPC reaches the maximum of 30.7 kN, exhibiting obvious strain-hardening characteristics and multicrack cracking behavior. Compared with the static mechanical properties, the dynamic impact resistance properties increase by 25%. Yoo et al. (2015) studied the flexural response of UHPC beams subjected to low-velocity impact load by drop hammer test. With the increase in fiber volume, the deflection of UHPC beams decreases, due to the increase in flexural and shear capacities of UHPC beams under dynamic load. Local flexural crack is found in plain UHPC and UHPC with a reinforcement ratio of 0.53%, while shear failure mode is shown in UHPC with a reinforcement ratio of 1.06%. The strain rate is more appropriate for analyzing the DIF of the flexural strength of UHPC considering the size effect, as shown in Figure 9.14. Hybrid fibers have a great influence on the flexural response of UHPC. The flexural strength of UHPC with 1.6% short PVA fiber and 0.4% long PVA fiber is better than that of UHPC with mono long or short PVA fibers (Kim et al. 2009).

Figure 9.15 illustrates the typically average impact force versus midspan displacement curves of UHPC with monofiber and hybrid fibers based on experiment and numerical simulation. These curves are similar and can be divided into three stages: (1) close-grained stage; (2) microcrack propagation stage; and (3) crack penetration stage. At the close-grained stage, some pores in UHPC specimens are broken or closed, and the fibers are in closer contact with the matrix. Fibers deform together with the matrix and transfer stress. It should be noted that more pores and cracks are too hard to be closed due to the dynamic impact force, and the failure of UHPC enters the next stage. At the microcrack propagation stage, microcracks develop in the surface and weak areas of UHPC; macrocrack occurs when impact force reaches the peak. Short straight fibers play a major role at this stage and hinder the extension of microcracks. At the crack penetration stage, the damage to UHPC specimens increases, and the macrofracture surface forms. Fibers in the fracture surface are pulled out gradually, and UHPC is destroyed completely. Ultimate strain is an important index to describe the deformation and fracture characteristics of UHPC. As the strain rate increases, the ultimate strain of concrete increases, as shown in Figure 9.16. The results are consistent with the previous works (Ahmad and Shah 1985; Rostasy and Hartwich 1985; Watstein 1953). However, some researchers

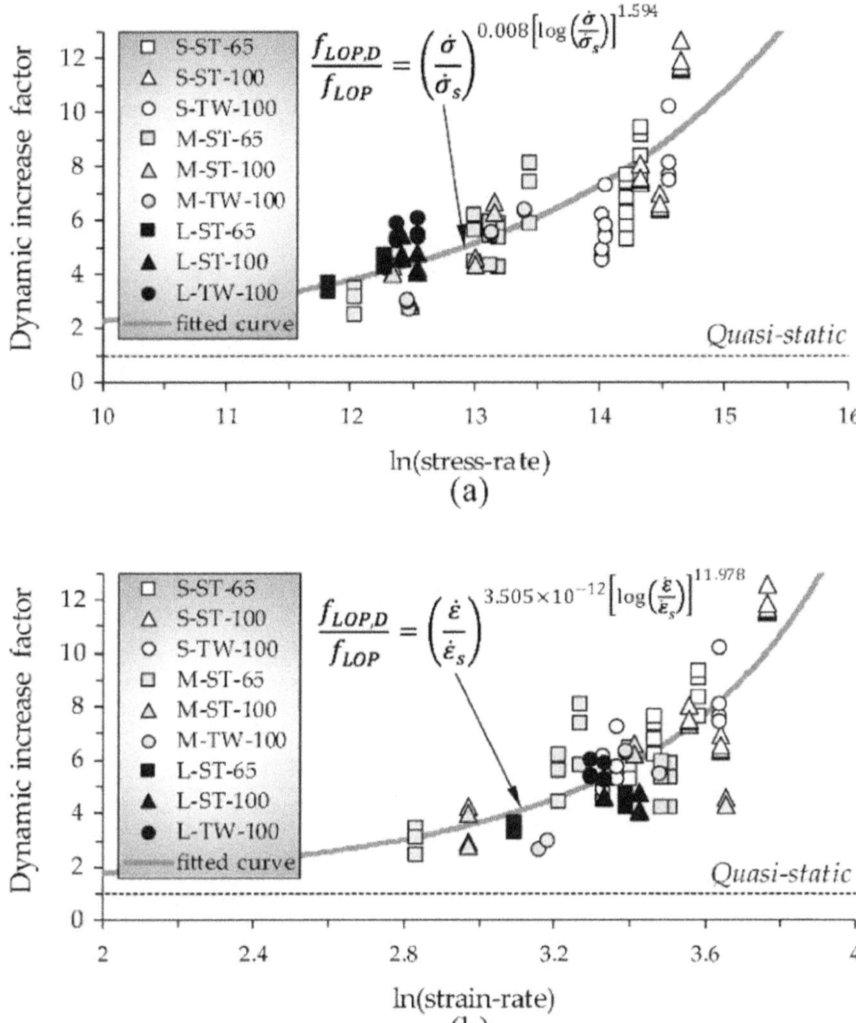

Figure 9.14 Relationship between DIF and flexural stress/strain rate of UHPC.

Source: Yoo et al. (2017).

(Mlakar et al. 1985) hold the opinion that the strain rate does not affect the ultimate strain of concrete, while others believe that the impact force increases but the ultimate strain decreases with the increase of strain rate (Dilger et al. 1984; Hughes and Watson 1978; Plauk 1982). The reason for

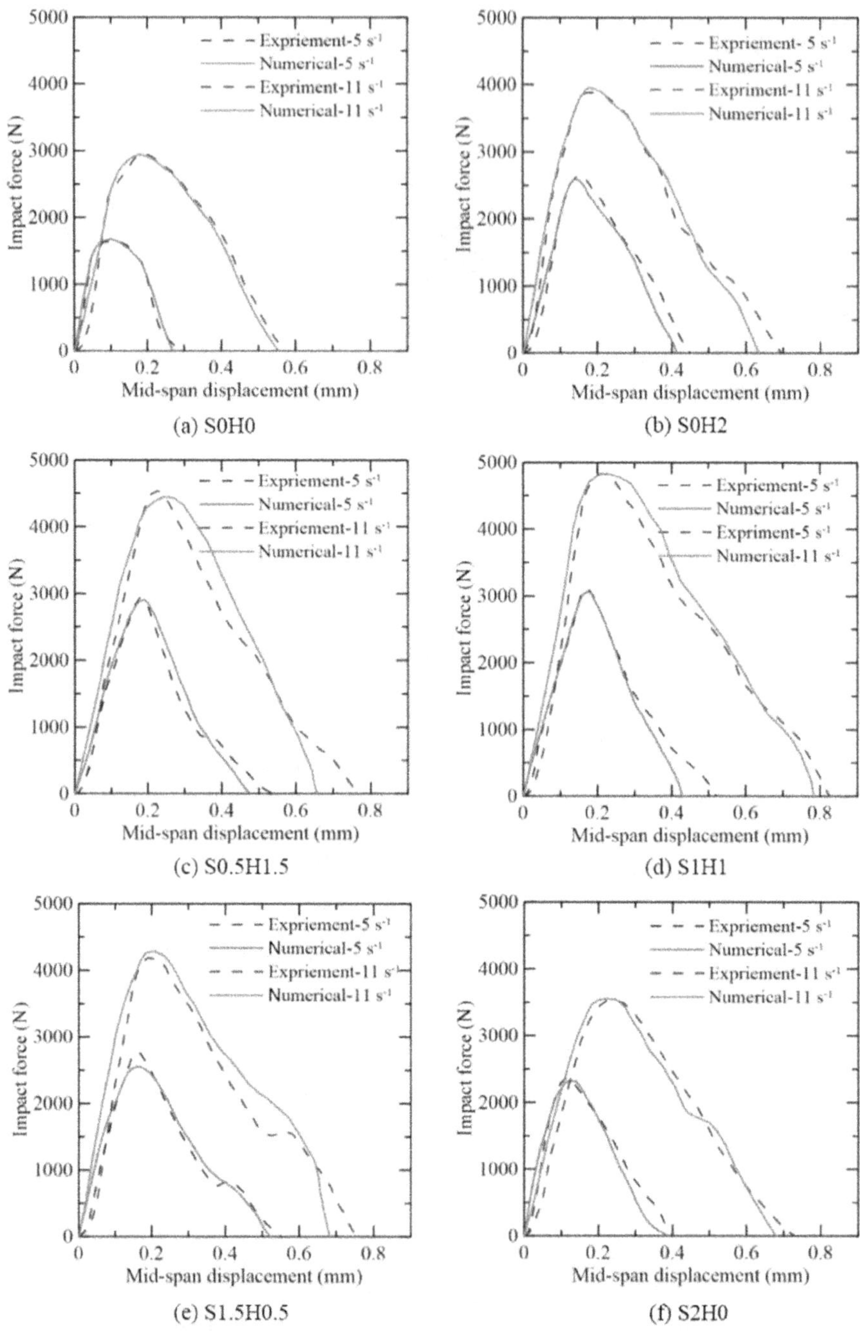

Figure 9.15 Impact force (N) versus mid-span displacement (mm) curves of UHPC.

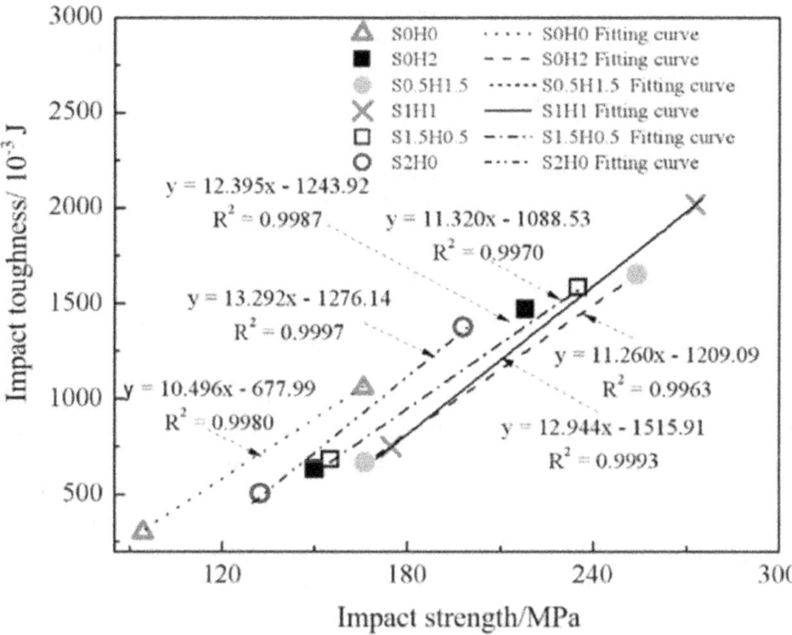

Figure 9.16 Relationship between impact energy and peak impact strength of UHPC.

these differences is that there are many factors affecting the ultimate strain of concrete, such as water-to-cement ratio, aggregate, fiber, curing condition, and test method. The coupling of these factors leads to the great dispersion of ultimate strain.

As shown in Figure 9.16, the impact toughness is linearly correlated with the impact strength. With the increase of the strain rate, the impact toughness and strength of UHPC increase, while the range of the enhancement of UHPC with hybrid fibers is greater than that of UHPC with a single type of fiber and without any fiber. Among them, the enhancement effect of UHPC with hybrid 1% short straight and hooked fibers is the best. It is indicated that the reinforcing effects of hybrid fibers on the impact toughness and impact strength of UHPC are better than that of single types of fibers.

9.4.4 Blast resistance of UHPC

UHPC is one of the building materials widely used in civil and national defense projects. Its projectile impact characteristics, such as penetration depth, perforation, and crack propagation, are far better than ordinary concrete and have great potential for protection and military application.

Generally, there are two indexes to evaluate the blast resistance capability of UHPC: (1) penetration depth of projectile and (2) velocity loss rate of projectile in UHPC matrix. Steel fiber is an important part to enhance the impact-resistance capability of high-performance concrete. It can effectively improve the stress transmission capacity and energy absorption capacity (Soufeiani et al. 2016; Cao et al. 2019). The "bridging effect" of steel fibers helps to restrain crack propagation, thus reducing the damage of UHPC and ensuring the integrity of concrete. Moreover, they can significantly reduce the scraps caused by scraping and peeling damage, thus reducing the secondary damage of UHPC. Yi et al. (2012) evaluated the blast resistance of normal-strength concrete (NSC), ultra-high-strength concrete (UHSC), and reactive powder concrete (RPC), and found that they both have excellent explosion resistance. The compressive strength, tensile strength, elastic modulus, and Poisson's ratio of UHSC and RPC are 3.0–7.9 times higher than those of NSC. Due to the short steel fiber used in RPC, the Poisson's ratio of UHSC is 1.2 times that of RPC. RPC has higher tensile strength than UHSC, while UHSC and RPC samples have higher explosion resistance than NSC samples. Wu et al. (2009) conducted a series of tests to investigate the blast resistances of slabs constructed with both nonfiberous UHPC matrix and UHPC, and slabs reinforced with externally bonded (EB) fiber-reinforced polymer (FRP) plates. The ordinary nonfiberous UHPC matrix slab has a similar blast resistance as the normal reinforced concrete (NRC) slab, and UHPC board is better than the others. The failure of these materials is shown in Figure 9.17. The ductility and blast resistance of reinforced concrete slabs are improved by adding EB carbon FRP plates to the compression surface of reinforced concrete slabs. Li et al. (2015) tested the response of a series of UHPC and NSC plates under blast. The numerical model was established in LS-DYNA. The feasibility and effectiveness of numerical prediction of UHPC slab response are proved by comparing the numerical results with test data.

9.5 REASONS FOR UHPC STRAIN RATE SENSITIVITY

There are four main explanations for the strain rate sensitivity phenomenon of UHPC: (1) Stefan effect; (2) crack propagation; (3) inertial effect; and (4) end-friction effect. The Stefan effect mainly works in the range of low strain rates. As shown in Figure 9.18, the governing equations are as follows:

$$F = 3\pi k r^4 2h^3 V \quad (9.2)$$

Figure 9.17 Cracks in different concretes (Wu et al. 2009). (a) Cracks in NRC specimens. (b) Cracks in FRP plate. (c) Cracks in plain UHPC slab. (d) Cracks in UHPC specimen.

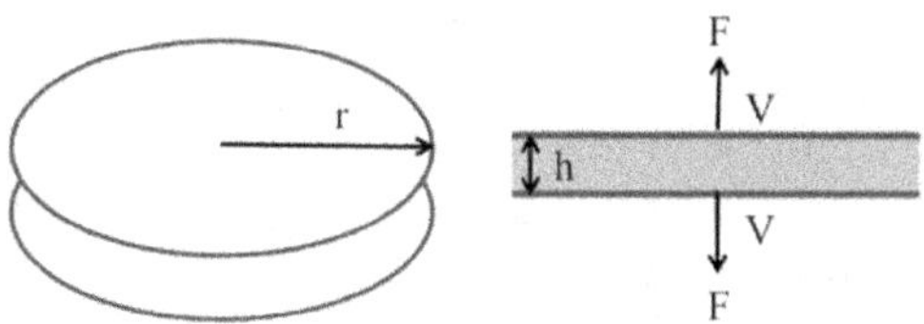

Figure 9.18 Stefan effect model.

where r is the radius of two parallel, circular plates that are separated by a Newton (incompressible) liquid; k is the viscosity; h represents the thickness; F represents the force; and V represents the velocity.

The mechanism is that the viscous force caused by free water in the concrete increases with the increase of the strain rate, which inhibits the expansion of micro- and macrocracks. At a high strain rate, cracks no longer propagate along with weak areas in concrete but directly pass through the aggregate, and crushing the concrete. This is the so-called crack propagation effect. The influence of the inertia effect is closely related to the ability of concrete to resist dynamic tensile load, which is helpful to the design of concrete structures in different environments. The inertia effect, also known as the stress wave propagation effect, is a kind of structural effect. It is the process of energy transmission of the stress wave in the concrete. For the end-friction effect, it can amplify the actual strain rate effect due to the "boundary constraint phenomenon" caused by the friction of concrete.

9.6 SUMMARY

This chapter describes the recent progress in experiments and numerical simulation studies on the dynamic mechanical properties of UHPC. The effects of different factors, including type and content of fiber, strain rate, and material composition, on dynamic compressive properties, dynamic tensile properties, and dynamic bending properties of UHPC are discussed and analyzed. The following conclusions can be drawn:

(1) UHPC is a strain rate-sensitive material with a strain rate strengthening effect. The peak load and failure strain increase with the increase in strain rate. The addition of steel fibers can effectively improve the dynamic mechanical properties of UHPC.
(2) The impact resistance of UHPC with a combination of hybrid fibers is better than that of UHPC with monofiber. The dynamic mechanical properties of UHPC can be greatly improved by adding profiled fibers.
(3) Incorporation of nanomaterials can improve the fiber-matrix bond properties, thereby improving the dynamic mechanical properties of UHPC.
(4) To obtain reliable simulation results, the inertia effect should be considered when the strain of UHPC is as large as 100 s^{-1}. A dynamic friction model needs to be adopted to simulate the friction between SHPB and UHPC specimens.
(5) The explicit dynamic finite element method is mesh based. The mesh distortion greatly affects the calculation results for solving the problem of high-speed impact of UHPC. Lattice models, in which only spring

networks are needed without meshes remeshing, lead to a high computational efficiency when studied UHPC subjected to low-velocity impact load. The particle-based method can better solve the problem of large deformation and damage behaviors of UHPC under extreme loadings, such as high-velocity impact load and ballistic load, etc., than mesh-based and lattice-based methods. Among the particle-based methods, SPH has higher computational efficiency and lower computational accuracy, while other particle methods have lower computational efficiency and higher computational accuracy.

To promote the wider application of UHPC in practical engineering, further studies are still needed:

(1) Reasonable optimization of the shape, size, and content of hybrid fibers can not only reduce the cost of UHPC but also optimize the performance of UHPC.
(2) Research on systematically studying the effects of fiber type on the dynamic mechanical properties of UHPC is very limited. There is a perceived lack of research to reveal the strengthening and toughening mechanisms of UHPC with different types of fibers.
(3) Compared with the mesh-based and lattice-based methods, the particle-based methods can better study the dynamic mechanical behaviors of UHPC. Because of its lower accuracy and efficiency, mesh-based methods are the most widely used methods at present. Therefore, it is necessary to develop a meshless method with high computation accuracy and efficiency to better investigate the dynamic mechanical behaviors of UHPC.

REFERENCES

Ahlborn, T. M., Misson, D. L., Peuse, E. J., Gilbertson, C. G. (2008). Durability and strength characterization of ultra-high performance concrete under variable curing regimes. In *Proc. 2nd Int. Symp. on Ultra High Performance Concrete*, Fehling, E., Schmidt, M., & Stürwald, S.(Eds.), Kassel, Germany, 197–204.

Ahmad, S. H., Shah, S. P. (1985). Structural properties of high strength concrete and its implications for precast prestressed concrete. *PCI J.* 30(6), 92–119.

Allena, S., Newtson, C. M. (2014). Materials specification needs for future development of ultra high performance concrete. *Adv. Civ. Eng. Mater.* 4(2), 17–37.

Alsalman, A., Dang, C. N., Prinz, G. S., Hale, W. M. (2017). Evaluation of modulus of elasticity of ultra-high performance concrete. *Constr. Build. Mater.* 153, 918–928.

Bardenhagen, S. G., Brackbill, J. U., Sulsky, D. (2000). The material-point method for granular materials. *Comput. Methods Appl. Mech. Eng.* 187(3–4), 529–541.

Bibora, P., Drdlová, M., Prachař, V., & Sviták, O. (2017). UHPC for blast and ballistic protection, explosion testing and composition optimization. *In IOP Conference Series: Materials Science and Engineering*. IOP Publishing. 251(1), 012004.

Bo C., Guoping Y., Junhao G., (2011). Modeling and simulation of gas-liquid united hydraulic impactor based on amesim. *Journal of Shanghai University of Engineering Science*, 25: 292–295.

Borrvall, T., Riedel, W. (2011). The RHT concrete model in LS-DYNA. In *Proceedings of The 8th European LS-DYNA user conference,* Strasbourg, France.

Cao, Y. Y. Y., Li, P. P., Brouwers, H. J. H., Sluijsmans, M., Yu, Q. L. (2019). Enhancing flexural performance of ultra-high performance concrete by an optimized layered-structure concept. *Compos. B: Eng.* 171, 154–165.

Chen, B., Yang, G. P., Gao, J. H. (2011). Modeling and simulation of gas-liquid united hydraulic impactor based on AMESim. *J. Shanghai Univ. Eng. Sci.* 25(4), 292–295.

Cotsovos, D. M., Pavlović, M. N. (2008). Numerical investigation of concrete subjected to compressive impact loading. Part 2: Parametric investigation of factors affecting behaviour at high loading rates. *Comput Struct.* 86(1–2), 164–180.

Cusatis, G., Mencarelli, A., Pelessone, D., Baylot, J. (2011b). Lattice discrete particle model (LDPM) for failure behavior of concrete. II: Calibration and validation. *Cem. Concr. Compos.* 33(9), 891–905.

Cusatis, G., Pelessone, D., Mencarelli, A. (2011a). Lattice discrete particle model (LDPM) for failure behavior of concrete. I: Theory. *Cem. Concr. Compos.* 33(9), 881–890.

Davies, E. D. H., Hunter, S. C. (1963). The dynamic compression testing of solids by the method of the split Hopkinson pressure bar. *J. Mech. Phys. Solids.* 11(3), 155–179.

Dilger, W. H., Koch, R., Kowalczyk, R. (1984). Ductility of plain and confined concrete under different strain rates. In *J. Proc.* 81(1), 73–81.

Dolbow, J., Belytschko, T. (1998). An introduction to programming the meshless Element Free Galerkin method. *Arch. Comput. Methods Eng.* 5(3), 207–241.

Habel, K., Gauvreau, P. (2008). Response of ultra-high performance fiber reinforced concrete (UHPUHPC) to impact and static loading. *Cem. Concr. Compos.* 30(10), 938–946.

Habel, K., Viviani, M., Denarié, E., Brühwiler, E. (2006). Development of the mechanical properties of an ultra-high performance fiber reinforced concrete (UHPFRC). *Cem. Concr. Res.* 36(7), 1362–1370.

Hughes, B. P., Watson, A. J. (1978). Compressive strength and ultimate strain of concrete under impact loading. *Mag. Concr. Res.* 30(105), 189–199.

Khosravani, M. R., Silani, M., Weinberg, K. (2018). Fracture studies of ultra-high performance concrete using dynamic Brazilian tests. *Theor. Appl. Fract. Mech.* 93, 302–310.

Kim, Y. W., Min, K. H., Yang, J. M., Yoon, Y. S. (2009). Flexural and impact resisting performance of HPUHPCCs using hybrid PVA fibers. *J. Korea. Concr. Inst.* 21(6), 705–712.

Kolsky H. (1949). An investigation of the mechanical properties of materials at very high rates of loading. *Proc. Phys. Soc.* Section B 62(11), 676.

Krahl, P. A., Gidrão, G. D. M. S., Carrazedo, R. (2018). Compressive behavior of UHPUHPC under quasi-static and seismic strain rates considering the effect of fiber content. *Constr Build Mater.* 188, 633–644.

Larsen, I. L., Aasbakken, I. G., O'Born, R., Vertes, K., Thorstensen, R. T. (2017). Determining the environmental benefits of ultra high performance concrete as a bridge construction material. In *IOP Conference Series: Materials Science and Engineering*. IOP Publishing. 245(5), 052096.

Lee, J. H., Kim, J. H., Choi, C. W., Kang, K. W. (2013). Simple approach to estimate residual flexural strength distribution of composite sandwich structure with impact-induced damage. *J. Renew. Sustain. Energy*. 5(5), 5052010.

Lehmann, C., Fontana, P., Müller, U. (2009). Evolution of phases and micro structure in hydrothermally cured ultra-high performance concrete (UHPC). In *Nanotechnology in Construction 3*. Springer, Berlin, Heidelberg, 287–293.

Li, J., Wu, C. (2018). Damage evaluation of ultra-high performance concrete columns after blast loads. *Int. J. Prot. Struct.* 9(1), 44–64.

Li, J., Wu, C., Hao, H. (2015). An experimental and numerical study of reinforced ultra-high performance concrete slabs under blast loads. *Mater. Des.* 82, 64–76.

Liu, G. R., Liu, M. B. (2003). Smoothed particle hydrodynamics: a meshfree particle method. World Scientific.

Lok, T. S., Zhao, P. J. (2004). Impact response of steel fiber-reinforced concrete using a split Hopkinson pressure bar. *J. Mater. Civ. Eng.* 16(1), 54–59.

Máca, P., Sovják, R., Konvalinka, P. (2014). Mix design of UHPUHPC and its response to projectile impact. *Int. J. Impact Eng.* 63, 158–163.

Markovich, N., Kochavi, E., Ben-Dor, G., 2011. An improved calibration of the concrete damage model. *Finite Elem. Anal. Des.* 47(11), 1280–1290.

Mlakar, P. F., Vitaya-Udom, K. P., Cole, R. A. (1985). Dynamic tensile-compressive ibehavior of concrete. In *J. Proc.* 82(4), 484–491.

Moës, N., Belytschko, T. (2002). Extended finite element method for cohesive crack growth. *Eng. Fract. Mech.* 69(7), 813–833.

Mohammed, H. (2015). Mechanical properties of ultra high strength fiber reinforced concrete (Doctoral dissertation, University of Akron).

Osman, B. H., Sun, X., Tian, Z., Lu, H., Jiang, G. (2019). Dynamic compressive and tensile characteristics of a new type of ultra-high-molecular weight polyethylene (UHMWPE) and polyvinyl alcohol (PVA) fibers reinforced concrete. *Shock. Vib.* 2019: 6382934.

Park, S. H., Kim, D. J., Kim, S. W. (2016). Investigating the impact resistance of ultra-high-performance fiber-reinforced concrete using an improved strain energy impact test machine. *Constr. Build. Mater.* 125, 145–159.

Plauk, G. (1982). Concrete structures under impact and impulsive loading.

Pyo, S. (2014). Characteristics of ultra high performance concrete subjected to dynamic loading. University of Michigan.

Ren, G. M., Wu, H., Fang, Q., Kong, X. Z. (2017). Parameters of Holmquist–Johnson–Cook model for high-strength concrete-like materials under projectile impact. *Int. J. Prot. Struct.* 8(3), 352–367.

Ren, G. M., Wu, H., Fang, Q., Liu, J. Z. (2018). Effects of steel fiber content and type on dynamic compressive mechanical properties of UHPCC. *Constr. Build. Mater.*, 164, 29–43.

Ren, X. T., Zhou, T. Q., Zhong, F. P., Hu, Y. L., Wang, W. P. (2011). Dynamic mechanical behavior of steel-fiber reactive powder concrete. *Explosion Shock Waves*. 31(5), 540–547.

Rong, Z., Sun, W. (2012). Experimental and numerical investigation on the dynamic tensile behavior of ultra-high performance cement based composites. *Constr. Build. Mater*. 31, 168–173.

Rostasy, F. S., Hartwich, K. (1985). Compressive strength and deformation of steel fibre reinforced concrete under high rate of strain. *Int. J. Cem. Compos. Lightweight Concr.*, 7(1), 21–28.

Schachinger, I., Hilbig, H., Stengel, T., Fehling, E. (2008). Effect of curing temperature at an early age on the long-term strength development of UHPC. In *2nd International Symposium on Ultra High Performance Concrete*. Kassel University, Kassel, Gemany, 10: 205–213.

Schwer, L. E., & Murray, Y. D., 2002. Continuous surface cap model for geomaterial modeling: A new LS-DYNA material type. In *Seventh International LSDYNA Users Conference*. Dearborn. Michigan, LSTC& ETA. 16–35.

Shi, C., Wu, Z., Xiao, J., Wang, D., Huang, Z., Fang, Z. (2015). A review on ultra high performance concrete: Part I. Raw materials and mixture design. *Constr. Build. Mater*. 101, 741–751.

Smith, J., Cusatis, G., Pelessone, D., Landis, E., O'Daniel, J., Baylot, J. (2014). Discrete modeling of ultra-high-performance concrete with application to projectile penetration. *Int. J. Impact Eng*. 65, 13–32.

Soufeiani, L., Raman, S. N., Jumaat, M. Z. B., Alengaram, U. J., Ghadyani, G., Mendis, P. (2016). Influences of the volume fraction and shape of steel fibers on fiber-reinforced concrete subjected to dynamic loading—A review. *Eng. Struct*. 124, 405–417.

Su, Y., Li, J., Wu, C., Wu, P., Tao, M., & Li, X. (2017). Mesoscale study of steel fibre-reinforced ultra-high performance concrete under static and dynamic loads. *Mater. Des*. 116, 340–351.

Tran, N. T., Tran, T. K., Jeon, J. K., Park, J. K., Kim, D. J. (2016). Fracture energy of ultra-high-performance fiber-reinforced concrete at high strain rates. *Cem. Concr. Res*, 79, 169–184.

Wang, D., Shi, C., Wu, Z., Xiao, J., Huang, Z., Fang, Z. (2015). A review on ultra high performance concrete: Part II. Hydration, microstructure and properties. *Constr. Build. Mater*. 96, 368–377.

Wang, J., Zhang, J. (2019). Research on high-power and high-speed hydraulic impact testing machine for mine anti-impact support equipment. *Shock. Vib*. 2019: 6545980.

Wang, Z. L., Li, Y. C., Shen, R. F., Wang, J. G. (2007). Numerical study on craters and penetration of concrete slab by ogive-nose steel projectile. *Comput. Geotech*. 34(1), 1–9.

Wang, Z., Wang, J. Q., Tang, Y. C., Liu, T. X., Gao, Y. F., Zhang, J. (2018). Seismic behavior of precast segmental UHPC bridge columns with replaceable external cover plates and internal dissipaters. *Eng. Struct*. 177, 540–555.

Watstein, D. (1953, April). Effect of straining rate on the compressive strength and elastic properties of concrete. In *J. Proc.* 49(4), 729–744.

Wilhelm, S. Curbach, M. (2016). Development of underwater housings for the deep sea made of ultra-high performance concrete (UHPC) and long-term testing at the Arctic Sea, in: *OCEANS 2016 MTS/IEEE Monterey*, Monterey, USA, IEEE. 1–6.

Wille, K., Kim, D. J., Naaman, A. E. (2011). Strain-hardening UHP-UHPC with low fiber contents. *Mater. Struct.* 44(3), 583–598.

Wille, K., Xu, M., El-Tawil, S., Naaman, A. E. (2016). Dynamic impact factors of strain hardening UHP-UHPC under direct tensile loading at low strain rates. *Mater. Struct.* 49(4), 1351–1365.

Wu, C., Oehlers, D. J., Rebentrost, M., Leach, J., Whittaker, A. S. (2009). Blast testing of ultra-high performance fibre and FRP-retrofitted concrete slabs. *Eng. Struct.* 31(9), 2060–2069.

Wu, H., Ren, G. M., Fang, Q., Liu, J. Z. (2018). Effects of steel fiber content and type on dynamic tensile mechanical properties of UHPCC. *Constr. Build. Mater.* 173, 251–261.

Wu, Z., Khayat, K. H., Shi, C. (2018). How do fiber shape and matrix composition affect fiber pullout behavior and flexural properties of UHPC? *Cem. Concr. Compos.* 90, 193–201.

Wu, Z., Shi, C., He, W., Wang, D. (2016). Uniaxial compression behavior of ultra-high performance concrete with hybrid steel fiber. *J. Mater. Civ. Eng.* 28(12), 06016017.

Yi, N. H., Kim, J. H. J., Han, T. S., Cho, Y. G., Lee, J. H. (2012). Blast-resistant characteristics of ultra-high strength concrete and reactive powder concrete. *Constr Build. Mater.* 28(1), 694–707.

Yoo, D. Y., Banthia, N. (2017). Size-dependent impact resistance of ultra-high-performance fiber-reinforced concrete beams. *Constr. Build. Mater.* 142, 363–375.

Yoo, D. Y., Banthia, N., Kim, S. W., Yoon, Y. S. (2015). Response of ultra-high-performance fiber-reinforced concrete beams with continuous steel reinforcement subjected to low-velocity impact loading. *Compos. Struct.* 126, 233–245.

Yu, R., Spiesz, P., Brouwers, H. J. H. (2015). Development of an eco-friendly Ultra-High Performance Concrete (UHPC) with efficient cement and mineral admixtures uses. *Cem. Concr. Compos.* 55, 383–394.

Yunsheng, Z., Wei, S., Sifeng, L., Chujie, J., Jianzhong, L., 2008. Preparation of C200 green reactive powder concrete and its static-dynamic behaviors. *Cem. Concr. Compos.* 30(9), 831–838.

Zienkiewicz, O. C., Taylor, R. L., Nithiarasu, P., Zhu, J. Z. (1977). The finite element method. London: McGraw-Hill.

Zhang, C. (2008). Experimental research on spalling strength of reactive powder concrete. Master dissertation, University of Hunan University.

Zhang, W.H., Liu, P.Y., Lv, L.J. (2019). Dynamic mechanical property of UHPCs: A Review. *Materials reports.* 33(10), 3257–3271.

Zheng, D., Li, Q. (2004). An explanation for rate effect of concrete strength based on fracture toughness including free water viscosity. *Eng. Fract. Mech.* 71(16–17), 2319–2327.

Zhu, Z., Li, B., Zhou, M. (2015). The influences of iron ore tailings as fine aggregate on the strength of ultra-high performance concrete. *Adv. Mater. Sci. Eng.* 2015: 412878.

Zukas, J. A., Nicholas, T., Swift, H. F., Greszczuk, L. B., Curran, D. R. (1982). Impact dynamics. Wiley New York.

Chapter 10

Durability

10.1 INTRODUCTION

Durability of cementitious materials is the capability to resist weathering actions, such as freezing and thawing, chemical attack, abrasion, electrochemical, or any other deterioration process. It is the most important concern for concrete structures since the deterioration of concrete can result in decreased mechanical properties, regular repair and maintenance, increased labor cost, and shortened service life. Different application needs for concrete require various degrees of durability depending on the exposure environment and properties desired. The motivations for greater strength and longer service life of concrete structures for use in hostile environments drive people to seek high-performance concrete (HPC) and ultra-high performance concrete (UHPC). Better durability of UHPC can be achieved along with greater strength and denser microstructure. What makes UHPC more durable than conventional concrete and HPC is its reduced porosity because of the dense particle packing and the low water-to-binder (w/b) ratio associated with the use of superplasticizer and improved homogeneity related to reduced particle size. The durability of UHPC in field cases is closely related to the long-term performance in particular environments and service conditions. Although the investigation of durability of UHPC is of great interest to researchers and there exists sufficient literature on this aspect, its field track record is rarely documented since it is a new material in the market with a relatively short service period.

This chapter describes various durability aspects of UHPC under physically or mechanically and chemically attacked conditions that are assessed using accelerated durability tests. The properties, including water permeability, chloride ion permeability, carbonation, corrosion of steel reinforcement, freezing-thawing resistance, deicer scaling resistance, chemical attack resistance, alkali-silica reaction, abrasion resistance, and fire resistance, are compared to those of conventional concrete and HPC. This chapter aims to

DOI: 10.1201/9781003203605-10

find out how durable it is for UHPC under different attacking environments and offers important tips for future design and applications.

10.2 WATER PERMEABILITY

Hardened concrete is a composite material comprising cement, fine and coarse aggregate, voids, and/or fibers. The presence of voids in concrete makes it permeable, which in turn allows water or gas to flow into it. The permeability of concrete, defined as the property that controls the rate of flow of fluids or gas into a porous solid, largely depends on the size and connectivity of pores, as well as the tortuousness of the transport path. The pores relevant to permeability are those connected ones with a minimum diameter of 0.01 or 10 μm. Isolated pores, pores filled with water, and pores with a narrow entrance are irrelevant to permeability. When concrete is subjected to an external chemical attack, lowering the porosity and the permeability of concrete to reduce or slow down as much as possible the penetration of the aggressive agents is the most effective way to reduce the intensity of external aggression. High water permeability of concrete can give chemical substances, such as chloride and sulfate ions, the privilege to permeate into concrete and eventually results in corrosion of steel rebars and/or fibers and sulfate attack.

Compared to conventional concrete and HPC, UHPC is characterized by much lower porosity and denser microstructure, which guarantees its superior permeability resistance (Abbas et al. 2015; Tayeh et al. 2012). Figure 10.1 compares the development of water absorption behaviors of C30, C80, RPC200, and RPC200C with exposure time (Roux et al. 1996). RPC200 shows a water absorption coefficient of 0.2 kg/m^2 after 7.5 h, which is significantly lower than 2.7 kg/m^2 of C30. The water permeability of concrete is mainly affected by w/b ratio, use of supplementary cementitious materials, pore diameter, and pore connectivity (Poon et al. 2006; Ghafari et al. 2012). As shown in Figure 10.2, the permeability coefficient is significantly reduced with the decrease in w/b ratio. The water absorption coefficient of concrete with w/b ratio of 0.4 is 0.04 at 14 d, and it decreases to 0.0025 at w/b ratio of 0.17. In addition, the permeability coefficient of concrete before 28 d decreases with the prolongation of curing age. The corresponding value of UHPC after 98 d is 0.0005, which is one-third that of conventional concrete. Dobias et al. (2016) found that the water absorption coefficient of UHPC at 90 d was approximately five times lower than that of conventional concrete. Li and Huang (2005) reported that the measured average water penetration height and relative seepage height were 7.2 and 2.2 × 10^{-8} mm, respectively, as determined by a single pressure method. The high permeability resistance of UHPC originates from its low w/b ratio and low porosity (Tam et al. 2012; Ghafari et al. 2012).

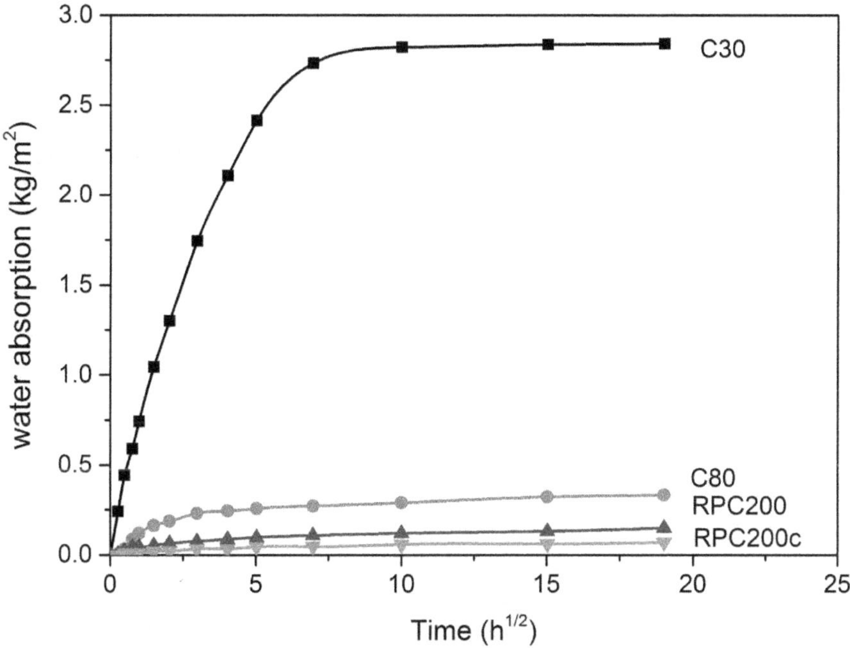

Figure 10.1 Water absorption of C30, C80, RPC200, and RPC200C.

Source: Roux et al. (1996).

10.3 CHLORIDE ION PERMEABILITY

Resistance to chloride ion penetration is a measure of the ability of concrete to retard the migration of chloride ions. The chloride ions in concrete come primarily from the use of deicing salts and seawater. By penetrating the concrete and attacking the reinforcing steel bars and fibers within the concrete, the corrosion of reinforcing steel in concrete would occur and the structure's integrity can be gradually destroyed. Chloride ions diffusing inside concrete can be either dissolved in the pore solution or chemically and physically bound to the hydration products (Yuan et al. 2009). Depending on the binding method, the chlorides in concrete can be divided into free and bound chlorides. Chemically bonded chloride ions can react with cement composition, especially unhydrated C_3A, to form monochloroaluminates ($3CaO{\cdot}Al_2O_3{\cdot}CaCl_2{\cdot}10H_2O$), leading to swelling stress. Free chloride ion penetration in steel fiber reinforced concrete induces depassivation of the steel rebars and/or fibers as well as the initiation of the corrosion process, leading to the degradation of concrete structures (Poupard et al. 2004; Chen et al. 2018). Steel reinforcements in concrete are protected from

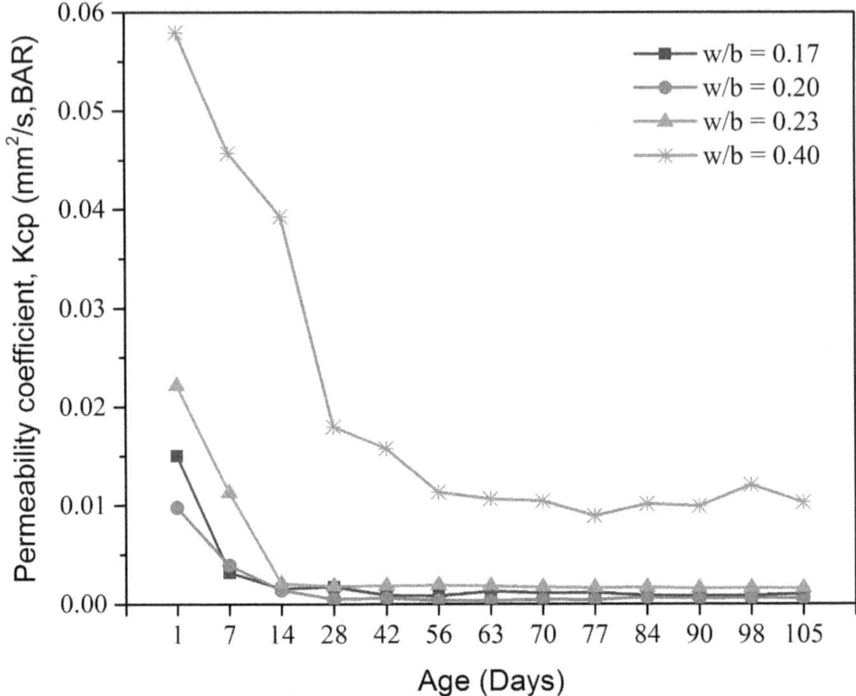

Figure 10.2 Water permeability coefficients of UHPC with different w/b ratios at various ages.

Source: Tam et al. (2012).

corrosion because of the passivation in high alkaline pore solution. Due to aggressive chloride ions and/or neutralization of the environment near the reinforcements, the passive layer on the steel surface will be damaged. This results in the corrosion of steel reinforcements and spalling of concrete due to the formation of rust, leading to the deterioration and reduced service life of concrete structures (Angst et al. 2011; Glass and Buenfeld 1997).

Chloride penetration is usually a slow process, which cannot be determined directly in a time frame that would be useful as a quality control measure. Hence, to evaluate chloride penetration, an accelerated test method is needed to render the determination of diffusion values in a reasonable time frame. Accelerated testing methods, including the rapid chloride permeability test (RCPT) according to ASTM C 1202 (2019) and Nordtest method NT Build 443 (1995), are commonly used to assess the chloride-ion permeability of concrete. The chloride ion diffusion coefficient serves as an evaluation indicator in the Nordtest method NT Build 443. Table 10.1

Table 10.1 Chloride ion diffusion coefficient of UHPC

Ref.	*Curing condition*	*w/b ratio*	*Steel fiber content*	*Medium solution*	*Testing age*	*Chloride ion diffusion coefficient ($\times 10^{-13}$ m^2/s)*
Roux et al. (1996)	Pressurized before and during setting at a pressure of 60 MPa and water-cured for 28 d at 20°C	0.14	2%	3% NaCl	1 y	0.2
Piérard and Cauberg (2009)	Moist curing (20°C at 95% relative humidity)	0.18	0	16% NaCl	56 d	4.0
Liu et al. (2009)	90°C steam for 1 d then 90°C hot water till room temperature	0.15	2.6%	10% NaCl	-	4.1
Thomas et al. (2012)	20°C for 2 d + 90°C curing for 2 d	0.12	2%	Seawater	28 d	1.3
Scheydt and Müller (2012)	Heat-treated at 90°C for 3 d	0.21	2.5%	3% NaCl	63 d	1.3
Piérard et al. (2012)	20°C and 95% RH for 90 d	0.21	2%	16% NaCl	3 m	2.3

compares the chloride permeability coefficients of UHPC in literature. The chloride ion diffusion coefficient of UHPC ranges from 0.2×10^{-13} to 4.1×10^{-13} m^2/s, depending on w/b ratio, curing regime, medium solution concentration, steel fiber volume, and testing age. Due to the different testing environments and mixture proportions used, it is difficult to quantitatively compare the chloride permeability of those UHPCs. It is worth mentioning that the chloride ion diffusion coefficient of UHPC is at least one order of magnitude lower than that of HPC and conventional concrete (Liu et al. 2009; Wang 2011; Chuang and Huang 2013). For instance, Roux et al. (1996) reported that the chloride ion diffusion coefficients of C30, C80, and UHPC were 1.1×10^{-12}, 0.6×10^{-13}, and 0.02×10^{-12} m^2/s, respectively. An et al. (2007) indicated that the chloride ion diffusion coefficients of UHPC and HPC were 2.2×10^{-13} and 15.4×10^{-13} m^2/s, respectively. Dobias et al. (2016) also reported that the chloride ion diffusion coefficients of UHPC were less than 1.4×10^{-13} m^2/s. This agrees well with the results reported by Voo and Foster (2010). Besides, Wei and Song (2005) showed that the chloride ion diffusion coefficient of UHPC was 1/5 to 1/6 of that of C60. Therefore, UHPC demonstrates superior chloride permeability resistance in comparison to HPC and conventional concrete. Although results from accelerated testing do not replicate actual conditions as concrete would experience in the field, it is a helpful quality control measure for structure design and service life prediction.

The RCPT method uses an applied electric potential across a specimen load cell to determine concrete conductance to check whether concrete is susceptible to chloride ingress or not. The calculated amount of Coulombs passed through tested specimens has been used as an indicator for concrete permeability. Bonneau et al. (1997) reported that the charge passing through UHPC specimens over 6 h when subjected to water curing at various ages and temperatures was less than 10 Coulombs. Graybeal (2006) indicated that standard curing and/or thermally treated UHPC had very low or negligible chloride ion penetration. Meanwhile, the use of steel fibers did not contribute to a short circuit effect during UHPC testing, owing to their short length and randomly discontinuous distribution (2006). Toutanji et al. (1998) revealed that adding 19.5 mm polypropylene fibers increased the permeability of concrete, while adding 12.7 mm shorter fibers reduced the permeability. El-Dieb (2009) reported that the chloride permeability of UHPC specimens increased with the increase in steel fiber volume. After 56 d standard curing, the total passed charge of UHPC specimens made with 0.08%, 0.12%, and 0.52% steel fibers were about 1.8, 2.9, and 3.6 times, respectively, greater than those of specimens without any steel fiber. On the contrary, Abbas et al. (2015) found that UHPC specimens with higher steel fiber content exhibited relatively low permeability. The total passed charges of specimens made with 1%, 3%, and 6% steel fibers were 22.9%, 38.6%, and 53.5%, respectively, lower than those without any steel fiber.

Table 10.2 summarizes the total passed charge for UHPC reported in the literature. The thermally cured UHPC shows a low passed charge of less than 100 Couloums, which is negligible. The standard-cured UHPC shows a relatively high passed charge up to 100 Couloums, which is at a very low level. The prolongation of standard curing from 28 to 56 d decreases the passed charge by 78.9% (2006). However, under an initial heat curing period, the passed charge of UHPC specimens exposed to different standard curing ages does not show much change (2006). Another observed general trend was that the charge passed through UHPC decreased with the increase in curing temperature (Ahlborn et al. 2008; Yu et al. 2013). This is because the increase in curing temperature accelerates the hydration of cement and promotes the pozzolanic reaction between supplementary cementitious materials and $Ca(OH)_2$ to form the strength-contributing product of C-S-H. A longer average chain of C-S-H and relatively greater pozzolanic reactivity of silica fume, slag, and fly ash are reported at high temperatures (Zhang et al. 2008; Zanni et al. 1996).

The porosity of UHPC decreases with an increase in curing temperature, as can be seen from Table 10.3. The increase in temperature from 20 to 180°C lowers the porosity by 40%–80%, depending on w/b ratio, mixture

Table 10.2 Total passed charge of UHPC

Ref.	*Curing condition*	*w/b ratio (w/c ratio*)*	*Steel fiber content by volume*	*Testing age (h)*	*Passed charge (Coulombs)*	*Chloride ion penetrability*
Bonneau et al. (1997)	7-d water-curing at 20°C followed by 3-d water-curing at 90°C	0.21	1.8%	6	<10	Negligible
Graybeal (2006)	2 d of 90°C steam curing and then standard curing for 28 d	0.12	2%	6	18	Negligible
	Standard curing for 28 d			6	360	Very Low
	Standard curing for 56 d			6	76	Negligible
	2 d of 60°C steam curing and then standard curing for 28 d			6	39	Negligible
	2 d of 60°C steam curing and then standard curing for 56 d			6	26	Negligible
	Standard curing for 15 d followed by 2 d of 90°C steam curing			6	18	Negligible
Ahlborn et al. (2008)	Air curing to 28 d	0.20*	2%	6	75	Negligible
	2 d of 90°C steam curing and then standard curing for 7 d			6	10	Negligible
	2 d of 90°C steam curing and then standard curing for 28 d			6	15	Negligible
Li et al. (2005)	3 d of 90°C hot water curing	0.20	2%	6	84	Negligible
Ye et al. (2006)	3 d of 90°C hot water curing	0.21	0	6	22.9	Negligible
Scheydt and Muller (2012)	3 d of 90°C hot water curing	0.21	2.5%	6	1.3	Negligible
Trinh (2011)	2 d of 90°C steam curing	0.22	2%	6	<100	Negligible
Tayeh (2012)	2 d of 90°C steam curing and then standard water curing for 7 d or 28 d	0.15	2%	6	< 60	Negligible

Table 10.3 Porosity of UHPC specimens subjected to different curing regimes

Ref.	*W/B*	*Curing condition*	*Porosity (%)*
Herold and Müller (2004)	0.16	20°C standard curing	10.5
		90°C hot water curing	6.4
Heinz and Ludwig (2004)	0.22	20°C and 93% relative humidity curing	8.2
		65°C and 93% relative humidity curing	5.3
		90°C and 93% relative humidity curing	4.2
		105°C and 93% relative humidity curing	4.0
		120°C and 93% relative humidity curing	3.5
		180°C and 93% relative humidity curing	2.9
Cwirzen (2007)	0.17	20°C and 93% relative humidity curing	5.8
		90°C hot water curing	≤1.3
Scheydt and Muller (2012)	0.21	28°C water curing (with fiber)	8.9
		90°C hot water curing	5.4
		28°C water curing (no fiber)	10.9

proportion, and curing temperature (Scheydt and Müller 2012; Cwirzen 2007; Heinz and Ludwig 2004). Such a great decrease in porosity can effectively densify and homogenize the microstructure, enhance aggregate-matrix interface quality, and eventually lead to great improvement in permeability resistance (Yang et al. 2000; Liu et al. 2005; Cheyrezy et al. 1995). It is worth noting that the accuracy of the laboratory apparatus for this testing method should be high enough with a specified error range. Otherwise, it will be a challenge to identify the precision of the low passed charge, especially for thermally cured UHPC.

The additions of supplementary cementitious materials and nanoparticles in UHPC greatly affect the permeability through the following effects: (1) pozzolanic effect; (2) filler effect; (3) seeding or nucleation effect; and additionally (4) chemical effect, in particular for $CaCO_3$ (Wu et al. 2017; Papadakis 2000; Wu et al. 2018). The addition of proper supplementary cementitious materials content can effectively decrease the permeability coefficient. The use of silica fume can render lower permeability, which may contribute to greatly reduced conductivity compared to other types of supplementary cementitious materials at the same content (Yu et al. 2013; Matte and Moranville 1999). However, the reduction in electrical conductivity of UHPC specimens is not proportional to silica fume content (Toutanji et al. 1998; Tafraoui et al. 2016). UHPC made with 58% fly ash, by mass of binder, indicates a chloride ion diffusion coefficient of 0.51×10^{-8} cm²/s, which is half the value of HPC of 1.1×10^{-8} cm²/s (Wei and Song 2005). In addition, the resistance to chloride ion permeability of UHPC gradually decreases with the decrease of silica fume content and the increase of slag content (Shi et al. 2012). From the above analysis, it can be concluded that the permeability results of UHPC vary with many factors, including testing method, specimen size, medium solution concentration, and mixture composition. Therefore, the results from RCPT testing method should reflect these claims and demonstrate its high resistance to chloride penetration.

10.4 CARBONATION

Carbonation is the chemical reaction of CO_2 in the environment with hydration products, such as calcium hydroxide $Ca(OH)_2$, in the presence of sufficient moisture, as shown in Eq. 10.1. This reaction produces $CaCO_3$ and lowers the pH value of concrete from 13.5 to around 9. The protective oxide layer surrounding the reinforcing steel breaks down at such pH value, leading to steel corrosion and eventually cracking of concrete structures (Bertos 2004). Carbonation can only occur in concrete in a certain condition. For example, if concrete is very dry, the carbonation rate will be slowed down because of a lack of water content to dissolve CO_2. However, in saturated concrete, too much moisture can serve as a barrier to the penetration of CO_2 and then slow down the carbonation rate. Therefore, the most favorable condition with sufficient moisture and CO_2 concentration should be secured for carbonation reactions (El-Turki 2007).

$$Ca(OH)_2 + CO_2 + H_2O = CaCO_3 + 2H_2O \qquad (10.1)$$

Porosity, w/b ratio, and compressive strength grade of concrete play significant roles in the carbonation rate. A greater w/b ratio and a higher porosity generally correspond to a faster carbonation rate (Liu et al. 2003; Park et al. 2007). Due to the very low w/b ratio and limited water content in UHPC, its microstructure is very dense, with a very small amount of the most probable pore size. It is stated that carbonation rarely happens on UHPC, either under standard or heat-curing regimes (Long et al. 2005). UHPC specimens subjected to 56 and 90 d of carbonation do not show any carbonation signs (Long et al. 2005; Sun and Lai 2009). The carbonation depth of UHPC after 6 months of exposure is 0.5 mm (Perry and Zakariasen 2004; Schmidt et al. 2003) and increases to 1.5–2.0 mm after a 1-year exposure to 1% CO_2 atmosphere (Piérard et al. 2012). Small-scale beams of UHPC placed in a carbonation chamber subjected to a flow of 5%–100% CO_2 for 2 years show no signs of carbonation (Andrade et al. 1996). After 3-year exposure, the carbonation depth of UHPC is 1.5–1.7 mm only, which is 2.5–4.5 times lower than those of HPC and conventional concrete (Alonso et al. 2006; Schmidt and Fehling 2005). To some extent, mechanically induced microcracks in UHPC are observed to be partly or completely filled by carbonation products, contributing to strength enhancement to some extent (Müller et al 2010).

10.5 CORROSION OF STEEL REINFORCEMENT

Corrosion of steel reinforcement may occur due to localized failure of the passive film on the steel reinforcement due to chloride ions ingress and carbonation. Inherently, the alkaline environment of concrete with a pH of 12–14 provides steel reinforcement with corrosion protection. Corrosion can occur when the passive layer is destroyed, when the alkalinity of the concrete is reduced, or when the chloride concentration in concrete is increased to a certain level. For the occurrence of corrosion, four basic elements, namely an electrolyte, a cathode, an anode, and a metal connection, must be present.

In the presence of chloride ions, Cl^-/OH^- ratio of concrete pore solution is recognized as a critical parameter for the corrosion initiation of steel bars. Depending on the Cl^-/OH^- ratio, the protective film may be destroyed even at pH values considerably above 11.5. Chloride ions activate the surface of the steel to form an anode, with the passivated surface being the cathode. The reactions involved are given as follows (Eqs. 10.2 and 10.3):

$$Fe^{2+} + 2Cl^- \rightarrow FeCl_2 \quad (10.2)$$

$$FeCl_2 + 2H_2O \rightarrow Fe(OH)_2 + 2HCl \quad (10.3)$$

Since the chloride ion permeability and carbonation in UHPC are extremely low, UHPC shows high resistance against corrosion. Ghafari et al. (2015) used the accelerated corrosion test to evaluate the corrosion rate of steel bars in HPC and UHPC. The time to cracking in HPC is less than half of the time in UHPC. The corrosion rate for steel bars in UHPC is reported to be 0.01 μm/year, which is much lower than the limited value of 1 μm/year (Roux et al. 1996). Fan et al. (2019) investigated the influence of brass and zinc-coated fibers on the electrical resistance of standard-cured UHPC and the corrosion resistance of embedded steel bars. The electrochemical test was conducted on UHPC specimens immersed in 3.5 wt.% NaCl solution for 1 year. The results showed that no chloride was detected and the corrosion current density was around 0.01 $\mu A/cm^2$, indicating no corrosion of the steel bar. Moreover, the ASTM C876 (2015) based on open circuit potential is found to be not suitable for determining the corrosion state of embedded steel bars in UHPC. Rafiee and Schmidt (2012) simulated the macro cell steel corrosion in cracked UHPC exposed to chloride solution up to 360 d. The predicted anode corrosion current and cross-section loss agreed well with the experimental data, which were in the ranges of 0.5×10^{-6}

to 2.5 × 10^{-5} A and 0 to 4.5%, respectively, depending on the exposure time and concrete cover. Zheng et al. (2022) evaluated the corrosion behavior of carbon steels embedded in UHPC contaminated with seawater. It was found that the initial high Cl^-/OH^- ratio greater than 0.6 induced the corrosion of steel after the casting of seawater mixed UHPC. However, it was immediately suppressed within the first 3 days due to the lack of water for the electrochemical process.

Nanoparticles are employed in UHPC to improve its durability by taking advantage of their unique physical and chemical properties. Faizal et al. (2016) evaluated the effect of nanoclay content (1%, 3%, and 5%, by weight of cement) on the corrosion potential of UHPC embedded with steel reinforcements. The specimens were subjected to a 3% sodium chloride solution up to 91 d. The corrosion potential decreased with the increase in nanoclay content, thus delaying the corrosion initiation. The addition of nanosilica also effectively reduced the corrosion rate of steel bars embedded in UHPC (Ghafari et al. 2015).

10.6 FREEZING-THAWING RESISTANCE

Freezing-thawing damage occurs in concrete when water molecules freeze and expand beyond the volume constraints of concrete. This leads to distress in concrete, especially when the pressure developed exceeds the tensile strength of concrete, eventually resulting in dilation and rupture of cavities. Typical deterioration of concrete from freezing-thawing actions includes random cracking, surface scaling, and joint deterioration due to D-cracking (Mu et al. 2002; Schwartz 1987; Bondar et al. 2015), which has a very adverse effect on the mechanical properties and permeability resistance of concrete.

The superior freezing-thawing durability of UHPC comes from its highly impermeable matrix with reduced capillary porosity. Several studies revealed that UHPC demonstrates no deterioration after 300 or 600 freezing-thawing cycles and has a durability factor of 100 or greater with nearly zero mass loss (Bonneau et al. 1997; Alkaysi et al. 2016). Even after 800 freezing-thawing cycles, no apparent deterioration phenomena are observed in some research (Liu et al. 2009). Typically, concrete subjected to freezing-thawing action can lose mass due to material spalling and undergo a decrease in relative dynamic modulus as the appearance of microcracking. The relative dynamic modulus is a value of the percent difference in the squares of the frequency at any cycle compared with the initial frequency before freezing-thawing cycling, as defined in the ASTM C666 test method (2019). However, Wang et al. (2017) found that the values of mass loss of UHPC were 0.18%, 0.50%, and 0.62%, respectively, after 500, 1000, and 1500 freezing-thawing cycles. Surprisingly, Lee et al. (2005) and Graybeal (2006) indicated that UHPC gained mass and increased relative dynamic modulus as the freezing-thawing cycles increased. A mass increase of 0.2% after 125 cycles was reported by Graybeal and Tanesi (2007). Magureanu et al. (2012) also found that UHPC specimens exhibited greater compressive strength, static modulus of elasticity, and dynamic modulus of elasticity after 1098 freezing-thawing cycles compared to the control specimens.

Table 10.4 summarizes the change in relative dynamic modulus of UHPC made with different w/b ratios, steel fiber contents, and curing regimes. Steam-cured UHPC specimens display an increase in elastic modulus even after 1500 freezing-thawing cycles (Liu et al. 2009). The relative dynamic modulus of UHPC with 28 d standard curing subjected to 1000 freezing-thawing cycles decreases, but it is still over 90% (Lee et al. 2007). It is also reported that the dynamic modulus values of UHPC with 28 d standard curing subjected to 180 freezing-thawing cycles are decreased but do not drop below 95% of the baseline (Smarzewski and Bernat-Hunek 2017). The relative dynamic modulus is still over 98% after 1500 freezing-thawing cycles (Wang et al. 2017).

The increases in mass and relative dynamic modulus may be attributed to the absorbed water and/or continued hydration during the freezing-thawing

Table 10.4 Change in relative dynamic modulus of UHPC subjected to freezing-thawing condition

Ref.	*w/b ratio (w/c ratio*)*	*Steel fiber content (%)*	*Curing condition*	*Freeze-thaw cycles*	*Relative dynamic modulus (%)*
Shaheen and shrive (2006)	0.13*	0%	150°C	300	101.0
		1.5%			101.6
Gao et al. (2006)	0.17	7%	90°C steam curing	800	100.0
Ahlborn et al. (2008)	0.20*	6%	Air curing	300	101.6
			90°C hot curing for 2 d	300	100.3
Lee et al. (2007)	0.14	3%	Saturated limestone water for 28 d	100	103
				300	96
				600	92
				1000	90
Liu et al. (2009)	0.21*	2%	90°C steam curing for 1 d	1500	101.0
Magureanu et al. (2012)	0.12	0%	90°C hot curing for 5 d	1095	100.8
			20°C curing for 5 d	1095	100.4
		2.5%	90°C hot curing for 5 d	1095	100.4
			20°C curing for 5 d	1095	100.6

Note: * means water-to-cement ratio.

process. It is revealed that UHPC specimens can self-heal the microcracks when they are submerged in water after or during deterioration because of the presence of a high amount of unhydrated cement particles (Jacobsen and Sellevold 1996). This can recover the mechanical properties of UHPC from deterioration due to freezing-thawing action given sufficient immersed water. The self-healing phenomenon of UHPC is proved by Granger et al. (2007) and is further investigated by Graybeal (2006). In Graybeal's research (2006), changes in mass and mechanical behavior of untreated, air-treated, and steam-treated UHPC specimens submerged in water without freezing-thawing conditions were investigated. The reference UHPC specimens are reported to gain 0.25%–0.35% mass, while those air-treated specimens obtain a mass gain of 0.09%–0.18% (Graybeal 2006). Moreover, the compressive strengths of untreated and steam-treated UHPC specimens increase by 12% and 3%, respectively, when compared to that of the air-treated specimens.

Graybeal (2006) also investigated the freezing-thawing resistance of UHPC that was treated with different curing regimes, as illustrated in Figure 10.3. The mass loss and relative dynamic modulus were studied. Test results showed that all the UHPC specimens treated by four different curing regimes exhibited mass increases throughout the testing. The untreated specimens displayed the greatest mass increase of 0.2% after 125 cycles. The relative dynamic modulus of tempered steam and delayed steam-treated specimens was only sightly changed. However, such value slightly decreased in the steam-treated regime and significantly increased in the untreated regime. On the contrary, Cwirzen et al. (2008) revealed that a generally more extensive internal damage developed in the case of the heat-treated UHPC specimens after freezing-thawing cycling combined with deicing salts, which is attributed to the variation of the very low internal relative humidity after heat treatment.

10.7 DEICER SCALING RESISTANCE

Deicers are frequently used in cold climates for snow and ice control to ensure safe and efficient roadway and/or aircraft operation. However, scaling due to freezing in the presence of deicers is a big concern for portland cement concrete pavements. ASTM C672 (2012) is often used for the evaluation of scaling resistance of concrete in the United States. UHPC mixtures indicate very low amounts of surface scaling after 50 cycles (Bonneau et al. 1997; Cwirzen et al. 2008). The scaling rate of UHPC after 56 cycles is 100 g/m^2, which is relatively low compared to the normal acceptance limit of 1500 g/m^2 after 28 cycles (Schmidt et al. 2003). Graybeal (2006) reported that UHPC specimens did not show surface scaling after 215 freezing-thawing cycles. Cwirzen et al. (2008)

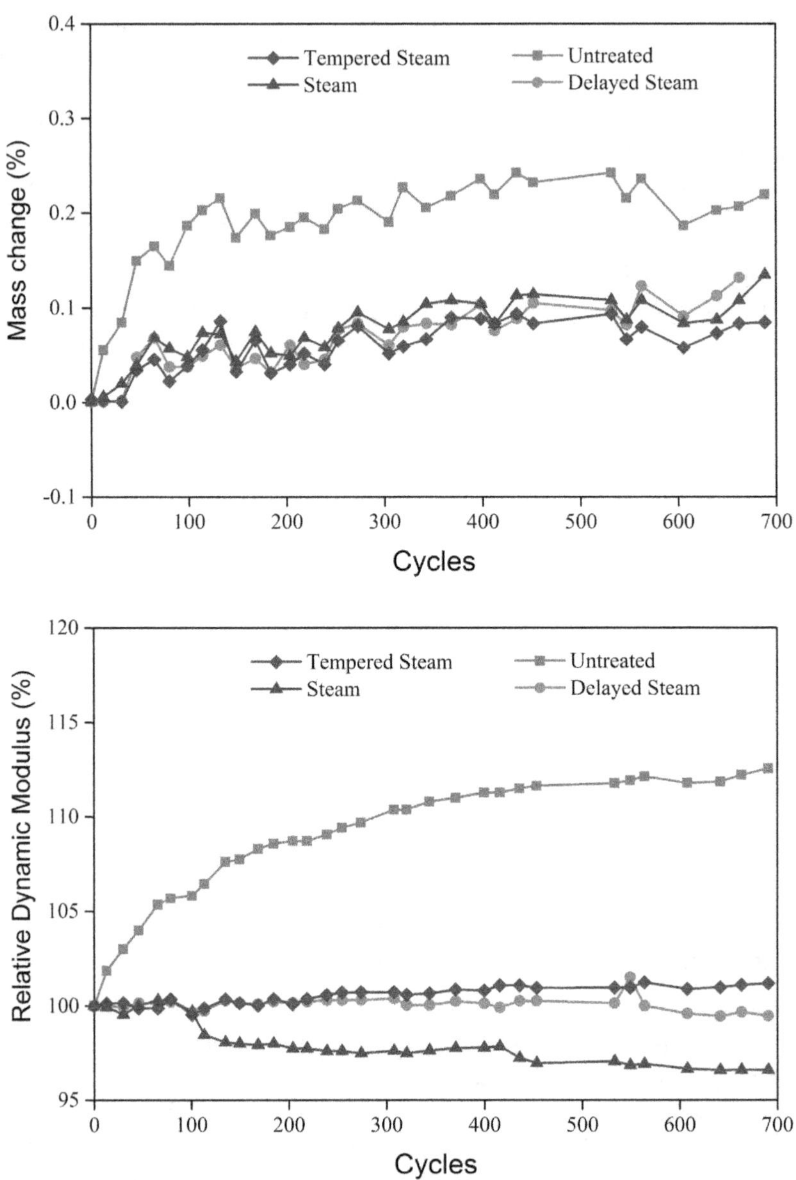

Figure 10.3 Changes in mass loss and relative dynamic modulus of UHPC subjected to freezing-thawing action under different curing regimes (Graybeal 2006). (Untreated regime: standard laboratory environment; steam curing: 90°C and 95% RH for 48 h; tempered steam: 60°C and 95% RH for 48 h; delayed steam: initiated steam curing mentioned above until the 15th day after casting. (a) Mass change. (b) Relative dynamic modulus.)

reported that heat-treated UHPC specimens prepared with a w/b ratio of 0.2 showed 500 g/m^2 surface scaling after 150 cycles and the relative dynamic modulus dropped below 50% after 200 cycles. The presence of steel fibers restrains the internal damage but causes higher surface scaling due to enhanced water ingress from the interfacial transition zone between the fiber and matrix as observed by the backscattered scanning electron microscope images (Cwirzen et al. 2008).

10.8 CHEMICAL ATTACK RESISTANCE

UHPC structures exposed to marine environments suffer from chemical attacks, such as sulfates and chlorides, leading to corrosion and spalling. Different successive altered layers, including carbonation, magnesium attack, and sulfate attack with the formation of gypsum, are found in a concrete structure exposed to seawater for several years (Aïtcin 2003). It takes time to show obvious chemical attack signs in UHPC due to its extremely dense structure. Yang and Huang (2003) found that the compressive and flexural strengths of UHPC specimens after 180 d of immersion in tap water and artificial seawater were greater than the original strengths. This is attributed to the continued hydration of unhydrated cement particles in concrete with the extension of soaking time. However, Tang et al. (2016) revealed that the imitative UHPC has an extremely shallow depth of invasion, and the factor of antierosion retained above 1.1 after 120 d of sulfate leaching attack. El-Dieb (2009) reported that the reduction in the compressive strength was around 12% after 12 months of immersion in high sulfate exposure conditions. Besides, Ye et al. (2006) investigated the behavior of high-strength concrete (HSC) and UHPC subjected to different chemical solutions, including 5% H_2SO_4, 20% Na_2SO_4, 20% $(NH_4)_2SO_4$, and artificial seawater. The relative compressive strength, which is the percentage of compressive strength of specimens after exposure to solutions to the original value before immersion, was used as an anticorrosion index. The results showed that the relative compressive strength of UHPC specimens was 1%–37% greater than that of the HSC specimens, depending on the solution type. Franke et al. (2008) compared the behaviors of UHPC specimens exposed to sulfuric acid, lactic acid, ammonium nitrate, and sodium solutions to those of conventional concrete. The results indicated that the corrosion depth of UHPC subjected to sulfuric acid with a pH value of 3 was the same as that of the reference mortar in sulfuric acid with a pH value of 4. It was also reported that there was not much reduction in mass and compressive strength (9.2% max) when UHPC were immersed in various concentrations of sulfuric acid solution, sodium chloride solution, and sulfate solution for 60 d (Reju and Jacob 2012). Thus, UHPC is a good candidate for applications in harsh marine environments.

10.9 ALKALI-SILICA REACTION

Alkali-silica reaction refers to a swelling reaction between alkali cations and hydroxyl ions in the pore solution of hydrated cement paste and certain reactive forms of silica from used aggregates in concrete in the presence of sufficient moisture. For conventional concrete with a w/b ratio of less than 0.5, it is necessary to have a 65% relative humidity at 38°C to develop significant expansions greater than 0.04% per year using reactive aggregates, such as Spratt limestone and Beauceville tuff (Bérubé et al. 2002). The critical relative humidity increases up to 85% in the presence of less reactive aggregates. Since UHPC usually has a high cement content and a low water content, more alkalis per cubic meter exist, and these alkalis have less residual water to dissolve. Consequently, an interstitial solution that is rich in alkalis with a high pH can be reached. On the other hand, 10% or more silica fume is often added in UHPC, which can decrease the pH value and mitigate alkali-silica reaction to some extent owing to the pozzolanic reaction with portlandite.

Graybeal (2006) evaluated the potential alkali-silica reaction ability of UHPC specimens with a size of 25 × 25 × 280 mm. The specimens were subjected to a sodium hydroxide solution for 2–4 weeks at 80°C. The expansion values of untreated and thermally treated specimens at 28 d were 0.012% and 0.002%, respectively. Moser et al. (2009) found that the intentionally predamaged UHPC specimens showed an expansion value of 0.02% after 603 d, which is below the threshold limit of 0.04%. Waste glass is recycled as aggregate in UHPC to replace quartz sand, but its application is often limited due to the deleterious effect of alkali-silica reaction expansion of the waste glass. Soliman and Tagnit-Hamou (2017) investigated the alkali-silica reaction expansion of UHPC made with 0%, 50%, and 100% glass sand according to ASTM C 1260 (2021). It was reported that the maximum expansion of UHPC at 16 d was 0.03%, which is much lower than the specified limit of 0.1%. Thus, the alkali-silica reaction is not a concern for UHPC under any curing regime given its low permeability (Vernet 2003; Graybeal 2007).

10.10 ABRASION RESISTANCE

Abrasion resistance refers to the ability of a surface to resist being worn away by rubbing or friction. It is a critical design factor for concrete structures, including highway pavements, floors, and some parts of hydraulic structures such as spillways, pipes, outlets of dams, and bridge piles subjected to the action of water-containing sediments. The abrasion resistance of concrete is directly related to its strength, which is affected by mixture proportions, aggregate-paste bond, placing, compaction, materials and surface finishing,

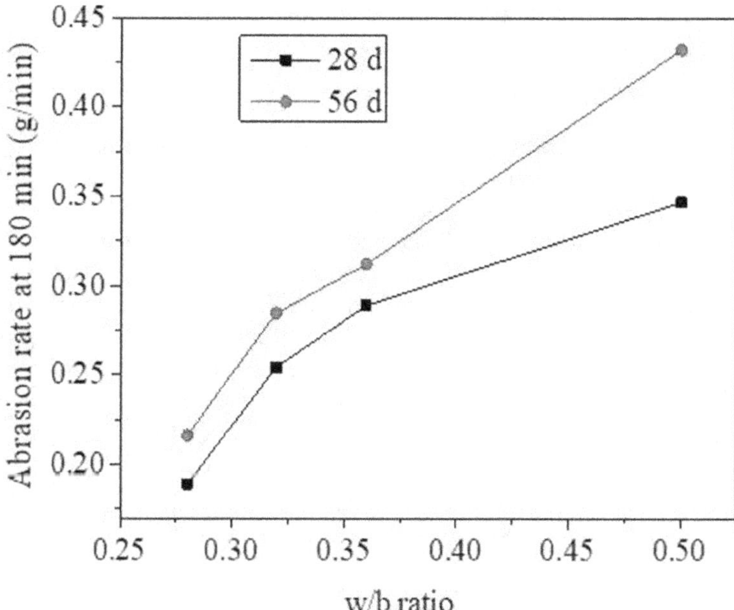

Figure 10.4 Relationship between w/b ratio and abrasion rate of concrete at 180 minutes (modified based on website https://ccppa.ca/abrasion-resistance-of-concrete-pipe/).

aggregate hardness, and curing. A reduction in w/b ratio increases the abrasion resistance of concrete, as illustrated in Figure 10.4.

Graybeal (2006) investigated the abrasion resistance of three UHPC surfaces, including cast against a steel mold, sand-blasted, and ground, following ASTM C944 (2019). The results indicated that steam-cured specimens showed much higher abrasion resistance when compared to the untreated specimens. Under steam-curing, the specimens with surfaces cast against the steel mold demonstrated higher abrasion resistance than those of the sand-blasted or ground surfaces. Roux et al. (1996) studied the abrasion resistance of C30, C80, and RPC200 using the Compagnie Nationale du Rhone test. RPC200 showed an abrasion coefficient of 1.3, comparable to 1.2 of a mortar cast with corundum aggregate. The abrasion coefficients of C30 and C38 were 4.0 and 2.8, respectively. Lee et al. (2007) investigated the abrasion resistance of conventional concrete, HPC, and UHPC with w/b ratio of 0.14, conforming to ASTM C131 (2020). It is indicated that conventional concrete displayed a weight reduction of 32% and 67%, respectively, after 500 and 1000 cycles of abrasion. The corresponding values of

UHPC were 5% and 8%, respectively, which were approximately 8 and 4 times greater than those of conventional concrete and HPC, respectively.

10.11 FIRE RESISTANCE

Concrete is an excellent fire-resistant construction material because of its low thermal conductivity and high thermal capacity. The behavior of concrete in a fire depends on its mixture proportions, specimen thickness, exposed heating rate, permeability, pore structure, and moisture content. Three different mechanisms, namely restrained thermal stress theory, build-up of pore pressure, and combined thermal stress and pore pressure, are referred to in explaining explosive spalling (Khoury 2000). Spalling of concrete under fire exposure is the breaking up of fragments and is mainly caused by the accumulation of pore pressure within concrete mass, which is generated as a consequence of presented water transformed into steam. When this built-up pore pressure becomes greater than the decreasing tensile strength of concrete, layers or pieces of concrete disintegrate from concrete members. This can lead to loss of concrete cover and overall cross-sectional area, thus resulting in lower fire resistance. The physicochemical processes of concrete during heating at different temperature ranges are summarized in Table 10.5. These changes gradually affect the mechanical properties of UHPC.

Conventional concrete and HPC microstructurally follow similar trends when heated, but UHPC behaves differently. Conventional concrete retains much of its strength and exhibits minimal spalling after exposure to elevated temperatures. The explosive spalling in conventional concrete is unlikely to occur if its free moisture content is below 3%–4% (Zhukov 1976). The very high compressive strength and superior durability of UHPC arise mainly

Table 10.5 Physicochemical processes of concrete during heating

Temperature	*Reaction and characteristics*
30–105°C	• Hydrothermal reaction • Loss of free water and Chemical bonded water
110–300°C	• Decomposition of gypsum and ettringite • Loss of water from parts of carboaluminate hydrate; • Loss of chemically bound water from the decomposition of C-S-H and carboaluminate hydrates.
300–600°C	• Dehydroxylation of calcium hydroxide • Transformation of quartz in aggregates from α to β form with volumetric expansion
600–900°C	• Dissociation of calcium hydroxide • Decomposition of C-S-H
900–1400°C	Melting

from its increased packing density and decreased porosity. This can result in increased susceptibility to explosive spalling due to the movement of the moisture restrained by the limited porosity. It is stated that the virtual absence of capillary pores with the low amount of free water in UHPC makes it more prone to explosive spalling due to increasing vapor pressure and nonuniform thermal gradient during heating (Abid et al. 2017; Chan et al. 2000). The temperature of explosive spalling of UHPC is lower than that of conventional concrete. The following sections discuss the relationships between the elevated temperature and the mechanical properties of UHPC.

10.11.1 Compressive strength

Abid et al. (2017) summarized that the residual compressive strength vs. temperature relationship of UHPC is characterized by three following distinct stages: (1) initial stabilizing and regaining stage from 150 to 350°C; (2) strength loss stage ranging from 350 to 400 to 800°C; and (3) total strength loss stage starts after 800°C. The initial increase in compressive strength is due to dry hardening and continued hydration of unhydrated cement particles from heating (Rashad et al. 2012; Phan 1996). The strength loss in the second stage is due to the decomposition of $Ca(OH)_2$ and the breakage of C-S-H bond and volumetric expansion associated with quartz transformation from α to β form around 571°C (Rashad et al. 2012; Abrams 1971). The strength loss after 800°C is due to the complete decomposition of C-S-H and partial decomposition of $CaCO_3$, which finally results in severe cracking (So et al. 2015; Canbaz 2014).

Figure 10.5 compares the changes in compressive strengths of conventional concrete, HPC, and UHPC heated in a fixed temperature chamber of 500 ± 50°C. The compressive strengths of conventional concrete, HPC, and UHPC show strength losses of 41.5%, 53.3%, and 37.8%, respectively, after 60 minutes of heating, as shown in Figure 10.5 (Liu and Huang 2009). The corresponding values are 47.3%, 65.4%, and 44.4% after 120 min of heating duration. Studies reveal that compressive strength loss of UHPC specimens ranged from 30% to 50% after 1000°C heating (Sobia et al. 2015; Liang et al. 2018).

The increase in steel fiber content can decrease strength loss due to the restraining of cracking propagation. Tai et al. (2011) reported that the compressive strengths of UHPC made with 1%, 2%, and 3% steel fibers were increased by about 24%, 4%, and 7%, respectively, after 200°C heating. After 800°C heating, the compressive strengths dropped by about 82%, 78%, and 77%, respectively, which were approximately 20% of the original strength. Zheng et al. (2013) also found that compressive strengths of UHPC made with 1%, 2%, and 3% steel fibers dropped by about 28.1%, 26.1%, and 18.8%, respectively, after 100°C heating. After 600°C, the compressive strengths of specimens made with 1%, 2%, and 3% steel

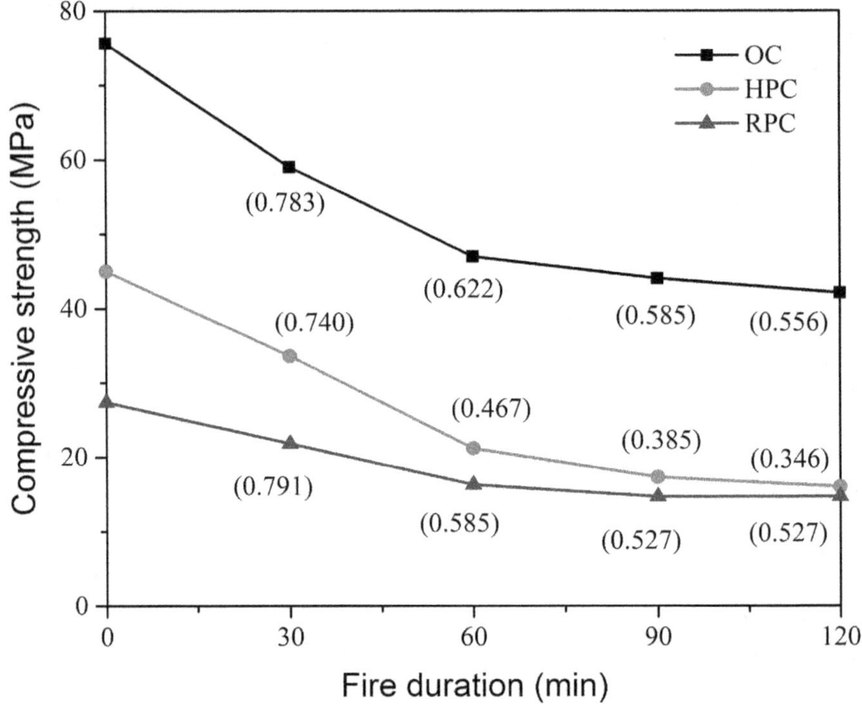

Figure 10.5 Changes in compressive strength with fire duration of ordinary concrete (OC), HPC, and UHPC.

Source: Liu and Huang (2009).

fibers were approximately 37.9%, 43.0%, and 33.5%, respectively, of the original strength. After 800°C heating, UHPC suffers severe compression deformation, which results in a decrease in performance. Besides, it was reported that the mass losses during heating from 150 to 800°C of UHPC specimens with different steel fiber volumes did not exceed 12% of the initial mass (2017). Compared to conventional concrete and HPC, UHPC specimens displayed a lower temperature difference between the outer and inner layers after the same exposure temperature and duration. This led to a slight spalling of UHPC and lower mass loss after 800–900°C heating (Sanchayan and Foster 2016; Feylessoufi et al. 1997).

10.11.2 Elastic modulus

Generally, the loss of elastic modulus increases as the water-to-cement ratio increases. Xiong et al. (2016) found that the elastic modulus of UHPC decreases with increasing temperatures due to the built-up internal pressure by the evaporation of free water. The secant modulus between the stress equal to 40% of peak stress and the stress corresponding to strain of 5 × 10^{-5} following ASTM C469 (2002) was defined as the elastic modulus. Compared to normal-strength concrete (NSC) and high-strength concrete (HSC), the elastic modulus of nonfibrous UHPC matrix was less affected by the elevated temperatures. Banerji and Kodur (2022) investigated the elastic modulus of UHPC at a temperature range of 20–750°C. The average elastic moduli were 58%, 32%, 8%, and 4% at 200, 400, 600, and 750°C, respectively, compared to that under room temperature. The loss of elastic modulus up to 400°C is due to microcracking and microstructure damage of concrete, resulting from moisture loss and shrinkage of cement paste. With temperature ranging from 400 to 750°C, the thermal mismatch and onset of decompositions of $Ca(OH)_2$ and C-S-H lead to a reduction in aggregate-paste bond and eventually the decrease in elastic modulus. Data generated from experiments were utilized to propose empirical relations of temperature-dependent mechanical properties, including compressive strength, elastic modulus, and tensile strength (Banerji and Kodur 2022). The predicted elastic modulus of UHPC was compared to those of NSC and HSC according to the ASCE manual (1994), Kodur et al. (2002), ACI (2014), and Eurocode (2004). As shown in Figure 10.6, UHPC experienced faster degradation in elastic modulus in comparison to NSC. However, its relative rate of modulus loss was comparable to that of HSC. Ahmad et al. (2019) examined the effects of fiber content and duration of thermal exposure on the elastic modulus of UHPC mixtures at a prespalling elevated temperature of 300°C. It was found that UHPC specimens showed 2%–5% decrease in elastic modulus after 60 min of exposure. The total decrease in elastic modulus of UHPC after 300 min of exposure was in the range of 27%–31%.

10.11.3 Flexural and tensile strengths

Flexural and tensile strengths of UHPC are much greater than those of NSC and can be efficiently utilized in achieving higher capacity of concrete. Under fire conditions, flexural and tensile strengths are important properties because they can resist crack propagation in concrete. For HSC, flexural and tensile strengths are critical properties as they help to overcome fire-induced spalling to some extent (Khaliq and Kodur 2011; 2018). Splitting tensile strength is often employed to evaluate the tensile properties of UHPC due to its ease of execution and more accurate obtained results (Banerji and Kodur 2022).

Banerji and Kodur (2022) evaluated the splitting tensile strength of UHPC at a temperature range of 20–750°C. The inclusion of steel fibers in UHPC helps in slowing down the strength loss with increasing temperature. The

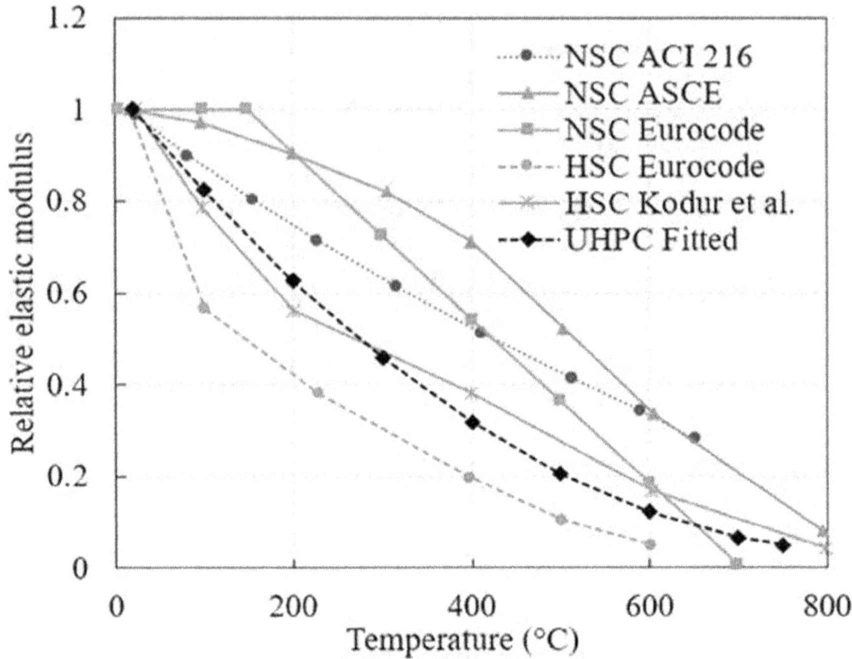

Figure 10.6 Variations in relative elastic modulus of UHPC, NSC, and HSC.

Source: Banerji and Kodur (2022).

average splitting tensile strength retentions of UHPCs at 200, 400, 600, and 750°C were 70%, 55%, 40%, and 20% of those at room temperature, respectively. Ahmad et al. (2019) found that UHPC at a prespalling elevated temperature of 300°C showed 13%–27% decrease in flexural strength. Li et al. (2020) investigated the flexural behavior of UHPC reinforced with hybrid polyethylene and steel fibers under exposed temperatures of 300 and 600°C. After 300°C heat exposure, the flexural performance of UHPC decreased significantly due to the melting of polyethylene fibers and reduced bond strength between steel fibers and matrix. The polyethylene fiber is not efficient in preventing spalling due to its lower coefficient of thermal expansion.

Park et al. (2019) studied the flexural and tensile performance of UHPCs incorporated with polyvinyl alcohol fiber (PVA), polyethylene fiber (PE), steel fiber (S), polypropylene fiber (PP), and nylon fiber (Ny) after 2-h duration of fire exposure up to 1100°C according to KSF 2257 (2014) using a high-temperature electric furnace. Compared to PP and PE fibers, the addition of PVA fibers was the most effective in enhancing the fire resistance

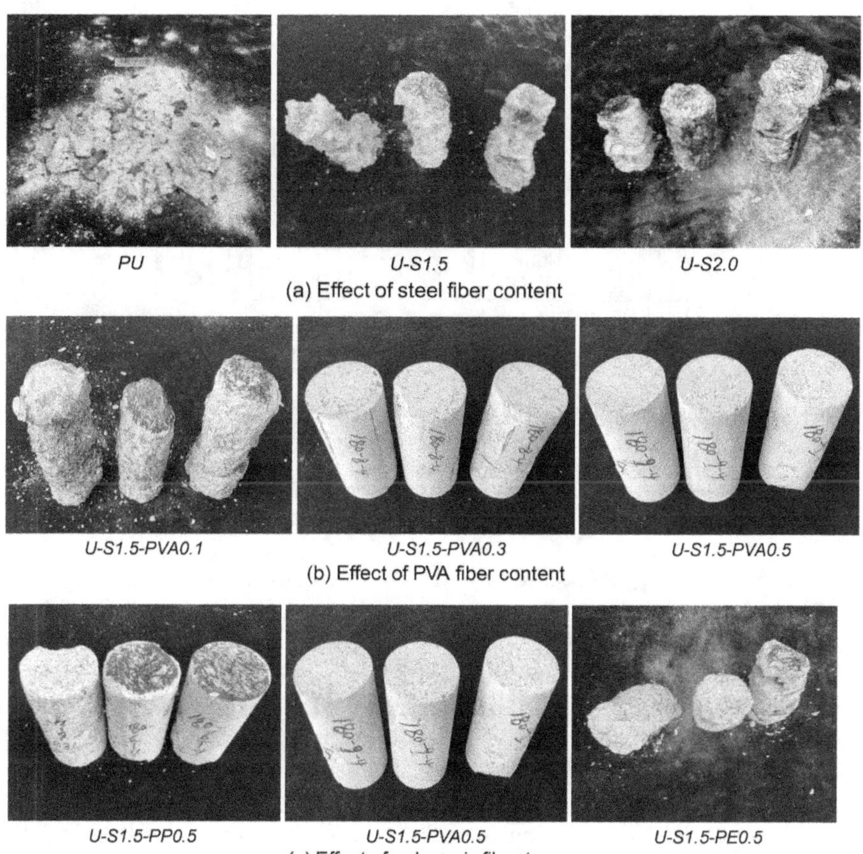

Figure 10.7 Pictures of explosive spalling of UHPC specimens after 2-h exposure to the ISO fire (Park et al. 2019). (a) Effect of steel fiber. (b) Effects of hybrid PVA and steel fibers. (c) Effects of hybrid polymer-based and steel fibers.

and residual tensile and flexural performances of UHPC, while the PE fiber was ineffective in enhancing the fire resistance (Figure 10.7). Liang et al. (2019) evaluated the coupled effect of temperature and impact loading on the tensile strength of UHPC, which retains 69% of its original compressive strength after exposure to 1000°C. The splitting tensile strength decreased sharply beyond 400°C but still retained 41% of its original strength at 800°C. Temperature and combined action of elevated temperature and impact loading had different effects on splitting tensile strength.

Qin et al. (2021) evaluated the effects of curing regimens, including standard curing (ST), 20°C water curing (CW), 90°C hot water curing

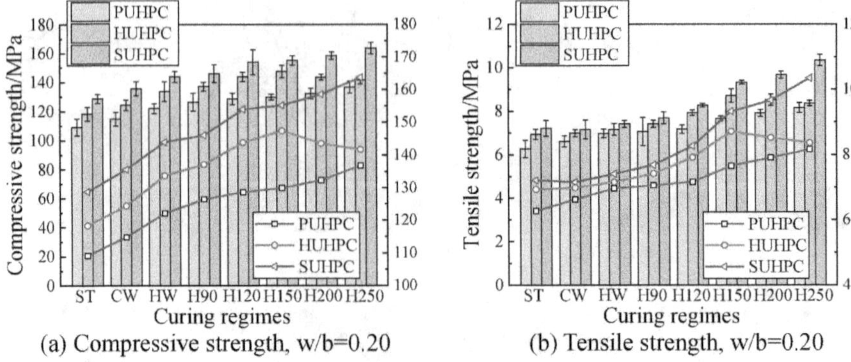

Figure 10.8 Compressive and tensile strengths of UHPC under various curing regimes (Qin et al. 2021). Notes: PUHPC represents plain specimens without fiber; SUHPC means specimens with 2% steel fiber volume; HUHPC is a mixture mixed with 2% steel fibers and 0.2% PP fibers.

(HW), steam curing (AC), and two-stage hot air curing (H90, H120, H150, and H250), on 2-h fire resistance of UHPC according to the ISO 834 fire standard. H90, H120, H150, and H250 present specimens initially cured at 50°C in hot air with a humidity of 95% curing for 24 h and then heated with a heating rate of 30°C/h in 3 h to 90, 120, 150, and 250°C in hot air for another 24 h, respectively. Figure 10.8 illustrates the compressive and tensile strengths of UHPC under various curing regimes. UHPC subjected to two-stage hot air curing showed higher compressive and tensile strengths than those under other curing regimes. Compared to standard curing, the compressive and tensile strengths of H150 were increased by 21.35% and 32.56%, respectively. However, the mechanical properties of H200 with PP fibers decreased due to the melting of the PP fibers.

10.11.4 Mass loss

Mass loss can be used as an effective parameter to describe the water evaporation and gas escape from UHPC being heated (Liang et al., 2018). UHPC mixtures have a small mass loss below 200°C, indicating a very compact structure that hinders the vapor from escaping. Being heated to 200°C, UHPC specimens experience an average mass loss of 1 g or 0.34% in percentage, indicating that steel fiber has little influence on the mass loss of UHPCs exposed to elevated temperatures. The mass loss of UHPC containing PP fibers increases sharply between 200 and 400°C because of the melting of PP fibers at 165°C. With the temperature further increasing

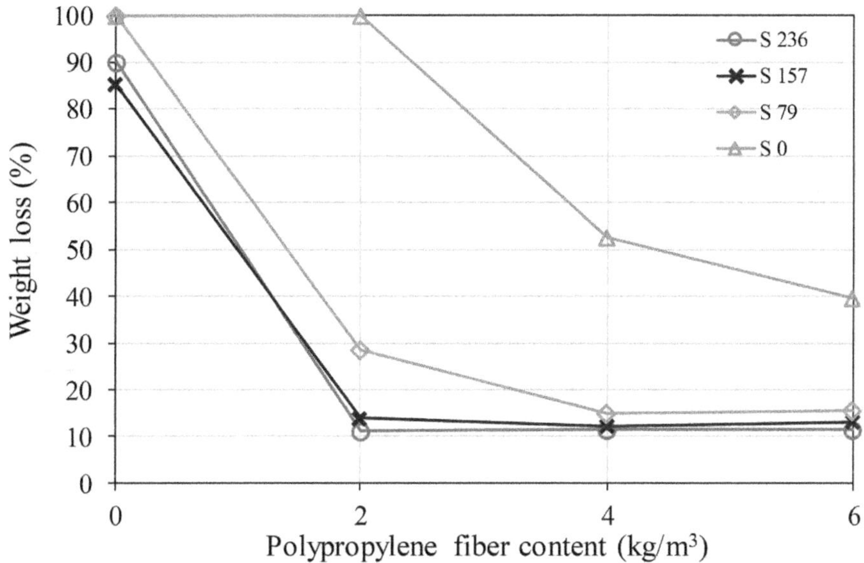

Figure 10.9 Weight loss of UHPC specimens made with different PP and steel fiber contents after spalling testing.

Source: Li et al. (2019).

from 400 to 1000°C, the mass loss increases slightly and reaches a peak value at around 8% (Liang et al. 2018).

Li et al. (2019) evaluated the synergetic effects of hybrid polypropylene (PP) and steel fibers on explosive spalling prevention of ultra-high performance concrete (UHPC) at elevated temperature. In this research, weight loss was examined and calculated by the weight of the most significant remaining part of UHPC after spalling divided by the original weight, including moisture. As can be seen from Figure 10.9, the control specimens without any fiber show the highest weight loss of nearly 100%, while UHPC with only steel fibers has a weight loss of more than 80%. The weight loss of UHPC containing only PP fibers is reduced with increasing fiber content. At a PP fiber content of 6 kg/m^3, the residual weight of specimens is 60.3%. With the hybridization of PP and steel fibers, the weight loss of UHPC is dramatically reduced. At a higher combination content, spalling is suppressed with a weight loss of 11.1%–15.5%, which is mainly attributed to the loss of moisture and decomposition of hydration products. According to Heo et al. (2010), 20% weight loss (inclusive of moisture loss) is taken as the threshold for severe spalling of UHPC.

10.11.5 Strategies to alleviate spalling of UHPC under elevated temperature

To alleviate the spalling of UHPC, several preventive measures, including uses of polypropylene fibers, air-entraining agent, protective lining and thermal barrier, two-stage hot air curing, steel fibers, and low expansion and small-size aggregates, can be employed. Table 10.6 summarizes the effectiveness, working mechanism, and characteristics of these preventive measures. The use of proper content of synthetic fiber with a low melting

Table 10.6 Preventive measures of spalling of UHPC

Method	*Effectiveness*	*Mechanism*	*Comments*
Polypropylene fibers	Very effective	Formation of pores and channels from melted fibers to release vapor and reduce build-up of pore pressure	Reducing strength due to the introduction of pores Poor workability due to hydrophobic nature
Air-entraining agent	Very effective	Introducing pores and channels to release vapor and reduce the build-up of pore pressure	Increase of 1% entrained air voids can reduce compressive strength by 5%
Protective lining and thermal barrier	Very effective	Reducing concrete temperature and heating rate on concrete surface	Additional required work and increase in labor cost Effectiveness depends on its thickness and cohesion with concrete Limited design criteria for application
Use of two-stage hot air curing	Very effective	Reducing moisture content	Formed a dense structural framework in the first curing stage and then promoted the further hydration and pozzolanic reaction to consume considerable free water and optimize pore structures
Steel fibers	Effective	Restraining cracking and spalling	Higher material cost
Low expansion and small-size aggregates	Effective	Improving thermal compatibility with cement paste	Porous aggregates with stored moisture lead to severe violent spalling

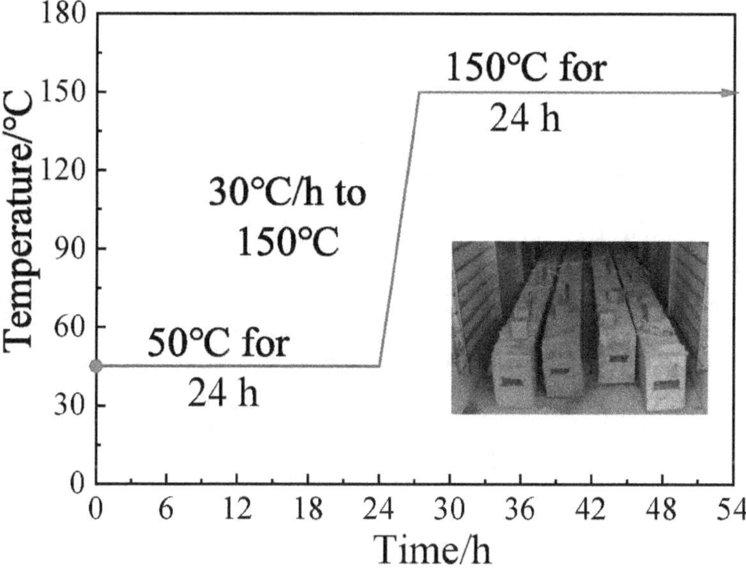

Figure 10.10 Two-stage hot air curing regime of H150.

Source: Qin et al. (2021).

point can alleviate explosive spalling. This is attributed to the formation of pores and channels from the melting of synthetic fiber around 150°C heating, which can release vapor and reduce the build-up of pore pressure (Sanchayan and Foster 2016; Zheng et al. 2012). Besides, good bond properties between the polyvinyl alcohol fiber and cementitious matrix, together with their higher elastic modulus than that of polypropylene fiber, can lead to better structural performance (Arisoy and Wu 2008; Qiu et al. 2016). Zheng et al. (2012) evaluated the heating temperature on the compressive strength of UHPC made with steel and polypropylene fibers. The use of polypropylene fiber exerted an adverse effect on the compressive strength of UHPC when exposed to a lower temperature but showed a positive influence at a higher temperature. Heinz et al. (2004) investigated the fire resistance of UHPC prepared with either steel fibers or a combination of steel and polypropylene fibers under loading. It is concluded that UHPC cylinders without polypropylene fibers exhibited spalling after a few minutes and were destroyed beyond recognition after 90 min. UHPC specimens made with 3.05% steel fibers and 0.6% polypropylene fibers provided the best results. Peng et al. (2012) found that 0.9% of polypropylene fibers was the critical content to enhance the fire resistance of UHPC. Therefore, incorporating steel and polypropylene fiber can obviously minimize or completely prevent spalling under fire exposure and effectively decrease strength loss because of cracking (Zheng et al. 2012; Li and Liu 2016; Chen and Liu 2004). As revealed from microstructural analysis, UHPC with steel and polypropylene fibers shows increased permeability and improved connectivity due to the formation of pores and channels associated with the melting of polypropylene fibers (Li et al. 2009). Low thermal expansion aggregates would be useful in combating thermal stress spalling (Khoury 2000). However, high moisture-conditioned aggregates in UHPC, such as saturated lightweight aggregate, can lead to severe violent spalling because of the additional introduction of water.

Qin et al. (2021) proposed a new curing regime, two-stage hot air curing (Figure 10.10), to inhibit the explosive spalling of UHPC in a high-stress state and under an extreme heating rate. It is found that severe spalling occurred in UHPC beams cured under standard curing. However, the two-stage hot air curing could reduce the moisture content and completely prevent spalling from occurring in small-size specimens and full-scale structural members under fire exposure. UHPC formed a dense structural framework in the first stage, with curing at 50°C. Subsequently, further hydration and pozzolanic reaction could be promoted in the second stage with curing at 150°C in hot air, resulting in the formation of dense and complete C-S-H gels. This improved the mechanical properties, consumed considerable free water, and optimized the pore structures.

10.12 SUMMARY

The superior durability of UHPC relies on its ultra-high strength, dense structure, and decreased capillary pores. The chloride ion diffusion coefficient of UHPC was found to be at least one order of magnitude lower than that of conventional concrete or HPC. UHPC can gain increases in mass, compressive strength, and relative dynamic modulus over several hundreds of freezing-thawing cycles up to 1000 cycles due to absorbed water and/or continued hydration. Accelerated tests indicated that corrosion of steel reinforcement and alkali-silica reaction are not concerns for UHPC under any curing regime given its low permeability. Those superior performances render UHPC an outstanding candidate for infrastructure systems to obtain long service life, especially those exposed to harsh climatic environments.

However, because of the direct implementation of UHPC into the structural applications without proper design codes, inefficient material use and function exertion need to be considered. Besides, various testing methods, especially traditional methods and threshold values of the performance evaluation used for conventional concrete or HPC, are used to evaluate the durability of UHPC. This may make the durability results of UHPC not credible and comparable.

REFERENCES

Abbas, S., Soliman A. M., Nehdi, M. L. (2015). Exploring mechanical and durability properties of ultra-high performance concrete incorporating various steel fiber lengths and dosages. *Constr. Build. Mater.* 75, 429–441.

Abid, M., Hou, X., Zheng, W., Hussain, R. R. (2017). High temperature and residual properties of reactive powder concrete–A review. *Constr. Build. Mater.* 147, 339–351.

Abrams, M. S. (1871). Compressive strength of concrete at temperatures to 1600 F. American Concrete Institute, Temperature and concrete 1971, ACI SP 25, Detroit (Michigan), 33–58.

ACI 216.1. (2014). Code requirements for determining fire resistance of concrete and masonry construction assemblies. Farmington Hills, MI: American Concrete Institute.

Ahmad, S., Rasul, M., Adekunle, S. K., Al-Dulaijan, S. U., Maslehuddin, M., Ali, S. I. (2019). Mechanical properties of steel fiber-reinforced UHPC mixtures exposed to elevated temperature: Effects of exposure duration and fiber content. *Compo. Part B: Eng.* 168, 291–301.

Ahlborn, T.M., Misson, D.L., Peuse, E.J., Gilbertson, C.G. (2008). Durability and strength characterization of ultra-high performance concrete under variable curing regimes. In Proc. 2nd Int. Symp. on Ultra High Performance Concrete, Fehling, E., Schmidt, M., Stürwald, S.(Eds.), Kassel, Germany, 197–204.

Aïtcin, P. C. (2003). The durability characteristics of high performance concrete: a review. *Cem. Concr. Compos.* 25 (4–5). 409–420.

Alkaysi, M., El-Tawil, S., Liu, Z., Hansen, W. (2016). Effects of silica powder and cement type on durability of ultra high performance concrete (UHPC). *Cem. Concr. Compos.* 66, 47–56.

Alonso, C., Castellote M., Llorente I., Andrade, C. (2006). Ground water leaching resistance of high and ultra high performance concretes in relation to the testing convection regime. *Cem. Concr. Res.* 36(9), 1583–1594.

Angst, U.M., Elsener, B., Larsen, C.K., Vennesland, Ø. (2011). Chloride induced reinforcement corrosion: electrochemical monitoring of initiation stage and chloride threshold values. *Corr. Sci.* 53 (4), 1451–1464.

An, M., Yang, X., Wang, J., Cui, N. (2007). Research on durability of RPC. *Architecture Technology* 38(5), 367–368.

Andrade, M.C., Frías, M., Aarup, B. (1996). Durability of ultra high strength concrete: compact reinforced composite (CRC). Proceedings of the Fourth International Symposium on the Utilization of High-Strength/High-Performance Concrete, 29–31 May, Paris, France, Ed., de Larrard, F. and Lacroix, R., 2, 529–534.

Arisoy, B., Wu, H. C. (2008). Material characteristics of high performance lightweight concrete reinforced with PVA. *Constr. Build. Mater.* 22(4), 635–645.

ASTM C 131/131M. (2020). Standard test method for resistance to degradation of small-size coarse aggregate by abrasion and impact in the Los Angeles Machine. ASTM International: West Conshohocken, PA, USA.

ASTM C 666. (2019). Standard test method for resistance of concrete to rapid freezing and thawing. ASTM International: West Conshohocken, PA, USA.

ASTM C 672/672M. (2012). Standard test method for scaling resistance of concrete surfaces exposed to deicing chemicals. ASTM International: West Conshohocken, PA, USA.

ASTM C 876. (2015). Standard Test Method for Corrosion Potentials of Uncoated Reinforcing Steel in Concrete. ASTM International: West Conshohocken, PA, USA.

ASTM, C 944. (2019). Standard test method for abrasion resistance of concrete or mortar surfaces by the rotating-cutter method. In American Society for Testing and Materials (pp. 1–4). ASTM International: West Conshohocken, PA, USA.

ASTM C1202, (2019). Electrical indication of concrete's ability to resist chloride ion penetration, Annual Book of American Society for Testing Materials Standards, vol. C04.02.

ASTM C1620. (2021). Standard test method for potential alkali reactivity of aggregates (mortar-bar method). ASTM International: West Conshohocken, PA, USA.

ASTM C 469. (2002). Standard test method for static modulus of elasticity and Poisson's ratio of concrete in compression. ASTM International: West Conshohocken, PA, USA.

Banerji, S., Kodur, V. (2022). Effect of temperature on mechanical properties of ultra-high performance concrete. *Fire Mater.* 46(1), 287–301.

Bertos, M.F., Simons, S.J.R., Hills, C.D., Carey, P.J. (2004). A review of accelerated carbonation technology in the treatment of cement-based materials and sequestration of CO_2. *J. Hazard. Mater.* 112 (3), 193–205.

Bérubé, M.A., Dorion, J.F., Rivest, M (2002). Laboratory assessment of alkali contribution by aggregates to concrete and application to concrete structures affected by ASR. *Cem. Concr. Res.* 32(8), 1215–1227.

Bondar, D., Lynsdale, C. J., Milestone, N. B., Hassani, N. (2015). Sulfate resistance of alkali activated pozzolans. *International Journal of Concrete Structures and Materials* 9, 145–158.

Bonneau, O., Lachemi, M., Dallaire, E., Dugat, J., Aïtcin, P.C. (1997), Mechanical properties and durability of two industrial reactive powder concretes. *ACI Mater. J.* 94(4), 286–290.

Canbaz, M. (2014). The effect of high temperature on reactive powder concrete. *Constr. Build. Mater.* 70(70), 508–513.

Chan, Y. N., Luo, X., Sun, W. (2000). Compressive strength and pore structure of high-performance concrete after exposure to high temperature up to 800 °C. *Cem. Concr. Res.* 30(2), 247–251.

Chen, B., Liu, J. (2004). Residual strength of hybrid-fiber-reinforced high-strength concrete after exposure to high temperatures. *Cem. Concr. Res.* 34(6), 1065–1069.

Chen, Y., Yu, R., Wang, X., Chen, J., Shui, Z. (2018). Evaluation and optimization of Ultra-High Performance Concrete (UHPC) subjected to harsh ocean environment: towards an application of layered double hydroxides (LDHs). *Constr. Build. Mater.* 177, 51–62.

Cheyrezy, M., Maret, V., Frouin, L. (1995). Microstructural analysis of RPC (reactive powder concrete). *Cem. Concr. Res.* 25(7), 1491–1500.

Chuang, M. L., Huang, W. H. (2013). Durability Analysis Testing on Reactive Powder Concrete. *Adv. Mater. Res.* 811, 244–248.

Cwirzen, A. (2007). The effect of the heat-treatment regime on the properties of reactive powder concrete. *Adv. Mater. Res.* 19 (1), 25–34.

Cwirzen, A., Penttala, V., Vornanen, C. (2008). Reactive powder based concretes: Mechanical properties, durability and hybrid use with OPC. *Cem. Concr. Res.* 38(10), 1217–1226.

Cwirzen, A., Penttala, V., Cwirzen K. (2008). The effect of heat treatment on the salt freeze-thaw durability of UHSC. In *Proceedings of the 2nd International Symposium on Ultra High Performance Concrete*, Kassel, Germany (221–230). March.

Dobias, D., Pernicova, R., Mandlik, T. (2016). Water transport properties and depth of chloride penetration in ultra high performance concrete. *Key Eng. Mater.* 711, 137–142.

El-Dieb, A. S. (2009). Mechanical, durability and microstructural characteristics of ultra-high-strength self-compacting concrete incorporating steel fibers. *Mater. Des.* 30(10), 4286–4292.

El-Turki, A., Ball, R.J., Allen, G.C. (2007). The influence of relative humidity on structural and chemical changes during carbonation of hydraulic lime. *Cem. Concr. Res.* 8 (37), 1233–1240.

EN 1992-1-2. (2004). Eurocode 2: Design of Concrete Structures—Part 1–2: General Rules-Structural Fire Design. Brussels, Belgium: European Committee for Standardization.

Fan, L., Meng, W., Teng, L., Khayat, K. H. (2019). Effect of steel fibers with galvanized coatings on corrosion of steel bars embedded in UHPC. *Compos. Part B: Eng.* 177, 107445.

Faizal, M. M., Hamidah, M. S., Norhasri, M. M., Noorli, I. (2016). Effect of clay as a nanomaterial on corrosion potential of steel reinforcement embedded in ultra-high performance concrete. In InCIEC 2015 (679-687). Springer, Singapore.

Fehling, E., Schmidt, M., Teichmann, T., et al. (2005). Entwicklung, Dauerhaftigkeit und Berechnung Ultra hochfester Betone (UHPC). Kassel: Kassel Universitz Press GmbH.

Feylessoufi, A., Crespin, M., Dion, P., Bergaya, F., Van Damme, H., Richard, P. (1997). Controlled rate thermal treatment of reactive powder concretes. *Advanced cement based materials* 6(1), 21–27.

Franke, L., Schmidt, H., Deckelmann, G. (2008). Behavior of ultra-high performance concrete with respect to chemical attack. In *Proceedings of the 2nd International Symposium on Ultra-High Performance Concrete*, Kassel, Germany (pp. 453–460), March.

Gao, R., Liu, Z. M., Zhang, L.Q., Stroeven, P. (2006). Static properties of plain reactive powder concrete beams. *In Key Engineering Materials* 302, 521–527, Trans Tech Publications.

Ghafari, E., Costa, H., Júlio, E., Portugal, A., Durães, L. (2012). Enhanced durability of ultra high performance concrete by incorporating supplementary cementitious materials. In The 2nd international conference microdurability Delft, Amsterdam, Netherland (pp 86–94), April.

Ghafari, E., Costa, H., Júlio, E., Portugal, A., Durães, L. (2014). The effect of nanosilica addition on flowability, strength and transport properties of ultra high performance concrete. *Mater. Des.* 59, 1–9.

Ghafari, E. Arezoumandi, M., Costa, H., Julio, E. (2015). Influence of nano-silica addition on durability of UHPC. *Constr. Build. Mater.* 94, 181–188.

Glass, G. K., Buenfeld, N. R. (1997). The presentation of the chloride threshold level for corrosion of steel in concrete. *Corr. Sci.* 39(5), 1001–1013.

Graybeal, B. A. Material property characterization of ultra-high performance concrete (No. FHWA-HRT-06-103), 2006.

Granger, S., Loukili, A., Pijaudier-Cabot, G., Chanvillard, G. (2007). Experimental characterization of the self-healing of cracks in an ultra high performance cementitious material: Mechanical tests and acoustic emission analysis. *Cem. Concr. Res.* 37, 519–527.

Graybeal, B. (2006). Material property characterization of ultrahigh performance concrete. In FHWA-HRT-06-103, U.S. Department of Transportation (p. 176).

Graybeal B., Tanesi J. (2007). Durability of an ultra high-performance concrete. *J. Mater Civil Eng.* 19(10), 848–854.Heinz, D., Ludwig, H.M. (2004). Heat treatment and the risk of DEF delayed ettringite formation in UHPC. In Proceedings of the International Symposium on UHPC, Kassel, Germany, 717–730.

Heinz, D., Dehn, F., Urbonas, L. (2004). Fire resistance of ultra high performance concrete (UHPC)-Testing of laboratory samples and columns under load. In International Symposium on Ultra High Performance Concrete, Kassel, Germany, 703–715.

Heo, Y. S., Sanjayan, J. G., Han, C. G., Han, M. C. (2010). Synergistic effect of combined fibers for spalling protection of concrete in fire. *Cem. Concr. Res.* 40(10), 1547–1554.

Herold, G., Müller, H. S. (2004, September). Measurement of porosity of ultra high strength fibre reinforced concrete. In Proceedings of the International Symposium on Ultra-High Performance Concrete, Kassel, Germany, 685–694.

Imm, P. Abrasion Resistance of Concrete Pipe. Ontario Concrete Pipe Association.

Jacobsen, S., Sellevold, E. J. (1996). Self-healing of high strength concrete after deterioration by freeze/thaw. *Cem. Concr. Res.* 26(1) 55–62.

Khaliq, W., Kodur, V.K.R. (2011). Effect of high temperature on tensile strength of different types of high-strength concrete. *ACI Mater. J.* 108(4), 394–402.

Khaliq, W., Kodur, V. (2018). Effectiveness of polypropylene and steel fibers in enhancing fire resistance of high-strength concrete columns. *J. Struct. Eng.* 144(3), 04017224.

Khoury, G. A. (2000). Effect of fire on concrete and concrete structures. *Progress in Structural Engineering and Materials* 2(4), 429–447.

Kodur, V., Naser, M. (2020) Structural fire engineering. London, UK: McGraw Hill Professional.

KS F 2257-1. (2014). Method of fire resistance test for elements of building construction - general requirements. Korean Standard Association, Seoul, Korea.

Lee, M. G., Chiu, C. T., Wang, Y. C. (2005). The study of bond strength and bond durability of reactive powder concrete. *J. ASTM Int.* 2(7), 1–10.

Lee, M. G., Wang, Y. C., Chiu, C. T. (2007). A preliminary study of reactive powder concrete as a new repair material. *Constr. Build. Mater.* 21(1), 182–189.

Li, Z., Huang, L. (2005). Study on durability of fiber-reinforced reactive powder concrete. *China Concr. Cement Prod.* 23(3), 42–43.

Li, H., Liu, G. (2016). Tensile properties of hybrid fiber-reinforced reactive powder concrete after exposure to elevated temperatures. *Int. J. Concr. Struct. Mater.* 10(1), 29–37.

Li, Y., Pimienta, P., Pinoteau, N., & Tan, K. H. (2009). Effect of aggregate size and inclusion of polypropylene and steel fibers on explosive spalling and pore pressure in ultra-high-performance concrete (UHPC) at elevated temperature. *Cem. Concr. Compos.* 99, 62–71.

Li, Y., Tan, K. H., Yang, E. H. (2019). Synergistic effects of hybrid polypropylene and steel fibers on explosive spalling prevention of ultra-high performance concrete at elevated temperature. *Cem. Concr. Compos.* 96, 174–181.

Li, Y., Yang, E. H., Tan, K. H. (2020). Flexural behavior of ultra-high performance hybrid fiber reinforced concrete at the ambient and elevated temperature. *Constr. Build. Mater.* 250, 118487.

Liang, X., Wu, C., Yang, Y., Wu, C., Li, Z. (2019). Coupled effect of temperature and impact loading on tensile strength of ultra-high performance fibre reinforced concrete. *Compo. Struct.* 229, 111432.

Lie T.T. (1992). Structural Fire Protection. New York, NY: American Society of Civil Engineers.

Liu, J., Song, S., Wang, L. (2009). Durability and micro-structure of reactive powder concrete. *J. Wuhan Univ. Technol. Mater.* 24(3), 506–509.

Liu, B., Xie, Y., Li, J. (2005). Influence of steam curing on the compressive strength of concrete containing supplementary cementing materials. *Cem. Concr. Res.* 35(5), 994–998.

Liu, S., Sun, W., Lin, W., Lai, J. (2003). Preparation and durability of a high performance concrete with natural ultra-fine particles. *J. Chinese Ceram. Soc.* (Chinese), 31, 1080–1085.

Liu, C. T., Huang, J. S. (2009). Fire performance of highly flowable reactive powder concrete. *Constr. Build. Mater.* 23(5), 2072–2079.

Liang, X., Wu, C., Su, Y., Chen, Z., Li, Z. (2018). Development of ultra-high performance concrete with high fire resistance. *Constr. Build. Mater.* 179, 400–412.

Long, G. C., Xie, Y. J., Wang, P. M., Jiang, Z. W. (2005). Properties and micro/macrostructure of reactive powder concrete. *Journal of the Chinese Ceramic Society* 33 (4), 456–461.

Magureanu, C., Sosa, I., Negrutiu, C., Heghes, B. (2012). Mechanical properties and durability of ultra-high-performance concrete. *ACI Mater J.* 109(2), 177–184.

Matte, V., Moranville M. (1999). Durability of reactive powder composites: influence of silica fume on the leaching properties of very low water/binder pastes. *Cem. Concr. Compos.* 21(1), 1–9.

Mehta, P.K., Monteiro, P. (2013). Concrete: Structure, Properties and Materials. McGraw Hill; 4th edition, 13 (4), 499.

Moser, B., Pfeifer, C., Stark, J. (2009). Durability and microstructural development during hydration in ultrahigh performance concrete. 87–88, London, UK: Taylor and Francis Group.

Müller, C., et al. (2010). Durability of Ultra-High Performance Concrete (UHPC). Proceedings of the Third International fib Congress and Exhibition

Incorporating the PCI Annual Convention and National Bridge Conference, Washington, DC, May 29 to June 2, Compact Disc, Paper 135.

Mu, R., Miao, C., Luo, X., Sun, W. (2002). Interaction between loading, freeze-thaw cycles, and chloride salt attack of concrete with and without steel fiber reinforcement. *Cem. Concr. Res.* 32(7), 1061–1066.

Nordtest Method NT Build 443. (1995). Concrete. Hardened: accelerated chloride penetration. Nordtest, Finland.

Papadakis, V.G. (2000). Effect of supplementary cementing materials on concrete resistance against carbonation and chloride ingress. *Cem. Concr. Res.* 30 (2), 291–299.

Park, J. J., Koh, K. T., Ryou, G. S., Kim, S. W. (2007). Evaluation on durability of ultra-high strength cementitious composites. *J. Korean Soc. Civil Eng.* 27(2A), 257–263.

Park, J. J., Yoo, D. Y., Kim, S., Kim, S. W. (2019). Benefits of synthetic fibers on the residual mechanical performance of UHPFRC after exposure to ISO standard fire. *Cem. Concr. Compos.* 104, 103401.

Perry, V., Zakariasen, D. (2004). First use of ultra-high performance concrete for an innovative train station canopy. *Conc. Technol. Today* 25(2), 1–2.

Peng, G. F., Kang, Y. R., Huang, Y. Z., Liu, X. P., Chen, Q. (2012). Experimental research on fire resistance of reactive powder concrete. *Adv. Mater. Sci. Eng.* 2012: 860303.

Peng, Y. Z., Chen, K., Hu, S. G. (2011). Durability and microstructure of ultra-high performance concrete having high volume of steel slag powder and ultra-fine fly ash. *Advanced Mater. Res.* 255, 452–456. Trans Tech Publications.

Phan, L. T. (1996). Fire performance of high strength concrete. A report of the state-of-the-art building and fire research laboratory. National Institute of Standards and Technology, Maryland.

Piérard, J., Cauberg, N., Remy, O. (2009). Evaluation of durability and cracking tendency of ultra-high performance concrete. In: *Creep, shrinkage and durability mechanics of concrete and concrete structures*, Taylor and Francis Group. London, UK, 695–700.

Piérard, J., Dooms, B., Cauberg, N. (2012). Evaluation of durability parameters of UHPC using accelerated lab tests. In Proceedings of the 3rd International Symposium on UHPC and Nanotechnology for High Performance Construction Materials, Kassel, Germany, 371–376.

Poon, C.S., Kou, S.C., Lam, L. (2006). Compressive strength, chloride diffusivity and pore structure of high performance metakaolin and silica fume concrete. *Constr. Build. Mater.* 20(10), 858–865.

Poupard, O., Aıt-Mokhtar, A., Dumargue, P. (2004). Corrosion by chlorides in reinforced concrete: Determination of chloride concentration threshold by impedance spectroscopy. *Cem. Concr. Res.* 34(6), 991–1000.

Qin, H., Yang, J., Yan, K., Doh, J. H., Wang, K., Zhang, X. (2021). Experimental research on the spalling behaviour of ultra-high performance concrete under fire conditions. *Constr. Build. Mater.* 303, 124464.

Qiu, J., Lim, X. N., Yang, E. H. (2016). Fatigue-induced deterioration of the interface between micro-polyvinyl alcohol (PVA) fiber and cement matrix. *Cem. Concr. Res.* 90, 127–136.

Rafiee, A., Schmidt, M. (2012). Computer modeling and investigation on the chloride induced steel corrosion in cracked UHPC. In Ultra-High Performance Concrete and Nanotechnology in Construction. Proceedings of Hipermat 2012. 3rd International Symposium on UHPC and Nanotechnology for High Performance Construction Materials (No. 19, p. 357). Kassel University Press GmbH.

Reju, R., Jacob, G. J. (2012). Investigations on the chemical durability properties of ultra high performance fibre reinforced concrete. In Proceedings of International Conference on Green Technologies, New York, USA, 181–185.

Rashad, A. M., Bai, Y., Basheer, P. A. M., Collier, N. C., Milestone, N. B. (2012). Chemical and mechanical stability of sodium sulfate activated slag after exposure to elevated temperature. *Cem. Concr. Res.* 42(2), 333–343.

Rabehi, B., Ghernouti, Y., Boumchedda, K., Li, A., Drir, A. (2017). Durability and thermal stability of ultra high-performance fibre-reinforced concrete (UHPFRC) incorporating calcined clay. *Eur. J. Environ. Civil Eng.* 21(5), 594–611.

Roux, N., Andrade, C., Sanjuan, M.A. (1996). Experimental study of durability of reactive powder concretes. *J. Mater. Civil Eng.* 8(1), 1–6.

Sanchayan, S., Foster, S. J. (2016). High temperature behaviour of hybrid steel-PVA fibre reinforced reactive powder concrete. *Mater. Struct.* 49 (3), 769–782.

Scheydt, J.C., Muller, H.S. (2012). Microstructure of ultra high performance concrete (UHPC) and its impact on durability. In Proceedings of the 3rd International Symposium on UHPC and Nanotechnology for High Performance Construction Materials, Kassel, Germany, 349–356.

Schmidt, M., Fehling, E., Bornemann, R., Bunje, K., Teichmann, T. (2003). Ultra-high performance concrete: Perspective for the precast concrete industry. *Betonwerk Und Fertigteiltechnik*, 69(3), 16–29.

Schmidt, M., Fehling, E. (2005). Ultra-high-performance concrete: research, development and application in Europe. *ACI Special Publication* 228, 51–78.

Schwartz, D.R. (1987). D-cracking of concrete pavements. NCHRP Synthesis of Highway Practice No. 134.

Shi, T., Wei, S., Shen, J. Y., Ye, Q. (2012). Preparation of slag reactive powder concrete and the research on its resistance to chloride ion permeability. *In Adv. Mater. Res.* 391, 1189–1194, Trans Tech Publications Ltd.

Sun, W., Lai, J. Z. (2009). Dynamic mechanical behaviour and durability of ultra-high performance cementitious composite. *Key Eng. Mater.* 400–402, 3–15.

Smarzewski, P., Barnat-Hunek, D. (2017). Effect of fiber hybridization on durability related properties of ultra-high performance concrete. *Int. J. Conc. Struct. Mater.* 11(2), 315–325.

Shaheen, E., Shrive, N. G. (2006). Optimization of mechanical properties and durability of reactive powder concrete. *ACI Mater. J.* 103 (6), 444.

Soliman, N. A., Tagnit-Hamou, A. (2017). Using glass sand as an alternative for quartz sand in UHPC. *Constr. Build. Mater.* 145, 243–252.

So, H. S., Jang, H. S., Khulgadai, J., So, S. Y. (2015). Mechanical properties and microstructure of reactive powder concrete using ternary pozzolanic materials at elevated temperature. *KSCE J. Civil Eng.* 19 (4), 1050–1057.

Sobia, A. Q., Hamidah, M. S., Azmi, I., Rafeeqi, S. F. (2015). Elevated temperature resistance of ultra-high-performance fibre-reinforced cementitious composites. *Mag. Conc. Res.* 67 (17), 923–937.

Tafraoui, A., Escadeillas, G., Vidal, T. (2016). Durability of the ultra high performances concrete containing metakaolin. *Constr. Build. Mater.* 112, 980–987.

Tai, Y. S., Pan, H. H., Kung, Y. N. (2011). Mechanical properties of steel fiber reinforced reactive powder concrete following exposure to high temperature reaching 800 °C. *Nucl. Eng. Des.* 241(7), 2416–2424.

Tam, C.M., Tam, V.W., Ng, K.M. (2012). Assessing drying shrinkage and water permeability of reactive powder concrete produced in Hong Kong. *Constr. Build. Mater.* 26(1), 79–89.

Tang, X. G., Xie, Y. J., Long, G. C. (2016). Experimental study on performance of imitative RPC for sulphate leaching. *Journal of Asian Ceramic Societies* 4, 143–148.

Tayeh, B.A., Bakar, B. H., Johari, M. A., et al. (2012). Mechanical and permeability properties of the interface between normal concrete substrate and ultra high performance fiber concrete overlay. *Constr. Build. Mater.* 36, 538–548.

Thomas, M., Green, B., O'Neal, E., Perry, V., Hayman, S., Hossack, A. (2012). Marine performance of UHPC at Treat Island. In Proceedings of the 3rd International Symposium on UHPC and Nanotechnology for High Performance Construction Materials, Kassel, Germany, 365–370.

Toutanji, H., McNeil, S., Bayasi, Z. (1998). Chloride permeability and impact resistance of polypropylene-fiber-reinforced silica fume concrete. *Cem. Concr. Res.* 28(7), 961–968.

Trinh, B. P., Chanh, N. V. (2011). Study on the properties of ultra high performance fiber reinforced concrete. *Adv. Mater. Res.*. 383–390, 3305–3312.

Vernet, C. (2003). UHPC microstructure and related durability performance laboratory assessment and field experience examples. Proc., Int. Symp. on High Performance Concrete, Orlando, Florida.

Voo, Y. L., Foster, S. J. (2010). Characteristics of ultra-high performance "ductile" concrete and its impact on sustainable construction. *IES J. Part A Civil Struct. Eng.* 3(3), 168–187.

Wang, Z. (2011). Chloride-ion permeability of reactive powder concrete. Maser thesis, Beijing Jiaotong University.

Wang, Y., An, M. Z., Yu, Z. R., Han, S., Ji, W. Y. (2017). Durability of reactive powder concrete under chloride-salt freeze-thaw cycling. *Mater. Struct.*50, 1–9.

Wei, C., Song, S. (2005). Study on durability of high content fly ash active powder concrete. *New Building Mater.* 09, 27–29.

Wu, Z., Khayat, K.H., Shi, C. (2017). Effect of nano-SiO_2 particles and curing time on development of fiber-matrix bond properties and microstructure of ultra-high strength concrete. *Cem. Concr. Res.* 95, 247–256.

Wu, Z., Shi, C., Khayat, K. H. (2018). Multi-scale investigation of microstructure, fiber pullout behavior, and mechanical properties of ultra-high performance concrete with nano-$CaCO_3$ particles. *Cem. Concr. Compos.* 86, 255–265.

Xiong, M. X., Liew, J. R. (2015). Spalling behavior and residual resistance of fibre reinforced ultra-high performance concrete after exposure to high temperatures. *Mater. Const.* 65(320), 071–071.

Xiong, M. X., Liew, J. R. (2016). Mechanical behaviour of ultra-high strength concrete at elevated temperatures and fire resistance of ultra-high strength concrete filled steel tubes. *Mater. Des.* 104, 414–427.

Yang, Q., Zhang, S., Huang, S., He, Y. (2000). Effect of ground quartz sand on properties of high-strength concrete in the steam-autoclaved curing. *Cem. Concr. Res.* 30 (12), 1993–1998.

Yang, W., Huang, Z. (2003). Study on durability of reactive powder concrete. *China Concrete Cement Prod.* 1, 19–20.

Ye, Q., Zhu, J., Ma, C., Shi, T. (2006). Investigation on durability of reactive powder concrete. *New Building Mater.* 06, 33–36.

Yuan, Q., Shi, C., De Schutter, G., Audenaert, K., Deng, D. (2009). Chloride binding of cement-based materials subjected to external chloride environment–a review. *Constr. Build. Mater.* 23 (1), 1–13.

Yu, Z., Gao, K., An, M., Han, S. (2013). Influence of micro-structure on the strength and resistance to chloride ion permeability of reactive powder concrete. *J. Xi'an Uni. Arch. Technol.* 45(1), 31–37.

Zanni, H., Cheyrezy, M., Maret, V., Philippot, S., Nieto, P. (1996). Investigation of hydration and pozzolanic reaction in reactive powder concrete (RPC) using 29Si NMR. *Cem. Concr. Res.* 26(1), 93–100.

Zhang, Y., Wei, S., Liu, S., Jiao, C., Liu, J. (2008). Preparation of C200 green reactive powder concrete and its static–dynamic behaviors. *Cem. Concr. Compos.* 30(9), 831–838.

Zheng, H., Lu, J., Shen, P., Sun, L., Poon, C. S., Li, W. (2022). Corrosion behavior of carbon steel in chloride-contaminated ultra-high-performance cement pastes. *Cem. Concr. Compos.* 128, 104443.

Zheng, W., Luo, B., Wang, Y. (2013). Compressive and tensile properties of reactive powder concrete with steel fibres at elevated temperatures. *Constr. Build. Mater.* 41(2), 844–851.

Zheng, W., Li, H., Wang, Y. (2012). Compressive behaviour of hybrid fiber-reinforced reactive powder concrete after high temperature. *Mater. Des.* 41, 403–409.

Zheng, W., Li, H., Wang, Y. (2012). Compressive stress-strain relationship of steel fiber-reinforced reactive powder concrete after exposure to elevated temperatures. *Constr. Build. Mater.* 35, 931–940.

Zhukov, V. V. (1976). Reasons for the explosive spalling of heated concrete. Beton I Sjeljesobeton 3, 26–28 (in Russian).

Chapter 11

Self-healing properties

11.1 INTRODUCTION

Concrete structures in harsh and complex environments can deteriorate over time due to volume instability like dry and autogenous shrinkage, environmental actions, or mechanical stresses, such as fatigue, creep, or structural loads, during the service life (Sohail et al. 2021; Xue et al. 2020). The presence of cracks can impair mechanical properties, decrease durability, weaken the serviceability of concrete structures, and eventually cause failure of structures. Although regular means are employed to maintain and repair the infrastructure, they are always expensive and dangerous (Ersan et al. 2018). Therefore, there is a strong demand for self-healing cement-based materials, which drives the development of sustainable and long-lasting cement-based materials.

Self-healing is an intrinsic property of living organisms, which can successfully deal with various damage or injuries experienced during their lifetime without external intervention. Many ancient Roman architectures made of limestone survived because of their self-healing capacities when cracks expose limestone to moisture/water from the external environment. During the last few decades, innovative strategies for self-healing of cement-based materials have emerged and been investigated. The self-healing process within cement-based materials can be divided into two categories, namely autonomous and autogenous self-healing, as summarized in Figure 11.1. Autonomous self-healing is an engineered healing process which often involves the use of nonconventionally additives into cementitious materials to enhance the self-healing capability. It can be further divided into active and passive modes, depending on if there is a human intervention to activate it or not. Typically, it includes polymer-based, thermal-based, biological-based, chemobased, and electrobased self-healing. Autogenous self-healing is a natural/intrinsic phenomenon of used materials, where cracks can be healed by their generic composition through further hydration and/or calcium carbonate precipitation.

DOI: 10.1201/9781003203605-11

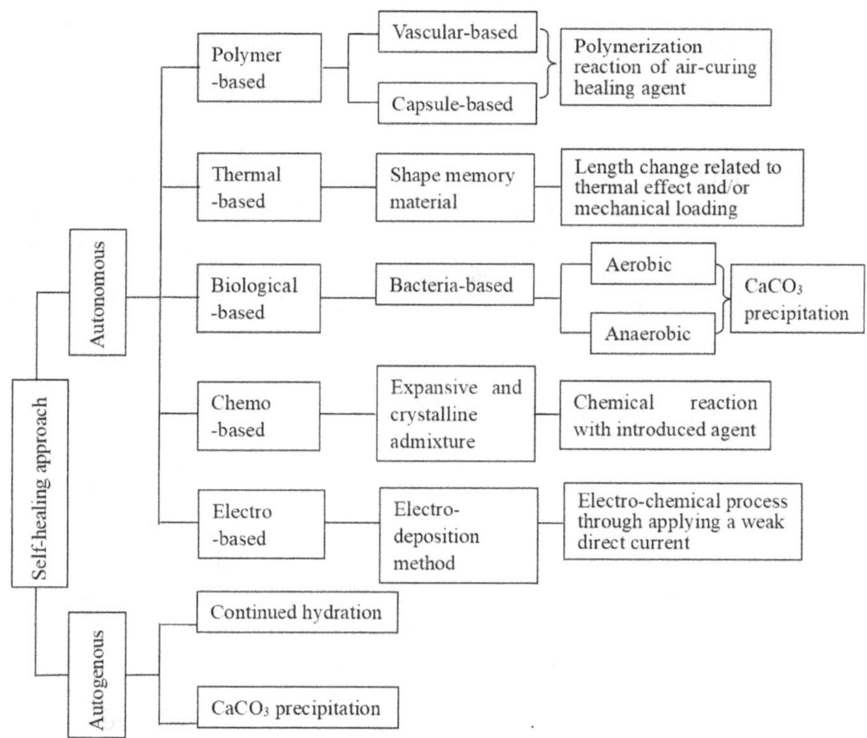

Figure 11.1 Self-healing approach for cement-based materials.

UHPC shows a great self-healing potential due to high content of cementitious materials and low water-to-binder (w/b) ratio (Granger et al. 2007). A lot of unhydrated cement and reactive pozzolanic materials are present in UHPC. It is reported that the hydration degree of UHPC with 0.16 w/b ratio at 28 d standard curing is only 28%–30% (Courtial et al. 2013). The hydration degree of UHPC at 91 d is reported to be 30%–40% (Van Tuan et al. 2011). Once water percolates inside the internal structure of UHPC, it will react with unhydrated cement particles to form hydration products, resulting in the closure of microcracks and then having the potential to recover mechanical properties and durability (Xue et al. 2023).

Exposure conditions play an essential role in the healing capacity of UHPC (Granger et al. 2007). Poor healing capacity is found on UHPC specimens under air curing, while better healing efficiency can be obtained under cryogenic cooling and 3-d air curing (Kim et al. 2019). Crack width is an essential factor for the self-healing of concrete. For micro-cracks that exist in UHPC, a few healing products generated from continuous hydration of unhydrated cement particles and pozzolanic activity of supplementary cementitious materials can lead to full crack closure. This is due to less crack void space in the microcracks (Skibsted and Snellings 2019; Yoo et al. 2020). Recently, alumina nanofibers (Cuenca et al. 2021) and silica microcapsules containing healing agents (García Calvo et al. 2017) were added to UHPC for surface crack sealing and performance restoration.

This chapter describes self-healing mechanisms and efficiency evaluation of UHPC. Surface crack sealing and healing of mechanical and transport properties of UHPC are analyzed and summarized. It aims at providing insights into the improvement of performance of UHPC during its service life.

11.2 HEALING MECHANISM

There are three mechanisms responsible for the self-healing behavior of cement-based material, which can be explained from the physical, chemical, and mechanical perspectives, as follows (Rooij et al. 2013). Firstly, the swelling of the hydrated cementitious matrix in cracks, which is mainly attributed to the water absorbed into calcium silicate hydrate (C-S-H). Thus, the crack widths are reduced to some extent (Figure 11.2(a)). The chemical mechanism for self-healing includes the reprecipitation of portlandite and ettringite because of the leaching of Ca^{2+} and SO_4^{2+} from portlandite and AFm, respectively, in bulk paste (Figure 11.2(b)), the ongoing hydration of unhydrated clinker particles (Figure 11.2(c)) and precipitation of $CaCO_3$ due to the carbonation of the products of hydration (Figure 11.2(d)). Substances of C-S-H gels, $Ca(OH)_2$, and $CaCO_3$ are often formed to close the cracks. Thirdly, mechanical sealing of debris in water and loose paste particles derived from the crushed crack surface (Figure 11.2(e)) and particle sedimentation in the water (Figure 11.2(f)). In the presence of fibers, they can exert certain crack-restraining effects to prevent crack propagation.

Several publications (Huang et al. 2016; Snoeck and De Belie 2019; Suleiman and Nehdi 2018) reported that the chemical reactions dominated the healing effect of natural healing. The recovery of mechanical properties of engineered cementitious composites is mainly related to the accumulation of C-S-H gels. The variability of transport capacity is more affected by the formation of $CaCO_3$ precipitation (Yıldırım et al. 2015). Once unhydrated cement particles reacted with water, space would be occupied by calcium silicate hydrate and calcium hydroxide (Huang and Ye 2012). Some researchers developed a novel computational model to investigate the autogenous self-healing behavior of concrete. Some authors (Lv and Chen 2012; Lv et al. 2011) investigated the self-healing efficiency related to the hydration reaction of unhydrated cement nuclei. They used splitting crack

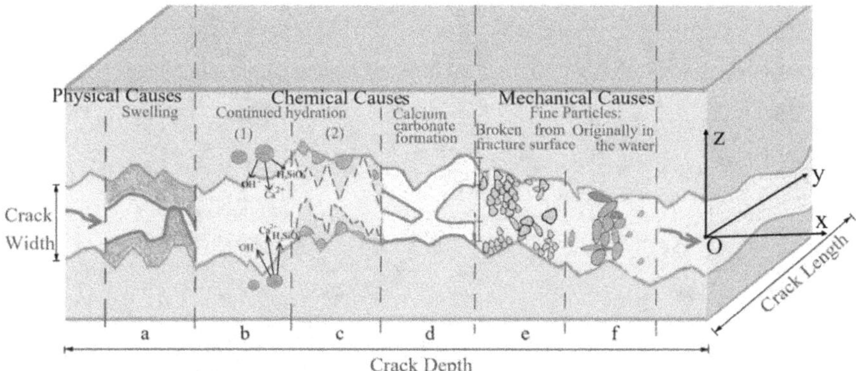

Figure 11.2 Schematic diagram of autogenous healing mechanisms in cementitious materials (modified based on Huang and Ye 2012; Rooij et al. 2013; Schlangen and Sangadji 2013).

and dome-like crack modes to model the cracking in the unhydrated nuclei. The self-healing efficiency was presented in terms of geometrical probability. They showed that the degree of self-healing efficiency is characterized based on the crack modes. Moreover, Huang and Ye (2012) and Huang et al. (2013) proposed a self-healing model by providing extra water on unhydrated cement for further hydration. They assumed that an amount of water is stored in capsules and can be passed by cracks and unhydrated cement particles. These unhydrated particles were exposed on both crack surfaces and embedded inside the cement paste. If carbon dioxide exists, $CaCO_3$ would be the final healing product (Kim et al. 2019; Suleiman and Nehdi 2018). The $CaCO_3$ precipitation process is mainly related to temperature, reactant concentration, and pH of the pore solution (Figure 11.2(a)) and can be summarized as follows (Edvardsen 1999):

$$CO_2\,(aq.)+H_2O \rightleftharpoons H_2CO_3 \rightleftharpoons H^{+}+HCO_3^{-} \rightleftharpoons 2H^{+}+CO_3^{2-} \tag{11.1}$$

$$Ca^{2+}+CO_3^{2-} \rightleftharpoons CaCO_3\,(pH_{water}>8) \tag{11.2}$$

$$Ca^{2+}+HCO_3^{-} \rightleftharpoons CaCO_3+H^{+} \quad (7.5<pH_{water}<8) \tag{11.3}$$

The cracks will be filled due to $CaCO_3$ precipitation, resulting in a partial or even fully closed crack (Aliko-Benítez et al. 2015). Sisomphon et al.

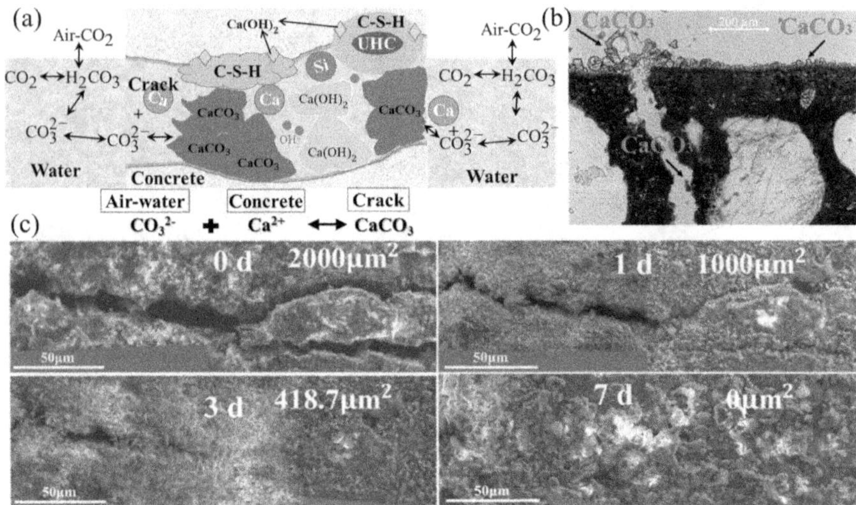

Figure 11.3 Self-healing process of $CaCO_3$ precipitation: (a) Media, interfaces, and chemical reactions in $CaCO_3$-CO_2-H_2O system (Hou et al. 2022); (b) Precipitation of $CaCO_3$ at different locations (Sisomphon et al. 2012); and (c) Crack sealing phenomena observed in UHPC at different ages (crack surface areas are calculated using Software Image J) (Kim et al. 2019).

(2012) demonstrated that crack opening absorbs more $CaCO_3$ since water containing CO_3^{2-} and bicarbonates ions (HCO_3^-) first reacts with Ca^{2+} near the crack mouth, which is illustrated in Figure 11.3(b). Kim et al. (2019) reported that the observed fully closed cracks in UHPC are attributed to the carbonation of crystallization, as shown in Figure 11.3(c). With further hydration of unhydrated cement particles, more $Ca(OH)_2$ is formed. When CO_2 dissolves into water, the consequent carbonation is stimulated to first form calcite in the internal crack. However, some different phenomena were observed by Rajczakowska et al. (2019). They demonstrated that the natural healing effect of cracked UHPC depends on both the pozzolanic reaction and the carbonation process. It is reported that the leached Ca^{2+} from the matrix primarily reacts with CO_3^{2-} so that the first calcite precipitates on the crack mouth due to a gradient of CO_3^{2-} concentration formation. The increment of ion concentration then leads to precipitation of hydration products with fewer carbonation products in the closed crack. This is in line with the results stated by Sisomphon et al. (Sisomphon et al. 2012; Snoeck et al. 2016). It is reported that the closed opening mouth of cracks prevents the CO_2 and water from the ingressing into the internal crack.

11.3 TECHNIQUES FOR EVALUATING HEALING EFFICIENCY

Techniques for assessing the self-healing efficiency of cement-based materials can be divided into three categories: (1) microstructure characterization; (2) mechanical properties recovery; and (3) durability evaluation. These consist of destructive and nondestructive techniques. Advanced material characterization techniques, such as scanning electron microscopy (SEM), thermal gravimetric analysis (TGA), and X-ray microcomputed tomography, can be employed to characterize chemical composition of hydration products and visualize the development of cracks during self-healing process. Mechanical properties and durability recovery indexes of precracked specimens after a certain self-healing regime can be obtained by comparing them to the original values of specimens. Table 11.1 summarizes the test methods for evaluating self-healing of cement-based materials.

11.3.1 Mechanical properties

The test methods used for evaluating the healing efficiency of UHPC regarding mechanical performance can be divided into destructive tests and nondestructive tests. Destructive tests include the three-point bending test, four-point bending test, compression test, and bond properties test. Nondestructive tests contain acoustic emission and ultrasonic diagnostics.

11.3.1.1 Destructive tests

11.3.1.1.1 Four-point bending performance

Cuenca et al.(2021) carried out a four-point bending test to evaluate the healing capacity of mechanical performance of UHPC using strength recovery (ISR) (as Eq. 11.4) and stiffness/damage recovery (IDaR) as indicators (as Eq. 11.5). The authors followed a loading/unloading cyclic protocol at 0.5 mm crack opening deflection intervals (Figure 11.4) and a monotonic load path to evaluate stiffness recovery and the scalar of variable damage. The input parameters corresponding to values of strength and stress in ISR and IDaR are illustrated in Figure 11.5. After 6 months of healing in geothermal water or a moist room, there is a similar trend of the index of crack sealing (ICS) of surface crack to ISR. They found when the specimens were treated following the specific program (i.e., cracking, 3-month healing, recracking, 3-month healing, and reload to failure), the cracked specimens showed the highest recovery effect. These results are in line with the results of Huang et al. (2016) and Jiang et al. (2019), that a faster full closure of surface crack can be obtained to prevent the ingress of water. The unhydrated particles reacted with water to heal the deeper crack when the precracked specimens were recracked. Besides, the IDaR increased

Table 11.1 Techniques and test methods used to evaluate self-healing efficiency of cement-based materials

Test methods		*Parameters and/or possibilities*	*References*
Mechanical properties	Four-point bending	Regain in strength, stiffness, and toughness of precrack specimens after a certain curing time	Cuenca et al. (2021); Kim et al. (2019)
	Three-point bending		García Calvo et al. (2017); Granger et al. (2007); Granger et al. (2009); Hilloulin et al. (2014)
	Compressive strength		Beglarigale et al. (2021)
	Bonding strength		Beglarigale et al. (2019)
	Direct tensile		Guo et al. (2019); Shin and Yoo (2021); Yoo et al. (2020)
Nondestructive tests	Acoustic emission	Stiffness over time	Granger et al. (2007); Guo et al. (2019)
	Ultrasonic diagnostics	Stiffness over time	Granger et al. (2009)
Transport properties	Water absorption	Water absorption capacity	Cuenca et al. (2021)
	Air permeability	Airflow capability from crack	Beglarigale et al. (2021); Guo et al. (2019); Kunieda et al. (2012)
	Chloride ion permeability	Resistance against chloride ingress	Doostkami et al. (2021); Jiang et al. (2019); Ma et al. (2016)
	Water permeability	Water flow capability from crack	Beglarigale et al. (2021); Kunieda et al. (2012); Lo Monte and Ferrara (2021); Ma et al. (2016)
Microstructure characterization	Microscopy	Morphology of precipitated products and crack visualization	Kunieda et al. (2012)
	X-ray tomography	Crack visualization	García Calvo et al. (2017)
Crack sealing	Surface crack width (area) testing	Crack width and area evaluation	Beglarigale et al. (2021); Beglarigale et al. (2019); Cuenca et al. (2021); Cuenca et al. (2021); Guo et al. (2019); Jiang et al. (2019)

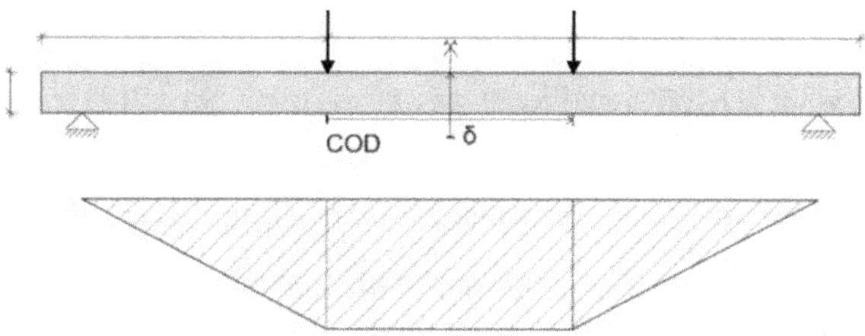

Figure 11.4 Schematic of the four-point bending test setup.

Source: Cuenca et al. (2021).

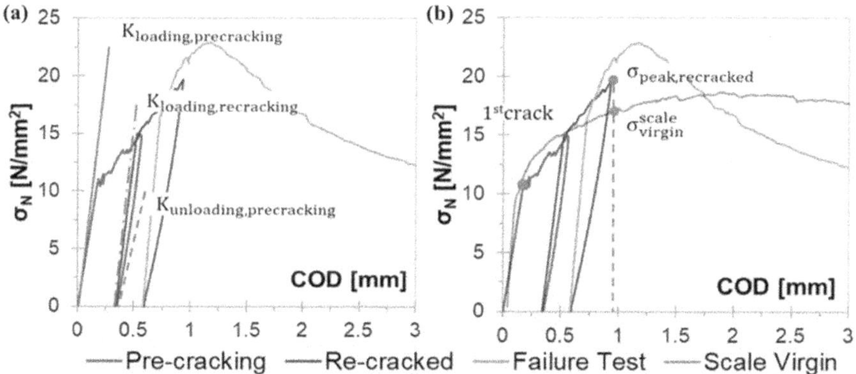

Figure 11.5 Schematic and definition of healing indices related to the recovery of stress capacity (ISR) and stiffness/damage (IDaR).

Source: Cuenca et al. (2021).

faster than that of ISR, which means that the recovery of stiffness of cracked specimens is slower than the recovery of the loading bearing capacity. It is highlighted that the ISR and IDaR of specimens with narrow cracks can be better healed, when the alumina nanofibers were added into mixtures.

Kim et al. (2019) evaluated the healing performance of UHPC with 2% steel fibers under cryogenic temperature using four-point bending testing. S65, S97.5, and S100 specimens produced with straight fibers at specific ratios of 65, 97.5, and 100, respectively, and T100 specimens with twisted fiber at an aspect ratio of 100 were put into a cryogenic temperature of –165°C. The specimens were then returned to room temperature for precracking and postcondition. The precracking value was 0.5 times the sum of the initial crack strength and ultimate strength. The bending bearing capacity of specimens with S65 was fully recovered after 3-d air curing, while the S97.5 and S100 specimens still showed apparent cracks. The bending property of cracked specimens with straight fibers was fully healed after 28-d water immersion, but the ones with twisted fibers were partially recovered. Thus, the addition of straight steel fiber can effectively improve the bending performance of UHPC after ultra-low temperature erosion. UHPC with low aspect ratio fiber indicated a better self-healing effect than that with high aspect ratio one. Lo Monte and Ferrara (2021) found that UHPC specimens with lower ISR can hardly obtain the initial stiffness, proving that it is difficult to achieve a full stiffness recovery. This is because effective bonding products, such as C-S-H and ettringite, need to be formed between fiber and matrix (Rajczakowska et al. 2019).

$$\mathrm{ISR}=\frac{\sigma_{\mathrm{peak,recracked}}-\sigma_{\mathrm{virgin}}^{scale}}{\sigma_{\mathrm{virgin}}^{scale}} \tag{11.4}$$

$$\mathrm{IDaR}=\frac{\mathrm{K}_{\mathrm{peak,post\text{-}conditioning}}-\mathrm{K}_{\mathrm{unloading,pre\text{-}cracking}}}{\mathrm{K}_{\mathrm{loading,pre\text{-}cracking}}-\mathrm{K}_{\mathrm{unloading,pre\text{-}cracking}}} \tag{11.5}$$

11.3.1.1.2 Three-point bending performance

Granger et al. (2007) precracked the prismatic specimens and unloaded them when reaching a load of 2 kN. Some specimens were under immediate reloading, and others were reloaded after 3 weeks in air curing or water immersion. It was found that the bending capacity of UHPC specimens cured in the air is similar to that with secondary loading immediately after precracking. The bending capacity of UHPC specimens cured in water increased by about 40% when the precrack width is 20 μm. Hilloulin et al. (2014) attempted to describe the partial recovery of the mechanical properties of healed UHPC under three-point bending testing using a hydro-chemo-mechanical model, as shown in Figure 11.6. They found that when a relatively small zone with 1/10 the total height was healed to achieve the initial mechanical properties, UHPC showed a great improvement in bending capacity. The hydration speed in small cracks can be faster than the speed of origin hydration. Although this model can reproduce the decreasing process of the stiffness and crack width of UHPC, it cannot reflect the hysteresis behavior of the specimen with slight stiffness recovery due to the friction at the crack tip.

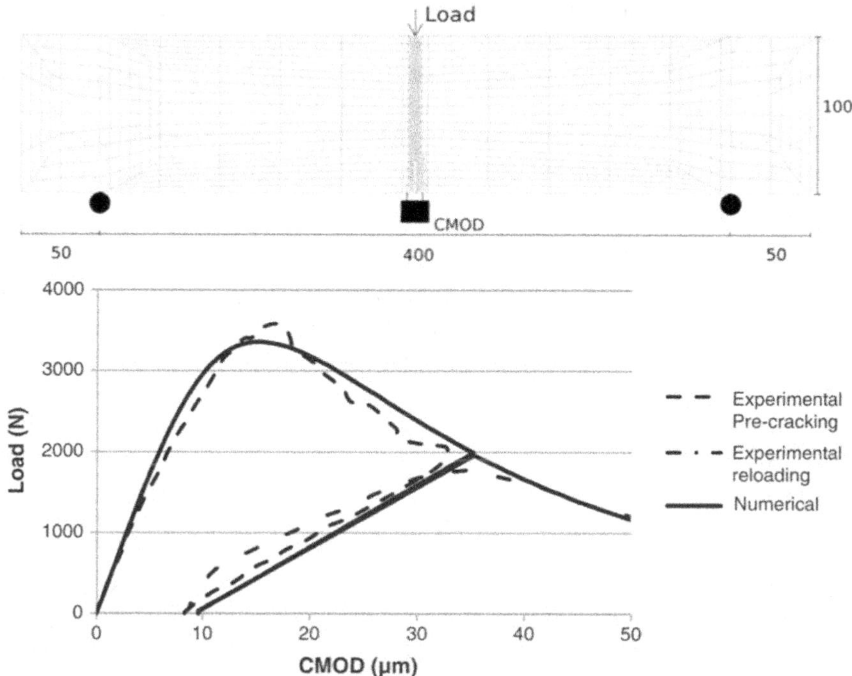

Figure 11.6 Experimental and numerical relationships between crack mouth opening displacement (CMOD) and loading of notched beams under three bending tests.

Source: Hilloulin et al. (2014).

11.3.1.1.3 Compressive performance

Beglarigale et al. (2021) applied 100% compressive strength on different specimens, including control, FA-CM, and GGBS-CM, to induce initial damage. The self-healing efficiency values for FA-CM and GGBS-CM after 7-d immersion in saturated lime water were 68% and 76%, respectively, which are 44% higher than that of the control mixtures. The pozzolanic reaction could be ignored due to a short time immersion, but it would play an important role in the long-term performance after 90 d and result in higher healing efficiency.

Since there are many original microcracks in UHPC specimens at early age, further hydration can help to fill the flaws (Fanella and Krajcinovic 1988). Ge et al. (2016) found that the compressive strengths of UHPC specimens obtained obvious increment after 90-d immersion in 60°C water. The compressive strengths of UHPC specimens with w/b ratios of 0.17 and

0.2 were increased by 41% and 36%, respectively. It is explained that the microcracks were healed because of formed hydration products from further hydration. If the volume of further hydration products exceeds the volume of microcrack space, expansion stress will adversely affect the compressive strength. Wang et al. (2020) found that the compressive strength of UHPC with 0.2 w/b ratio was decreased by 14.5% after immersing in 20°C water for 2 years.

11.3.1.1.4 Bond properties

Beglarigale et al. (2019) investigated the bond properties of fracture surfaces in different UHPC mixtures made with fly ash and ground granulated blast furnace slag (GGBS). The bond loads are measured by a direct tension test using a novel mold. The split specimens are then put together and cured in water for 30 d. The bonding load of the reference mixture without any supplementary cementitious materials was 81 N. With the use of 33% fly ash or GGBS, the bond performance of UHPC mixture was significantly improved. UHPC mixture with fly ash showed a bond load of 308 N compared to 201 N of specimens with GGBS, which is attributed to more unhydrated particles in the presence of fly ash. However, with the supplementary cementitious materials content reaching 66%, the hydration processes in the UHPC matrix were delayed, and the 30-d bond load was decreased to approximately 50 N.

11.3.1.1.5 Direct tensile performance

Guo et al. (2019) conducted direct tensile testing on UHPC to induce precracking with 1% strain. The precracked specimens were then stored in a water and humid room with 80% or 45% relative humidity. It is found that the stiffness of cracked specimens after 28 d water immersion was recovered to 46.6 GPa, compared to the original value of 48.0 GPa. However, after recuring under 80% and 45% relative humidity, the reloading stiffness values were 29.8 and 23.5 GPa, respectively, which are much lower than the initial stiffness. In addition, the recovery rates of initial cracking strength of specimens cured in water and 80% relative humidity room were 44% and 16%, respectively. However, there was no recovery in specimens stored in a 45% relative humidity room. Thus, the higher relative humidity can lead to a higher healing efficiency. It was also found that the recovery rates of stiffness and initial cracking strength of specimens after 3-d steam curing were higher than those after 28-d water immersion curing. The stiffness recovery can be attributed to the recovery of fiber-matrix bond strength.

Yoo et al. (2020) found that the corrosion of steel fiber in self-healed UHPC increased the roughness of fiber surface, which is beneficial for bond strength and direct tensile stiffness. The surface roughness parameter (Ra) of steel fibers of cracked specimens immersed in NaCl solution was 173.4 nm, which is much greater than that of plain specimens at 56.2 nm. Once the steel fiber corrosion initiates, it will get worse in long-term exposure of 20 weeks under an aggressive environment due to partially filled cracks. Shin and Yoo (2021) reported that if the crack width is smaller than 20 μm, the self-healing is capable of resisting the ingress of aggressive agents into the inner structure of matrix. However, when a wider crack width (50 μm) is induced, the deterioration of tensile bearing capacity is accelerated. This is because insufficient healing products are formed to seal the crack. Thus, self-healing plays a slight role in preventing the ingress of toxic substances.

11.3.1.2 Nondestructive tests

Guo et al. (2019) conducted acoustic emission and direct tensile testing on UHPC to identify crack generation. The mechanical responses can be received after new healing products in crack space from autogenous healing. The crack would be healed and bridged by healing products after water immersion. When the healed specimens were reloaded, and the closed cracks were reopened, the cracking events can be recorded by acoustic emission. The recorded events of cracked and healed specimens can be compared and used to evaluate healing efficiency. Guo et al. (2019) found that the events of UHPC specimens were 14 after first loading and ones under reloading were 11 after 28-d water immersion. This means that newly formed crystals bridge the microcracks during the healing process, but the bounding ability of the crystals is weaker than that of the original matrix. However, as for the specimens cured in 80% relative humidity, the events under first loading and reloading were 11 and 1, respectively. It demonstrated that water immersion can help cracked UHPC obtain a higher healing efficiency. This is in line with the results reported by Granger et al. (2007), as shown in Figure 11.7(a), in which the total response map is recorded after the initial cracking. The cracked specimens cured in water showed higher healing efficiency because there are more events in Figure 11.7(b), compared to Figure 11.7(c). In addition, Granger et al. (2009) monitored the cracking and healing of UHPC loaded under three bending tests using the time-reversal technique. It is reported that the energy and amplitude always decreased during the cracking process but regained during the healing process, as illustrated in Figure 11.8. They also found that the testing technique under the bending load with time reversal is sensitive to the variation of stiffness of UHPC specimens because of crack propagation and crack healing.

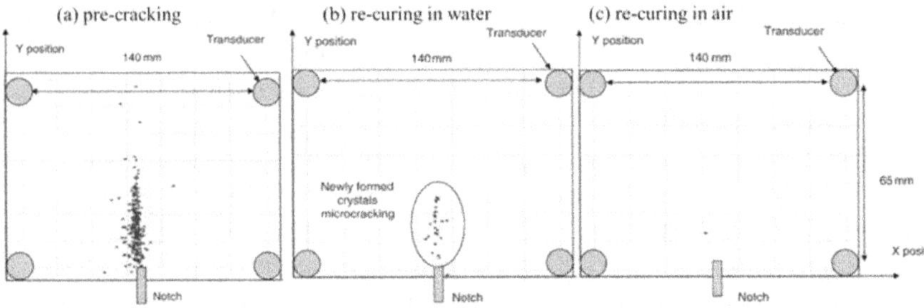

(a) pre-cracking
(b) re-curing in water
(c) re-curing in air
Y position
Transducer
140 mm
Newly formed crystals microcracking
65 mm
X position
Notch

Figure 11.7 Microcracking map: (a) Precracking phase and reloading phase; (b) Aging in water; and (c) Aging in air.

Source: Granger et al. (2007).

Energy
Amplitude
Amplitude (mV)
Energy (aJ)
cracking
healing

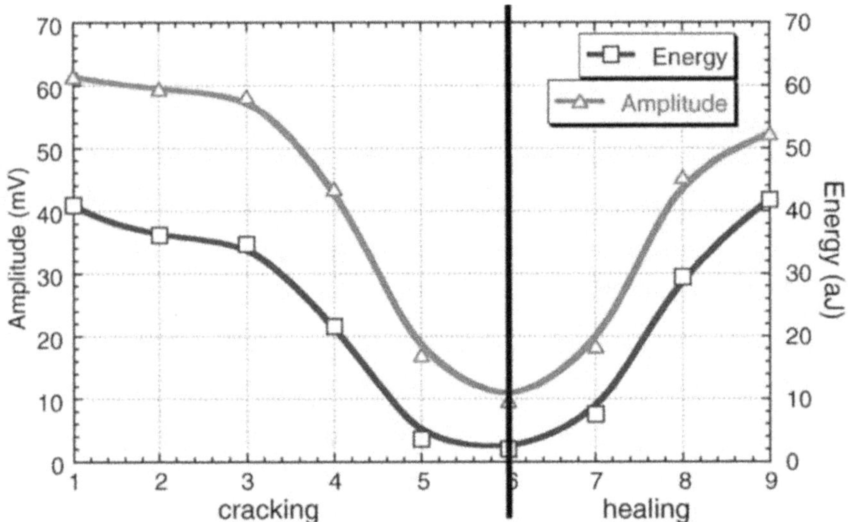

Figure 11.8 Evolution of energy and amplitude of the focusing signals of UHPC detected at different states after time reversal during cracking and healing processes.

Source: Granger et al. (2009).

11.3.2 Crack sealing properties

The index of crack sealing (ICS) (Eq. 11.6) is selected to describe the autogenous healing performance of concrete (Wiktor and Jonkers 2011).

$$ICS = 1 - \frac{A_{crack,t}}{A_{crack,0}} \quad (11.6)$$

where ICS is the index of crack sealing and $A_{crack,t}$ and $A_{crack,0}$ represent areas of healed crack and initial crack, respectively.

11.3.2.1 Internal crack sealing

Internal crack sealing is closely related to the mechanical performance of concrete (Pei et al. 2013; Wang et al. 2020). However, the evidence of the healed phenomena is difficult to obtain. X-ray computed tomography (X-CT) can be used to detect the internal crack distribution (García Calvo et al. 2017; Suleiman and Nehdi 2018). García Calvo et al.(2017) used this technique to observe crack patterns, including crack path and boundary, in different depths of UHPC with and without healing agent. As shown in Figure 11.9, the crack path and boundary in three different depths of UHPC without a healing agent were observed, reflecting a continuous crack. However, continuous cracks were observed in UHPC with a 5% healing agent, suggesting that the crack path was cut down by healing products. The fact is that the healing products formed in the cracks due to autogenous self-healing in concrete are always uneven (Huang et al. 2016). The healing products mainly accumulated near the crack mouth, and only fewer healing products can be found near the crack core (Fan and Li 2015; Suleiman and Nehdi 2018).

11.3.2.1 Surface crack sealing

Crack width is an essential factor for ICS (Qiu et al. 2016; Wiktor and Jonkers 2011). Proper addition of supplementary cementitious materials in concrete can contribute to crack closure. Different cracked UHPC groups

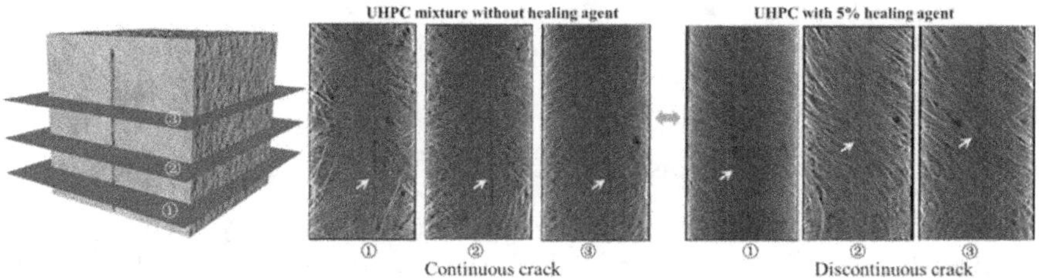

Figure 11.9 X-CT images of internal crack closure in UHPC with and without healing agent.

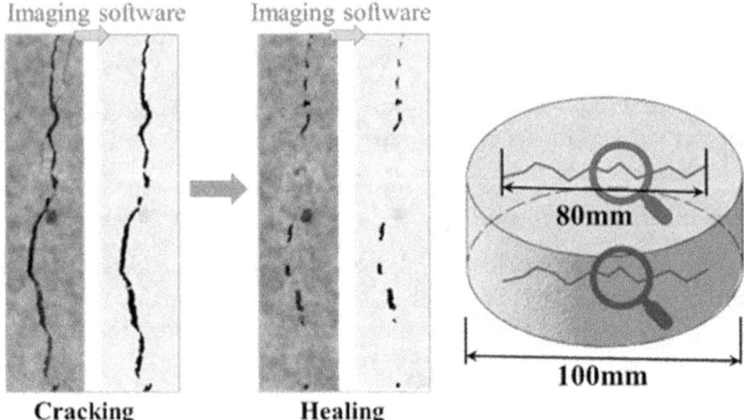

Figure 11.10 Visual imaging of surface crack closure.

Source: Cuenca et al. (2021).

containing different mass ratios of fly ash or GGBS were cured in water for 30 d (Beglarigale et al. 2019). Compared to the control group without any supplementary cementitious materials, ICS is greater for UHPC mixtures with fly ash or GGBS replacement of 33%. However, the ICSs reduce when the mass ratio of fly ash or GGBS is up to 66%. This is because fewer healing products are formed because of lower cement and $Ca(OH)_2$ contents, thus lowering the self-capability of UHPC. After 90-d water curing, a higher ICS in UHPC with 33% fly ash or GGBS was observed (Beglarigale et al. 2021). This suggests that more pozzolanic action products are formed with prolonged curing time. Kim et al. (2019) found that the healing product mainly consists of calcium carbonate after 28-d water curing. Cuenca et al. (2021) employed visual imaging to evaluate surface crack closure (Figure 11.10). It was found that ICSs of specimens cured in geothermal water were higher than those in the moist room due to the presence of sulfate ions.

11.3.3 Transport properties

11.3.3.1 Water absorption

Water sorptivity of cracked UHPC can reflect the ability of microcracks to absorb and transfer water by capillary suction (Ma et al. 2016). The sorptivity healing index (ISH), as shown in Eq. (11.7), is selected to describe the crack self-sealing performance. Ma et al. (2016) found that a wider crack area was observed in UHPC specimens under a greater load, leading to higher water absorption. However, effective crack closure of specimens reduced the water absorption after 28-d water curing, even for those with relatively big cracks. This is consistent with the results reported by García Calvo et al. (2017). The water absorption coefficients of cracked UHPC drop dramatically after 7 d and 28 d. The repair agent of silica microcapsules containing epoxy can provide further autonomous healing, leading to lower water absorption. However, the continuous increase in repair agent content will not continue to increase the healing performance of UHPC. To some extent, it will adversely affect the original performance of UHPC matrix.

$$ISH = 1 - \frac{\text{Sorptivity coefficient}_{\text{after healing}}}{\text{Sorptivity coefficient}_{\text{after pre-cracking}}} \tag{11.7}$$

Beglarigale et al. (2021) found the use of 33% fly ash or GGBS can reduce water absorption. The ISH of the control mixture was 15.5%, and it increased to 56.3% and 56.3% for FA-CM and GGBS-CM, respectively. Cuenca et al. (2021) reported that UHPC incorporated with nanoconstituents and crystalline admixture showed reduced water absorption after wet-dry cycles.

11.3.3.2 Air permeability

Slight changes in crack width can be accurately detected by air permeability after cracking and healing of microcracks because of the superior airtightness of UHPC matrix (Li et al. 2020). Kunieda et al. (2012) found that the air permeability coefficient of UHPC specimens increased dramatically after tensile load. A torrent permeability tester was used, as illustrated in Figure 11.11. The air permeability of cracked specimens reduces to an extremely low value after 20-d recuring in water, and it is similar to that of the original value after 1-year water immersion. However, the reduction rate of air permeability of UHPC in air curing is much lower than that in water curing. Guo et al. (2019) found that the water content under curing conditions plays an important role in the sealing of interconnected cracks. The sound UHPC specimens were applied to the direct tensile load up to 0.1% strain so that the air permeability coefficients increased from 0.0022–0.0068 to 0.003–0.1283 Ln(mbar)/min. As for cracked UHPC specimens with the same damage degree, cracked specimens healed under 45% relative humidity showed no decrement of air permeability. However, the air permeability index of ones healed under 80% relative humidity and water immersion significantly decreased with the curing period. Especially, the air permeability of cracked specimens can decrease and reach the same level as intact ones after 7-d water immersion. It is in line with the results of

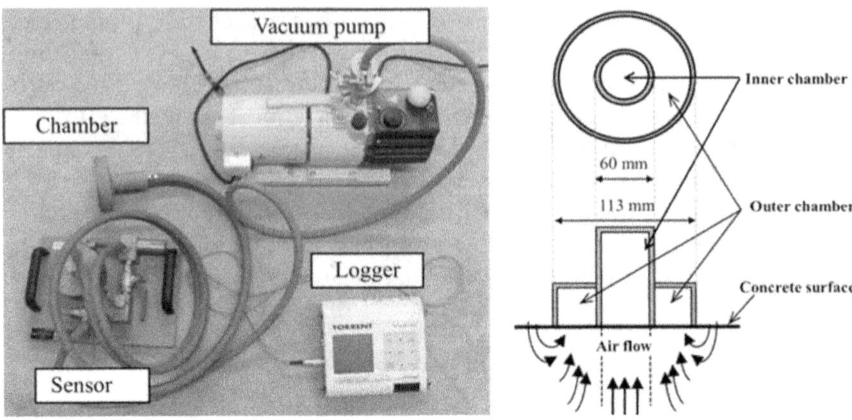

Figure 11.11 Torrent permeability tester.

Source: Kunieda et al. (2012).

Granger et al. (2007), which reflects that water is an essential factor in the autogenous healing process.

11.3.3.3 Chloride ion permeability

Chloride ion permeability of UHPC is very low due to its dense structure (Li et al. 2020). However, microcracks can be inducted and propagated in UHPC due to mechanical and environmental factors (Graybeal 2008). The cracks make chloride ions permeate into the inner structure of UHPC, which can induce corrosion of steel fibers. Since the volume of corrosion products expanded twice, the expansion will lead to cracks and further deterioration of concrete composites (Marcotte and Hansson 2006). The self-healing properties of UHPC with crack widths of 10, 30, and 50 μm were simulated based on the diffusion coefficient of chloride ions. A greater load always leads to more severe damage and higher transport capacity. Ma et al. (2016) found that the greater loads bore by UHPC specimens would result in larger cracks and higher chloride ion permeability. After 28-d recuring in the saturated $Ca(OH)_2$ solution, the microcracks induced from lower loading obtained full closure, leading to recovered chloride ion permeability of the healed specimen to that of the virgin. It is in line with the findings from Doostkami et al. (2021). In addition, Shin and Yoo (2021) found that the tensile strength of UHPC with a crack width of 20 μm did not deteriorate under a chloride environment, but the tensile strength decreased when the crack width was up to 50 μm. Fortunately, UHPC can control the crack width to a small range, usually 10 μm of the low strain level and

25–50 μm of the high strain level, which improves the self-healing effect of UHPC and effectively prevents the invasion of chloride ions (Doostkami et al. 2021). Even though the steel fibers in UHPC can be slightly eroded in the chloride environment, the tensile strength increases due to the increased surface roughness of steel fibers (Yoo et al. 2020).

11.3.3.4 Water permeability

There are two test methods for evaluating the water permeability coefficient (K) of cracked (K_0) and healed (K_t) specimens (Ferrara et al. 2018; Hou et al. 2022; Lepech and Li 2009): (1) measurement of the decreased water head after a certain period and (2) evaluation of water flow passing through a crack during a period. The index of permeability healing (IPH), as shown in Eq. (11.8), can be used to describe crack sealing after the self-healing effect.

$$\mathrm{IPH} = 1 - \frac{K_t}{K_0} \tag{11.8}$$

The first approach is selected to measure the IPH of cracked UHPC by researchers such as Beglarigale et al. (2021), Cuenca et al. (2021), Kunieda et al. (2012), and Lo Monte and Ferrara (2021). Kunieda et al. (2012) reported that greater water permeability reduction is closely associated with a lower w/b ratio in ultra-high-performance strain hardening cementitious composites. Cuenca et al. (2021) found that a full recovery of the permeability performance can be observed in a short healing time due to crack healing in surface and inner of 3D crack, when the initial crack opening is within 50 μm. The hydrophilic nature of the alumina nanofibers resulted in better and faster healing in a narrow crack. Beglarigale et al. (2021) reported that the IPH values of 7-d specimens are higher than that of 90-d specimens. The maximum IPH of 7-d FA-CM specimens is 89%, while the IPH values of the 7-d control and GGBS-CM specimens are 74% and 81%, respectively.

11.4 SUMMARY

This chapter describes self-healing mechanisms and efficiency evaluation of UHPC. Surface crack sealing and healing of mechanical and transport properties of UHPC are analyzed and summarized. Based on the discussion from this chapter, the following conclusions can be summarized.

(1) Because of ultra-low w/b ratio of the matrix, there are a large number of unhydrated cementitious materials and pozzolanic materials inside

UHPC. This provides UHPC with vigorous secondary hydration and pozzolanic reaction capabilities, thus a high potential for self-healing ability to regain its initial properties.

(2) The autogenous healing efficiency is closely related to exposure conditions of cracked UHPC. When enough water content is accessible to cracked specimens, a higher healing efficiency can be obtained. In addition, at a lower damage level or a smaller crack width, a higher crack healing efficiency can be observed.

(3) Further hydration products can fill the original flaws in UHPC. Steam curing and hot water curing can provide a faster chemical reaction to close flaws and cracks in UHPC, leading to denser microstructure and a higher recovery degree of mechanical properties.

(4) Binder type and content play an essential role in the self-healing of UHPC. Appropriate addition of GGBS or fly ash is beneficial for the self-healing performance of UHPC.

REFERENCES

Aliko-Benítez, A., Doblaré, M., Sanz-Herrera, J. A. (2015). Chemical-diffusive modeling of the self-healing behavior in concrete. *Int. J. Solids Struct.* 69–70, 392–402.

Beglarigale, A., Eyice, D., Tutkun, B., Yazıcı, H. (2021). Evaluation of enhanced autogenous self-healing ability of UHPC mixtures. *Constr. Build. Mater.* 280, 122524.

Beglarigale, A., Vahedi, H., Eyice, D., Yazıcı, H. (2019). Novel Test Method for Assessing Bonding Capacity of Self-Healing Products in Cementitious Composites. *J. Mater. Civ. Eng.* 31(4), 04019028.

Courtial, M., de Noirfontaine, M. N., Dunstetter, F., Signes-Frehel, M., Mounanga, P., Cherkaoui, K., Khelidj, A. (2013). Effect of polycarboxylate and crushed quartz in UHPC: Microstructural investigation. *Constr. Build. Mater.* 44, 699–705.

Cuenca, E., D'Ambrosio, L., Lizunov, D., Tretjakov, A., Volobujeva, O., Ferrara, L. (2021). Mechanical properties and self-healing capacity of ultra high performance Fibre Reinforced Concrete with alumina nano-fibres: Tailoring Ultra High Durability Concrete for aggressive exposure scenarios. *Cem. Concr. Compos.* 118, 103956.

Cuenca, E., Mezzena, A., Ferrara, L. (2021). Synergy between crystalline admixtures and nano-constituents in enhancing autogenous healing capacity of cementitious composites under cracking and healing cycles in aggressive waters. *Constr. Build. Mater.* 266, 121447.

Doostkami, H., Roig-Flores, M., Serna, P. (2021). Self-healing efficiency of Ultra High-Performance Fiber-Reinforced Concrete through permeability to chlorides. *Constr. Build. Mater.* 310, 125168.

Edvardsen, C. (1999). Water permeability and autogenous healing of cracks in concrete. *ACI Mater. J.* 96, 448–454.

Ersan, Y. C., Palin, D., Yengec Tasdemir, S. B., Tasdemir, K., Jonkers, H. M., Boon, N., De Belie, N. (2018). Volume Fraction, Thickness, and Permeability of the Sealing Layer in Microbial Self-Healing Concrete Containing Biogranules. *Front. Built Environ.* 4(70).

Fan, S., Li, M. (2015). X-ray computed microtomography of three-dimensional microcracks and self-healing in engineered cementitious composites. *Smart Mater. Struct.* 24(1), 015021.

Fanella, D., Krajcinovic, D. (1988). A micromechanical model for concrete in compression. *Eng. Fract. Mech.* 29(1), 49–66.

Ferrara, L., Van Mullem, T., Alonso, M. C., Antonaci, P., Borg, R. P., Cuenca, E., Jefferson, A., Ng, P.-L., Peled, A., Roig-Flores, M., Sanchez, M., Schroefl, C., Serna, P., Snoeck, D., Tulliani, J. M., De Belie, N. (2018). Experimental characterization of the self-healing capacity of cement based materials and its effects on the material performance: A state of the art report by COST Action SARCOS WG2. *Constr. Build. Mater.* 167, 115–142.

García Calvo, J. L., Pérez, G., Carballosa, P., Erkizia, E., Gaitero, J. J., Guerrero, A. (2017). Development of ultra-high performance concretes with self-healing micro/nano-additions. *Constr. Build. Mater.* 138, 306–315.

Ge, X., Liu, D., Wang, Y., Tian, O., Shen, J. (2016). Effect of rehydration of unhydrate cement on ultra high performance concrete. *J. Civ.,Archit. Environ. Eng.* 38, 40–45.

Granger, S., Loukili, A., Pijaudier-Cabot, G., Chanvillard, G. (2007). Experimental characterization of the self-healing of cracks in an ultra high performance cementitious material: Mechanical tests and acoustic emission analysis. *Cem. Concr. Res.* 37(4), 519–527.

Granger, S., Pijaudier Cabot, G., Loukili, A., Marlot, D., Lenain, J. C. (2009). Monitoring of cracking and healing in an ultra high performance cementitious material using the time reversal technique. *Cem. Concr. Res.* 39(4), 296–302.

Graybeal, B. A. (2008). Flexural behavior of an ultrahigh-performance concrete I-girder. *J. Bridge Eng.* 13(6), 602–610.

Guo, J.-Y., Wang, J.-Y., Wu, K. (2019). Effects of self-healing on tensile behavior and air permeability of high strain hardening UHPC. *Constr. Build. Mater.* 204, 342–356.

Hilloulin, B., Grondin, F., Matallah, M., Loukili, A. (2014). Modelling of autogenous healing in ultra high performance concrete. *Cem. Concr. Res.* 61–62, 64–70.

Hou, S., Li, K., Wu, Z., Li, F., Shi, C. (2022). Quantitative evaluation on self-healing capacity of cracked concrete by water permeability test—A review. *Cem. Concr. Compos.* 127, 104404.

Huang, H., Ye, G. (2012). Simulation of self-healing by further hydration in cementitious materials. *Cem. Concr. Compos.*, 34(4), 460–467.

Huang, H., Ye, G., Damidot, D. (2013). Characterization and quantification of self-healing behaviors of microcracks due to further hydration in cement paste. *Cem. Concr. Res.* 52, 71–81.

Huang, H., Ye, G., Qian, C., Schlangen, E. (2016). Self-healing in cementitious materials: Materials, methods and service conditions. *Mater. Des.* 92, 499–511.

Jiang, J., Zheng, X., Wu, S., Liu, Z., Zheng, Q. (2019). Nondestructive experimental characterization and numerical simulation on self-healing and chloride ion transport in cracked ultra-high performance concrete. *Constr. Build. Mater.* 198, 696–709.

Jiang, Z., Yuan, Z., Li, W. (2019). Acoustic emission analysis of characteristics of healing products in steam-cured cementitious materials with mineral additives. *Constr. Build. Mater.* 201, 807–817.

Kim, S., Yoo, D.-Y., Kim, M.-J., Banthia, N. (2019). Self-healing capability of ultra-high-performance fiber-reinforced concrete after exposure to cryogenic temperature. *Cem. Concr. Compos.* 104, 103335.

Kunieda, M., Choonghyun, K., Ueda, N., Nakamura, H. (2012). Recovery of protective performance of cracked ultra high performance-strain hardening cementitious composites (UHP-SHCC) due to autogenous healing. *J. Adv. Concr. Technol.* 10(9), 313–322.

Lepech, M. D., Li, V. C. (2009). Water permeability of engineered cementitious composites. *Cem. Concr. Compos.* 31(10), 744–753.

Li, J., Wu, Z., Shi, C., Yuan, Q., Zhang, Z. (2020). Durability of ultra-high performance concrete – A review. *Constr. Build. Mater.* 255.

Lo Monte, F., Ferrara, L. (2021). Self-healing characterization of UHPFRCC with crystalline admixture: Experimental assessment via multi-test/multi-parameter approach. *Constr. Build. Mater.* 283, 122579.

Lv, Z., Chen, H. (2012). Modeling of self-healing efficiency for cracks due to unhydrated cement nuclei in hardened cement paste. *Pro. Eng.*, 27, 281–290.

Lv, Z., Chen, H., Yuan, H. (2011). Quantitative solution on dosage of repair-agent for healing of 3D simplified cracks in materials: short capsule model. *Mater. Struct.* 44(5), 987–995.

Lv, Z., Chen, H., Yuan, H. (2011). Quantitative solution on dosage of repair agent for healing of cracks in materials: short capsule model vs. two-dimensional crack pattern. *Sci. Eng. Comp. Mater.* 18, 13–19.

Ma, Z., Zhao, T., Yao, X. (2016). Influence of Applied Loads on the Permeability Behavior of Ultra High Performance Concrete with Steel Fibers. *J. Adv. Concr. Technol.* 14(12), 770–781.

Marcotte, T. D., Hansson, C. M. (2006). Corrosion products that form on steel within cement paste. *Mater. Struct.* 40(3), 325–340.

Pei, R., Liu, J., Wang, S., Yang, M. (2013). Use of bacterial cell walls to improve the mechanical performance of concrete. *Cem. Concr. Compos.* 39, 122–130.

Qiu, J., Tan, H. S., Yang, E.-H. (2016). Coupled effects of crack width, slag content, and conditioning alkalinity on autogenous healing of engineered cementitious composites. *Cem. Concr. Compos.* 73, 203–212.

Rajczakowska, M., Nilsson, L., Habermehl-Cwirzen, K., Hedlund, H., Cwirzen, A. (2019). Does a High Amount of Unhydrated Portland Cement Ensure an Effective Autogenous Self-Healing of Mortar? *Materials (Basel)*, 12(20), 3298.

Rooij, M. d., Tittelboom, K. V., Belie, N. D., Schlangen, E. (2013). Self-Healing Phenomena in Cement-Based Materials. In *Self-Healing Phenomena in Cement-Based Materials: State-of-the-Art Report of RILEM Technical Committee 221-SHC: Self-Healing Phenomena in Cement-Based Materials*, M. de Rooij, K. Van Tittelboom, N. De Belie, and E. Schlangen, eds., Springer Netherlands, Dordrecht, 65–117.

Schlangen, E., Sangadji, S. (2013). Addressing Infrastructure Durability and Sustainability by Self Healing Mechanisms—Recent Advances in Self Healing Concrete and Asphalt. *Proc. Eng.* 54, 39–57.

Shin, W., Yoo, D.-Y. (2021). Tensile properties of cracked reactive powder concrete in corrosive environment—effects of crack width and exposure duration. *Constr. Build. Mater.* 272.

Sisomphon, K., Copuroglu, O., Koenders, E. A. B. (2012). Self-healing of surface cracks in mortars with expansive additive and crystalline additive. *Cem. Concr. Compos.* 34(4), 566–574.

Skibsted, J., Snellings, R. (2019). Reactivity of supplementary cementitious materials (SCMs) in cement blends. *Cem. Concr. Res.* 124, 105799.

Snoeck, D., and De Belie, N. (2019). Autogenous Healing in Strain-Hardening Cementitious Materials With and Without Superabsorbent Polymers: An 8-Year Study. *Front. Mater.* 6, 48.

Snoeck, D., Dewanckele, J., Cnudde, V., De Belie, N. (2016). X-ray computed microtomography to study autogenous healing of cementitious materials promoted by superabsorbent polymers. *Cem. Concr. Compos.* 65, 83–93.

Sohail, M. G., Kahraman, R., Al Nuaimi, N., Gencturk, B., Alnahhal, W. (2021). Durability characteristics of high and ultra-high performance concretes. *J. Build. Eng.* 33, 101669.

Suleiman, A. R., Nehdi, M. L. (2018). Effect of environmental exposure on autogenous self-healing of cracked cement-based materials. *Cem. Concr. Res.* 111, 197–208.

Van Tuan, N., Ye, G., van Breugel, K., Copuroglu, O. (2011). Hydration and microstructure of ultra high performance concrete incorporating rice husk ash. *Cem. Concr. Res.* 41(11), 1104–1111.

Wang, R., Yu, J., Gu, S., He, P., Han, X., Liu, Q. (2020). Investigation of self-healing capability on surface and internal cracks of cement mortar with ion chelator. *Constr. Build. Mater.* 236, 117598.

Wiktor, V., Jonkers, H. M. (2011). Quantification of crack-healing in novel bacteria-based self-healing concrete. *Cem. Concr. Compos.* 33(7), 763–770.

Xue, C., Tapas, M. J., Sirivivatnanon, V. (2023). Cracking and stimulated autogenous self-healing on the sustainability of cement-based materials: a review. *J. Sust. Cem. Based Mater.* 12(2), 184–206.

Xue, J., Briseghella, B., Huang, F., Nuti, C., Tabatabai, H., Chen, B. (2020). Review of ultra-high performance concrete and its application in bridge engineering. *Constr. Build. Mater.* 260, 119844.

Yıldırım, G., Keskin, Ö. K., Keskin, S. B., Şahmaran, M., Lachemi, M. (2015). A review of intrinsic self-healing capability of engineered cementitious composites: Recovery of transport and mechanical properties. *Constr. Build. Mater.* 101, 10–21.

Yoo, D.-Y., Shin, W., Chun, B., Banthia, N. (2020). Assessment of steel fiber corrosion in self-healed ultra-high-performance fiber-reinforced concrete and its effect on tensile performance. *Cem. Concr. Res.* 133, 106091.

Chapter 12

Seawater and sea sand UHPC

12.1 INTRODUCTION

The corrosion risk of steel reinforcements in marine environments and the lack of freshwater and river and manufactured sand are two significant challenges for reinforced concrete structures in marine and coastal applications. To eliminate the corrosion risk of steel reinforcements and address the challenges arising from the shortages of freshwater and river and manufactured sand for making concrete onsite, ultra-high-performance seawater sea-sand concrete (UHPSSC) has been developed. UHPSSC refers to a new type of UHPC prepared using seawater and sea sand. Generally, the use of seawater as mixing water in concrete can lead to the gain of early age strength, reduced setting time, and possible deteriorating effect due to the existence of chloride ions, sulfate ions, sodium cations, and potassium cations from salts (Nishida et al. 2015; Kalousek et al. 1970). More importantly, seawater can escalate the corrosion risk of the embedded reinforcements in reinforced concrete (Xiao et al. 2017). Besides, a high amount of salts and coral/sea shell particles in sea sand have detrimental effects on the workability of concrete and may affect the elastic modulus and mechanical strength of concrete (Richardson and Fuller 2013; Yang et al. 2005).

UHPC is an ideal material that can be confidently used in concrete structures exposed to aggressive environments due to its superior durability. Roux et al. (1996) reported that the chloride ion diffusion coefficients of C30, C80, and UHPC were 1.1×10^{-12}, 0.06×10^{-12}, and 0.02×10^{-12} m^2/s, respectively. An et al. (2007) indicated that the chloride ion diffusion coefficients of UHPC and HPC were 2.2×10^{-13} and 15.4×10^{-13} m^2/s, respectively. Therefore, the chloride ion diffusion coefficient of UHPC was found to be at least one order of magnitude lower than that of conventional concrete and HPC. The detrimental effects of salt ions from seawater and sea sand of UHPSSC can be expected to be much lower than that of normal concrete. Thus, UHPSSC has a great potential application in coastal and marine environments.

DOI: 10.1201/9781003203605-12

This chapter discusses the properties of seawater and sea sand and their influence on the hydration, microstructure, and properties of UHPSSC. The salinity and chemical compositions of seawater are summarized, and their influences on the properties of UHPSSC are revealed. The influences of sea sand from different sea areas are also discussed. The effects of cement type, supplementary cementitious materials type, and curing condition on fresh and hardened properties of UHPSSC are described. This chapter summarizes the recent advances in UHPSCC, which aims at promoting its potential applications in the marine environment to relieve the shortages of freshwater and river/manufactured sand for onsite preparation.

12.2 SEAWATER AND SEA SAND

It is necessary to characterize the properties of seawater and sea sand and determine a suitable mixture proportion of UHPSSC according to the performance requirements of concrete structures.

12.2.1 Chemical composition of seawater

Table 12.1 presents the chemical compositions of seawater employed in the literature. The seawater used can be either artificial or natural. It can be seen that the salinity values of seawater in different sea areas and literature are close to ASTM D1141 (2013) with two exceptions. The concentration of Cl^- ranges from 3000 to 94,120 mg/l, while the total salinity varies from 5615 to 94,120. As reported in the literature (Cwirzen et al. 2014), the salinity of seawater naturally fetched from the Baltic Sea is much lower than others due to the temperate maritime climate of rainy, foggy cloudy, and low temperatures. The pH value of seawater is between 7.5 and 8.4, with an average value of about 8.2 (ASTM D11412013). The change in the salinity of seawater can significantly influence the properties of UHPSSC (Kaushik and Islam, 1995).

The main reactive ions in seawater that affect the properties of UHPSSC and potentially threaten the durability of concrete structures include Cl^-, SO_4^{2-}, and Mg^{2+}. The high concentration of Cl^- can accelerate steel corrosion, as expressed by Eqs. (12.1–12.3), whereas the presence of SO_4^{2-} can accelerate the formation of gypsum and ettringite. The latter may result in the expansion and cracking of concrete (Wegian et al. 2010; Shi et al. 2018), as shown in Eq. (12.4). The Mg^{2+} in the pore solution of cement-based materials can react with calcium hydroxide to produce brucite, thus decreasing the salinity of pore solution and destabilizing C-S-H gels. Besides, the chemical ions in seawater may also interact with other ions in the system to affect the solid phases within concrete (Al-Amoudi et al. 1994; Frias et al. 2013). The combined effects of Cl^- and SO_4^{2-} or other cation ions, such as Na^+, K^+, Ca^{2+}, and Mg^{2+}, on the microstructure and durability of concrete need to

Table 12.1 Chemical compositions of seawater in different regions used in literature

Major ions	*Water type*	Cl^- *(mg/l)*	Na^+ *(mg/l)*	SO_4^{2-} *(mg/l)*	Mg^{2+} *(mg/l)*	Ca^{2+} *(mg/l)*	K^+ *(mg/l)*	*Total salinity*	*Difference* (%)*
Li et al. (2018)	Artificial	16,476	11,044	2772	302	436	53	31,083	−13.09
Cwirzen et al. (2014)	Natural	3000	1800	410	240	98	67	5615	−84.30
Fraternali et al. (2014)	Artificial	50,060	31,560	9660	2480	180	180	94,120	163.18
Nishida et al. (2015)	Artificial	19,720	11,111	1325	1319	422	364	34,261	−4.20
Moukwa (1990)	Natural	17,035	10,231	2800	1006	327	-	31,399	−12.20
Solution et al. (1996)	Natural	19,130	10,750	1890	1370	320	380	33,840	−5.38
Mohammed et al. (2004)	Artificial	17,087	9290	2378	1167	356	346	30,624	−14.37
Ganjian and Sadeghi (2005)	Natural	22,330	11,400	3070	1600	450	397	39,247	9.74
Province (2014)	Artificial	19,753	11,109	2765	1328	422	399	35,776	0.04
Weerdt and Justne (2014, 2015)	Natural	19,400	9500	258	1100	350	350	30,958	−13.44
Huang et al. (2018)	Artificial	19,744	11,052	1325	1328	422	369	34,240	−4.26
Rashad et al. (2018)	Artificial	21,584	11,993	3601	1458	311	480	39,427	10.25
Wang et al. (2018)	Natural	26,000	15,000	3700	2300	500	520	48,020	34.27
Cheng et al. (2018)	Artificial	19,837	11,032	2765	1314	418	397	35,763	0
Typical Seawater (Cotruvo 2005)	Natural	18,980	10,556	2649	1262	400	380	34,227	−4.29
Eastern Mediterranean (Cotruvo 2005)	Natural	21,200	11,800	2950	1403	423	463	38,239	6.92
Arabian Gulf (Cotruvo 2005)	Natural	23,000	15,850	3200	1765	500	460	44,775	25.20
Red Sea (Cotruvo 2005)	Natural	22,219	14,255	3078	742	225	210	40,729	13.89
ASTM D1141 (2021)	Artificial	19,837	11,032	2765	1314	418	397	35,763	-

Note: *Denotes the difference percentage amount of value between ASTM D1141 and previous literature on the salinity of seawater.

be considered when using seawater with different chemical compositions to prepare UHPSSC.

$$Fe^{2+} + 2Cl^- \rightarrow FeCl_2 \tag{12.1}$$

$$FeCl_2 + 2H_2O \rightarrow Fe(OH)_2 + 2Cl^- \tag{12.2}$$

$$6FeCl_2 + O_2 + 6H_2O \rightarrow Fe_3O_4 + 12H^+ + 12Cl^- \tag{12.3}$$

$$C_3A + 3CaSO_4 \cdot 2H_2O + 26H_2O \rightarrow C_3A \cdot 3CaSO_4 \cdot 32H_2O \tag{12.4}$$

12.2.2 Sea sand

The chemical and physical properties of sea sand vary greatly with sea area due to the various climate zones where the sea sand resides (Wang et al. 2003; Etxeberria et al. 2016). Sea sand mainly originates from dunes, shorelines, and offshore zones (Dhondy et al. 2020). Figure 12.1 shows that clear cubic salt crystals and high salinity are observed in offshore sea sand, while sea sand sourced from the shoreline displays an even distribution of salts resulting from tide rising and ebbing. The sea sand collected from the dunes with more soil content has lower chloride content and greater calcium content than those in shoreline and offshore areas (Dhondy et al. 2020). As far as offshore sea sand is concerned, the sea sand from the surface disposal layer has the highest chloride and fine contents (Limeira et al. 2010). The variations in the chemical and physical properties of sea sand in different climate zones should be considered before being applied in construction.

The main element compositions of sea sand and river sand are compared in Figure 12.1. Sea sand contains higher chloride content than river sand, especially for the sea sand from the offshore zone. The silica element content of sea sand varies with location due to the different contents of shells and possible organic matters in sea sand, while other chemical elements are comparable. After washing with freshwater, such as the sea sand in dunes, the salt content of sea sand can be significantly reduced to be below 0.015% (Dhondy et al. 2020; Sun et al. 2016). Sea sand in the dunes contained more calcite ($CaCO_3$) than river sand (Limeira et al. 2011; Dhondy et al. 2020). According to the X-ray diffraction (XRD) analysis, sea sand from the coastal zone possesses similar mineral compositions to river sand (Liu et al. 2016). Their main minerals are quartz (SiO_2) and feldspar ($K_2O{\cdot}Al_2O_3{\cdot}6SiO_2$, $Na_2O{\cdot}Al_2O_3{\cdot}6SiO_2$, and $CaO{\cdot}Al_2O_3{\cdot}2SiO_2$), as can be seen from Figure 12.1. However, the high contents of shells and organic substances in sea sand can decrease the strength and durability of concrete.

Figure 12.1 SEM observations of main element composition within tested river sand and sea sand.

Source: Dhondy et al. (2020).

Therefore, their contents need to be limited based on their grain size and the target compressive strength of concrete.

The particle size distributions of sea sand vary with locations from the physical characteristics. However, some studies suggest that they are generally located within the limits of Grade II sand as specified in JGJ206 (Dhondy et al. 2020) and ISO 7033:1987 (2014). Typically, sea sand has a smoother surface texture and a lower fineness modulus than river sand under the scouring effect of seawater with higher density (Limeira et al. 2010; 2011). Sea sand with an angular shape and higher surface area can decrease the bond strength by reducing the available cement paste for coating aggregate, thus decreasing the compressive strength of concrete (Ning et al. 2012). However, the usage of sea sand with rough surface can increase the compressive strength of concrete due to the enhanced mechanical properties of the interfacial transition zone (Cui et al. 2014). Sea sand with lower surface porosity and finer particles in less coagulated conditions can improve the mechanical properties of sea sand concrete (Limeira et al. 2011).

12.3 HYDRATION AND MICROSTRUCTURE

12.3.1 Hydration

Like ordinary cement paste, the hydration process of cement paste mixed with seawater can also be divided into five stages: dissolution, induction, acceleration, deceleration, and diffusion control stages. The comparisons of heat flow rate and accumulative hydration heat of cement pastes with ultra-low w/b ratio hydrated in seawater and freshwater during the first 3 d of hydration are shown in Figure 12.2. The hydration rate of the cement paste with seawater is different from that mixed with freshwater under the effect of salinity, especially during the dissolution and acceleration stages (Sikora et al. 2019; Li et al. 2015). The use of seawater significantly improves the height of the first peak due to the faster dissolutions of calcium aluminate (C_3A) and gypsum phases (Wang et al. 2018). The second peak at the acceleration stage is also accelerated by the relatively faster hydration reaction of silicate hydration products (mainly C_3S). The second hydration peak of pure cement paste mixed with seawater is accelerated by 2–5 h compared to those cast with freshwater. With the addition of supplementary cementitious materials, the peak can be retarded up to 5–12 h.

The acceleration mechanism of seawater on C_3S hydration is due to the fact that calcium hydroxide reacts with the soluble ions in seawater, which leads to a greater pH and greater amounts of gypsum. This step is followed by the reaction of the sodium hydroxide with salts in seawater, such as calcium chloride, resulting in the formation of additional calcium hydroxide. In addition, due to the presence of Cl^-, Ca^{2+}, Mg^{2+}, and SO_4^{2-}, seawater can decrease the induction stage of cement paste. The second peak shoulder is identified in the heat flow of seawater paste as a result of secondary hydration of C_3A and C_4AF induced by Cl^- in seawater. In the deceleration stage, around 40–60 h, a small peak is observed in the heat flow curve of seawater cement paste due to Friedel's salt formation through the transformation of the AFt to AFm phases (Wang et al. 2018). At this stage, sulfate in the aqueous is exhausted to form ettringite, and the free chloride concentration in the pore solution of seawater cement paste is decreased (Jensen and Pratt, 1987).

The total hydration heat of cement paste mixed with seawater is higher than that mixed with freshwater at a high w/b ratio of 0.5 (Sikora et al. 2019). However, the role of seawater played in cement paste with various w/b ratios is different. This is because the change in w/b ratio substantially alters the hydration characteristics of cement in seawater. Li et al. (2018) found that the acceleration effect of hydration can be further improved by seawater at a 0.7 w/b ratio compared to the counterpart at a 0.5 w/b ratio. In other words, the effect of seawater on cement hydration is declined with the decrease of w/b ratio. The influence of seawater on the first exothermic peak is not obvious when a 0.2 w/b ratio is used. Still, seawater mitigates the hydration heat release during the second stage and the total hydration heat (Li et al. 2018). The low activity of chemical ions and low extent of secondary hydration of C_3A in seawater with a low w/b ratio is considered to be the reason for the retarded cement hydration (Li et al. 2018).

The hydration process of supplementary cementitious materials blended paste is also changed by seawater. The pozzolanic reaction of fly ash is slowed down using seawater at a later age of 28 d (Jensen and Pratt 1987). This is probably because the dissolution of the glassy phases is initially accelerated so that little reactive material is left within fly ash particles to take part in the pozzolanic reaction at the later hydration stages. On the contrary, seawater accelerates the hydration process of cement paste containing ground granulated blast furnace slag (GGBS) and silica fume, as shown in Figure 12.2. The accelerating effect of seawater on the hydration process of GGBS blends is more remarkable than that of silica fume specimens as a result of the accelerated chemical reactions between Al-rich phases in GGBS and chemical ions (Cl^- and SO_4^{2-}) and the alkaline environment provided by seawater (Jensen and Pratt 1987). The occurrence of the main hydration peak of specimens containing 50% of GGBS is accelerated for almost 10 h with the use of seawater (Figure 12.2(a)). The corresponding value for specimens containing 30% silica fume is about 5 h compared to specimens mixed with freshwater. However, the effect of seawater on the hydration rate of cement paste without any supplementary cementitious material is not obvious (Li et al. 2018). Additionally, the addition of 1 wt.% nanosilica accelerates the cement hydration of seawater-mixed cement paste by about 15% at the early age of 1–7 d, while 3 wt.% nanosilica would delay the hydration rate (Sikora et al. 2019). However, for a ternary binder system with several types of supplementary cementitious materials at a low w/b ratio, the influential mechanism of seawater is still not clear.

12.3.2 Microstructure

Seawater reduces the porosity and refines the pore sizes to result in less capillary pores, especially the larger ones. The flake-like Friedel's salt produced by the reaction between chlorides and aluminum is a typical hydration product in seawater cement paste (Cheng et al. 2018). It can fill the pores of cement paste to densify the microstructure. Figure 12.3 shows XRD patterns of control and slag-containing pastes at 3 d. The use of seawater promotes the formation of ettringite, portlandite, and Friedel's salt. It also enhances the hydration degree of cement paste with an ultra-low w/b ratio. Supplementary cementitious materials and chemical admixtures, such as calcium nitrate and nanosilica, can greatly decrease the portlandite content by pozzolanic reactions. High Al-phase supplementary cementitious materials, such as GGBS and metakaolin, promote Friedel's salt formation, while low Al-phase supplementary cementitious materials decrease Friedel's salt content. The high calcium chemical admixtures, such as calcium nitrate, facilitate the ettringite formation.

The formation of ettringite and Friedel's salt using seawater in UHPC paste can densify the microstructure (Yuan et al. 2009). The pore structures of freshwater and seawater mixed specimens with pure cement, 25% slag, and 15% silica fume are analyzed using nitrogen adsorption and desorption (Figure 12.4). The pore size distribution of SW3 specimens containing 25% GGBS exhibits lower pore volume at any pore range. This is related to the formation of a high Ca/Si ratio of C-S-H and the formation of Friedel's

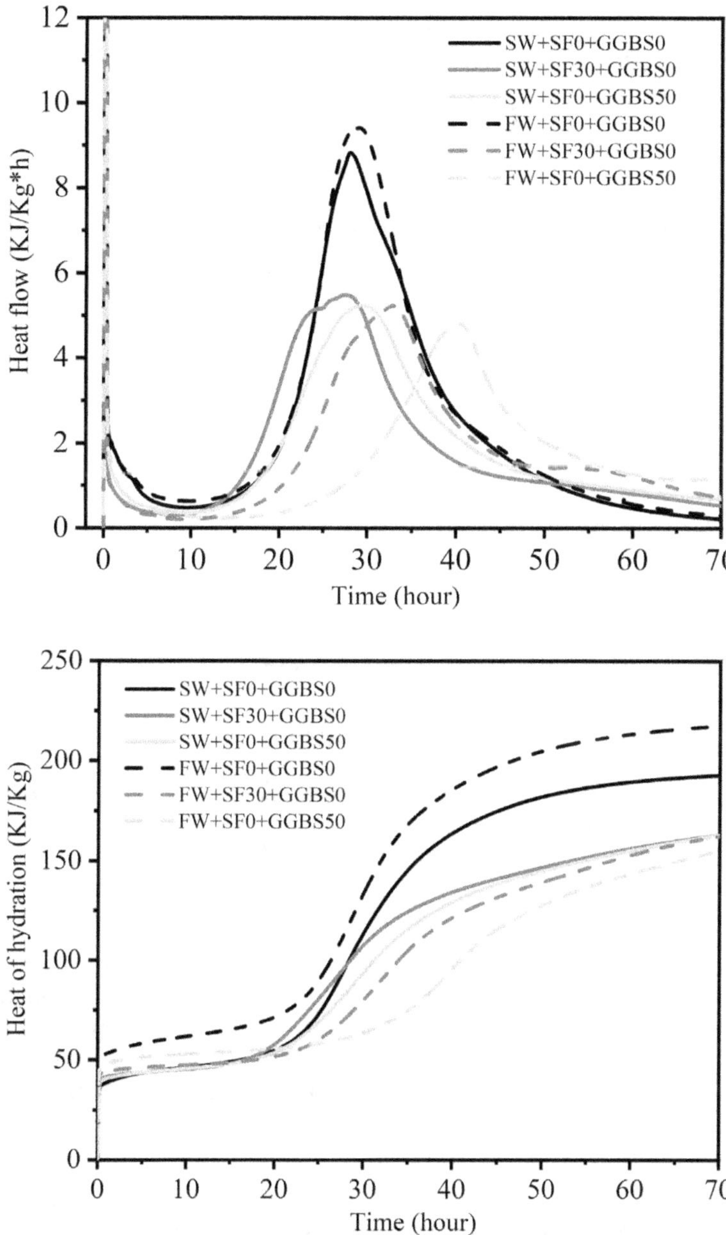

Figure 12.2 Heat flow rates and total heat of hydration for different cement pastes mixed with either freshwater or seawater at 0.2 w/b ratio (Li et al. 2018). [SW + SF30 + GGBS0 means seawater mixing cement paste made with 30% silica fume (SF) and 0% replacement of GGBS, at 2% superplasticizer dosage]. (a) Heat flow. (b) Total heat of hydration.

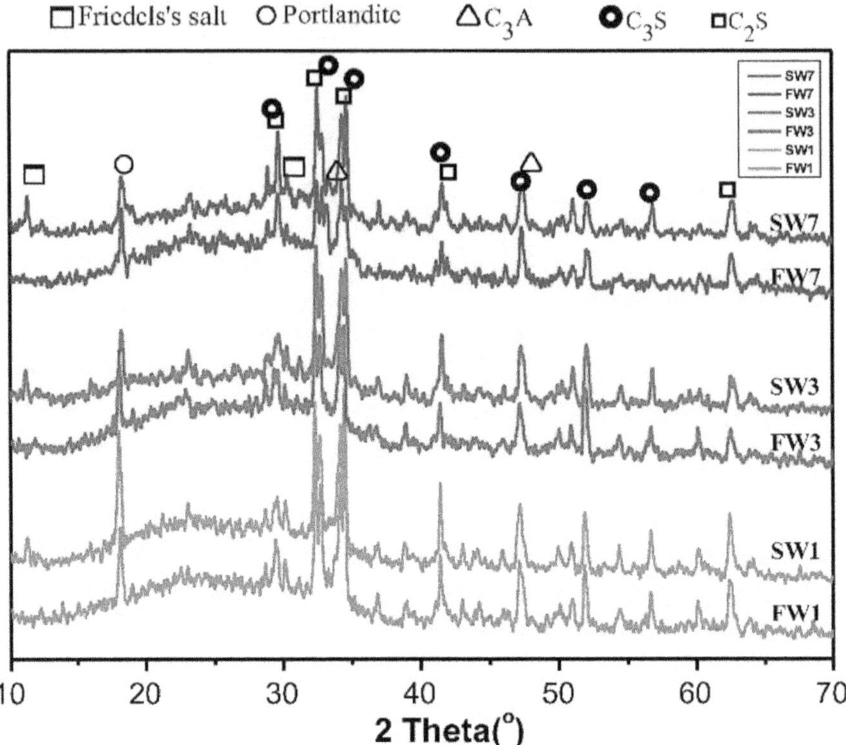

Figure 12.3 XRD patterns of control and slag-containing pastes at 3 d (Li et al. 2018). (SW1 and FW1 are cement paste groups; SW3 and FW3 are paste groups containing 25% GGBS; and SW7 and FW7 present paste groups containing 50% GGBS).

salt, as shown in Figure 12.3. On the contrary, the addition of seawater cement paste (SW1) and paste added with 15% silica fume (SW2) renders the capillary pores refined to smaller nanopores. The specimen SW1 exhibits a higher pore fraction with a pore size below 4 nm than that of FW1. However, within the pore size ranging from 4 to 128 nm, the pore fraction of SW1 decreases. The same trend is observed for the silica fume-containing specimens, especially in larger pores.

12.4 FRESH AND HARDENED PROPERTIES

12.4.1 Workability

The use of seawater can reduce the workability of UHPSSC due to the accelerated hydration associated with chemical ions in seawater. Thus, the

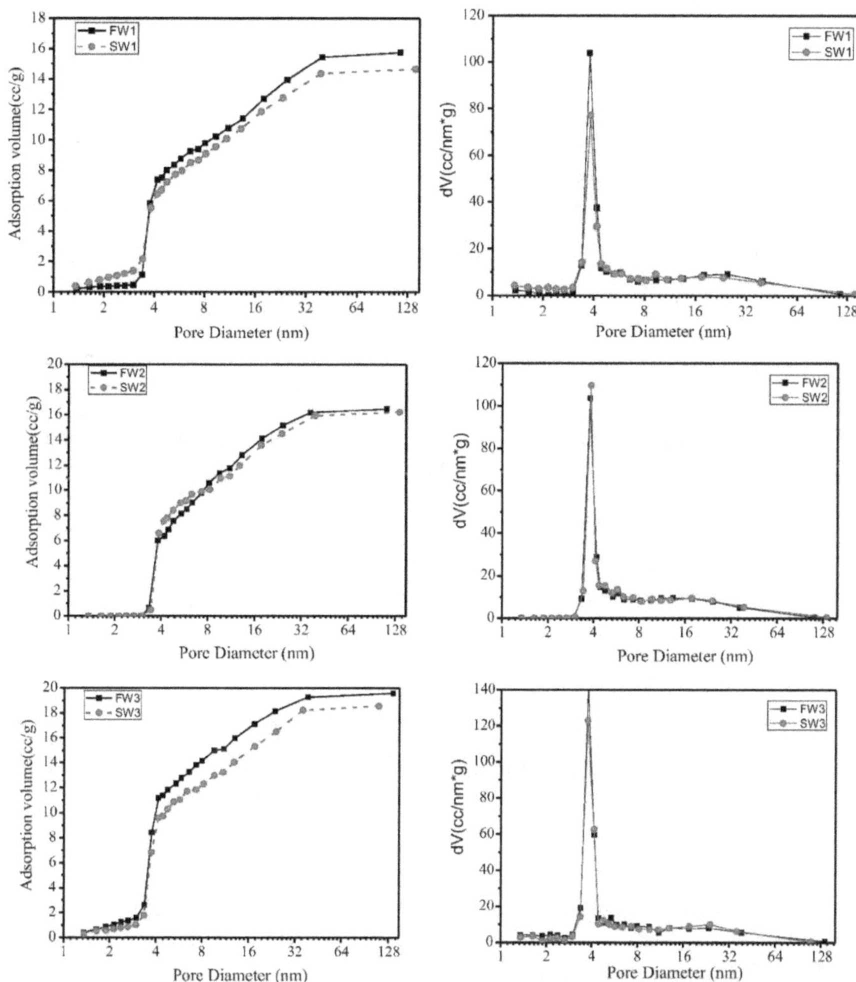

Figure 12.4 Pore structure distribution of freshwater and seawater mixed pastes at 3 d (Li et al. 2018). ("FW1 and SW1" denote freshwater and seawater mixed cement paste; "FW2 and SW2" mean freshwater and seawater mixed cement pastes containing 15% silica fume; and "FW3 and SW3" imply freshwater and seawater mixed cement pastes containing 25% GGBS).

salinity of mixing water has an important influence on the workability of UHPSSC. Besides, the workability of UHPSSC is also related to packing density and good compatibility. The slump spreads of UHPSSC prepared with different mixture compositions obtained from free mini-slump spread tests are summarized in Table 12.2.

Table 12.2 Workability and densities of UHPSSC and ordinary UHPC (Teng et al. 2019)

Group number	*Mixture number*	*Code*	*Slump spread (mm)*	P_{1d} *(kg/m³)*	P_{28d} *(kg/m³)*	P_{90d} *(kg/m³)*
1	1	QS-TW-WC-QP	267.50	2350	2379	2388
	2	QS-50ASW-WC-QP	227.50	2352	2363	2380
	3	QS-100ASW-WC-QP	175.00	2317	2341	2334
	4	QS-150ASW-WC-QP	195.00	2320	2331	2345
	5	QS-200ASW-WC-QP	130.00	2313	2300	2312
2	6	uRS-TW-WC-QP	315.00	2423	2437	2445
	7	RS-TW-WC-QP	338.75	2405	2435	2454
	8	RS-TW-WC-FA	342.50	2419	2430	2443
	9	RS-TW-OPC-QP	316.67	2428	2443	2447
	10	RS-TW-OPC-FA	330.00	2430	2453	2461
3	11	uSS-SW-WC-QP	270.00	2316	2330	2346
	12	SS-SW-WC-QP	324.17	2409	2424	2433
	13	SS-SW-WC-FA	331.25	2403	2407	2422
	14	SS-SW-OPC-QP	301.67	2426	2444	2459
	15	SS-SW-OPC-FA	322.50	2434	2444	2455

Note: WC = white cement; OPC = ordinary Portland cement; QP = quartz powder; FA = Class C fly ash; QS = quartz sand; uRS = unwashed river sand; RS = river sand; uSS = unwashed sea sand; SS = sea sand; TW = tap water; 50ASW = artificial seawater whose salinity is half that of typical natural seawater (TNSW); 100ASW = artificial seawater whose salinity is the same as that of TNSW; 150ASW = artificial seawater whose salinity is 1.5 times greater than that of TNSW; 200ASW = artificial seawater whose salinity is twice more than that of TNSW; SW = natural seawater.

The workability of UHPSSC generally decreases with the salinity of mixing water (Group 1, Mixtures 1–5, Table 12.2). The slump spread of UHPSSC using seawater with a salinity of 72 g/l is only 50% of UHPC made with tap water (Mixtures 5 and 1). In addition, the existence of $CaCl_2$ in artificial seawater can decrease the workability by the accelerated formation of C-S-H and heat release in the hydration process (Juenger et al. 2005). By comparing Group 2 (Mixtures 6–10) and Group 3 (Mixtures 11–15), seawater and sea sand generally lead to decreased slump spread. The decreasing degree also depends on other raw materials used in the mixture. Besides, the accelerated hydration due to salts and the finer particles results in greater surface areas of sea sand and lower workability of UHPSSC (Hasdemir et al. 2016).

The effects of various raw materials on workability are evident by comparing Mixtures 6–10 in Group 2. The desilting of sand and the replacement of quartz powder with fly ash leads to increased slump spread. The replacement of white cement with ordinary Portland cement negatively affects the workability. This can be due to the fineness or specific surface areas of the raw materials. Especially, the desilting of sand significantly reduces its clay amount, which consists of very fine particles (Fernandes et al. 2007). Compared with white cement, ordinary Portland cement has a greater specific surface area. In addition, fly ash can cause potentially pozzolanic reactions and may reduce frictions between aggregate particles because of its spherical shape of particles, which both contribute to increased slump spreads (Hemalatha and Ramaswamy 2017).

Mixtures 11–15 in Group 3 in Table 12.2 also reveal the influence of the properties of raw materials on the workability of UHPSSC. Notably, the effect of the desilting process seems to be much more pronounced for sea sand than for river sand. This is probably due to the greater clay content in the sea sand (5.46%) than that in the river sand (0.61%).

12.4.2 Density

Generally, the use of seawater can increase the density of UHPSSC, similar to ordinary concrete, due to the salinity contained in seawater. In contrast, the use of sea sand can cause a decrease in density of UHPSSC due to the influence of sea shell contained in sea sand. The densities of UHPSSC at different ages are also summarized in Table 12.2. These results are obtained using samples subjected to tap water curing at room temperature.

In general, the density increases with curing age for all the mixtures because of the continuous water absorption process of the concrete when immersed in water (Table 12.2). The effects of mixture proportion parameters on the density appear to be similar to workability. The density generally decreases with the salinity in seawater for Mixtures 1–5 in Group 1. At the same time, the use of seawater and sea sand generally leads to a decrease in density, as can be seen in Groups 2 and 3. It is evident that the density linearly increases with the slump spread.

12.4.3 Strength

The use of seawater and sea sand can improve the compressive strength of UHPSSC at early age due to the accelerated hydration and improved microstructure. Mixture proportion, curing conditions, and testing procedures affect the strength of UHPSSC. Table 12.3 summarizes the cube compressive strengths of mixtures at various ages. The compressive strength of UHPSSC gradually increases with age from 1 to 28 d, similar to UHPC. After 28 d, the compressive strength remains stable. The 28-d cube compressive strength of UHPSSC can reach 184 MPa.

Table 12.3 Cube compressive strengths of different UHPSSC mixtures (Teng et al. 2019)

Group number	*Mixture number*	$f_{cu,1d}$ *(MPa)*	$f_{cu,7d}$ *(MPa)*	$f_{cu,14d}$ *(MPa)*	$f_{cu,28d}$ *(MPa)*	$f_{cu,90d}$ *(MPa)*	$f_{cu,H\text{-}24hr}$ *(MPa)*
1	1	95.80	130.13	156.48	170.57	179.28	177.83
	2	96.60	142.07	158.77	178.90	183.35	179.94
	3	87.38	137.04	155.72	164.07	165.19	165.53
	4	96.85	138.95	149.87	160.83	164.15	164.96
	5	85.31	129.27	142.28	160.84	160.91	168.06
2	6	103.11	146.84	152.60	171.15	181.03	185.92
	7	105.60	146.26	164.33	177.89	185.18	178.91
	8	105.45	146.06	165.78	185.14	189.42	186.85
	9	78.46	147.31	169.17	184.66	190.82	182.72
	10	73.89	142.98	164.34	186.40	191.47	183.78
3	11	109.10	132.09	150.16	167.27	171.17	173.00
	12	112.98	152.47	174.71	184.04	194.86	193.71
	13	107.57	145.86	161.90	176.41	186.53	187.01
	14	74.37	135.53	160.46	171.21	181.80	186.83
	15	90.42	141.66	159.66	174.03	184.98	191.48

Note: Same mixture proportions used as shown in Table 12.2.

12.4.3.1 Effect of salinity

Figure 12.5 presents the effect of salinity of seawater on compressive strength of UHPSSC Mixtures 1–5 in Group 1. The use of saltwater generally leads to a higher early strength of concrete. An optimal salinity of mixing water, equal or close to that of Mixture 2 with a salinity of 18 g/L, may exist for the compressive strength of UHPSSC. When the salinity of seawater exceeds a specific value, it may have a slightly negative effect on the long-term compressive strength of concrete. It is believed that the slightly higher early strength of UHPSSC with salt water is due to the formation of Friedel's salt ($3CaO{\cdot}Al_2O_3{\cdot}CaCl_2{\cdot}10H_2O$) and Kuzel's salt ($3CaO{\cdot}Al_2O_3{\cdot}0.5CaSO_4{\cdot}0.5CaCl_2{\cdot}11H_2O$) (Weerdt et al. 2014). However, the decomposition of these salts with time is believed to affect the long-term strength of concrete (Suryavanshi and Swamy 1996).

12.4.3.2 Effect of seawater and sea sand

Figure 12.6 compares the effects of seawater and sea sand on the development of compressive strength of UHPSSC by comparing the results of Groups 2 and 3. The Group 2 used river sand and tap water, while the Group 3 used sea sand and seawater. Due to the use of seawater and sea sand, the early strength of UHPSSC is likely to increase, but the strengths at and above 7 d decrease. Nevertheless, the differences at various ages between the two mixtures in each of the four subfigures are all within 8%

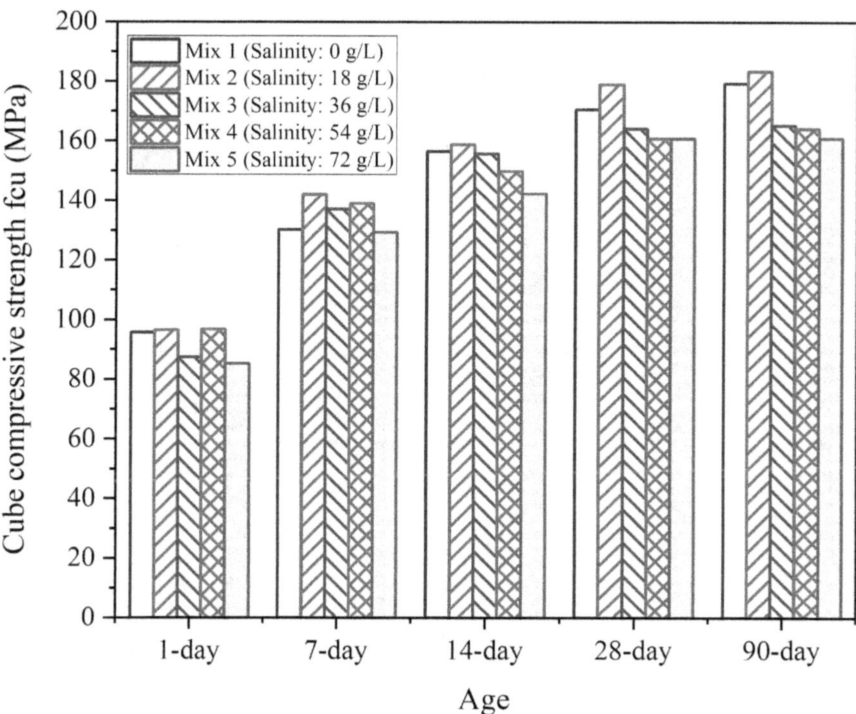

Figure 12.5 Effect of salinity on compressive strength of UHPSCC.

Source: Data from Teng et al. (2019).

except for the 1-d strength (Figure 12.6(d)), suggesting that the uses of seawater and sea sand to replace tap water and river sand, respectively, have negligible effects on the compressive strength of UHPSSC. However, the sea sand containing sea shell may lead to the deterioration of the strength of UHPSSC. Figure 12.7 presents the effect of sea shell content on the pores distribution and the compressive strength. As can be seen, the number of macroscopic pores continually increases with the sea shell content, beginning at the content of 10%. It is believed that incorporating much sea shell can lead to a serious air-entraining effect, which induces an increase in the number of macroscopic pores (from 23% to 31% with 20% replacement). Increasing the number of macroscopic pores (d > 50 nm) can decrease the strength. The microstructural examination results also demonstrate that mixing a large amount of sea shell can weaken the compressive strength (Figure 12.7(b)).

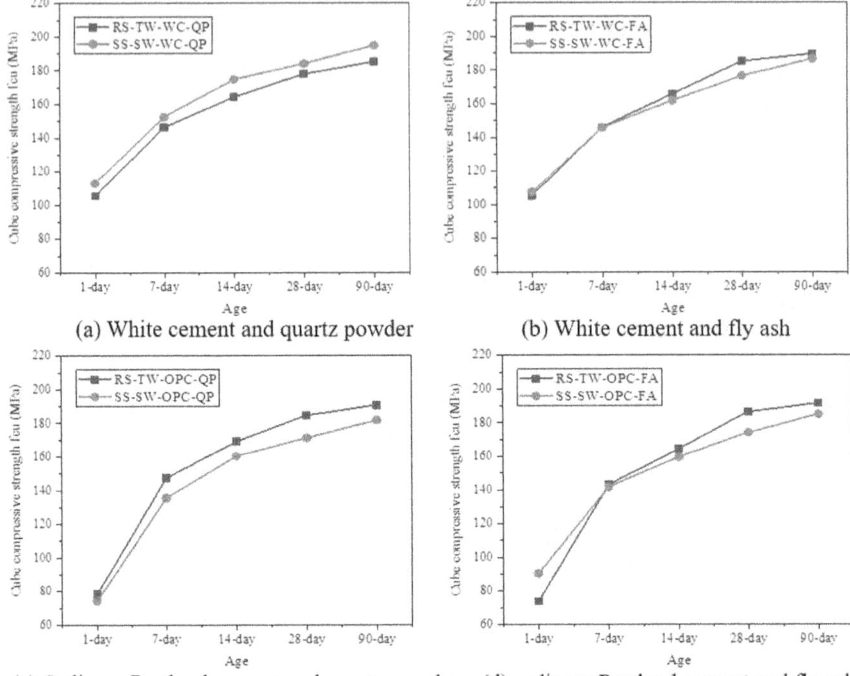

(a) White cement and quartz powder (b) White cement and fly ash

(c) Ordinary Portland cement and quartz powder (d) ordinary Portland cement and fly ash

Figure 12.6 Effects of seawater and sea sand on strength developments of UHPSSC prepared with: (a) White cement and quartz powder, (b) White cement and fly ash, (c) Ordinary Portland cement and quartz powder, and (d) Ordinary Portland cement and fly ash.

Source: Data from Teng et al. (2019).

12.4.3.3 Effect of supplementary cementitious material and cement types

The use of various types of supplementary cementitious materials and cement significantly affects the strength of UHPSSC. The effect of using Class C fly ash to replace quartz powder is illustrated in Figure 12.8 by comparing four mixtures. The only difference between the two mixtures is the fine supplemental material (i.e., fly ash or quartz powder). The mixtures with fly ash have similar strengths to those made with quartz powder at 7 d or above. Fly ash is known to have the potential for pozzolanic reactions (Hemalatha and Ramaswamy 2017), which is beneficial to the strength development of concrete. Besides, the high content of free calcium oxide of Class C fly ash negatively affects the concrete strength, especially with sulfate ions (Tikalsky and Carrasquillo 1989).

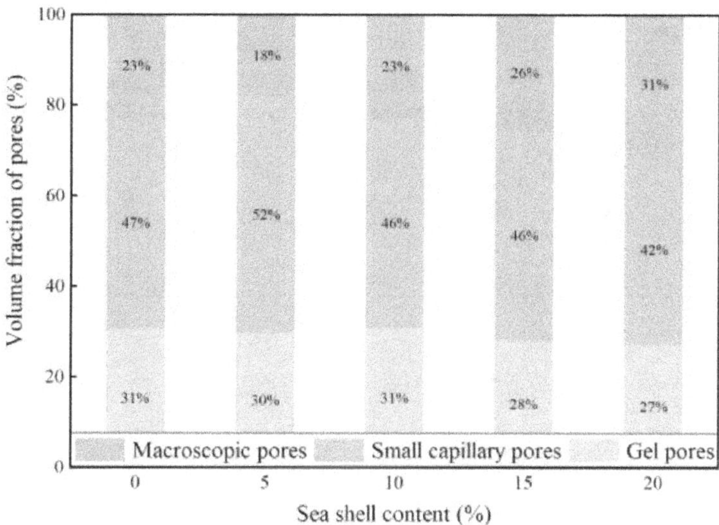

(a) Accumulation percentage of the different pore

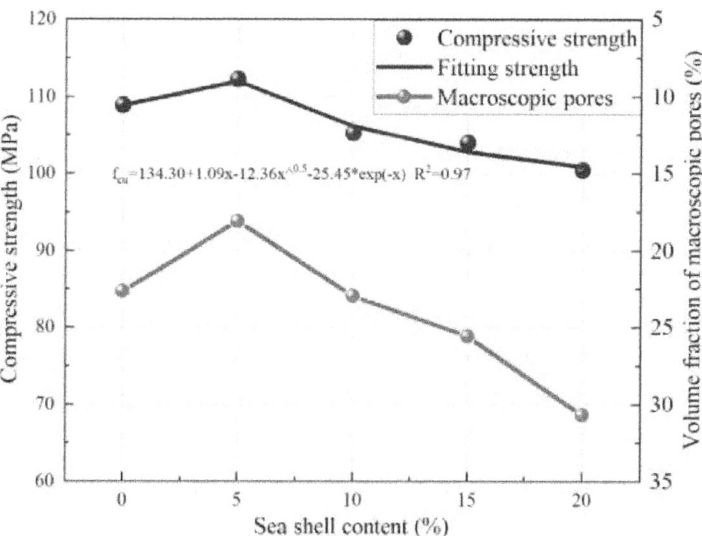

(b) Compressive strength of matrix versus volume fraction of macroscopic pores

Figure 12.7 Pore structure of UHPSSC matrix with different sea shell contents. (a) Accumulation percentage of the different pore. (b) Compressive strength of matrix versus volume fraction of macroscopic pores.

Source: Data from Jiang et al. (2022).

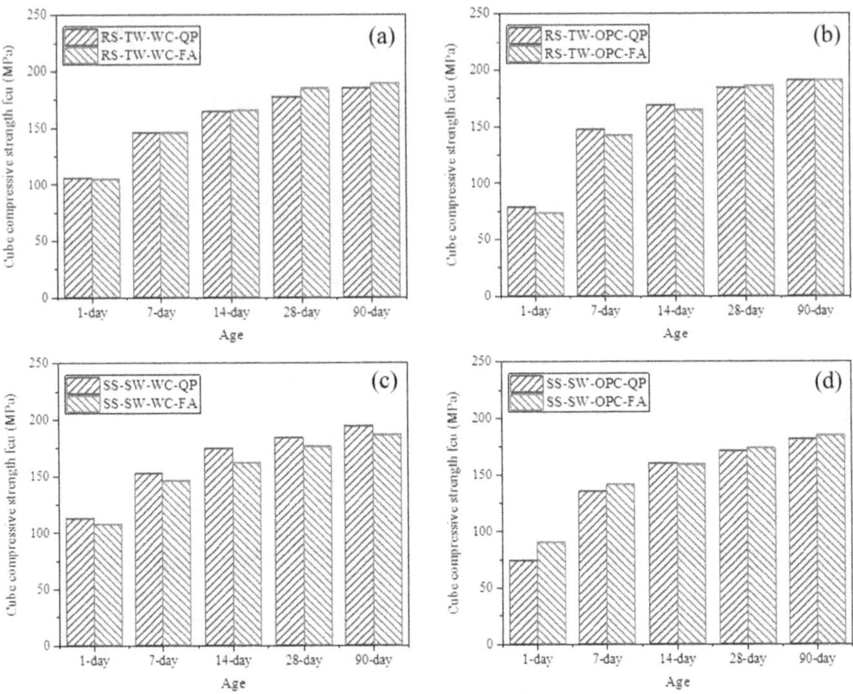

Figure 12.8 Effect of supplementary cementitious material type on strength development: (a) Tap water-river sand UHPC with white cement; (b) Tap water-river sand UHPC with ordinary Portland cement; (c) UHPSSC with white cement; and (d) UHPSSC with ordinary Portland cement.

Source: Data from Teng et al. (2019).

The observation illustrated in Figure 12.8 is believed to result from counteracting effects of many factors, including salinity, seawater, and sea sand. Further research involving analysis of material structure of UHPSSC is needed to clarify these effects.

Figure 12.9 illustrates the effects of cement type on strength development of UHPSSC. The only difference between the two mixtures in each subfigure of Figure 12.9 is the type of cement (i.e., white cement and ordinary Portland cement). The use of ordinary Portland cement to replace white cement generally leads to lower early-age strengths, especially the 1-d strength. However, the 28-d and 90-d strengths depend on other mixed constituents. For Group 2 with river sand and tap water, the mixtures with ordinary Portland cement have higher 28-d and 90-d strengths, but the opposite is found for Group 3 with seawater and sea sand.

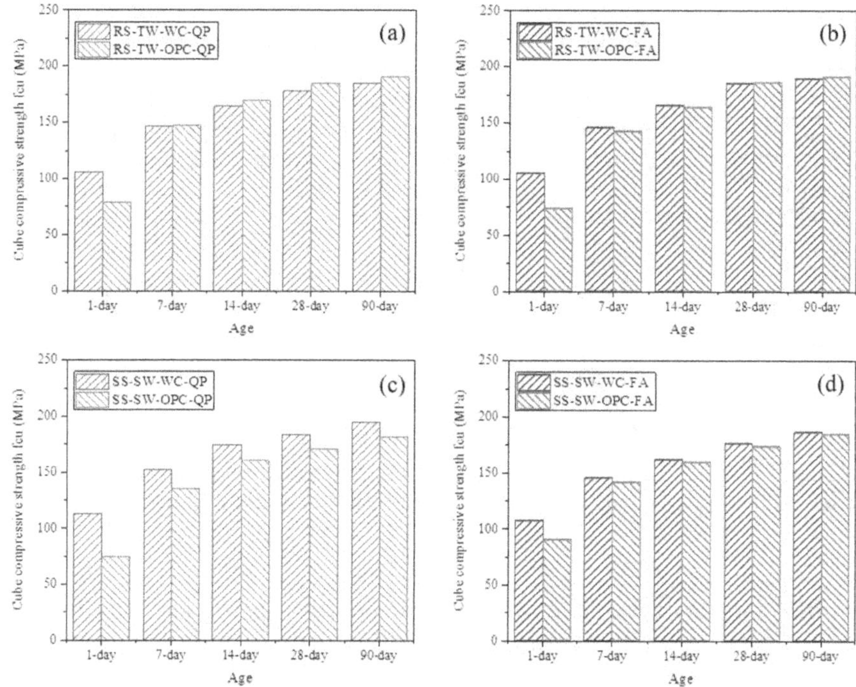

Figure 12.9 Effect of cement type on strength development: (a) Tap water-river sand UHPC with quartz powder; (b) Tap water-river sand UHPC with fly ash; (c) UHPSSC with quartz powder; and (d) UHPSSC with fly ash.

Source: Data from Teng et al. (2019).

12.4.3.4 Effect of curing method

Figure 12.10 depicts the influence of curing methods, including seawater curing and tap water curing, on the development of compressive strength of UHPSSC. Compared with tap water curing, seawater curing leads to evident reductions in the compressive strength of UHPSSC. The reduction is up to 15% at 90 d, and such a reduction appears to increase with age. The seawater curing has a limited adverse effect on the elastic modulus of UHPSSC, but this effect is not as pronounced as the effect on strength (see Table 12.4). The above observations are believed to be at least partially due to the existence of magnesium sulfate when seawater is used for curing (Ragab et al. 2016).

In Table 12.3, the compressive strengths of UHPSSC after 28 d and 90 d of 22 ± 3°C tap water immersion curing ($f_{cu,28d}$ and $f_{cu,90d}$) are shown against the strengths of the corresponding UHPSSC after 24 h of 90 ± 1°C heat

Table 12.4 Effect of curing method on properties of UHPSSC (Teng et al. 2019)

Mix name	*SS-SW-WC-QP (Mix 12/Group 3)*		*SS-SW-WC-FA (Mix 13/Group 3)*	
Curing method	*Tap water curing*	*Seawater curing*	*Tap water curing*	*Seawater curing*
ρ_{1d} (kg/m^3)	2405	2409	2419	2403
ρ_{28d} (kg/m^3)	2435	2433	2430	2449
ρ_{90d} (kg/m^3)	2454	2446	2443	2446
$f_{cu,1d}$ (MPa)	112.98	110.91	107.57	115.01
$f_{cu,7d}$ (MPa)	152.47	145.98	145.86	142.47
$f_{cu,14d}$ (MPa)	174.71	159.91	161.9	158.43
$f_{cu,28d}$ (MPa)	184.04	172.26	176.41	167.39
$f_{cu,90d}$ (MPa)	194.86	165.22	186.53	163.19
f_{co} (MPa)	197.59	185.57	185.87	194.10
E_c (GPa)	51.024	50.924	50.132	50.499
υ	0.20	0.21	0.21	0.21
$\varepsilon_{co}(\times 10^{-6})$	4381	3870	4636	4613
$\varepsilon_{lo}(\times 10^{-6})$	1165	968	1497	1495

Note: WC = white cement; QP = quartz powder; FA = Class C fly ash; QS = quartz sand; SS = sea sand.

curing ($f_{cu,H\text{-}24hr}$). It is evident from Table 12.3 that $f_{cu,H\text{-}24hr}$ is generally close to $f_{cu,28d}$, but is lower than $f_{cu,90d}$ of UHPC prepared with tap water and river sand. However, for the UHPSSC, both $f_{cu,28d}$ and $f_{cu,90d}$ are lower than $f_{cu,H\text{-}24hr}$. It can be concluded that it takes more time for UHPSSC cured at room temperature to develop the same strength as that subjected to heat curing.

12.4.3.5 Effect of fiber content

Polymer fiber, as an alternative material, is tried to be used to prepare UHPSSC (Huang et al. 2021). The effect of polymer fiber content on the compressive strength of UHPSSC is shown in Figure 12.11. As the polymer fiber content increases, the compressive strength of UHPSSC slightly decreases. The average compressive strengths of UHPSSC with 0%, 0.5%, 1.0%, and 1.5% polymer fibers are 140.9, 140.8, 137.7, and 131.7 MPa, respectively. This indicates that the compressive strength of UHPSSC is sensitive to the matrix compactness, which is caused by the addition of polymer fiber. In addition, polymer fiber content can also affect the bond strength. The addition of 0.5% polymer fibers in UHPSSC can lead to a 3%–10% increase in bond strength (Zeng et al. 2022). However, a further increase in polymer fiber content to 1% may impair the strength of UHPSSC. Minimal information has been reported on the influence of steel fiber content and shape on UHPSSC.

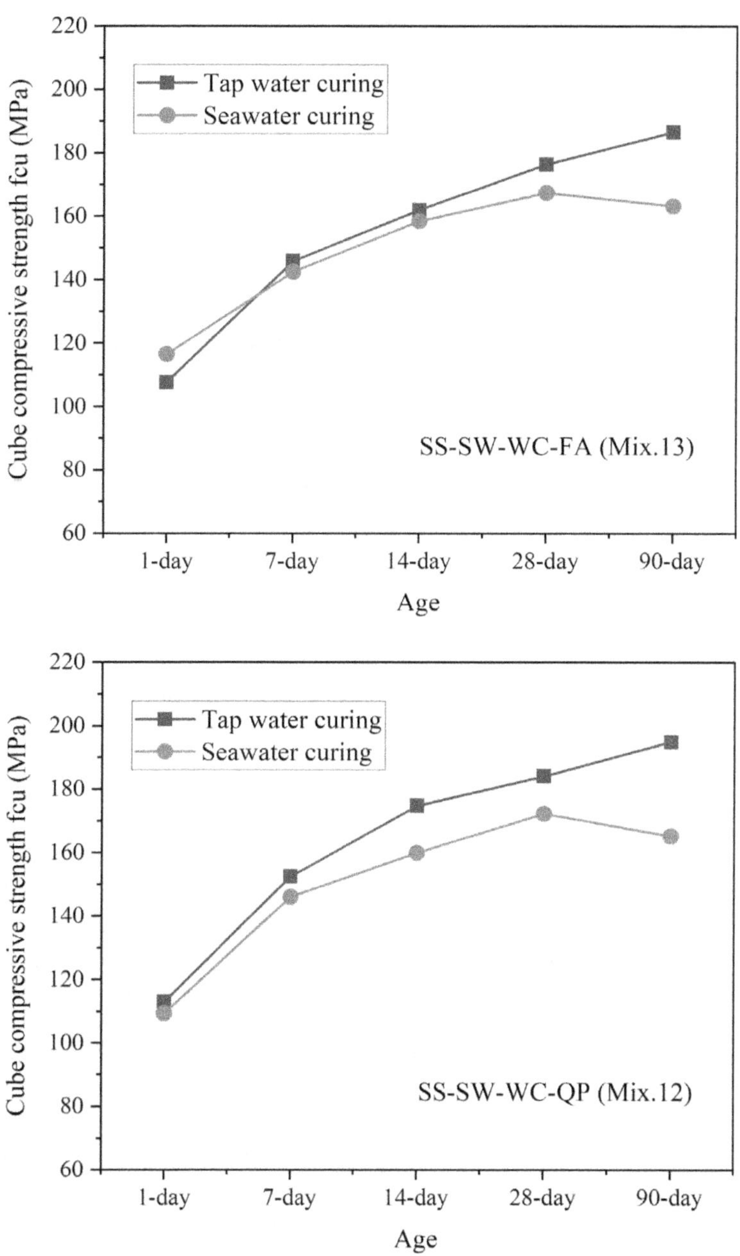

Figure 12.10 Effect of seawater curing on strength development of UHPSSC.

Source: Data from Teng et al. (2019).

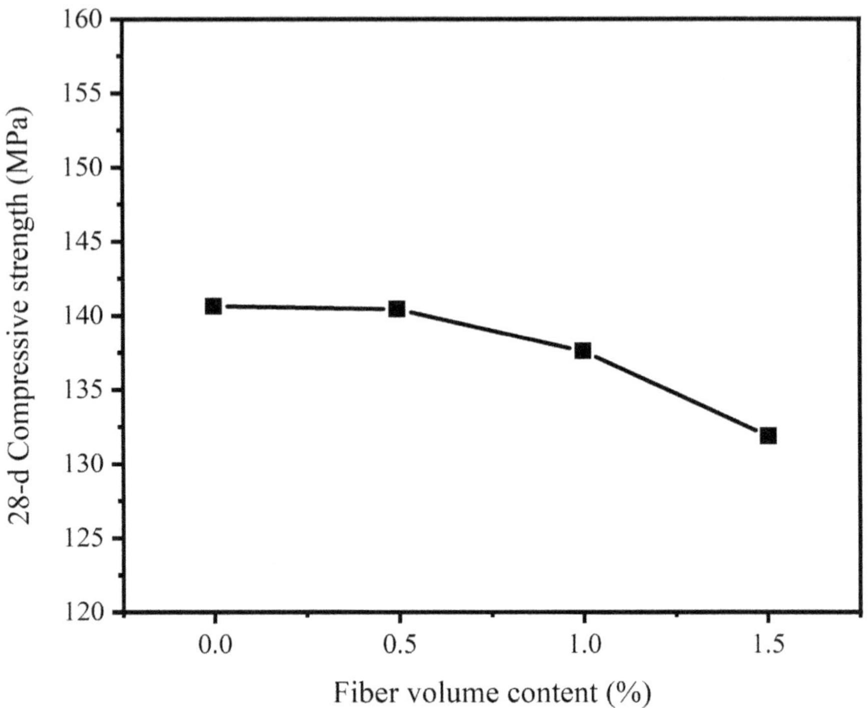

Figure 12.11 Effect of polymer fiber content on compressive strength of UHPSSC.

Source: Data from Huang et al. (2021).

Figure 12.12 illustrates tensile strength and strain capacity of UHPSSC. The tensile strength of UHPSSC increases from 4 to 8 MPa, with polymer fiber content increasing from 0.5% to 1.5%, as presented in Figure 12.12(a). Additionally, the tensile strain capacity also significantly increases with the increase of polymer fiber content, as can be seen from Figure 12.12(b). The tensile strain capacities of UHPSSC with 1.0% and 1.5% polymer fibers are 1.8% and 4.6%, respectively, which are 3.5 and 9.2 times greater than those of UHPSSC with 0.5% polymer fibers. It should be pointed out that excess fiber content in UHPSSC can lead to difficulty in fiber dispersion.

12.4.4 Durability

The effect of seawater and sea sand on the durability of UHPSSC mainly depends on the characteristics of seawater and sea sand, similar to ordinary concrete. Generally, for ordinary concrete, the use of seawater has little

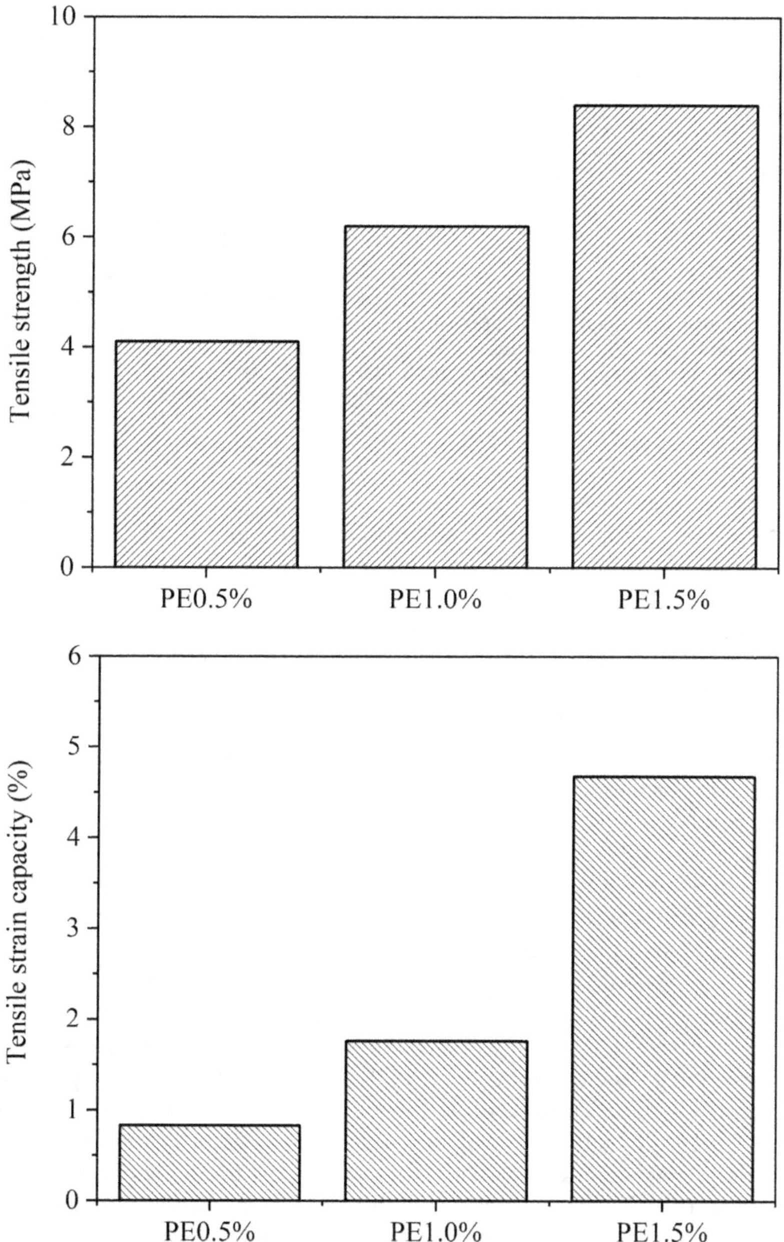

Figure 12.12 Tensile strength and strain capacity of UHPSSC: (a) Strength and (b) Strain capacity.

Source: Data from Huang et al. (2021).

Table 12.5 Laboratory test results of durability of UHPC and UHPSSC (Li et al. 2020)

Code	*Chloride ion diffusion coefficient (RCM method)*	*Chloride ion permeability (electricity method)*	*Carbonization*	*Frost resistance*	*Resistance sulfate attack*
UHPSSC(s)	1.54×10^{-12} m²/s	131 C	0 mm	100%	103%
UHPSSC(h)	1.28×10^{-12} m²/s	85 C	0 mm	100%	105%
UHPC(s)	1.66×10^{-12} m²/s	155 C	0 mm	100%	101%
UHPC(h)	1.33×10^{-12} m²/s	91 C	0 mm	100%	101%

Note: UHPSSC(s) refers to specimens made with simulated seawater and sea sand under standard curing; UHPSSC(h) presents specimens made with simulated seawater and sea sand that cured in 85°C hot water for 48 h; UHPC(h) denotes specimens mixed with deionized water and river sand that cured in 85°C hot water for 48 h; and UHPC(s) represents the specimens prepared using deionized water and river sand under standard curing.

effect on durability except for the risk of steel corrosion and alkali-silica reaction. In contrast, the sea sand harms durability due to the existence of seashells. The effects of seawater and sea sand on the durability of UHPC are different from ordinary concrete due to its dense microstructure. The durability of UHPC and UHPSSC subjected to standard curing and 85°C hot water for 48 h is shown in Table 12.5. The indexes of durability of UHPSSC(s) and UHPSSC(h) are at the same level as those of UHPC(s) and UHPC(h). The chloride ion diffusion coefficient of UHPSSC(h) is 1.28×10^{-12} m²/s, while the electric flux is 85 Coulomb. Its carbonization depth of accelerated carbonization of UHPSSC at 28 d is 0, and the relative elastic modulus is 100% after 1000 freezing-thawing cycles. The UHPSSC has good durability due to its dense microstructure. Therefore, UHPSSC can be used in the marine environment. In addition, steel corrosion is a severe problem for ordinary concrete prepared with seawater and sea sand. However, it has been identified that the embedded steel reinforcements in UHPSSC are safe enough against corrosion under drying-wetting cycle conditions (Wang et al., 2021).

The use of supplementary cementitious materials can offer a significant enhancement in the durability of UHPSSC due to disconnected pore structure and pore size modification through the formation of Friedel's salt. Figure 12.13 shows the effect of use of slag and silica fume on the nonsteady state migration coefficient. The 28-day migration coefficient of SS-50 is measured to be 0.2×10^{-12} m²/s, which is 19 times lower than that of SS-0. The silica fume can act as the filler to optimize the pore structure by filling effect. The use of slag increases the content of aluminum that can react with chloride to form Friedel's salt. It densifies the microstructure further. It is noteworthy that the chloride diffusion resistance continued to improve in slag and silica fume-based UHPCs for up to 90 days, whereas normal

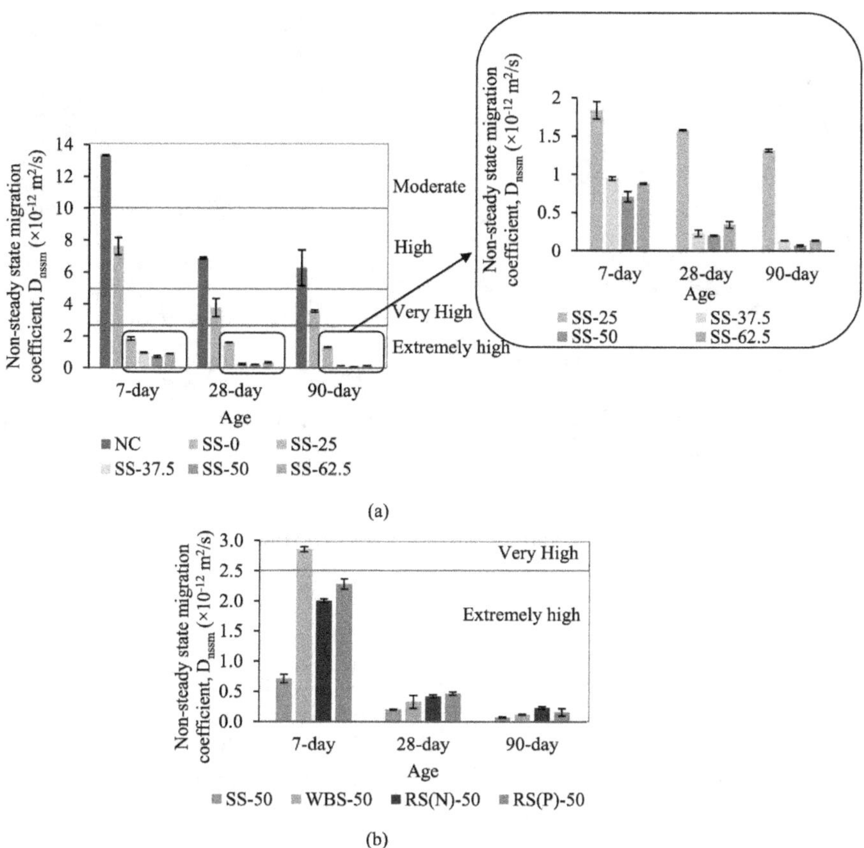

Figure 12.13 Chloride migration test results (according to NT Build 492) of UHPSSC mixes with different (a) OPC replacement ratio and (b) Aggregate sources (Saleh et al. 2023). ("NC" indicates the ordinary concrete, "SS50" means the UHPSSC mixes containing 37.5% slag and 12.5% silica fume, "WBS" indicates the UHPC mixes prepared by washed beach sand and tap water, and "RS" means the UHPC mixes prepared by river sand and tap water.)

strength concrete and control SWSS-UHPC tend to stabilize after 28 days of curing. This is caused by the pozzolanic reactivity of silica fume and slag that continues to disconnect the pores, thus increasing the mesopores (6–50 nm pore dia.) and decreasing the capillary pores (>50 nm pore dia.).

The durability of UHPSSC in marine environments is excellent due to its dense microstructure and the decreased difference in chloride concentration between UHPSSC and the seawater solution where specimens are exposed. Figure 12.14 shows the surface appearance of UHPSSC specimens before and after the exposure testing in the marine environment. After

Figure 12.14 Surface appearance of UHPSSC: (a) Before serving; (b) After serving and before cleaning; and (c) After serving and cleaning.

Source: Li et al. (2020).

1 year of serving in the marine environment, the surface of UHPSSC shows no cracking. The structure is complete, and the surface is attached with a layer of transparent and smelly marine biological humus, as shown in Figure 12.14(b). There is no crack on the surface, and no cement paste fell off after washing the concrete, as shown in Figure 12.14(c).

12.4.5 Shrinkage

Shrinkage, especially autogenous shrinkage, is a critical issue for UHPC due to the use of highly cementitious materials. However, the autogenous shrinkage of UHPSSC can be significantly mitigated by the use of seawater (Li et al. 2018). In addition, the effect of seawater on autogenous shrinkage is more remarkable when used with silica fume. This could be attributed to the combined effect of seawater on hydration and the very large specific surface area of silica fume particles.

12.5 SUMMARY

UHPSSC has a faster hydration rate, lower density, and more compact microstructure than conventional UHPC due to chloride existing in seawater and sea sand. UHPSSC also has high strength and good durability. The properties of seawater and sea sand have an important influence on the performance of UHPSSC. Generally, the shells and organic substances in sea sand have adverse effects on the properties of UHPSSC. Hence, the qualities of seawater and sea sand deserve careful consideration before the mixture design of UHPSSC. The strength of UHPSSC varies with salinity. An optimal salinity of mixing water exists for the compressive strength of UHPSSC. Supplementary cementitious materials and cement with high Al-phase content have beneficial effects on the properties of UHPSSC due to the pozzolanic effect and high chloride binding capacity.

UHPSSC presents excellent durability performance in the marine environment. There is only a possibility of corrosion for steel fibers close to the surface area due to the extremely low chloride and oxygen permeability. In addition, the microstructure and Vickers hardness near the surface of UHPSSC become more compact due to the filling effect of rehydration products and corrosion products. Thus, the corrosion effect of seawater on UHPSSC slows down after serving for a period.

REFERENCES

Al-Amoudi, O. S. B., Rasheeduzzafar, M., Maslehuddin, S. N., Abduljauwad. (1994). Influence of chloride ions on sulphate deterioration in plain and blended cements. *Mag. Concr. Res.* 46 (167), 113–123.

An, M., Yang, X., Wang, J., Cui, N. (2007). Research on durability of RPC. *Archit. Technol.* 38(5), 367–368.

ASTM D 1141 (2021). Standard practice for preparation of substitute ocean water. ASTM International: West Conshohocken, PA, USA.

Cheng, S., Shui, Z., Sun, T., Huang, Y., Liu, K. (2018). Effects of seawater and supplementary cementitious materials on the durability and microstructure of lightweight aggregate concrete. *Constr. Build. Mater.* 190, 1081–1090.

Cotruvo, J. A. (2005). Water desalinization processes and associated health and environmental issues. *Water Conditioning and Purification* 47(1), 13–17.

Cui, M., Mao, J., Jia, D., Li, B. (2014). Experimental study on mechanical properties of marine sand and seawater concrete. *Int. Conf. Mech. Civ. Eng.* 106–111.

Cwirzen, A. P., Sztermen, P., Habermehl-Cwirzen, K. (2014). Effect of Baltic seawater and binder type on frost durability of concrete. *J. Mater. Civ. Eng.* 26(2), 283–287.

De Weerdt, K., Justnes, H., Geiker, M. R. (2014). Changes in the phase assemblage of concrete exposed to sea water. *Cem. Concr. Compos.* 47, 53–63.

De Weerdt, K., Justnes, H. (2015). The effect of sea water on the phase assemblage of hydrated cement paste. *Cem. Concr. Compos.* 55, 215–222.

Dhondy, T., Remennikov, A., M., Neaz Sheikh, M.(2020). Properties and application of sea sand in sea sand-seawater concrete. *J. Mater. Civil Eng.* 32(12), 04020392.

Etxeberria, M., Fernandez, J.M., Limeira, J. (2016). Secondary aggregates and seawater employment for sustainable concrete dyke blocks production: case study. *Constr. Build. Mater.* 113, 586–595.

Fernandes, V. A., Purnell, P., Still, G. T., Thomas, T. H. (2007). The effect of clay content in sands used for cementitious materials in developing countries. *Cem. Concr. Res.* 37(5), 751–758.

Fraternali, F., Spadea, S., Berardi, V. P. (2014). Effects of recycled PET fibres on the mechanical properties and seawater curing of Portland cement-based concretes. *Constr. Build. Mater.* 61, 293–302.

Frias, M., Goñi, S., García, R., De La Villa, R. V. (2013). Seawater effect on durability of ternary cements. Synergy of chloride and sulphate ions. *Compos. Part B* 46, 173–178.

Ganjian, E., Sadeghi, H. (2005). Effect of magnesium and sulfate ions on durability of silica fume blended mixes exposed to the seawater tidal zone. *Cem. Concr. Res.* 35(7), 1332–1343.

Hasdemir, S., Tugrul, A., Yilmaz, M. (2016). The effect of natural sand composition on concrete strength. *Constr. Build. Mater.* 112, 940–948.

Hemalatha, T., Ramaswamy, A. (2017). A review on fly ash characteristics–Towards promoting high volume utilization in developing sustainable concrete. *J. Clean. Prod.* 147, 546–559.

Huang, B. T., Wang, Y. T., Wu, J. Q., Yu, J., Dai, J. G., Leung, C. K. (2021). Effect of fiber content on mechanical performance and cracking characteristics of ultra-high-performance seawater sea-sand concrete (UHP-SSC). *Adv. Struct. Eng.* 24(6), 1182–1195.

Huang, Y., He, X., Wang, Q., Sun, Y. (2018). Mechanical properties of sea sand recycled aggregate concrete under axial compression. *Constr. Build. Mater.* 175, 55–63.

ISO 7033: 1987. (2014). Fine and coarse aggregates for concrete—Determination of the particle mass-per-volume and water absorption-Pycnometer method. The International Organization for Standardization.

Jensen, H. U., Pratt, P. L. (1987). The effect of fly ash on the hydration of cements at low temperature mixed and cured in seawater. *MRS Online Proc. Lib.Arch.*, 113, 279–289.

JGJ 206-2010 (2010). Technical code for application of sea sand concrete. China, Architecture and Building Press (in Chinese).

Jiang, K., Wang, X., Chen, Z., Ding, L., Peng, Z., Wu, Z. (2022). Effect of constituent content on mechanical behaviors of ultra-high performance seawater sea-sand concrete, *Construct. Build. Mater.* 351, 128952.

Juenger, M. C. G., Monteiro, P. J. M., Gartner, E. M., Denbeaux, G. P. (2005). A soft X-ray microscope investigation into the effects of calcium chloride on tricalcium silicate hydration. *Cem. Concr. Res.* 35(1), 19–25.

Kalousek, G.L., Benton, E.J. (1970). Mechanism of seawater attack on cement pastes. *Am. Concr. Inst. J. Proc.* 67(2), 187–192.

Kaushik, S.K., Islam, S. (1995). Suitability of sea water for mixing structural concrete exposed to a marine environment. *Cem. Concr. Compos.* 17(3), 177–185.

Kawamura, M., Takeuchi, K. (1996). Alkali-silica reaction and pore solution composition in mortars in sea water. *Cem. Concr. Res.* 26(12), 1809–1819.

Li, H., Farzadnia, N., Shi, C. (2018). The role of seawater in interaction of slag and silica fume with cement in low water-to-binder ratio pastes at the early age of hydration. *Constr. Build. Mater.* 185, 508–518.

Li, T., Liu, X., Zhang, Y., Yang, H., Zhi, Z., Liu, L., Ma, W., Shah, P. S., Li, W. (2020). Preparation of sea water sea sand high performance concrete (SHPC) and serving performance study in marine environment. *Constr. Build. Mater.* 254, 119114.

Li, Q., Geng, H., Shui, Z., Huang, Y. (2015). Effect of metakaolin addition and seawater mixing on the properties and hydration of concrete. *Appl. Clay Sci.* 115, 51–60.

Limeira, J., Agullo, L., Etxeberria, M. (2010). Dredged marine sand in concrete: an experimental section of a harbor pavement. *Constr. Build. Mater.* 24(6), 863–870.

Limeira, J., Etxeberria, M., Agulló, L., Molina, D. (2011). Mechanical and durability properties of concrete made with dredged marine sand. *Constr. Build. Mater.* 25(11), 4165–4174.

Liu, W., Cui, H., Dong, Z., Xing, F., Zhang, H., Lo, T. Y. (2016). Carbonation of concrete made with dredged marine sand and its effect on chloride binding. *Constr. Build. Mater.* 120, 1–9.

Mohammed, T. U., Hamada, H. (2004). Long-term performance of alumina cement concrete mixed with tap water and seawater. *Mater. J.* 101(3), 226–232.

Mohammed, T. U., Hamada, H., Yamaji, T. (2004). Performance of seawater-mixed concrete in the tidal environment. *Cem. Concr. Res.* 34(4), 593–601.

Moukwa, M. (1990). Deterioration of concrete in cold sea waters. *Cem. Concr. Res.* 20(3), 439–446.

Ning, B., Ouyang, D., Wen, X.L. (2012). Experimental study on sea sand high-performance concrete. *Concrete* 28(1), 88–92.

Nishida, T., Otsuki, N., Ohara, H., Garba-Say, Z. M., Nagata, T. (2015). Some considerations for applicability of seawater as mixing water in concrete. *J. Mater. Civ. Eng.* 27(7), B4014004.

Province, A. (2014). Viscosity and antiwashout properties of cement mortar at low temperature seawater environment. *Asian J. Chem.* 26(6), 1595–1598.

Pyo, S., Koh, T., Tafesse, M., Kim, H. K. (2019). Chloride-induced corrosion of steel fiber near the surface of ultra-high performance concrete and its effect on flexural behavior with various thickness. *Constr. Build. Mater.* 224, 206–213.

Pyo, S., Tafesse, M., Kim, H., Kim, H. K. (2017). Effect of chloride content on mechanical properties of ultra high performance concrete. *Cem. Concr. Compos.* 84, 175–187.

Ragab, A. M., Elgammal, M. A., Hodhod, O. A., Ahmed, T. E. (2016). Evaluation of field concrete deterioration under real conditions of seawater attack. *Constr. Build. Mater.* 119, 130–144.

Rashad, A.M., Ouda, A.S., Sadek, D.M. (2018). Behavior of alkali-activated metakaolin pastes blended with quartz powder exposed to seawater attack. *J. Mater. Civ. Eng.* 30(8), 1–8.

Richardson, A.E., Fuller, T. (2013) Sea shells used as partial aggregate replacement in concrete. *Struct. Surv.* 31(5), 347–354.

Roux, N., Andrade, C., Sanjuan, M. (1996). Experimental study of durability of reactive powder concretes. *J. Mater. Civ. Eng.* 8(1), 1–6.

Saleh, S., Mahmood, A. H., Hamed, E., Zhao, X. L. (2023). The mechanical, transport and chloride binding characteristics of ultra-high-performance concrete utilizing seawater, sea sand and SCMs. *Constr. Build. Mater.* 372, 130815.

Shi, L., Liu, J., Liu, J. P., Gao, X. L., Zhang, W. N., Jiang, Q. (2018). Transmission and binding behavior of chloride ion in concrete under sulfate ion coupling effect in marine environment. *Concrete* 348(10), 67–71.

Sikora, P., Cendrowski, K., Abd Elrahman, M., Chung, S. Y., Mijowska, E., Stephan, D. (2019). The effects of seawater on the hydration, microstructure and strength development of Portland cement pastes incorporating colloidal silica. *Appl. Nanosci.* 1–12.

Sun, W., Liu, J.Z., Yan, J.L., Dai, Y.H. (2016). Study on the influence of chloride ions content on the sea sand concrete performance. *Am. J. Civ. Eng.* 4(2), 50–54.

Suryavanshi, A. K., Swamy, R. N. (1996). Stability of Friedel's salt in carbonated concrete structural elements. *Cem. Concr. Res.* 26(5), 729–741.

Teng, J. G., Xiang, Y., Yu, T., Fang, Z. (2019). Development and mechanical behaviour of ultra-high-performance seawater sea-sand concrete. *Adv. Struct. Eng.* 22(14), 3100–3120.

Tikalsky, P. J., Carrasquillo, R. L. (1989). The effect of fly ash on the sulfate resistance of concrete Research report 481–5. Austin, TX: Center for Transportation Research, The University of Texas at Austin.

Wang, J., Liu, E., Li, L. (2018). Multiscale investigations on hydration mechanisms in seawater OPC paste. *Constr. Build. Mater.* 191, 891–903.

Wang, S., Liu, X., Dai, D., Li, X., Chen, H. (2003). Distribution characteristic of marine aggregate resources and potential prospect in China. *Mar. Geol. Quat. Geol.* 23(3), 83–89 (in Chinese).

Wang, Z., Wang, J., Zhou, Y., Song, P., Guan, Q., Zhou, Z. (2021). The ability of UHPC prepared with unpurified sea sand and seawater to protect embedded steel against corrosion. In E3S Web of Conferences 293: 01024.

Wegian, F.M. (2010). Effect of seawater for mixing and curing on structural concrete. *IES J. Part A Civ. Struct. Eng.* 3(4), 235–243.

Xiao, J., Qiang, C., Nanni, A., Zhang, K. (2017). Use of sea-sand and seawater in concrete construction: current status and future opportunities. *Constr. Build. Mater.* 155, 1101–1111.

Yang, E. I., Yi, S. T., Leem, Y. M. (2005) Effect of oyster shell substituted for fine aggregate on concrete characteristics: part I. Fundamental properties. *Cem. Concr. Res.* 35(11), 2175–2182.

Younis, A., Ebead, U., Suraneni, P., Nanni, A. (2018). Fresh and hardened properties of seawater-mixed concrete. *Constr. Build. Mater.* 190, 276–286.

Yuan, Q., Shi, C., De Schutter, G., Audenaert, K., Deng, D. (2009). Chloride binding of cement-based materials subjected to external chloride environment—a review. *Constr. Build. Mater.* 23(1), 1–13.

Zeng, J. J., Liao, J., Zhuge, Y., Guo, Y. C., Zhou, J. K., Huang, Z. H., Zhang, L. (2022). Bond behavior between GFRP bars and seawater sea-sand fiber-reinforced ultra-high strength concrete. *Eng. Struct.* 254, 113787.

Chapter 13

Applications of UHPC and case studies

13.1 INTRODUCTION

UHPC is developed initially for large and specialized civil engineering applications that can benefit from its high strength and durability under extreme environmental conditions. After more than 30 years of research and use experience, it has been commercially produced and used in various construction projects in Europe, North America, Asia, Australia, etc. Because the mixing operation, casting technique, and curing procedure require highly specialized equipment and controls unique to UHPC, products are usually delivered as finished and precast elements. During the last several years, precast UHPC bridges, architectural UHPC, precast UHPC members for utilities, urban furnishings, and interior decoration elements have been installed to meet usage purposes. UHPC is also a promising material for cast-in-situ applications, including bridge deck overlays, strengthening slabs, joint fill-in bridge decks to connect two adjacent prefabricated elements, and rehabilitation of existing deteriorated concrete elements.

This chapter describes specific applications and case studies of UHPC in infrastructure projects, including roads and bridges, structural/architectural members, bridge deck overlays, crash barrier walls, structural rehabilitation and strengthening, urban finishing, and decorative arts.

13.2 ROADS AND BRIDGES

13.2.1 Precast footbridges

13.2.1.1 Prefabricated pedestrian, Sherbrooke, Canada, 1997

The first UHPC structure, a precast footbridge, was built in Sherbrooke, Canada, in 1997 (Figure 13.1). The structural concept consists of a space truss with a top UHPC chord as the riding surface, two UHPC bottom chords, and truss diagonals that slope in two directions. Each diagonal consists of UHPC confined in 152 mm diameter stainless steel tubes. The

DOI: 10.1201/9781003203605-13

Figure 13.1 Sherbrooke prefabricated pedestrian bridge, Canada, 1997.

Source: Provided by Dr. Kamal H. Khayat.

bridge has a span of 60 m and a width of 3.3 m with 200 MPa compressive strength and 25 MPa flexural strength. The erection of the bridge only took 4 days since it is composed of six prefabricated segments.

13.2.1.2 Sakata-Mirai footbridge, Sakata, Japan, 2002

Sakata-Mirai footbridge was built in Sakata, Japan, in 2002 (Figure 13.2). The designed footbridge consists of pretensioned box girder segments that were posttensioned together to form a single span of 49.2 m. The bridge is exceptionally light with a self-weight of 20% to 25% of a conventional reinforced concrete structure, and its cost savings are about 10%. Two external prestressing tendons, consisting of 31 strands with a diameter of 15.2 mm, are used. Six precast segments and no conventional steel reinforcing bars are needed. The load-carrying capacity of this footbridge is achieved through external prestressing.

13.2.1.3 Celakovice footbridge, Northeast Prague, Czech Republic, 2014

Celakovice footbridge is located in the northeast of Prague, Czech Republic, which crosses the Rabe River and connects the comprehensive park on the right bank and the urban street on the left bank. The superstructure of the Celakovice footbridge was completed in December 2013, and the bridge was opened to the public in April 2014. This is a continuous three-span cable-stayed bridge with a length of 242 m, a main span of 156 m, a main

Figure 13.2 Sakata-Mirai footbridge, Japan, 2002 (Tanaka et al. 2009): (a) View of footbridge. (b) Segments being erected on temporary supports.

beam width of about 3.6 m, a beam height, and a middle plate thickness of only 6 cm only. The bridge deck is located on the arch with a radius of 777 m, and the bridge tower is made of steel structure. The main beam is constructed by UHPC with a compressive strength of 150 MPa, and the precast segment length is 12 m.

13.2.1.4 Beichen Delta footbridge, Changsha, China, 2016

The first fully prefabricated UHPC bridge was built in the Beichen Delta of Changsha, Hunan, China, in 2016 (Figure 13.3). It is the first box girder UHPC bridge constructed by accelerated building construction technology in China. The total length of the footbridge is 70.8 m with three-span arrangements of 27.6, 36.8, and 6.4 m of a cantilever. The main girders and slats of the superstructure are prefabricated using UHPC with a compressive strength of 150 MPa (Fang et al. 2016). The main girder employs a continuous single-box, three-chamber section prefabricated prestressed UHPC box girder. UHPC integral prefabricated elements with a compressive strength of 100 MPa and a minimum section size of 60 cm are used as substructure pier. The prefabricated girder is equipped with ten prestressing bundles of through-length steel strands, and the prestressing channel is injected using densified small particles with compressive strength of 100 MPa. The box girders were prefabricated in the factory and were assembly erected on site (Fang et al. 2016). The main features of the project include:

(1) Improved structural and long-term performance.
(2) Green and rapid construction of urban bridges, which is realized by prefabrication of short girders, assembly of long girders, and overall prefabrication and hoisting of bridge piers (Figure 13.3(b)).
(3) Excellent durability, which can realize maintenance-free during the service life of the main structure.
(4) Lightweight and high aesthetic value, a weight reduction of nearly 1/3 compared to ordinary concrete structures using one main span of 36.8 m crossing the street.

13.2.2 Road bridges

13.2.2.1 Bourg-lès-Valence Bridges, Occitanie, France, 2001

Overpasses OA4 and OA6 were built at Bourg-lès-Valence in South-Eastern France in 2001 (Hajar et al. 2003). They are the oldest road bridges made of UHPC in the world. The material properties of UHPC were determined according to a prototype application that issued AFGC recommendations.

(a) Initially completed footbridge

(b) On-site assembling

(c) Precast section of box girder

(d) Prefabricated panel installation

Figure 13.3 Precast UHPC bridge, China, 2016 (Provided by Dr. Zhi Fang): (a) Initially completed footbridge; (b) Onsite assembling; (c) Precast section of box girder; (d) Prefabricated panel installation.

Typical compressive strength of 180 MPa is achieved (AFGC-Sétra, 2002). Each span of the decks was made of five precast parallel Pi-shaped pretensioned beams keyed onsite longitudinally. Since its opening to traffic in 2002, no significant maintenance operation has been carried out (Toutlemonde et al. 2013).

13.2.2.2 Batu 6 Bridge, Gerik-Perak, Malaysia, 2016

The segmental box-girder bridge at Batu 6 in Gerik Perak, Malaysia, is a typical bridge using UHPC materials. It has a 100 m long span. The use of UHPC material greatly reduces the deadweight of the bridge, which is only 50% of the ordinary concrete bridge, and increases the span capacity. The lighter girder section can reduce the requirements of hoisting equipment, machinery, and temporary facilities and make the construction more convenient and faster (Binard 2017).

13.2.2.3 The fifth Yangtze River Bridge, Nanjing, China, 2017

The Yangtze River Fifth Bridge is the world's first all-steel-composite cable-stayed bridge, which was built in Nanjing, Jiangsu, China, in 2017 (Figure 13.4). It connects the Jiangnan area and the Jiangbei area, located between the first and the third Nanjing Yangtze River Bridges. This two-way bridge has six lanes and a total length of about 10.33 km. It includes a 4.4 km main branch bridge across the river, a 1.8 km branching channel tunnel, and a 4.1 km cross-bank connection. The bridge project was completed in December 2020.

Figure 13.4 The Fifth Nanjing Yangtze River Bridge, China, 2018 (Provided by Sobute New Materials): (a) View of bridge; (b) Part of bridge; (c) Onsite assembling.

Its cable towers are steel-shell-concrete composite structures, and its main girders are steel-composite composite girders incorporated with coarse aggregate UHPC bridge decks. The incorporation of coarse aggregate in UHPC decreases binder content, increases elastic modulus, and restrains shrinkage. The high-performance lightweight steel-composite beam of the bridge employs a flat and streamlined integral box-type steel-composite beam with good overall torsion resistance. The steel-composite beam has a total width of 35.3 m and is divided into three internal chambers. The beam height is 3.6 m at the centerline of the steel-composite beam, and the standard section length is 14.6 m. The main girder bottom plate, including the inclined bottom plate and web, is made of steel structure. The carrier plate is made of coarse aggregate reactive powder concrete slab and is connected with the main steel girder through shear nails. It is a part of the main girder and serves as a lane slab.

13.2.2.4 The Second Dongting Lake bridge, Yueyang, China, 2017

The Second Dongting Lake bridge has a total length of 2390.18 m and a main span of 1480 m (Figure 13.5). It is located at the junction of Dongting Lake and Yangtze River, which starts from Qili Mountain in Yueyang City in the east of Hunan Province, crosses Dongting Lake, and then connects the Junshan Mountain in Yueyang City in the west. It is recognized as the first long-span steel truss girder suspension bridge in China and the second long-span one in the world after its completion in 2017. The orthotropic light composite deck of the bridge is composed of a 12 mm thick steel panel,

Figure 13.5 The Second Dongting Lake Bridge, China, 2017.

Source: Provided by Dr. Xudong Shao.

a 45 mm thick Super Toughness Concrete (STC) plate, and a 40 mm SMA asphalt abrasion layer. Its local stiffness is equivalent to that of a 38 mm thick steel panel.

The STC used in this bridge is a modified and strengthened UHPC with excellent mechanical properties and durability, which can meet the severe stress requirements of the steel bridge deck to achieve a long life without major repair. The construction process of the bridge mainly includes water tank counterweight construction, STC casting, vibration and leveling, moisture curing, high-temperature steam curing, and sandblasting of STC surface. Compared to traditional asphalt pavement steel decks, this light composite bridge deck structure can increase the local stiffness of steel deck by 40 times and reduce the average stress of steel deck by 50%. This helps to solve the two global problems of orthotropic steel bridge panels, i.e., the bridge deck pavement is vulnerable to breakage and steel structures are prone to fatigue cracking, finally greatly reducing the life cycle cost.

13.2.2.5 Songpu Bridge, Shanghai, China, 2018

Songpu Bridge, which stretches across the Huangpu River, is a double-decker railway bridge consisting of G320 Cheting Highway and Jinshan Branch Railway. The superstructure is unable to meet the needs of service after many years of use. The new upper highway bridge deck uses the steel-concrete combination orthotropic bridge deck system with a main bridge span of 96 + 112 m (Figure 13.6). The bridge width is widened from 12 to 24.5 m, and the concrete layer of the bridge deck employs 8 cm thick UHPC made with coarse aggregate to improve the stiffness of the combined bridge deck, reduce the local stress concentration caused by traffic load and fatigue damage, and prolong the service life of the bridge deck. The upper deck is divided into 54 sections, which consist of three types of TA, TB, and TC, with a width of 23.5 m. The length of the TA section (TD1) is 3.805 m, and the length of the TB segment (TD2 to TTD 26) is 8 m. The length of the TC segment (TD27) is 5.84 m.

13.2.2.6 Nanxun Bridge, Huzhou, China, 2019

Nanxun Bridge, located in Nanxun District, Huzhou City, Zhejiang Province, China, is a road reconstruction project. The main bridge employs a double-cable self-anchored suspension bridge with a span of 3 + 45 + 100 + 45 + 3 m (Figure 13.7). The cast-in-situ section of the main tower within 15.5 m is fabricated using UHPC made with coarse aggregate. The prepared UHPC shows good working performance, low shrinkage, high elastic modulus, and tensile strain-hardening behavior. The concrete bridge panel has a thickness of 25 cm. The load-bearing capacity and

Figure 13.6 Shanghai Songpu Bridge, China, 2018.

Source: Provided by Sobute New Materials.

antifatigue performance of bridge panels are greatly improved with the use of UHPC. The total shrinkage of coarse aggregate UHPC is limited to 200 με.

13.2.2.6 Qinglongzhou Bridge, Yiyang, China, 2020

Qinglongzhou Bridge, the new landmark architecture of Yiyang City, Hunan Province, China, is officially open to traffic at the end of 2020 (Figure 13.8). It is the first twin-tower self-anchored suspension bridge made of steel-UHPC composite girder in China. The bridge has a total length of 1636 m, a bridge width of 36.5 m, and six lanes in two-way with

Figure 13.7 Nanxun Bridge, China, 2019.

Source: Provided by Sobute New Materials.

a design speed of 80 km/h. The steel-UHPC composite beam structure is composed of a main beam, a middle beam, a box beam, a small beam, and a UHPC deck. The rise span ratio of the main cable is 1/5, the ratio of the main cable side to the middle span is 0.423, and the steel UHPC composite beam is used for the stiffening beam along the whole length of the main bridge. The traditional orthotropic steel bridge deck is replaced by UHPC low-rib bridge decks with an average thickness of 13.3 cm. The UHPC bridge deck is prefabricated in a factory, and the joints are zero-welded onsite. The total height of the main girder bridge deck is 22 cm. The UHPC solid bridge deck is employed for both sides of the box girder, and the low-ribbed UHPC bridge deck is adopted on the carriageway to reduce the dead weight of the bridge deck. In a word, the low-ribbed UHPC bridge deck has the advantages of high stiffness, excellent fatigue resistance, and low cost.

To speed up the construction speed and improve the construction quality, the deck slab of the bridge is prefabricated in a factory, hoisted onsite, and poured wet joint onsite to connect the deck slab with the main beam below. Compared with conventional steel-concrete composite bridge structures, the average thickness of UHPC bridge deck is reduced from more than 28 to 13.3 cm. The dead weight of the bridge deck is reduced by more than 50%, which significantly decreases the dead weight of the main girder. Besides, the prefabricated UHPC bridge deck eliminates the shrinkage cracking risk after steam curing. Moreover, the transportation and hoisting of the UHPC bridge deck are very convenient due to its lightweight. After hoisting in place, new stepped joints are constructed between precast slabs to ensure no risk of cracking.

(a) Real view of Qinglongzhou Bridge

(b) Moisture conservation of precast slab

(c) Precast slab hoisting to the real bridge

Figure 13.8 Qinglongzhou Bridge, China, 2018 (Provided by Dr. Xudong Shao): (a) Real view of Qinglongzhou Bridge; (b) Moisture conservation of precast slab; (c) Precast slab hoisting to the real bridge.

13.3 STRUCTURAL/ARCHITECTURAL UHPC MEMBERS

Within the last several years, architectural UHPC has been installed on buildings, including foreign consulates and courthouses, university buildings, museums, airports, commercial office buildings, and hotel and residential high-rise developments.

13.3.1 Cladding panels and facades

UHPC makes it possible to design and produce thin, complex shapes, curvatures, and customized textures due to its superior properties. These precast elements made by UHPC are durable, have low porosity and high ductility, and require less energy consumption and maintenance over time. The following section presents various projects made with UHPC systems completed in recent years.

13.3.1.1 MuCEM Museum, Marseille, France, 2013

The Museum of European and Mediterranean Civilisations, also known as MuCEM, is located in Marseille, France. Encompassing a total gross floor area of over 55,000 m^2, the MuCEM is one of the world's largest ethnographic museums with extensive use of UHPC (Figure 13.9). This building includes the application of UHPC in its columns, 115 m and 69 m pedestrian bridges, brackets, and, most innovatively, its tree-like facade. It is the first building in the world to make such extensive use of UHPC. The MuCEM consists of two adjacent buildings located on Marseille's waterfront, i.e., the historical Fort Saint-Jean and a new iconic building that opened in 2013. The two buildings are connected by a 115-m-long UHPC pedestrian bridge.

Surrounding the 18 m high building in concrete lace, the UHPC latticework enshrouds the building, acting as sunshades and spreading out over the facade and roof of the MuCEM. Designed to shield the southern and western facades from the strong Mediterranean sun, they cast a shadow along the peripheral walkway and onto the roof terrace. These panels can support extreme loads, especially the facades that are exposed to prevailing winds.

13.3.1.2 The Roof, Shanghai, China, 2020

The Roof was designed by the famous French architect Jean Nouvel and is located in Xintiandi of Shanghai, China (Figure 13.10). The key factors that need to be considered in material selection include mechanical properties, toughness, compactness, volume stability, durability, pollution resistance, and easy manufacturing. Therefore, UHPC is the best choice. A total of 450

Figure 13.9 MuCEM museum, France, 2013.

Figure 13.10 The Roof Project, China, 2020.

Source: Provided by Dr. Xiong Ke.

tons of UHPC premix was used to prefabricate the flower bowls and curtain wall panels.

The Roof is composed of four buildings and was completed in 2020. The facade is composed of 2500 UHPC flower pots and UHPC curtain wall panels of different sizes. The whole building is covered with plants to form a horizontal and vertical three-dimensional garden. In addition to the self-weight, the flower bowl also bears the weight of soil, water, and plants, as well as the stress caused by wind loads and installation. The curtain wall system adopts large plates, of which the first-floor plate is a high-angle integral plate with a height of greater than 4 m. Using UHPC to manufacture thin-wall flower bowls can reduce self-weight and meet the load demand and firm fixation. It is also suitable for manufacturing large-size curtain wall panels.

13.3.1.3 Museum of Zijing network information security, Zhengzhou, China, 2020

Museum of Zijing network information security is the core venue of the National Cyber Security Awareness Week in 2020 and the permanent site of the Strong Network Security Challenge. It is the first museum of network security science and technology in China, and the future new landmark building of Zhengzhou, which displays high-tech fields and cyberspace. The facades of the building and two surrounding hotels are fabricated using UHPC to achieve a high perforation rate, rich texture, visual impact, high stain resistance, and antifouling durability (Figure 13.11). The use of high flow and high thixotropy UHPC realizes the self-compaction and shotcrete

Figure 13.11 Museum of Zijing network information security and hotel, China, 2020 (Provided by Subote Materials).

technology and eventually achieves different textures of the same component. Its low shrinkage of less than 300 με reduces the risk of cracking of large thin-walled members.

13.3.1.4 Future Garden Hotel at Jiangsu Garden Expo, Nanjing, China, 2021

Future Garden Hotel in Jiangsu Garden Expo is located in an abandoned limestone mine in Tangshan, Nanjing, Jiangsu, China (Figure 13.12). The length of the mine is about 1100 m from east to west and 200 m from north to south. The elevation of the mountainside main road ranges from 95 to 110 m. The site retains the natural cliff-wall texture and presents a cut platform state. The architect chose UHPC, which is plastic and rich in texture, to achieve the integration of the exterior wall of the hotel and the feature formed by quarrying. It successfully copied the stone gouges on the cliff wall to the exterior wall of the hotel.

Different from the molding process of conventional UHPC panels, the facade of this project has a large size and complex three-dimensional shape. The facade needs to be fully fitted with the balcony, the partition elevation, and the exterior wall surface. The grooved surface of the axe and the folding plate from parts of the smooth surfaces are formed entirely to improve the sense of integrity and reduce the risk of leakage. The largest section uses more than 2.8 t premixed UHPC materials, which proposes extremely high requirements for the fluidity and thixotropy of the material. It also needs to coordinate the contradiction between strength development and slow retarding and washing processes.

Figure 13.12 Future Garden Hotel at Jiangsu Garden Expo, 2021, China (Provided by Subote Materials).

13.3.1.5 Chongqing Greenland Xinli autumn moon platform, Chongqing, China, 2021

Chongqing Greenland Xinli autumn moon platform is located at the former Autumn Moon Gate station in Fuling Old Town of Chongqing, China, which has an overall assembly rate of 65%. The facade of the whole building is manufactured by UHPC to take advantage of its ultra-high strength, toughness, and durability to achieve the simplification of the installation and connection points, fine and stable texture, antifouling durability, creative design, etc. (Figure 13.13). UHPC can meet the special needs of engineering due to the following two aspects. First, its good rheological performance with time can meet the texture manufacturing requirements of different structural parts of the same plate. Secondly, the low shrinkage and high crack resistance of UHPC can reduce the risk of shrinkage cracking of special-shaped, thin-walled large structures. The developed UHPC showed a reduction in viscosity by over 50%, which realizes the high flow and self-leveling characteristics and ensures the appearance and compactness of special-shaped decorative components. The shrinkage of UHPC materials is less than 300 $\mu\varepsilon$.

13.3.2 Roof components

UHPC enables the design of structural elements with a high load-bearing capacity and smaller slenderness compared to conventional concrete elements. Besides, UHPC has a very high resistance to environmental influences. These features allow UHPC to be advantageous for the roof covering or claddings (Holý 2017).

Figure 13.13 Precast facade for Chongqing Greenland Xinli autumn moon platform, 2021, China (Provided by Sobute New Materials).

13.3.2.1 Shawnessy Light Rail Transit (LRT) Station in Calgary, AB, Canada, 2003

The Shawnessy Light Rail Transit (LRT) Station in Calgary, AB, Canada, has the world's first thin-shelled UHPC canopy roof system, constructed during the fall of 2003 and winter of 2004 (Perry and Zakariasen 2004). Twenty-four unique and thin-shelled canopies (5.1 × 6 m and just 20 mm thick) are supported on single columns to protect commuters from the elements at Calgary's new Shawnessy LRT Station. The shells provide a dual function at night, diffusely reflecting the artificial light to the platform below and animating the facade through the louvered windows. Furthermore, the canopies are highly durable and easy to clean and very little maintenance is required. In addition to the canopies, the components include struts, columns, beams, and gutters. The volume of material used is 80 m^3.

13.3.2.2 Millau Viaduct Toll Gate, France, 2004

A spectacular example of architecture taking advantage of the remarkable benefits of UHPC is the toll gate of the Millau Viaduct in France. The elegant roof looks like an enormous twisted sheet of paper or an aircraft wing. The design is based on a succession of identical precast segments, and hollow-core thin-shell elements assembled by posttensioning (Thibaux et al. 2004). It is made of match-cast prefabricated 2 m wide segments connected by an internal longitudinal prestressing. The roof is 98 m long and 28 m wide, with a maximum thickness of 85 cm at the center. The UHPC has a compressive strength of 165 MPa, and the use of steel fiber ensures ductile behavior under tension and dispenses with the need for traditional passive reinforcement (Resplendino 2004).

13.3.2.3 G8WAY DC, Washington DC, USA, 2013

The G8WAY DC is located in Washington DC, constructed during the summer of 2013. It is a 4877-m^2 open-air and multiuse facility that spreads over almost a hectare on the east campus of St Elizabeth's Hospital, which was built in 1852. It is one of the speediest fast-track projects. It involved the production of 181 UHPC panels that would cover a total area of 2884 m^2, each averaging 15.9 m^2, in just 19 days. The panels range between 2.1 and 4.3 m in length, with a gradual dimensional change in width from 2.7 to 0.6 m over the entire structure. This pavilion is unique because of its thin, 44-mm lightweight UHPC roof, providing a large, landscaped seating area above and expansive views of the US Capitol.

13.3.2.4 Pérez Art Museum, Miami, USA, 2013

UHPC was used in ultra-thin cladding at the Pérez Art Museum in Miami to provide aesthetics and strength with specific consideration for hurricane resistance (Sayaffi et al. 2018). To meet the aesthetic and resilience requirements, Ductal® UHPC was chosen to produce approximately 100 long-span, precast vertical mullions to blend with the structure's cast-in-place elements and support the large curtain wall glazing which surrounds the building. The museum overlooks Biscayne Bay, where frequent tropical storms and exposure to salt/sea air can cause severe problems for buildings in the area. The Pérez Art Museum, designed by award-winning architects Herzog & de Meuron, combined concrete and glass with elegance and was open to the public in December 2013. The museum has 200,000 ft^2 of space for works of art, educational activity, relaxation, and dining.

13.4 BRIDGE DECK OVERLAYS

The high strength and toughness together with very low permeability and high resistance to freezing-thawing damage make UHPC a good candidate for bridge deck overlay. There are three possible options when a UHPC overlay may be applied to bridge deck overlays: (1) an overlay for a new deck; (2) a protective overlay for an existing deck; and (3) a structural overlay for an existing deck. The following section shows several examples of bridge deck overlays.

13.4.1 Chillon Viaduct, Switzerland, 2015

The Chillon viaduct was built in the late 1960s and is located in Switzerland. The reinforcing bars were corroded and alkali-aggregate reactions were detected in 2012. With an ever-increasing number of vehicles and to ensure structural safety for future traffic demands, the use of a UHPC layer of 40 mm reinforced with steel rebars was decided to strengthen the slab in 2015 (Martín-Sanz et al. 2020). Strain-hardening UHPC with low permeability is used as a waterproofing layer protecting the slabs from the water to reduce the alkali-aggregate reaction rate. UHPC has excellent mechanical properties and acts as the slab's external tensile reinforcement, increasing its bending and shear resistance and extending the fatigue life (Brühwiler et al. 2015).

13.4.2 Mafang Bridge, Zhaoqing, China, 2011

Mafang Bridge is located in Sihui County, Zhaoqing City, Guangdong Province, which crosses the Beijiang River and is also known as Mafang Beijiang Bridge. The bridge is divided into upper and lower sections, and the

Figure 13.14 Mafang Bridge, China, 2011 (Provided by Dr. Xudong Shao): (a) View of bridge; (b) Bridge after rehabilitation.

section towards Sihui was built in 1984 (Figure 13.14(a)). It has 14 spans and each span is a simply supported steel box girder bridge with a length of 64 m and a full width of the cross-section of 12.1 m.

Mafang Bridge is the first bridge paved with steel decks in China. A polychloroprene latex-modified asphalt paving scheme with a thickness of 70 mm was initially employed. Cracks occurred after 3 months of operation with visible steel plates. The asphalt pavement layer of the whole bridge was renovated due to the serious damage to the steel deck pavement, starting from June to July 2001. The double-layer asphalt concrete of 40-mm thick SAC10 in the lower layer and 40-mm thick SAC13 in the upper layer was employed. In 2010–2011, the steel deck of the whole bridge was rehabilitated again due to damage to the pavement layer. Different solutions were used for each span of the bridge. However, fatigue cracking of the steel deck still appeared.

Under the support of the Guangdong Department of Transportation and Guangdong Humen Bridge Co., Ltd., a steel-thin layer STC light composite deck structure was used on the 11th span of the Mafang Bridge in 2011 (Figure 13.14(b)). The STC layer is arranged with HRB400 grade steel mesh, the center distance of steel bars is 32 mm, and the transverse bridge steel bars are located on the upper layer. The stud size is $\Phi 13 \times 40$ mm, the spacing along the transverse bridge is 300–325 mm, and the spacing along the longitudinal bridge is 250 mm. No shrinkage cracks were found in the STC layer after high-temperature steam curing.

13.4.3 Mauves Sur Loire Bridge, Pays de la Loire, France, 2020

Mauves Sur Loire bridge was rehabilitated by substituting the existing deck (3500 m^2) made of brick vaults with a thin deck made of thin precast UHPC slabs, completed in November 2020. The fibers used in this project were 14/0.2 mm steel brass coated. Particular methods and additional rebar were implemented to increase the bonding in the joints between the slabs. The project's main objective is to upgrade this historic landmark by increasing its bearing capacity and allowing the addition of cantilevered footbridges.

13.5 CRASH BARRIER WALL, SWITZERLAND, 2008

Crash barrier walls on new and existing bridges are usually cast on site. Macrocrack occurrence favors the corrosion of reinforcing bars due to restrained early-age deformations. UHPC has been used to strengthen the concrete elements that are locally subjected to severe environmental exposure, such as the splash zone of deicing salts and high mechanical loading. This is mainly due to its unique combination of extremely low permeability, high strength with significant tensile strain hardening, and excellent rheological properties in the fresh state, allowing for easy casting of the self-compacting UHPC with conventional concreting equipment.

Figure 13.15 shows the crash barrier wall with a UHPC layer covering the areas subjected to splash exposure. The fresh self-compacting UHC was fabricated in a conventional ready-mix concrete plant, transported to the site by a truck, and successfully filled into the thin slot to realize the UHPC coating. The required mechanical properties and the protective function of the UHPC layer have been confirmed by in-situ air permeability tests and laboratory tests on specimens cast on site. After 4 months of rehabilitation, no crack was observed (Bruhwiler and Denarie 2008).

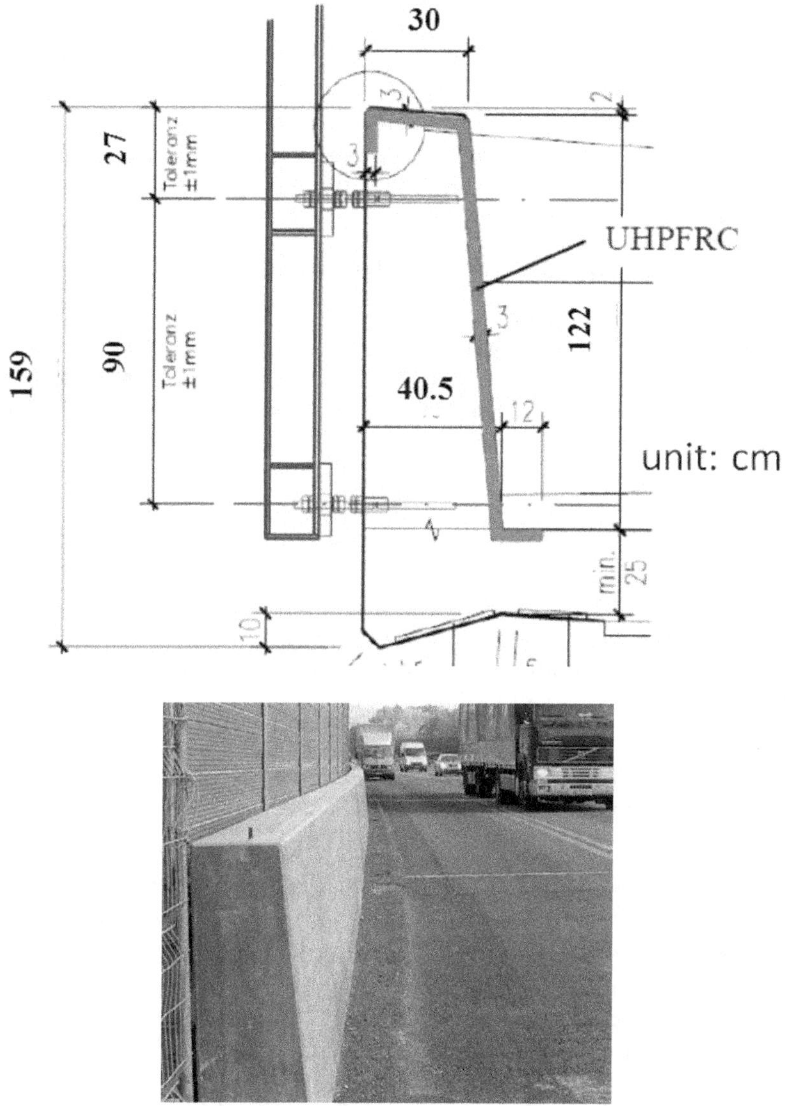

Figure 13.15 Typical cross-section of the crash barrier wall and view after rehabilitation, Switzerland, 2008.

Source: Bruhwiler and Denarie (2008).

13.6 STRUCTURAL REHABILITATION AND STRENGTHENING

UHPC material is also used for repair and strengthening techniques and traditional fiber-reinforced polymer (Zhu et al. 2020; Hallaq et al. 2017; Garner 2011). Reinforced concrete slabs strengthened with UHPC postpone cracks growth and increase the ultimate load capacity (Yin et al. 2017). The following are several cases using UHPC to reinforce concrete structures.

13.6.1 Mission Bridge, British Columbia, Canada, 2014

The Mission Bridge, opened to traffic in 1973, is a 1126 m long bridge on Hwy 11 that crosses the Fraser River between Mission and Abbotsford, British Columbia. The project was completed in 2014 and required approximately 18 m^3 of material. The existing fiber-reinforced polymer wrap was removed, and the surface was roughened. The pedestal was widened with traditional concrete to force the plastic hinge at this location under seismic load. Additional steel dowels were embedded in the existing column. A reinforcing rebar cage was secured to the dowels and new 225 mm thick UHPC jackets were used to increase and stiffen confinement up to a height of approximately 3 m. Compared to traditional pile compaction techniques, using a UHPC jacket provides savings of $1.5 million. It would allow for a high seismic deformation capacity using a thin jacket, eliminating underground costs and risks from piling.

13.6.2 Single-Lane Suspension Bridge, New Hazelton, Canada, 2015

The single-lane suspension bridge, with a main span of 140 m that hovers 75 m above the water below, is located in New Hazelton, BC. The main span was built in 1931. The bent legs of the bridge approaches had developed local corrosion at their base. Because of access limitations, congestion, and a small zone available to transfer concentrated load from the bent leg to its concrete pedestal, the solution demanded a very high-strength concrete with excellent flow-ability during placement, virtually no shrinkage, virtually no permeability, and good tensile capacity. UHPC material fit those characteristics and was used to encase the steel bent legs in 2015. The design demanded a minimum compressive strength of 100 MPa.

13.6.3 Yicheng Hanjiang Bridge, Yicheng, China, 2020

The UHPC wet joint reinforcement project of Yicheng Hanjiang Bridge was completed by China Railway Qiaoyan Technology Co., Ltd. The bridge, with a total length of 1886.95 m, was built in 1986. In 2017, the technical condition of the bridge was assessed as Class 3, and the technical condition of the deck pavement was assessed as Class 4. The maintenance and reinforcement work started in 2020. To improve the joint bearing capacity of 50 and 30 m prestressed concrete T-beam of east and west approach bridges, respectively, the original joint layer was chiseled out and UHPC wet joints were used. The joints have widths of 80 and 40 cm with a thickness of 10 cm.

13.6.4 Fuhe Bridge, Wuhan, China, 2020

The bridge is an important passage to Wuhan airport (Figure 13.16). Under the coupling action of river abrasion and dynamic load, the concrete reinforcement covers of four piers have severe abrasion and cracks, which need to be repaired and strengthened. The materials used in this

Figure 13.16 Rehabilitated Fuhe Bridge, China, 2020. (Provided by Dr. Qingjun Ding): (a) View of bridge; (b) Pier after rehabilitation.

project were UHPC with low shrinkage and high abrasion resistance. This type of UHPC material has a 28-d compressive strength of 144.3 MPa, a flexural tensile strength of 23.7 MPa, an impact wear strength of 157 h/(kg/m^2), a 28 d wear rate of only 1.7%, a final crack impact energy of 1610 J, and an interfacial bond strength with C40 concrete of 3.2 MPa. It can improve the abrasion resistance of piers and prolong the service life of the bridge.

13.6.5 Yichang Yangtze River Highway Bridge, Yichang, China, 2021

Yichang Yangtze River Highway Bridge is a super-large bridge that crosses the Yangtze River in Yichang City, Hubei Province, China. It is a double-tower single-span suspension bridge with a span of 960 m and a distance between the two main cable centers of 24.4 m (Figure 13.17). The bridge was opened to traffic in September 2001. After more than 20 years of service, problems, such as cracking of asphalt pavement, upheaval, rutting, stripping, and fatigue cracking of orthotropic steel bridge panels, successively appear in the steel deck system due to heavy load traffic. Given the fact that a large number of cracks in the steel bridge deck are difficult to be repaired and have a high risk of recracking in the short term, the Hubei Department of Transportation decided to comprehensively upgrade the bridge. The new UHPC reinforcement technology of steel bridge deck invented by Prof. Xudong Shao at Hunan University was employed to rehabilitate the bridge.

The project began on August 5, 2021, and was completed and opened to traffic on December 4, 2021. The upper and lower decks were separately cast, and the whole bridge was closed to traffic for 7 d during the period

Figure 13.17 Rehabilitated Yichang Yangtze River Highway Bridge, China, 2021 (Provided by Yichang Bridge Group).

(a) Welding bolts
(b) Attaching steel plates
(c) Assembling reinforcement
(d) UHPC casting
(e) UHPC curing
(f) Open to traffic soon

Figure 13.18 Rehabilitation process of Yichang Yangtze River Highway Bridge (Provided by Dr. Xudong Shao): (a) Welding bolts; (b) Attaching steel plates; (c) Assembling reinforcement; (d) UHPC casting; (e) UHPC curing; (f) Open to traffic soon.

from the beginning of UHPC casting to the end of steam curing. During the other construction processes, one-half of the deck is open to traffic. Figure 13.18 shows the construction process of the bridge deck. The UHPC was symmetrically poured from the middle of the span to both ends and achieved a new record of efficient construction of 10,250 m^2 in a single pour in about 20 h for the first time.

Detailed information regarding the construction process of the bridge deck is as follows: (1) milling the existing asphalt paving layer of the steel bridge deck; (2) sandblasting and removing the rust of the steel bridge panel; (3) anticorrosion coating of the steel panel; (4) welding bolts of the steel bridge panel; (5) attaching steel plates; 6) recoating of anticorrosion painting; (7) laying steel mesh and local reinforcement of the guardrail; (8) casting UHPC layer; (9) moisture curing of UHPC using membrane; (10) high-temperature steam curing of UHPC; (11) surface blasting of UHPC layer; (12) scattering TPO binder; (13) grave spreading; and (14) open to traffic.

13.7 URBAN FURNISHINGS AND DECORATIVE ARTS

UHPC can be used for the production of practical, unique street furnishings in urban settings. It can provide functionality, durability, aesthetics, and a wide range of colored or textured options. Its impact and weather-resistant qualities, durability, and low maintenance make it an excellent alternative to traditional materials, such as steel, cast iron, aluminum, plastic, or wood. Besides, the low permeability makes it suitable for corrosive environments, such as underground utilities and products for wastewater treatment plants. Its strength and ductility properties allow it to compete against materials, such as stainless steel, for wastewater treatment plant troughs.

Figure 13.19 shows a Substation Pavillion, a UHPC public art installation located in Vancouver, BC, Canada. It is installed in front of the Meccanica project by Cressey Developments, completed in 2014. The modular components are precast using dark grey UHPC, with a rough board-formed surface.

Figure 13.19 Meccanica/Substation Pavilion, Canada, 2014 (Provided by Szolyd Development).

13.8 SUMMARY

The application of UHPC in bridges, buildings, rehabilitation, and strengthening of existing concrete structures is attributed to its high mechanical properties and superior durability. The denser matrix of UHPC also creates feasible solutions for problems associated with concrete deterioration caused by inner steel reinforcement corrosion. Moreover, the use of UHPC materials in buildings and infrastructure can also effectively address the shortcomings of conventional construction work, which is rather labor intensive and time consuming. The excellent workability enables UHPC to be cast into any shape, which can be applied for structural and architectural precast works and some innovative urban furniture and decorative arts.

REFERENCES

AFGC-Sétra (2002). Ultra-high-performance fiber-reinforced concretes. Interim Recommendations.

Binard, J. P. (2017). UHPC: A game-changing material for PCI bridge producers. *PCI J.* 62(2): 34–46.

Bruhwiler, E., Denarié, E. (2008). Rehabilitation of concrete structures using ultra-high performance fiber reinforced concrete. In: Proceedings of the second international symposium on ultra high-performance concrete, Kassel, 1–8.

Brühwiler, E., Bastien-Masse, M., Mühlberg, H. (2015). Strengthening the Chillon viaducts deck slabs with reinforced UHPFRC, Structural Engineering: Providing Solutions to Global Challenges, Geneva, Switzerland. 23–25.

Fang Z., Zheng H., Zhengyu H., Wu L.M, Lv D.S., Shi Y.S., Ma, W. (2016). Design and construction of prestressed reactive powder concrete continuous box girder bridge. *Proceedings of the 22nd National Bridge Academic Conference. Guangzhou*, 543–551.

Garner, A. P. (2011). Strengthening reinforced concrete slabs using a combination of CFRP and UHPC. The University of New Mexico. Albuquerque, USA. M.S. thesis.

Hajar, Z., Simon, A., Thibaux, T., Wyniecki, P. (2004). Construction of an Ultra-High Performance Fibre-Reinforced Concrete thin-shell structure over the Millau Viaduct toll gates. In: fib symposium 2004: Concrete structures: the challenge of creativity, Avignon, France, p: 236–237.

Holý, M. (2017). Roof structural system study with use of UHPC. *Solid State Phenomena,* 259, 70–74.

Hallaq, A. L., Tayeh, A., Shihada, B. A. (2017). Investigation of the Bond Strength Between Existing Concrete Substrate and UHPC as a Repair Material. *Int. J. Eng. Adv. Technol.* 6, 210–217.

Martín-Sanz, H., Tatsis, K., Dertimanis, V. K., Avendaño-Valencia, L. D., Brühwiler, E., Chatzi, E. (2020). Monitoring of the UHPFRC strengthened Chillon viaduct under environmental and operational variability. *Struct. Infrastructure Eng.* 16(1), 138–168.

Perry, V., Zakariasen, D. (2004). First use of ultra-high performance concrete for an innovative train station canopy. Concr. Technol. Today, 25(2), 1–2.

Sayyafi, E. A., Chowdhury, A. G., Mirmiran, A. (2018). Innovative Hurricane-resistant UHPC roof system. *J. Archit. Eng.*, 24(1), 04017032.

Tanaka, Y., Maekawa, K., Kameyama, Y., Ohtake, A., Musha, H., Watanabe, N. (2009). Innovation and application of UFC bridges in Japan. In: Toutlemonde, F.and Resplendino, J., editors. Designing and building with UHPFRC. Chichester, UK: Wiley-ISTE. 149–187.

Thibaux, T., Hajar, Z., Simon, A., Chanut, S. (2004). Construction of an ultra-high-performance fiber-reinforced concrete thin-shell structure over the Millau viaduct toll gates. Proc. of 6th Int. RILEM Symposium on Fibre Reinforced Concrete (FRC), BEFIB 1183–1192.

Toutlemonde, F., Roenelle, P., Hajar, Z., Simon, A., Lapeyrère, R., Martin, R. P., Baron, L. (2013). Long-term material performance checked on world's oldest UHPFRC road bridges at Bourg-Lès-Valence. *In Proceedings of the RILEM-fib-AFGC Int. Symposium on Ultra-High Performance Fibre-Reinforced Concrete, UHPFRC,* Marseille, France, 265–274.

Yin, H., Teo, W., Shirai, K. (2017). Experimental investigation on the behaviour of reinforced concrete slabs strengthened with ultra-high performance concrete. *Constr. Build. Mater.* 155, 463–474.

Zhu, Y., Zhang, Y., Hussein, H. H., Chen, G. (2020). Flexural strengthening of reinforced concrete beams or slabs using ultra-high performance concrete (UHPC): A state of the art review. *Eng. Struct.* 205, 110035.

Index